110 Springer Series in Solid-State Sciences
Edited by Manuel Cardona

Springer
Berlin
Heidelberg
New York
Barcelona
Budapest
Hong Kong
London
Milan
Paris
Santa Clara
Singapore
Tokyo

Springer Series in Solid-State Sciences

Editors: M. Cardona P. Fulde K. von Klitzing H.-J. Queisser

Managing Editor: H. K.V. Lotsch

Volumes 1–89 are listed at the end of the book

90 **Earlier and Recent Aspects of Superconductivity**
Editors: J. G. Bednorz and K. A. Müller

91 **Electronic Properties of Conjugated Polymers III** Basic Models and Applications
Editors: H. Kuzmany, M. Mehring, and S. Roth

92 **Physics and Engineering Applications of Magnetism** Editors: Y. Ishikawa and N. Miura

93 **Quasicrystals** Editors: T. Fujiwara and T. Ogawa

94 **Electronic Conduction in Oxides**
By N. Tsuda, K. Nasu, A. Yanase, and K. Siratori

95 **Electronic Materials**
A New Era in Materials Science
Editors: J. R. Chelikowsky and A. Franciosi

96 **Electron Liquids** 2nd Edition By A. Isihara

97 **Localization and Confinement of Electrons in Semiconductors**
Editors: F. Kuchar, H. Heinrich, and G. Bauer

98 **Magnetism and the Electronic Structure of Crystals** By V.A. Gubanov, A.I. Liechtenstein, and A.V. Postnikov

99 **Electronic Properties of High-T_c Superconductors and Related Compounds**
Editors: H. Kuzmany, M. Mehring, and J. Fink

100 **Electron Correlations in Molecules and Solids** 3rd Edition By P. Fulde

101 **High Magnetic Fields in Semiconductor Physics III** Quantum Hall Effect, Transport and Optics By G. Landwehr

102 **Conjugated Conducting Polymers**
Editor: H. Kiess

103 **Molecular Dynamics Simulations**
Editor: F. Yonezawa

104 **Products of Random Matrices** in Statistical Physics By A. Crisanti, G. Paladin, and A. Vulpiani

105 **Self-Trapped Excitons**
2nd Edition By K. S. Song and R. T. Williams

106 **Physics of High-Temperature Superconductors**
Editors: S. Maekawa and M. Sato

107 **Electronic Properties of Polymers**
Orientation and Dimensionality of Conjugated Systems Editors: H. Kuzmany, M. Mehring, and S. Roth

108 **Site Symmetry in Crystals**
Theory and Applications 2nd Edition
By R. A. Evarestov and V. P. Smirnov

109 **Transport Phenomena in Mesoscopic Systems** Editors: H. Fukuyama and T. Ando

110 **Superlattices and Other Heterostructures**
Symmetry and Optical Phenomena 2nd Edition
By E. L. Ivchenko and G. E. Pikus

111 **Low-Dimensional Electronic Systems**
New Concepts
Editors: G. Bauer, F. Kuchar, and H. Heinrich

112 **Phonon Scattering in Condensed Matter VII**
Editors: M. Meissner and R. O. Pohl

113 **Electronic Properties of High-T_c Superconductors**
Editors: H. Kuzmany, M. Mehring, and J. Fink

114 **Interatomic Potential and Structural Stability**
Editors: K. Terakura and H. Akai

115 **Ultrafast Spectroscopy of Semiconductors and Semiconductor Nanostructures** By J. Shah

116 **Electron Spectrum of Gapless Semiconductors**
By J. M. Tsidilkovski

117 **Electronic Properties of Fullerenes**
Editors: H. Kuzmany, J. Fink, M. Mehring, and S. Roth

118 **Correlation Effects in Low-Dimensional Electron Systems**
Editors: A. Okiji and N. Kawakami

119 **Spectroscopy of Mott Insulators and Correlated Metals**
Editors: A. Fujimori and Y. Tokura

120 **Optical Properties of III–V Semiconductors**
The Influence of Multi-Valley Band Structures
By H. Kalt

121 **Elementary Processes in Excitations and Reactions on Solid Surfaces**
Editors: A. Okiji, H. Kasai, and K. Makoshi

122 **Theory of Magnetism**
By K. Yosida

123 **Quantum Kinetics in Transport and Optics of Semiconductors**
By H. Haug and A.-P. Jauho

124 **Relaxations of Excited States and Photo-Induced Structural Phase Transitions**
Editor: K. Nasu

E.L. Ivchenko G.E. Pikus

Superlattices and Other Heterostructures

Symmetry and Optical Phenomena

Translated by G.P. Skrebtsov

Second Edition
With 85 Figures

Springer

Professor Eougenious L. Ivchenko
Professor Grigory E. Pikus

A.F. Ioffe Physico-Technical Institute
Politechnicheskaya 26
St. Petersburg 194021, Russia

Library of Congress Cataloging-in-Publication Data.

Ivchenko, E. L. (Eougenious L.), 1946– Superlattices and other heterostructures: symmetry and optical phenomena / E. L. Ivchenko, G. E. Pikus; translated by G. P. Skrebtsov. – 2nd ed. p. cm. – (Springer series in solid-state sciences; 110) Includes bibliographical references and index. ISBN 3-540-62030-3 (hardcover: alk. paper) 1. Semiconductors–Optical properties. 2. Superlattices. 3. Heterostructures. 4. Symmetry (Physics) 5. Quantum wells. I. Pikus, Grigoriĭ Ezekielevich. II. Title. III. Series. QC611.6.06I93 1997 537.6'22–dc21 96-46761

ISBN-13: 978-3-642-64493-1 e-ISBN-13: 978-3-642-60650-2
DOI: 10.1007/978-3-642-60650-2

Typesetting: Macmillan India Ltd, Bangalore-25 and K&V Fotosatz, Beerfelden

SPIN: 10560353 54/3144 – 5 4 3 2 1 0 – Printed on acid-free paper

In Memoriam

our teacher, Professor

A.I. Anselm

1905–1988

Preface

Since the first edition of this book was published, the 22nd and 23rd International Conferences on the Physics of Semiconductors (ICPS) have been held, in Vancouver, Canada, and Berlin, Germany, respectively. Representing, as usual, the state of the art of semiconductor physics, the recent ICPSs have confirmed the significance of heterostructures and superlattices and showed the rapidly increasing role of small quantum systems. The main goal of this book is to demonstrate the efficiency of symmetry considerations in an analysis of optical phenomena in heterostructures with quantum wells and superlattices. With further progress in growth technology, the importance of the symmetry approach will become even more evident. We also expect this treatment to provide the framework within which future research into quantum wires and dots may develop. We hope that this monograph will prove useful for both researchers working in the field of semiconductors and semiconductor devices and students specializing in materials sciences.

In the preparation of this second edition we have corrected errors and misprints and added references to recent publications, including those on quantum microcavities and four-wave mixing in heterostructures.

St. Petersburg, Russia
January 1997

E.L. Ivchenko
G.E. Pikus

Preface to the First Edition

For a long time we had been contemplating the possibility of writing a book about optical phenomena in semiconductors in which the various optical phenomena would be considered from the standpoint of the theory of symmetry. We had planned to start with a short introduction into the theory of symmetry as the basis for expounding the methods for calculating the spectra of electrons, excitons and phonons in semiconductors. Using the results obtained we can then discuss the absorption and reflection of light in interband transitions including the exciton and polariton effects, electro- and magneto-optical phenomena, IR absorption and reflection, cyclotron and electron-spin resonance, light scattering by free and bound carriers and optical and acoustic phonons, polarized photoluminescence, optical spin orientation of electrons and excitons, electron alignment in momentum space, nonlinear optical and photogalvanic effects, with particular emphasis on the phenomena determined by crystal symmetry.

However, by the time the writing of such a book took place, the interest in the optics of semiconductors had shifted from bulk crystals to artificially produced low-dimensional systems. Having mastered the methods of fabricating complex synthetic structures, physicists imagined themselves capable of creating at will new objects with programmed properties and, predictably, could not resist the temptation to do so. As always, though, *Nature* proved more imaginative than *Man*, and what physicists had foreseen was only a part of what was to be revealed later.

In a book, it would hardly be possible to avoid describing the optical properties of these new objects. At the same time, an analysis of the properties of both low-dimensional and bulk crystals would require too extended a discussion. We decided therefore to modify the original idea, considering all of the above phenomena only for the quantum wells, superlattices and other heterostructures, and using them to illustrate the variety of polarization spectroscopic methods developed in the optics of semiconductors. Even after narrowing the topic, however, the number of works to be dealt with was obviously too large. Thus, we have restricted ourselves to a comparatively small number of studies which, in our opinion, illustrate most clearly the relation between the symmetry and optical phenomena, in order to follow consistently the road from the theory of symmetry to the description of particular physical phenomena; we hope that such a strategy will offer the reader a problem-solving tool. Hence, have followed, to a considerable extent, the pattern of the monograph *Symmetry and Strain-Induced Effects in Semiconductors* (by one of the authors together with G.L. Bir), but not expounding in detail on the group theory. Instead, we give only those results necessary for the understanding of subsequent chapters, including the required reference tables.

The literature used in the preparation of the book as well as publications which broaden and expand the material are collected partly in the list of references and partly under additional reading.

A special note concerns the nomenclature used in the monograph: the direction along the principal axis of a superlattice or a quantum well structure is denoted by the symbol $\|$, for instance, $a_\|$, $M_\|$, and the direction perpendicular to the axis, by $\perp$, for example, $k_\perp$, $m_\perp$. This is in agreement with the system of notation accepted for uniaxial crystals or many-valley semiconductors with anisotropic valleys. The reader should keep this in mind, since many publications on low-dimensional systems make use of the reverse nomenclature.

Finally, the authors wish to express their sincere gratitude to Dr. G.P. Skrebtsov, who undertook the not easy task of translating the manuscript.

St. Petersburg, Russia
August 1994

E.L. Ivchenko
G.E. Pikus

Contents

1 Quantum Wells and Superlattices

The Quantum Well (QW) is a system in which the electron motion is restricted in one direction thus producing quantum confinement; in other words, the spectrum in one of the quantum numbers changes from continuous to discrete. The quantum wells represent an example of systems with reduced dimensionality. Systems with electron motion restricted in two directions are called *quantum wires*, and those confined in all three directions were given the name of *quantum dots*. For quantum confinement to be observable, the size of a well must be less than the electron mean-free path. This requirement imposes constraints both on the geometric size of a well and on the quality of the sample and temperature determining the *mean-free-path length*.

The electron spectrum of quantum wells represents a series of subbands. The position of the bottom of each of them is determined by the conditions of confinement, the motion in the well plane being free.

The simplest subjects on which the quantum confinement effects were revealed and studied are silicon-based MOS structures with quantum wells formed at the semiconductor boundary (Fig. 1.1) and single GaAs/$Al_xGa_{1-x}As$ heterojunctions where the quantum well is created in the GaAs layer at the heterojunction boundary (Fig. 1.2). It is on these structures that the quantum-Hall effect, a totally unexpected and remarkable phenomenon occurring in a two-dimensional electron gas, was discovered. These structures provide the unique possibility of varying within a fairly broad range both the position of the quantum-well levels and the carrier concentration by properly varying the potential between the electrodes deposited at the outer boundaries.

Two closely lying heterojunctions make up a Single Quantum Well (SQW). If the separation between the heterojunctions is substantially less than the Debye length, the electron field will not distort the potential produced by the change in the lattice structure, and such a well will be practically rectangular (Fig. 1.3a). By varying properly the lattice composition, one can produce wells of another shape, e.g., parabolic, triangular, etc. (Fig. 1.3b,c).

A system of quantum wells separated by barriers thick enough to make them impenetrable for electrons is called the Multiple Quantum Well (MQW). While from the viewpoint of the electronic properties each of these wells is isolated, the presence of many wells affects noticeably the optical characteristics. As the barriers get thinner, carrier tunneling from one well to another becomes possible and, as a result, the levels of an isolated well smear out into one-dimensional

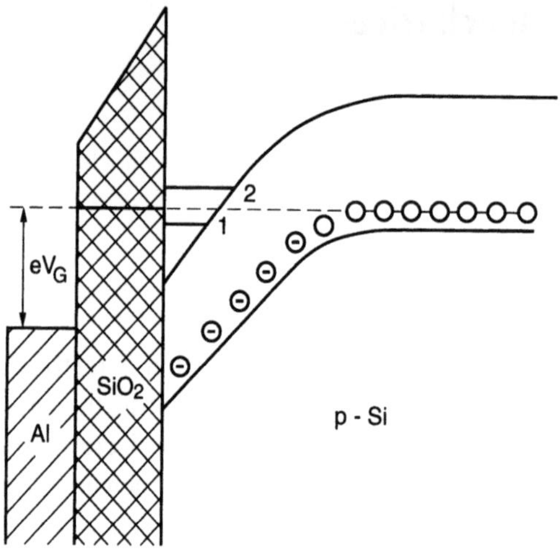

Fig. 1.1. Metal-insulator (SiO_2)-semiconductor (Si) structure

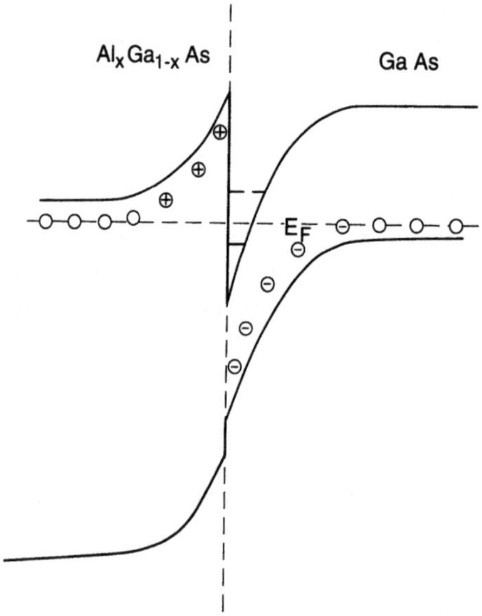

Fig. 1.2. Heterojunction

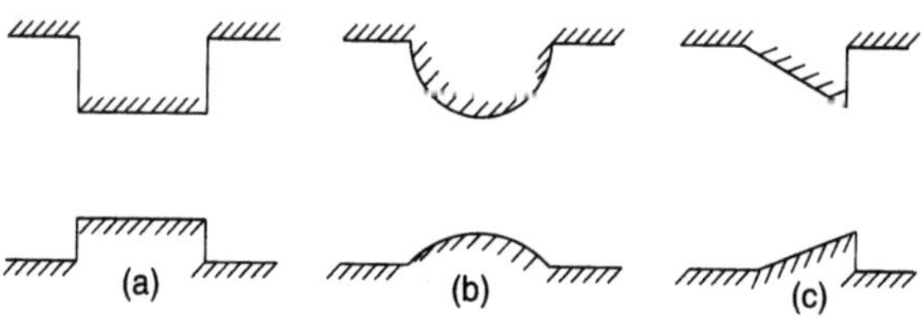

Fig. 1.3. Quantum wells: **a** rectangular, **b** parabolic, **c** triangular

minibands. For such subbands to form, the superlattice period d should be smaller than the mean free path in the corresponding direction.

As the barrier thickness continues to decrease, the minibands corresponding to different levels begin to overlap. Such UltraThin SuperLattices (UTSL) – with the thickness of wells and barriers being usually not in excess of three molecular layers – behave as an anisotropic bulk crystal whose properties, however, differ substantially from those of the original materials. The lattices where the superstructure potential is formed by properly varying the composition are called *compositional*. One distinguishes four types of compositional SuperLattices (SLs) according to the relative magnitude of the band gaps and the electron affinity (the distance from the vacuum level to the conduction band bottom of the original materials)[1]. In type-I lattices, the wells for the electrons and holes are located in the same layer, e.g., in the GaAs layers in the GaAs/$Al_xGa_{1-x}As$ structures with $x < 0.4$ (Fig. 1.4a). In this case, the well depth for the electrons is equal to the shift ΔE_c the bottom of the conduction bands undergoes as one crosses over from a GaAs layer to a AlGaAs layer, and that for the holes, to the corresponding displacement ΔE_v of the top of the valence band (Fig. 1.4a), with $\Delta E_v + \Delta E_c = \Delta E_g = E_{gA} - E_{gB}$, where E_{gA} and E_{gB} are the band gaps in AlGaAs and GaAs, respectively. The effective band gap of the superlattice, which determines the intrinsic concentration of the electrons and holes, can be written as $E_{g,eff} = E_{gB} + E_e + E_h$, where E_e and E_h are energies of the lowest electron and hole quantum confined states reckoned from the bottom of the corresponding wells. In type-II staggered lattices, the wells for the electrons and holes are located in different layers, the top of the valence band in layer B lying below the bottom of the conduction band in layer A (Fig. 1.4b). This structure is exemplified by the GaAs/AlAs lattice, where the bottom of the conduction band in AlAs lies below that in GaAs. Note that the effective band gap width $E_{g,eff} = E_{gB} - \Delta E_c + E_e + E_h = E_{gA} - \Delta E_v + E_e + E_h$. A specific feature of such structures is the large lifetime of the electron-hole pairs due to the electrons and holes being spatially separated. The top of the valence band in layer B in type-II misaligned lattices lies above the bottom of the conduction band in layer A i.e., $E_{g,eff} < 0$. In such structures, electrons transfer from the valence band of layer B to the conduction band of layer A (Fig. 1.4c). If the layers A and B are not too thin, the fields generated by these electrons and the remaining holes distort considerably the potential and, after the thickness has reached a certain value, the semimetal-semiconductor transition occurs, which manifests itself in the formation of regions without free electrons and holes (Fig. 1.4d). An example of such lattices are the GaSb/InAs structures. The lattices with one of the layers being a gapless semiconductor are called *type III* (Fig. 1.4e). Quantum confinement splits the levels of electrons and holes of the gapless semiconductor in quantum wells and superlattices, however the corresponding gap width is usually small. The electric properties of such superlattices depend to a considerable extent on the energy into which the

[1] Note that the electron affinity depends, in principle, not only on the material composition but on the interface orientation, i.e., on the growth direction, as well.

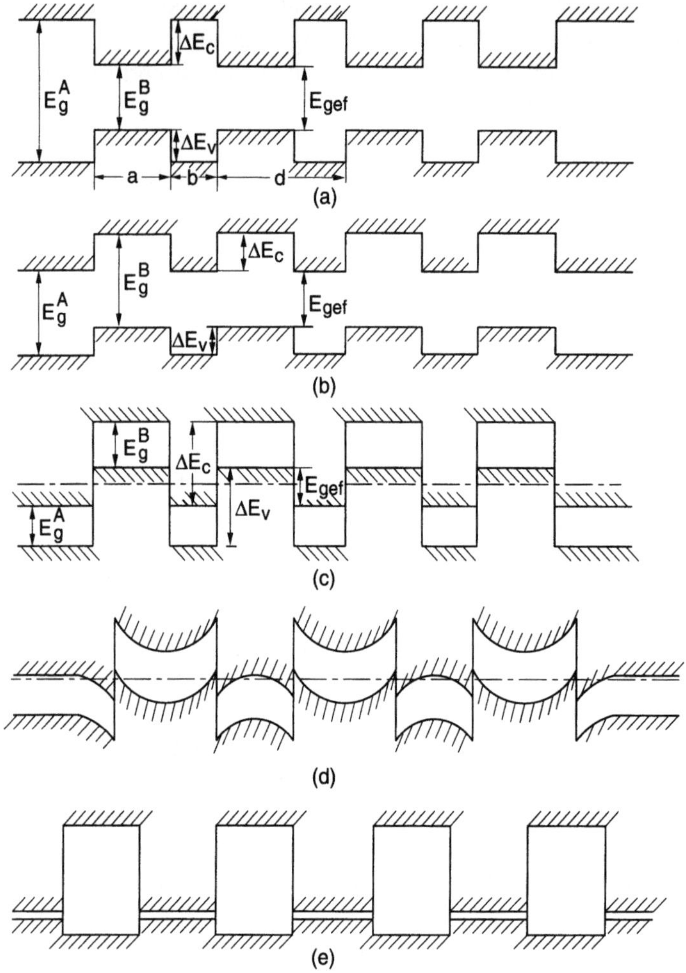

Fig. 1.4. Compositional superlattices: **a** type I, **b** type II–staggered, **c,d** type II–misaligned (for **c** smaller and **d** larger barrier thickness), **e** type III

bands of the gapless semiconductor merge relative to the bottom of the conduction band or the top of the valence band. Such a structure can be exemplified by the HgTe/CdTe lattice, in which the band merging point of HgTe lies inside the band gap of CdTe.

By varying smoothly the composition, one can produce lattices with wells of arbitrary shape, for instance, saw-tooth lattices with triangular wells. Besides periodic structures, it is possible to produce and study aperiodic superlattices; the Fibonacci lattice may serve as an example of such a structure, with the thicknesses of the layers A and B varying as $a_j = G_j a_0$, $b_j = G_{j-1} b_0$, and $G_j = G_{j-1} + G_{j-2}$. For instance, for $G_0 = G_1 = 1$, $G_2 = 2$, $G_3 = 3$, $G_4 = 5, \ldots$ In principle, aperiodic superlattices may possess very unusual properties; indeed, practically all biological objects have a fixed aperiodic structure. However, the methods for calculating the properties of such superlattices have yet to establish which structures could be of particular interest.

Quantum wells and superlattices were initially produced by choosing pairs with practically equal lattice constants, for example, the GaAs/GaAlAs pair. Such superlattices are called *lattice matched.* Progress in superlattice-growth technology has made it possible to obtain dislocation-free structures of materials with noticeably different lattice constants. In this case, lattices match because of internal stresses which result in a compression of one of the adjacent layers and a tension of the other. Such lattices are called *strained.*

Besides the compositional SLs produced by varying the composition, one can prepare superlattices by modulation doping with a donor and an acceptor impurity (modulation doped SLs). Such lattices represent a sequence of p-n or p-i-n junctions forming p- and n-layers which, in contrast to the compositional SLs, are separated by more or less thick i-layers (Fig. 1.5).

Compositional superlattices are obtained by means of Molecular-Beam Epitaxy (MBE) as well as liquid or gas-phase epitaxy (for instance, Metalloorganic Chemical Vapor Deposition, MOCVD). These techniques are capable of controlling the well and barrier thicknesses to within one atomic layer. Their description can be found in specialized literature [1.1]. One can also vary simultaneously the composition and concentration of the dopant. The application of these techniques, which have permitted, in particular, the production of perfect strained SLs, has broadened substantially the class of the semiconductor compounds suitable for making heterojunctions. This is illustrated by Fig. 1.6, which specifies the pairs used to produce superlattices and quantum wells.

The energy spectrum of electrons in a one-dimensional periodic potential and for one-dimensional quantum wells was calculated in many publications at the dawn of quantum mechanics, when such periodic structures were considered a one-dimensional model of the crystal, and isolated wells, a model of a defect in a solid. Most of these publications, as well as those containing calculations of spectra in thin layers and periodic layered structures, made before the practical realization of semiconductor superlattices in the 1970s, are now forgotten, their results being reproduced in later publications. A lucky exception is the Kronig-Penney model as well as Morse's model of a lattice with a periodic sine-shaped potential, for which Schrödinger's equation reduces to Mathieu's equation well known to mathematicians. These models are mentioned in many treatises on quantum mechanics and quantum theory of the solid state.

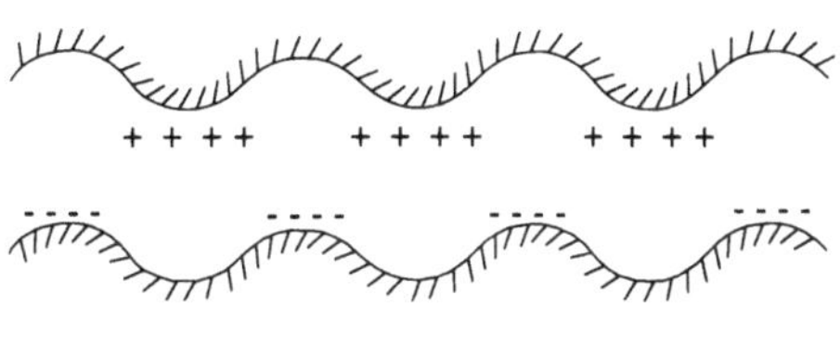

Fig. 1.5. Modulation-doped superlattices

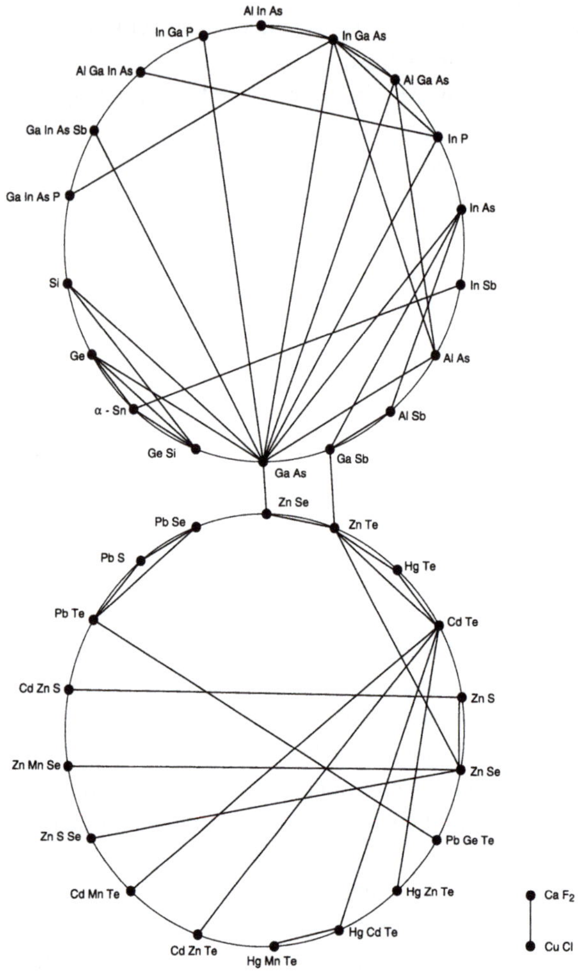

Fig. 1.6. Compound pairs used to produce superlattices and quantum wells (based on data reported at International conferences on semiconductor physics)

Initially, the interest in the preparation of superlattices was stimulated by the desire to obtain dropping I-V characteristics, i.e., Negative Differential Resistivity (NDR), and to use them to construct microwave generators or amplifiers.

Jones and *Zener* [1.2] showed as far back as 1934 that, in the absence of scattering and interband tunneling in an electric field, the electron placed in an ideal periodic structure performs periodic motion. *Zener* [1.3] demonstrated that in a one-dimensional lattice the electron is periodically reflected from the upper and lower band boundaries. Accordingly, the current under these conditions is zero. It is this factor that may result in a decrease of the current in strong electric fields. One cannot, however, observe this effect in conventional crystals, since for it to be observable, the allowed bands should be very narrow, and the forbidden ones quite the reverse, broad. Semiconductor superlattices offer the

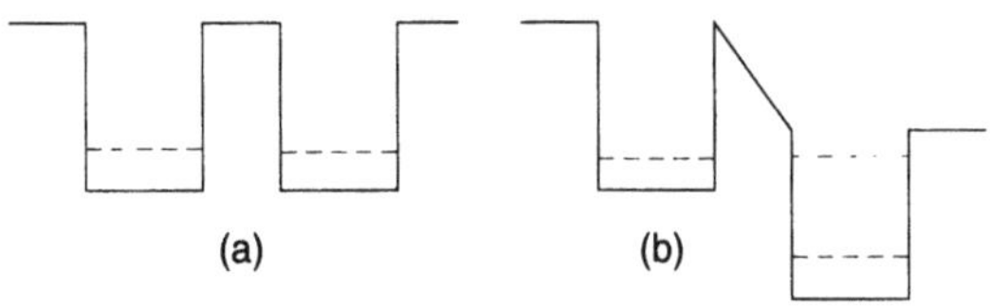

Fig. 1.7. Double-well structures with NDR

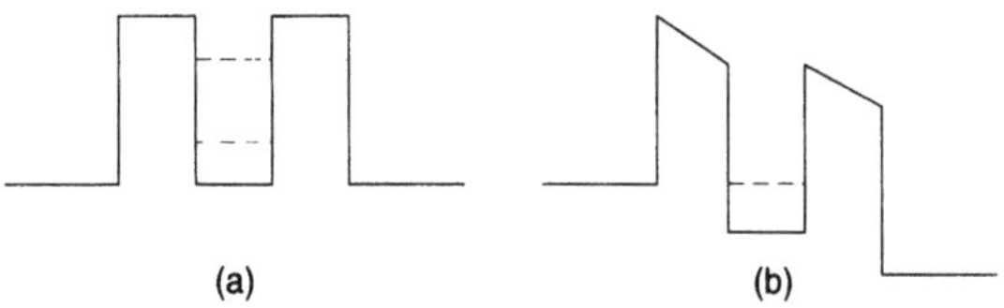

Fig. 1.8. Double-barrier structures with NDR

possibility to meet these conditions, since the width of the allowed bands in them can be reduced easily by increasing the barrier thickness, and the interband gap, by reducing the well width.

Keldysh [1.4] proposed to produce a superlattice by deforming a crystal with a strong acoustic wave. *Esaki* and *Tsu* [1.5] suggested instead composition modulation or selective doping, a method successfully implemented. One did not, however, succeed in making a domain free device which would allow for a uniform electric field along the NDR-based superlattice. More efficient has turned out to be a system of two wells (Fig. 1.7), or of two barriers (Fig. 1.8)[2] as proposed by *Kazarinov* and *Suris* [1.6]. In the first system, the conductivity reaches a maximum at such a bias that the lowest level of one of the wells coincides with one of the levels of the neighboring well (Figs. 1.7a,b), and in the second, when the bottom of the conduction band in region I coincides with one of the levels of the central well located between the two barriers (Fig. 1.8b). Resonant tunneling in such structures was observed by *Esaki* and *Chang* [1.10], and *Chang* et al. [1.11].[3]

As for the superlattices, the interest here focussed primarily on the study of their optical properties. One of the first significant results obtained was the observation of quantum confinement effects for the exciton levels and, in particular, of the splitting of the heavy and light exciton states [1.13].

Progress in the application of superlattices is associated mostly with optoelectronic devices. For example, using superlattices of a complex configuration for the construction of semiconductor lasers offers the possibility of reducing

[2] Similar structures with the metallic emitter and collector sandwiching two or more thin insulating barriers between which thin metallic layers are deposited were reported on as far back as 1963 – 1964 by *Davis* and *Hosack* [1.7], and *Johansen* [1.8,9].

[3] Note that as far back as 1966, *Lutskii* et al. [1.12] observed resistivity oscillations caused by quantum confinement in a thin Bi film when tunneling occurred from a thick film through the vacuum gap. In this experiment, the conductivity oscillations were caused not by a variation in transmission but rather by an increase of the density of states in the thin film with increasing bias, i.e., with increasing number of the quantum well states lying below the Fermi level of the thick film.

the threshold current and improving the device stability, as well as of controlling the radiation frequency by properly varying the lattice parameters. Another illustration is the highly efficient and low-consuming modulators based on the effect of electric field-induced shift of exciton levels in SQW or MQW structures (the excitonic optical Stark effect).

2 Crystal Symmetry

This chapter focusses attention on the theory of point- and space-group representations. Representations may be considered the basic mathematical formalism of the theory of symmetry used in the physics of the solid state.

Among the readers there may be physicists not acquainted with this term. Therefore, first we explain in some detail and, hopefully, clearly enough what are actually the groups, and reducible and irreducible group representations. It turns out to be possible to find, based on symmetry considerations only, all the irreducible representations of a given group and thus, to determine the possible degeneracy of states and the character of the electron wave functions or of the lattice vibrations allowed by symmetry.

Note that even for us, who have been dealing with applications of the theory of symmetry for many years, the methods of finding irreducible representations, particularly where this concerns the so-called *projective representations*, produce frequently the impression of an amazing mathematical trick. It is by no means accidental that these methods (one of the comparatively recent achievements in mathematics) owe their origin to I.I. Shur, an outstanding algebraist. Developed in the beginning of this century (1904–1911), the theory of projective representations fitted perfectly the description of space group and spinor representations taking into account the half-integral spin of the electron discovered a quarter of a century later. Although the theory of representations could appear to be a building specifically designed for the physics of the solid state, it soon became too involved. To include time reversal invariance, one had to cut additional corridors through it, which certainly did deprive it of some of its original elegance. This was the price for including, into the set of symmetry operations, the time inversion, an operation radically different from geometric transformations. But even here, the theory of representations turned out to be useful by having immediately shown one the way to the correct corridor.

The application of the theory of representations will be illustrated with a number of general problems such as the determination of selection rules for quantum-mechanical transitions and of the number of linearly independent components of material tensors describing different properties of crystals. More specific problems will be considered in the subsequent chapters.

2.1 Symmetry Operations, Groups

A distinctive feature of crystalline solids is the existence of symmetry elements, i.e., geometric transformations which bring a body into coincidence with itself. These symmetry operations may correspond both to real transformations which can be performed without destroying the body and are called *proper*, and to those that can be carried out only by a conceptual rearrangement of atoms and are called *improper*. Among the proper transformations we find translations, i.e., displacements of a body in a certain direction by a given amount, rotations about axes through a certain angle, and combinations thereof. Improper transformations include inversion about a point, reflection through a plane, mirror rotation, i.e., simultaneous rotation about an axis and reflection in a plane perpendicular to it, as well as combinations of these transformations and of translation. Usually rotation of nth order, i.e., rotation through an angle $2\pi/n$ is denoted by c_n, mirror rotations through $2\pi/n$ – by s_n, the symbol c_n^p or s_n^p referring to successive application of these operations p times, and c_n^{-1} or s_n^{-1} – to the corresponding rotation in the reverse direction. Inversion about a point is denoted by c_i (or i), reflection in a plane normal to the C_n or S_n rotation axis, by σ_h, and that by a plane containing the rotation axis, by σ_v, or by σ_d, if this plane lies midway between two two-fold axes perpendicular to the C_n or S_n axis. Successive application of symmetry operations also transforms a body into itself, i.e., it is a symmetry element.

A set of all symmetry elements forms a group. By definition of a group G one understands a set, finite or infinite, of elements g (which is denoted $g \in G$), for which:

1. A multiplication rule is defined. For symmetry elements, multiplication $g_i g_k$ is a successive application of the symmetry operations g_k and g_i. In the general case, $g_i g_k \neq g_k g_i$. If two groups, G and G', have an identical multiplication table differing only in the notation of the elements (e.g., in c_n^p being replaced by s_n^p), then these groups are called isomorphic;
2. Associative law of multiplication, $g_i(g_k g_l) = (g_i g_k)g_l$, holds, which is obvious for symmetry transformations;
3. Among the elements g_i, there exists one and only one identity element e, such that $ge = eg = g$ for all $g \in G$. For symmetry operations, the identity element is the identity transformation;
4. Each of the group elements g has an inverse element g^{-1}, such that $g^{-1}g = gg^{-1} = e$. Note that $(g_1 g_2)^{-1} = g_2^{-1} g_1^{-1}$. For instance, the inverse elements for the c_n or s_n rotations are $c_n^{-1} = c_n^{n-1}$ or s_n^{-1}, and, for the inversion c_i or reflection σ, the inverse element coincides with the direct one since $c_i^2 = e$ and $\sigma^2 = e$.

For finite groups, i.e., groups containing a finite number of elements h, one can perform the identity transformation, i.e., $g^m = e$, by successive application of any operation g a certain number of times m. The smallest possible number m which we will denote by ν is called the order of the element g, and the set of elements $g, g^2, \ldots, g^\nu$ is called the period of the element g and forms itself a group referred to as cyclic. Cyclic groups are commutative, i.e., for them $g_i g_k = g_k g_i$. It is the above postulates defining a group that correspond to the properties of the symmetry elements. And it is this that determines the role of group theory in physics.

If all rotation axes and reflection planes pass through one point, then the latter is not displaced under all the above symmetry operations. The set of the corresponding symmetry elements is called a point group. Point groups (together with the permutation groups) describe the symmetry of molecules and local centers in crystals.

2.2 Point-Group Classification

For classification of point groups one uses Schoenflies' and international notations. In Schoenflies' notation, the groups including only rotations about one axis are denoted by C_n. These are cyclic groups containing n elements c_n^k $(k = 1, \ldots n)$. The groups C_{nh} have, besides the axis C_n, a reflection plane σ_h and contain $2n$ elements c_n^k and $\sigma_h c_n^k$.

The groups containing mirror rotations s_n^k are denoted by S_n. Note that s_n^{2k} is the usual rotation through an angle $4\pi(k/n)$. For odd n, $s_n^n = \sigma_h$, i.e., this group coincides with C_{nh}. Groups which include only inversion c_i or mirror reflection σ are denoted, accordingly, by C_i or C_s, the group C_i coinciding with S_2.

The groups C_{nv} have, besides an n-fold axis, also n reflection planes passing through this axis, and the group D_n, an n-fold axis and n two-fold axes perpendicular to it. Rotations about these axes are denoted by u_2. The groups C_{nv} and D_n are isomorphic and contain $2n$ elements each.

The groups D_{nh} include, besides the elements of the group D_n, also a reflection plane σ_h perpendicular to the axis C_n, and n reflection planes σ_v passing through the n-fold axis and two-fold axes, and the groups D_{nd}-reflection planes passing between the two-fold axes. The groups D_{nh} and D_{nd} are isomorphic and contain $4n$ elements each, and for $n = 4$ or $n = 6$ these groups coincide and include inversion, i.e., they represent a direct product of the groups D_n and C_i. This implies that the elements of these groups are products of the elements of the groups D_n and of the elements e, c_i. Inversion is contained also in the group $D_{2h} = D_2 \times C_i$.

Besides the above groups, there are 5 cubic groups. Group T has 12 rotational symmetry elements of the tetrahedron; in addition to three two-fold symmetry axes of the group D_2, it contains also four three-fold axes and, accordingly, eight elements c_3 and c_3^2. There is also the group $T_h = T \times C_i$. The group O contains all 24 rotational cubic symmetry axes, and has three four-fold axes and four three-fold axes.

The group T_d is a complete tetrahedral symmetry group; besides elements of the group T, it contains 12 reflections in planes passing through the four-

and three-fold axes. It contains 24 elements and is isomorphic with the group O; indeed, the corresponding multiplication tables differ only in the elements u_2 and c_4^k $(k = 1, 3)$ of the group O being replaced by the elements σ_d and s_4^k of the group T_d. The group O_h is a complete group of cubic symmetry: $O_h = O \times C_i = T_d \times C_i$. In addition to these point groups, there are two groups of icosahedral symmetry which, just as the groups C_n, S_n, D_n with $n = 5$ and $n \geqslant 7$, cannot occur in crystals.

International notation specifies the symmetry elements determining a group, the number indicating the axis order, and the bar above it, the fact that the corresponding axis is an improper rotation. The group C_i is denoted by $\bar{1}$, the group C_s – by index m. The subscript index m indicates the existence of a reflection plane σ_h, perpendicular to this axis, and the same index in line with the number, the existence of reflection planes σ_v passing through the axis. Two indices mm indicate the existence of two systems of reflection planes passing through the axis and sent into one another under rotations c_n about this axis (in groups C_{nv} with even n).

One usually employs abbreviated international notations which specify only the generating elements, all the others being obtained by their successive multiplication. This principle, however, does not always hold. For instance, all elements of the group O can be obtained by multiplication of the elements c_4 and c_3, and those of the group T_d, by multiplication of the elements s_4 and c_3. However, the notations for these groups are 432 and $\bar{4}3m$, respectively. Besides the point finite groups describing the symmetry of molecules or crystals, there are also continuous groups. These are spherical symmetry groups describing, for instance, the symmetry of atoms, and groups of axial symmetry which describe the symmetry of diatomic molecules. There are two continuous spherical groups, namely, a group K including rotations through an arbitrary angle about any of the axes passing through one point, and a complete orthogonal group K_h which includes, besides the elements of the group K, inversion through a point and, accordingly, reflections by any of the planes passing through it. There exist also five groups of axial symmetry:

$$C_\infty, C_{\infty h}, C_{\infty v}, D_\infty, D_{\infty h} = D_\infty \times C_i = C_{\infty v} \times C_i.$$

The groups D_∞, and $C_{\infty h}$ include, respectively, rotations c_2 about any of the axes perpendicular to the axis C_∞, or reflections in any plane passing through this axis.

2.3 Space Groups

The space groups include, besides point group elements, also translations. The crystal has a specific feature of spatial periodicity, namely, it may be considered a structure obtained by translation of a primitive cell consisting of a finite, usually small, number of atoms or molecules, by an amount

$$\mathbf{t_m} = m_1\mathbf{a}_1 + m_2\mathbf{a}_2 + m_3\mathbf{a}_3, \tag{2.1}$$

where m_1, m_2, m_3 are arbitrary integers, $-\infty < m_i < +\infty$. The vectors $\mathbf{a}_i$ are called unit translations. The primitive cell represents here a parallelepiped based on vectors $\mathbf{a}_i$ (basic parallelepiped) and of volume $\Omega = (\mathbf{a}_1[\mathbf{a}_2\mathbf{a}_3])$, and the lattice proper may be represented as a structure built of identical primitive cells.

The vertices of these cells whose coordinates are determined by (2.1) are called Bravais sites, and the lattice formed by these sites, a Bravais lattice. It should be pointed out that the Bravais sites do not necessarily coincide with the sites of the crystal lattice, i.e., the location of one of the atoms in a given cell. The symmetry of the Bravais lattices is, as a rule, higher than that of the crystal lattice, in particular, inversion is always present as one of the elements of its symmetry. The symmetry elements of the Bravais lattice are a product of those of the corresponding point group determining the Bravais lattice symmetry or syngony, by the translation group elements defined by (2.1). Altogether, there are seven systems or syngonies specified in Table 2.1. To each syngony corresponds one or several types of Bravais lattices differing in the location of the sites, which are likewise given in this table. All these corresponding lattices are shown in Fig. 2.1. One has to keep in mind here that not all the cells shown in this figure are primitive; indeed, the sites of the main parallelepiped are located only at eight of its vertices and, accordingly, the volume of base-centered and body-centered cells (Fig. 2.1) is equal to two volumes of primitive cells, and that of face-centered cells, to four primitive cell volumes. The space-group elements include translations (2.1) denoted by $(e|\mathbf{t})$, elements of the corresponding point group denoted by $(r|0)$, and their products $(r|\mathbf{t})$. Besides the elements $(r|\mathbf{t})$, the space group can include rotations about screw axes, i.e., rotations c_n with simultaneous displacement along the axis by an amount $\tau \neq \mathbf{t}$, and glide reflections, i.e., reflections in glide planes with simultaneous displacement by an amount $\tau \neq \mathbf{t}$ in a direction in this plane. Note that here

$$\tau = \mu_1\mathbf{a}_1 + \mu_2\mathbf{a}_2 + \mu_3\mathbf{a}_3 \quad \text{with} \quad 0 \leqslant \mu_i < 1 \tag{2.2}$$

and the corresponding symmetry elements are denoted by $(r'|\tau)$. The elements $(r'|0)$, as well as $(e|\tau)$, are not elements of crystal symmetry. The vectors defined by (2.2) are called nontrivial translations, in contrast to the trivial translations t defined by (2.1). The set of vectors τ, i.e., of numbers μ_i in (2.2), is not arbitrary.

Since the elements r and r' themselves form a group, and $r'^{\nu} = e$, then, for instance, the condition $(r'|\tau)^{\nu} = (e|\mathbf{t})$ should be met, where $\mathbf{t}$ is one of the trivial translations.

Groups which do not contain nontrivial translations are called symmorphic. There are 230 different space groups altogether. 80 of them differ in that, under any transformation except for corresponding translations, one of the planes remains fixed. Excluding the translations which are perpendicular to this plane, we obtain 80 doubly periodic groups describing the symmetry of two-dimensional lattices. Of these 80 groups, 17 do not contain inversion or reflection in the plane. These 17 are called *surface groups*. The symmetry

Table 2.1. Types of crystal systems

System	Notation	Point group: By Schoenflies	Point group: International	Type of Bravais lattice	Notation Basic vectors
Triclinic	T	S_2	$\bar{1}$	Primitive	$T\,(\Gamma_{tr})$ Any $\mathbf{a}_1, \mathbf{a}_2, \mathbf{a}_3$
Monoclinic	M	C_{2h}	$2/m$	Primitive	$M\,(\Gamma_m)\,\mathbf{a}_3 \perp \mathbf{a}_1, \mathbf{a}_3 \perp \mathbf{a}_2$
				Base-centered	$M_c\,(\Gamma_m^b)\,2\mathbf{a}_3 - \mathbf{a}_1 \perp \mathbf{a}_1, 2\mathbf{a}_3 - \mathbf{a}_1 \perp \mathbf{a}_2$
Orthogonal (orthorhombic)	O	D_{2h}	$(2/m)\,(2/m)\,(2/m) \equiv mmm$	Primitive	$O\,(\Gamma_0)\,\mathbf{a}_1 \perp \mathbf{a}_2 \perp \mathbf{a}_3 \perp \mathbf{a}_1$
				Base-centered	$O_c\,(\Gamma_0^b)\,\mathbf{a}_1 \perp \mathbf{a}_2 \perp 2\mathbf{a}_3 - \mathbf{a}_1 \perp \mathbf{a}_1$
				Body-centered	$O_v\,(\Gamma_0^v)\,\mathbf{a}_1 \perp \mathbf{a}_2 \perp 2\mathbf{a}_3 - \mathbf{a}_1 - \mathbf{a}_2 \perp \mathbf{a}_1$
				Face-centered	$O_f\left(\Gamma_0^f\right)\mathbf{a}_1 \perp 2\mathbf{a}_2 - \mathbf{a}_1 \perp 2\mathbf{a}_3 - \mathbf{a}_1 \perp \mathbf{a}_1$
Tetragonal	Q	D_{4h}	$\left(\frac{4}{m}\right)(\bar{3})\left(\frac{2}{m}\right) \equiv 4/mmm$	Primitive	$Q\,(\Gamma_q)\,\mathbf{a}_1 \perp \mathbf{a}_2 \perp \mathbf{a}_3 \perp \mathbf{a}_1;\ a_1 = a_2$
				Body-centered	$Q_v\left(\Gamma_q^v\right)\mathbf{a}_1 \perp \mathbf{a}_2 \perp 2\mathbf{a}_3 - \mathbf{a}_1 - \mathbf{a}_2 \perp \mathbf{a}_1;\ a_1 = a_2$
Rhombohedral	R	D_{3d}	$\bar{3}\left(\frac{2}{m}\right) \equiv \bar{3}m$	Primitive	$R\,(\Gamma_{rh})\,a_1 = a_2 = a_3, \mathbf{a}_1\hat{\mathbf{a}}_2 = \mathbf{a}_1\hat{\mathbf{a}}_3 = \mathbf{a}_2\hat{\mathbf{a}}_3$
Hexagonal	H	D_{6h}	$\left(\frac{6}{m}\right)\left(\frac{2}{m}\right)\left(\frac{2}{m}\right) \equiv 6/mmm$	Primitive	$H\,(\Gamma_h)\,\mathbf{a}_3 \perp \mathbf{a}_1, \mathbf{a}_3 \perp \mathbf{a}_2;\ a_1 = a_2, \mathbf{a}_1\hat{\mathbf{a}}_2 = \frac{2\pi}{3}$
Cubic	C	O_h	$(4/m)\,(\bar{3})\,(2/m) \equiv m3m$	Primitive	$C\,(\Gamma_c)\,\mathbf{a}_1 \perp \mathbf{a}_2 \perp \mathbf{a}_3 \perp \mathbf{a}_1;\ a_1 = a_2 = a_3$
				Face-centered	$C_f\left(\Gamma_c^f\right)\mathbf{a}_i + \mathbf{a}_{i+1} - \mathbf{a}_{i+2} \perp \mathbf{a}_{i+1} + \mathbf{a}_{i+2} - \mathbf{a}_i;$ $a_1 = a_2 = a_3$
				Body-centered	$C_v\,(\Gamma_c^v)\,\mathbf{a}_1 \perp \mathbf{a}_2, \mathbf{a}_1 \perp 2\mathbf{a}_3 - \mathbf{a}_1 - \mathbf{a}_2 \perp \mathbf{a}_2;$ $a_1 = a_2 = \lvert 2\mathbf{a}_3 - \mathbf{a}_1 - \mathbf{a}_2 \rvert$

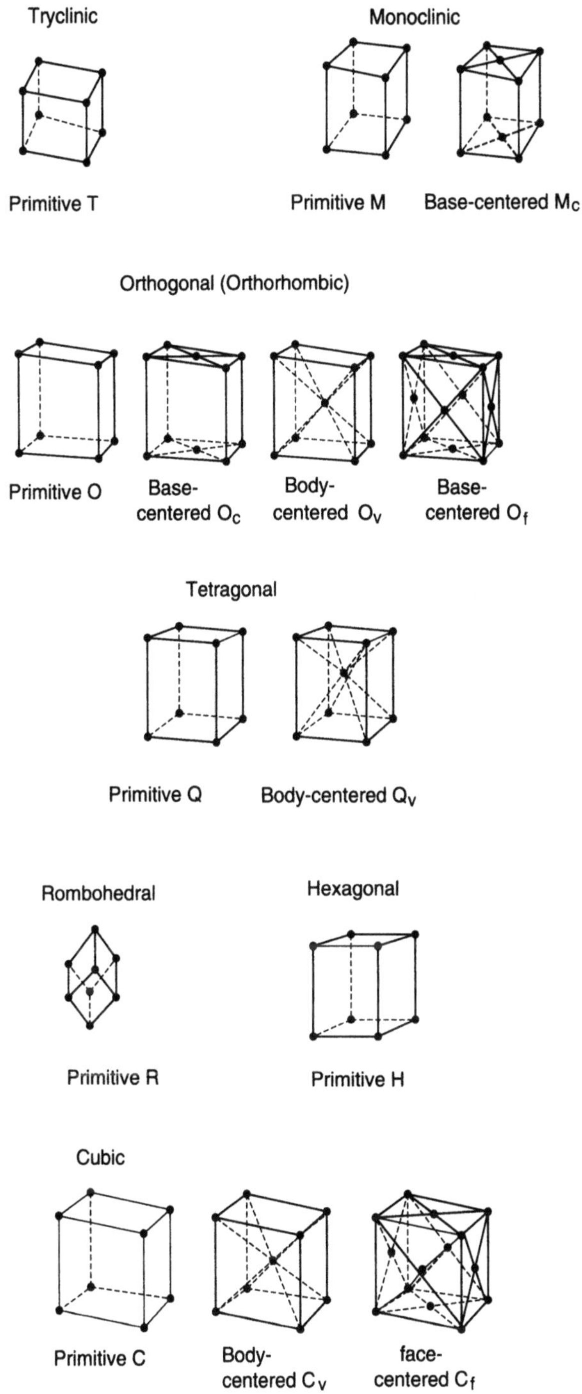

Fig. 2.1. Bravais lattices

Table 2.2. Crystal classes-distribution in Bravais lattice systems (Syngonies)

Syngony	Notation	Classes	
		Notation by Schoenflies	International notation
Triclinic	T	e; C_i	1; $\bar{1}$.
Monoclinic	M	C_2; C_s; C_{2h}	2; m; 2/m.
Orthogonal (orthorhombic)	O	C_{2v}; D_2; D_{2h}	2mm; 222; mmm.
Tetragonal	Q	C_4; S_4; C_{4h}; D_4; C_{4v}; D_{2d}; D_{4h}	4; $\bar{4}$; 4/m; 422; 4mm; $\bar{4}$2m; 4/mmm
Rhombohedral	R	C_3; S_6; C_{3v}; D_3; $D_{3d}(D_{3v})$	3; $\bar{3}$; 3m; 32; $\bar{3}$m.
Hexagonal	H	C_{3h}; D_{3h}; C_6; C_{6h}; C_{6v}; D_6; D_{6h}	$\bar{6}$, $\bar{6}$m2, 6; 6/m; 6mm; 622; 6/mmm
Cubic	C	T; T_h; T_d; O, O_n	23; m3; $\bar{4}$3m; 432; m$\bar{3}$m.

of quantum wells and superlattices is determined both by that of the original crystal and by the structure of the wells proper or superlattices. For nonsymmetric wells or aperiodic superlattices it is described by one of the 17 surface groups, and for symmetric wells, by the remaining 63 doubly periodic groups. For regular superlattices, the period in the corresponding direction is given by that of the superlattice.

The set of all rotation operations r and r' of a space group makes up a point group called the direction group and denoted by F. This group defines the crystal class. There are 32 crystal classes altogether, their notation coinciding with that of the corresponding point groups. It should be stressed that the direction group accounts for all the macroscopic properties of a crystal.

The crystal classes representing subgroups of the corresponding group which determines the Bravais lattice symmetry (or coinciding with it) can crystallize in any of the above seven systems. The crystal classes belong to the system of the lowest possible symmetry. The distribution of classes according to the systems is given in Table 2.2. The crystals of rhombohedral classes presented in the Table can have both rhombohedral and hexagonal Bravais lattices.

A set of all elements $(r|\tau)$ of a space group forms a factor group of the subgroup of translations. This means that all elements $(r|\tau+\mathbf{t})$ of the space group are obtained by multiplication of the factor group elements by the elements of the translation group $(0|\mathbf{t})$. A factor group is isomorphic with the point group of directions of the corresponding class, i.e., there is a one-to-one correspondence between the elements $(r|\tau)$ and r. However, this does not imply that the space groups relating to the same crystal class are themselves isomorphic.

2.4 Group Representations, Characters

Consider the solutions of Schrödinger's equation $\varphi(\mathbf{x})$ whose Hamiltonian $\hat{\mathcal{H}}$ possesses a symmetry, i.e., is invariant under symmetry transformations g of the group G. The wave functions themselves do not possess this symmetry, since the operations g over the coordinate system transform them into functions $\varphi(g^{-1}\mathbf{x})$ which in the general case differ from $\varphi(\mathbf{x})$. Applying all h operations g_k of the group G, we obtain h functions $\varphi(g_k^{-1}\mathbf{x})$, each being a solution of Schrödinger's equation with the same energy E. In the general case only some of these functions are linearly independent. Denote these linearly independent functions by $\varphi_i(\mathbf{x})$. Then all functions $\varphi_i(g^{-1}\mathbf{x})$ can be represented as linear combinations

$$\varphi_i(g^{-1}\mathbf{x}) = \sum_j D_{ji}(g)\varphi_j(\mathbf{x}). \tag{2.3}$$

One can readily see that if to the elements g_1 and g_2 correspond matrices $D(g_1)$ and $D(g_2)$, then to the element $g_3 = g_2 g_1$ corresponds a matrix

$$D(g_2 g_1) = D(g_2)D(g_1) \tag{2.4}$$

with matrix elements $D_{ji}(g_3) = \Sigma_k D_{jk}(g_2)D_{ki}(g_1)$. The matrices $D(g)$ satisfying these properties form a representation of the group. If the basis functions $\varphi_i(\mathbf{x})$ are constructed to be orthonormal, i.e., $\int \varphi_i^* \varphi_j \, dx = \delta_{ij}$, ($\delta_{ij} = 1$ for $i = j$, $\delta_{ij} = 0$ for $i \neq j$), then $D(g)$ are unitary matrices, i.e.,

$$D^{-1}(g) = D(g^{-1}) = D^{+}(g) = \tilde{D}^{*}(g).$$

Here the matrix $D^{-1}(g)$ is inverse to $D(g)$, i.e., it satisfies the condition $D(g)D^{-1}(g) = I$, where I is the identity matrix with elements $I_{ij} = \delta_{ij}$, $\tilde{D}$ is a transpose, $\tilde{D}_{ij} = D_{ji}$, D^* is a complex conjugate, and $D^+_{ij} = D^*_{ji}$.

If the basis functions $\varphi_i(\mathbf{x})$ are chosen such that no unitary transformation of them can divide simultaneously all h matrices $D(g)$ into quasidiagonal matrices of the type

$$D(g) = \begin{bmatrix} D_1(g) & 0 & 0 & 0 \\ 0 & D_2(g) & 0 & 0 \\ 0 & 0 & D_3(g) & 0 \\ 0 & 0 & 0 & D_4(g) \end{bmatrix} \tag{2.5}$$

including matrices $D_i(g)$ of a lower dimension, then this representation is called irreducible. Obviously, the dimension of a representation determines the degree of degeneracy of the corresponding term. It is here that the particular significance of irreducible representations lies. The actual form of the matrices $D(g)$ depends on the choice of the basis. Under unitary transformation S, i.e., as one transforms from a basis φ_i to $\varphi_i' = \Sigma_j S_{ji}\varphi_j$, the matrices $D(g)$ are sent into the matrices $D'(g) = S^{-1}D(g)S$. However, the traces of the matrices $D'(g)$ and $D(g)$ coincide for all g; indeed, since $(S^{-1}S)_{ik} = \delta_{ik}$, we have $\Sigma_i(S^{-1}DS)_{ii} = \Sigma_i D_{ii}$.

Obviously, for inequivalent representations which cannot be sent into one another under unitary transformation, the traces of the matrices $D(g)$ and $D'(g)$ do not coincide. This accounts for the particular significance of $\mathrm{Tr}\{D(g)\}$, which in the theory of representations is called the character of a representation and is denoted by $\chi(g)$. There are several important theorems related to the properties of irreducible representations and their characters. The most essential of them are given below.

The number of irreducible representations of a finite group is also finite. Denote this number by a symbol N, the corresponding irreducible representation, by an index μ, and its dimension, by n_μ.

By Burnside's theorem

$$\sum_{\mu=1}^{N} n_\mu^2 = h. \tag{2.6}$$

We recall that h is the number of elements in a group. The characters of inequivalent representations are orthogonal:

$$\sum_g \chi_\mu^*(g)\chi_\nu(g) = 0 \quad \text{for} \quad \mu \neq \nu, \tag{2.7a}$$

$$\sum_g |\chi_\mu(g)|^2 = h \quad \text{for} \quad \mu = \nu. \tag{2.7b}$$

If the characters of a representation are known, then (2.7) provide the possibility of determining whether this representation is irreducible, or, alternately, into what irreducible representations it can be decomposed. Indeed, if the representation μ occurs N_μ times in a representation with characters $\chi(g)$, then

$$\chi(g) = \sum_\mu N_\mu \chi_\mu(g)$$

whence, in accordance with (2.7), we obtain

$$N_\mu = \frac{1}{h}\sum_g \chi_\mu^*(g)\chi(g). \tag{2.8}$$

Irreducible representations always contain an identity representation whose basis function is an invariant, that is, it does not change under all transformations $g \in G$. For the identity representation $\chi(g) = D(g) = 1$. As follows from (2.8), the identity representation occurs N_0 times in a representation with the characters $\chi(g)$, where

$$N_0 = \frac{1}{h}\sum_g \chi(g). \tag{2.9}$$

In practical applications, it often becomes necessary to find the characters of the representation according to which products of the basis functions, each of which form basis functions of the known irreducible representations belong. If n_μ functions φ_i^μ transform according to a representation μ with the characters $\chi_\mu(g)$, and n_ν functions ψ_j^ν transform according to the representation ν with the characters χ_ν (g), then $n_\mu n_\nu$ functions $\psi_{ij}^{\mu\nu} = \varphi_i^\mu \psi_j^\nu$ transform according to the representation with the characters

$$\chi_{\mu\nu}(g) = \chi_\mu(g)\chi_\nu(g). \tag{2.10}$$

Using (2.8 and 10), one can readily determine the irreducible representations according to which the functions $\varphi_{ij}^{\mu\nu}$ transform. For instance, the basis functions transforming according to the identity representation are obtained from the products of the functions φ_i^{μ} and ψ_j^{ν} which transform according to the complex conjugate representations $D_\nu = D_\mu^*$. For $\mu = \nu$

$$\chi_{\mu\mu}(g) = \chi_\mu^2(g). \tag{2.11}$$

In this case, of n_μ^2 functions $\varphi_i^{\mu}\psi_j^{\mu}$, one can construct $1/2(n_\mu(n_\mu + 1))$ symmetrized products $\psi_{ij}^{s} = 1/2(\varphi_i\psi_j + \varphi_j\psi_i)$, and $1/2(n_\mu(n_\mu - 1))$ antisymmetrized products $\psi_{ij}^{a} = 1/2(\varphi_i\psi_j - \varphi_j\psi_i)$. For the symmetrized product, the character is

$$\left[\chi_\mu^2(g)\right]_{\text{symm}} = \frac{1}{2}\left[\chi_\mu^2(g) + \chi_\mu(g^2)\right] \tag{2.12}$$

and for the antisymmetrized one,

$$\left\{\chi_\mu^2(g)\right\}_{\text{antisymm}} = \frac{1}{2}\left[\chi_\mu^2(g) - \chi_\mu(g^2)\right]. \tag{2.13}$$

Classes of elements play a particular role in the representation theory. By the class of an element g one understands the set of all elements conjugate to g. By definition, elements g_1 and g_2 are conjugate if $g_2 = X^{-1}g_1X$, where X is any of the elements in the group G. By taking for X successively all elements $g \in G$, we obtain all elements of the class g_1, and by taking all elements g_1, we define all possible classes. Note that each element can occur only in one class ρ. If the number of elements of a given class is h_ρ, then each element of this class will occur the same number of times, $\nu = h/h_\rho$, in the sequence $X^{-1}gX$. For commutative groups $X^{-1}gX = gX^{-1}X = g$, i.e., each class contains only one element. It can be proved that the number of irreducible representations N is equal to that of classes, N_ρ,

$$N = N_\rho. \tag{2.14}$$

Indeed, the characters of all elements of one class coincide; the unitary nature of the matrices $D(g)$ implies that $\mathrm{Tr}\{D^{-1}(X)D(g)D(X)\} = Tr\{D(g)\}$. For elements g_1 and g_2 contained in the same class, one can write

$$\sum_{\mu=1}^{N_\rho} \chi_\mu^*(g_1)\chi_\mu(g_2) = \sum_\mu |\chi_\mu(g_1)|^2 = \frac{h}{h_\rho}. \tag{2.15a}$$

Summation here is performed over all irreducible representations μ.

For elements g_1 and g_2 belonging to different classes

$$\sum_{\mu=1}^{N_\rho} \chi_\mu^*(g_1)\chi_\mu(g_2) = 0. \tag{2.15b}$$

Since an identity element e forms a class in itself because $X^{-1}eX = e$, then for elements of all the other classes we will have, according to (2.15),

$$\sum_{\mu=1}^{N_\rho} \chi_\mu(g) = 0 \ (g \neq e). \tag{2.16}$$

2.5 Point-Group Representations

To construct representations of point groups, one has, first of all, to divide the elements of the groups into classes. This can be done by simple geometric reasoning. The rotations c_n^k (or improper rotations s_n^k) about different axes l or l' through the same angle belong to the same class if among the symmetry transformations there are ones which send the axis l into l'. In the same way reflections in planes σ and σ' belong to one class if there are symmetry elements which send plane σ into σ'. The rotations c_n^k and c_n^{-k} about the same axis belong to the same class if there is a two-fold axis perpendicular to C_n or a reflection plane passing through this axis. Such an axis C_n is called double-sided. If there are no such symmetry elements, the axis C_n is called one-sided. This rule holds also for the improper rotations s_n^k and s_n^{-k} (k is odd). Just as for the identity element e, inversion commutes with all elements of a group and forms a class by itself.

Using these rules, one can divide the elements of point groups into classes, find the number of the classes and, accordingly, the number of inequivalent irreducible representations. After this, (2.6) permits one to establish the dimension of the irreducible representations. The characters $\chi_\mu(g)$ can be found practically for all point groups by means of the orthogonality relations (2.7 and 15). Table 2.3 lists the characters of the representations of the above-mentioned point groups, as well as the basis functions transforming according to these representations. These functions can be readily determined by first finding the transformation matrices for the components of the polar or axial vectors for the generators; after that, knowing the corresponding characters, one can use (2.8) to find the irreducible representations according to which these components transform. This done, one can determine in a straightforward way, using (2.10–13 and 8), the representations according to which the transformation of their products occurs. Table 2.3 also contains multiplication tables for point groups which specify the irreducible representations contained in the products $D_\mu \times D_\nu^*$.

Table 2.3. Character tables of point groups

Group E

E	e
A	1

Groups C_2, C_i, C_s

C_2			e	c_2	C_2	C_i	C_s
	C_i		e	i			
		C_s	e	σ_h			
A	A^+	A^+	1	1	z	J_x, J_y, J_z	x, y
B	A^-	A^-	1	-1	x, u	x, y, z	z

Group C_3

C_3	e	c_3	c_3^2	
A	1	1	1	z
B_1	1	ε_3	ε_3^2	$x - \mathrm{i}y$
B_2	1	ε_3^2	ε_3	$x + \mathrm{i}y$
$\varepsilon_3 = \exp(2\pi \mathrm{i}/3)$				

Groups C_4, S_4

C_4		e	c_4	c_2	c_4^3	C_4	S_4
	S_4	e	s_4	c_2	s_4^3		
A	A	1	1	1	1	z	J_z
B_1	B_1	1	-1	1	-1	$x^2 - y^2$	z
B_2	B_2	1	i	-1	$-$i	$x - \mathrm{i}y$	$x - \mathrm{i}y$, $J_x + \mathrm{i}J_y$
B_3	B_3	1	$-$i	-1	i	$x + \mathrm{i}y$	$x + \mathrm{i}y$, $J_x - \mathrm{i}J_y$

Groups D_3, C_{3v}

D_3		e	$2c_3$	$3u_2$	D_3	C_3
	C_{3v}	e	$2c_3$	$3\sigma_v$		
A_1	A_1	1	1	1	$x^2 + y^2, z^2$	z
A_2	A_2	1	1	-1	z	J_z
E	E	2	-1	0	x, y	x, y, J_x, J_y

Groups D_d, C_{4v}, D_{2d}

D_4			e	c_2	$2c_4$	$2u_2$	$2u_2'$	D_4	C_{4v}	D_{2d}
	C_{4v}		e	c_2	$2c_4$	$2\sigma_v$	$2\sigma_v'$			
		D_{2d}	e	c_2	$2s_4$	$2u_2$	$2\sigma_d$			
A_1	A_1	A_1	1	1	1	1	1	$z^2, x^2 + y^2$	z	$z^2, x^2 + y^2$
A_2	A_2	A_2	1	1	1	-1	-1	z	J_z	J_z
B_1	B_1	B_1	1	1	-1	1	-1	$x^2 - y^2$	$x^2 - y^2$	$x^2 - y^2$
B_2	B_2	B_2	1	1	-1	-1	1	xy	xy	xy, z
E	E	E	2	-2	0	0	0	x, y	x, y, J_x, J_y	x, y, J_x, J_y

Group T

T	e	$3c_2$	$4c_3$	$4c_3^2$	
A	1	1	1	1	$x^2 + y^2 + z^2$
B_1	1	1	ε_3	ε_3^2	$x^2 + \varepsilon_3^2 y^2 + \varepsilon_3 z^2$
B_2	1	1	ε_3^2	ε_3	$x^2 + \varepsilon_3 y^2 + \varepsilon_3^2 z^2$
F	1	-1	0	0	x, y, z

Groups T_d, O

T_d		e	$8c_3$	$3c_2$	$6\sigma_d$	$6s_4$	T_d	O
	O	e	$8c_3$	$3c_2$	$6u_2$	$6c_4$		
A_1	A_1	1	1	1	1	1	$x^2+y^2+z^2$	$x^2+y^2+z^2$
A_2	A_2	1	1	1	-1	-1	$[J_xJ_yJ_z]_s$	xyz
E	E	2	-1	2	0	0	$x^2+\varepsilon_3y^2+\varepsilon_3^2z^2, x^2+\varepsilon_3^2y^2+\varepsilon_3z^2$	
F_1	F_1	3	0	-1	-1	1	J_x, J_y, J_z	x, y, z
F_2	F_2	3	0	-1	1	-1	$x, y, z; yz, zx, xy$	yz, zx, xy

Groups $D_2 = C_2 \times C_2'$, $C_{2h} = C_2 \times C_i$, $C_{2v} = C_2 \times C_s'$

D_2			e	c_{2z}	c_{2x}	c_{2y}	D_3	C_{2h}	C_{2v}
	C_{2h}		e	c_2	i	σ_h			
		C_{2v}	e	c_2	σ_v	σ_v'			
A_1	A^+	A^+	1	1	1	1	x^2, y^2, z^2	J_z	z
A_2	A^-	A^-	1	1	-1	-1	z	z	J_z
B_1	B^+	B^+	1	-1	1	-1	x	J_x, J_y	y, J_x
B_2	B^-	B^-	1	-1	-1	1	y	x, y	x, J_y

Groups $C_6 = C_3 \times C_2$, $C_{3h} = C_3 \times C_s$, $S_6 = C_3 \times C_i$

C_6			e	c_3	c_3^2	c_2	c_6^5	c_6	C_6	C_{3h}	S_6
	C_{3h}		e	c_3	c_3^2	σ_h	s_3	s_3^2			
		S_6	e	c_3	c_3^2	i	s_6^5	s_6			
A_1	A^+	A^+	1	1	1	1	1	1	z	J_z	J_z
A_2	A^-	A^-	1	1	1	-1	-1	-1	$(x \pm \mathrm{i}y)^3$	z	z
B_1	B_1^+	B_1^+	1	ε_3	ε_3^2	1	ε_3	ε_3^2	$(x-\mathrm{i}y)^2$	$x+\mathrm{i}y$	$J_x+\mathrm{i}J_y$
B_2	B_1^-	B_1^-	1	ε_3	ε_3^2	-1	$-\varepsilon_3$	$-\varepsilon_3^2$	$x+\mathrm{i}y$	$J_x+\mathrm{i}J_y$	$x+\mathrm{i}y$
B_3	B_2^+	B_2^+	1	ε_3^2	ε_3	1	ε_3^2	ε_3	$(x+\mathrm{i}y)^2$	$x-\mathrm{i}y$	$J_x-\mathrm{i}J_y$
B_4	B_2^-	B_2^-	1	ε_3^2	ε_3	-1	$-\varepsilon_3^2$	$-\varepsilon_3$	$x-\mathrm{i}y$	$J_x-\mathrm{i}J_y$	$x-\mathrm{i}y$

Groups $D_6 = D_3 \times C_2$, $C_{6v} = C_{3v} \times C_2$, $D_{3h} = D_3 \times C_s$

D_6			e	$2c_3$	$3u_2$	c_2	$2c_6$	$3u_2'$	D_6	C_{6v}	D_{3h}
	C_{6v}		e	$2c_3$	$3\sigma_v$	c_2	$2c_6$	$3\sigma_v'$			
		D_{3h}	e	$2c_3$	$3u_2$	σ_h	$2s_3$	$3\sigma_v'$			
A_1	A_1	A_1^+	1	1	1	1	1	1	x^2+y^2, z^2	z	x^2+y^2, z^2
A_2	B_2	A_1^-	1	1	1	-1	-1	-1	$(x+\mathrm{i}y)^3+$ $+(x-\mathrm{i}y)^3$	$(x+\mathrm{i}y)^3+$ $+(x-\mathrm{i}y)^3$	$\mathrm{i}z[(x+\mathrm{i}y)^3-$ $-(x-\mathrm{i}y)^3]$
A_3	A_2	A_2^+	1	1	-1	1	1	-1	z	J_z	J_z
A_4	B_1	A_2^-	1	1	-1	-1	-1	1	$\mathrm{i}[(x+\mathrm{i}y)^3-$ $-(x-\mathrm{i}y)^3]$	$\mathrm{i}[(x+\mathrm{i}y)^3-$ $-(x-\mathrm{i}y)^3]$	z
E_1	E_2	E^+	2	-1	0	2	-1	0	$(x+\mathrm{i}y)^2$, $(x-\mathrm{i}y)^2$	$(x+\mathrm{i}y)^2$, $(x-\mathrm{i}y)^2$	x, y
E_2	E_1	E^-	2	-1	0	-2	1	0	x, y	$x, y; J_x, J_y$	J_x, J_y

For the remaining eight groups

$C_{4h} = C_4 \times C_i, C_{6h} = C_6 \times C_i, D_{2h} = D_2 \times C_i, D_{3d} = D_3 \times C_i$

$D_{4h} = D_4 \times C_i, D_{6h} = D_6 \times C_i, T_h = T \times C_i, O_h = O \times C_i = T_d \times C_i$,

each representation of the original group D is associated with two representations $D^{\pm}$, with characters $\chi(ig) = \pm\chi(g)$ belonging to even and odd functions of the coordinates, respectively. All the functions $\varphi(J)$ transform according to even representations.

Multiplication table for T_d and O

	A_1	A_2	E	F_1	F_2
A_1	A_1	A_2	E	F_1	F_2
A_2	A_2	A_1	E	F_2	F_1
E	E	E	$A_1 + A_2 + E$	$F_1 + F_2$	$F_1 + F_2$
F_1	F_1	F_2	$F_1 + F_2$	$A_1 + E + F_1 + F_2$	$A_2 + E + F_1 + F_2$
F_2	F_2	F_1	$F_1 + F_2$	$A_2 + E + F_1 + F_2$	$A_1 + E + F_1 + F_2$

Basis functions for the groups T_d, O, O_h

Representations			Basis functions	
T_d	O	O_h	$\varphi(\mathbf{x})$	$\varphi(\mathbf{J})$
A_1	A_1	A_1^+	$x^2 + y^2 + z^2$; $x^4 + y^4 + z^4$	
A_2	A_2	A_2^+	$x^4(y^2 - z^2) + y^4(z^2 - x^2) +$ $+ z^4(x^2 - y^2)$	$[J_x J_y J_z]_s$
E	E	E^+	$x^2 + \varepsilon_3 y^2 + \varepsilon_3^2 z^2$, $x^2 + \varepsilon_3^2 y^2 + \varepsilon_3 z^2$; $x^4 + \varepsilon_3 y^4 + \varepsilon_3^2 z^4$, $x^4 + \varepsilon_3^2 y^4 + \varepsilon_3 z^4$	$J_x^2 + \varepsilon_3 J_y^2 + \varepsilon_3^2 J_z^2$, $J_x^2 + \varepsilon_3^2 J_y^2 + \varepsilon_3 J_z^2$
F_1	F_1	F_1^+	$yz(y^2 - z^2)$, $zx(z^2 - x^2)$, $xy(x^2 - y^2)$	J_x, J_y, J_z; J_x^3, J_y^3, J_z^3
F_2	F_2	F_2^+	yz, zx, xy; yzx^2, zxy^2, xyz^2	$[J_y J_z]_s$, $[J_z J_x]_s$, $[J_x J_y]_s$; $V_x = [J_x(J_y^2 - J_z^2)]_s$, V_y, V_z
A_1	A_2	A_1^-	xyz	
A_2	A_1	A_2^-	$xyz[x^4(y^2 - z^2) +$ $+ y^4(z^2 - x^2) + z^4(x^2 - y^2)]$	
E	E	E^-	$xyz(x^2 + \varepsilon_3 y^2 + \varepsilon_3^2 z^2)$, $xyz(x^2 + \varepsilon_3^2 y^2 + \varepsilon_3 z^2)$	
F_1	F_2	F_1^-	$x(y^2 - z^2)$, $y(z^2 - x^2)$, $z(x^2 - y^2)$; $x^3(y^2 - z^2)$, $y^3(z^2 - x^2)$, $z^3(x^2 - y^2)$	
F_2	F_1	F_2^-	x, y, z; x^3, y^3, z^3; x^5, y^5, z^5; xy^2z^2, yz^2x^2, zx^2y^2	

2.6 Spinor Representations

The basis functions of the above representations are the usual functions of coordinates x, y, z, which can be represented as functions Y_l^m with a distinct integer angular momentum l and its projection m on a given axis, or distinct superpositions of such functions. The wave functions of the electron depend not only on coordinates but on spin as well, and there are states with a half-integer moment l and half-integer m, or a superposition of such states. For the states Y_l^m, the component of the matrix (Dc_φ) associated with rotation about the quantization axis will be

$$D_{mm'}(c_\varphi) = \mathrm{e}^{im\varphi}\delta_{mm'}. \tag{2.17}$$

For $\varphi = 2\pi$, we obtain $D(c_{2\pi}) = 1$ for states with an integer m, and $D(c_{2\pi}) = -1$ for those with a half-integer m. Therefore, under operations g corresponding to a rotation through an angle $\varphi < 2\pi$ (or a rotation + inversion), the matrices of the spinor representations satisfy a relation different from (2.4):

$$D(g_2)D(g_1) = \omega_{12}D(g_2, g_1), \tag{2.18}$$

where $\omega_{12} = 1$ if g_2g_1 is rotation about an axis through an angle $\theta < 2\pi$, and $\omega_{12} = -1$ for $\theta \geqslant 2\pi$. Note that rotation through an angle $-\theta$ should be considered as rotation through $4\pi - \theta$.

The representations satisfying the condition (2.17) with $|\omega_{ij}| = 1$ are called projective, and the factors $\omega_{12} = \omega(g_2, g_1)$, a factor system. Spinor representations may be considered conventional representations of a double group containing, besides the elements g, also an element Qg, where $Q = c_{2\pi}$. This element Q commutes with all elements in the group. The above double group contains, besides the conventional (vector) representations, for which $D(Q) = D(e) = 1$, also spinor representations, with $D(Q) = -D(e) = -1$. This method of constructing projective representations was proposed by *Schur* in 1904 [2.1–3], and extended to spinor representations by *Bethe* in 1931 [2.4].

Table 2.4 specifies the characters of the spinor representations of point groups, as well as the corresponding basis functions. The characters of the representation $D_{1/2}$ according to which the spinors α, β with $m = \pm 1/2$ transform can be immediately derived from (2.17):

$$\chi(c_\varphi) = 2\cos(\varphi/2),$$
$$\chi(s_\varphi) = \chi(c_i c_{\pi+\varphi}) = -2\sin(\varphi/2), \chi(c_i) = 2. \tag{2.19}$$

Table 2.4. Character tables for spinor representation

Group E

E	e	Basic funct.	Mult. table
			A
A'	1	α, β	A'

Groups C_2, C_s

C_2	e	c_2	Basis	Mult. table	
C_s	e	σ_h	funct.	A	B
A'	1	i	α	A'	B'
B'	1	$-$i	β	B'	A'

Group C_3

C_3	e	c_3	c_3^2	Basis funct.	Mult. table		
					A	B_1	B_2
A'	1	-1	1		A'	B_1'	B_2'
B_1'	1	$-\varepsilon_3$	ε_3^2	β	B_1'	B_2'	A'
B_2'	1	$-\varepsilon_3^2$	ε_3	α	B_2'	A'	B_1'

Groups C_4, S_4

C_4	e	c_4	c_2	c_4^3	Basis funct.		Multipl. table			
S_4	e	s_4	c_2	s_4^3	C_4	S_4	A	B_1	B_2	B_3
A'	1	ω	i	ω^3	α		A'	B_1'	B_2'	B_3'
B_1'	1	$-\omega$	i	$-\omega^3$		β	B_1'	A'	B_3'	B_2'
B_2'	1	ω^3	$-$i	ω		α	B_2'	B_3'	B_1'	A'
B_3'	1	$-\omega^3$	$-$i	$-\omega$	β		B_3'	B_2'	A'	B_1'
$\omega = \exp(\pi \mathrm{i}/4)$										

Groups C_6, C_{3h}

C_6	e	c_6	c_3	c_2	c_3^2	c_6^5	Basis funct.		Multiplic. table					
C_{3h}	e	s_3^2	c_3	σ_h	c_3^2	s_3	C_6	C_{3h}	A_1	A_2	B_1	B_2	B_3	B_4
A_1'	1	i	-1	$-$i	1	i			A_1'	A_2'	B_1'	B_2'	B_3'	B_4'
A_2'	1	$-$i	-1	i	1	$-$i			A_2'	A_1'	B_2'	B_1'	B_4'	B_3'
B_1'	1	$\mathrm{i}\varepsilon_3^2$	$-\varepsilon_3$	$-$i	ε_3^2	$\mathrm{i}\varepsilon_3$	β		B_1'	B_2'	B_3'	B_4'	A_1'	A_2'
B_2'	1	$-\mathrm{i}\varepsilon_3^2$	$-\varepsilon_3$	i	ε_3^2	$-\mathrm{i}\varepsilon_3$		β	B_2'	B_1'	B_4'	B_3'	A_2'	A_1'
B_3'	1	$\mathrm{i}\varepsilon_3$	$-\varepsilon_3^2$	$-$i	ε_3	$\mathrm{i}\varepsilon_3^2$		α	B_3'	B_4'	A_1'	A_2'	B_1'	B_2'
B_4'	1	$-\mathrm{i}\varepsilon_3$	$-\varepsilon_3^2$	i	ε_3	$-\mathrm{i}\varepsilon_3^2$	α		B_4'	B_3'	A_2'	A_1'	B_2'	B_1'

Groups D_3, C_{3v}

D_3	e	c_3	c_3^2	$3u_2$	Basic funct.	Multipl. table		
C_{3v}	e	c_3	c_3^2	$3\sigma_v$		A_1	A_2	E
A_1'	1	-1	1	i		A_1'	A_2'	E'
A_2'	1	-1	1	$-$i		A_2'	A_1'	E'
E'	2	1	-1	0	α, β	E'	E'	$A_1' + A_2' + E'$

Groups D_2, C_{2v}

D_2	e	c_{2z}	c_{2x}	c_{2y}	Basis functions		Multiplication table			
C_{2v}	e	c_2	σ_v	σ_v'	D_2	C_{2v}	A_1	A_2	B_1	B_2
E'	2	0	0	0	α, β	α, β	E'	E'	E'	E'

Groups D_4, C_{4v}, D_{2d}

D_4	e	c_2	c_4	c_4^3	$2u_2$	$2u_2'$	Basis functions		Multiplication table				
C_{4v}	e	c_2	c_4	c_4^3	$2\sigma_v$	$2\sigma_v'$	D_4, C_{4v}	D_{2d}					
D_{2d}	e	c_2	S_4	S_4^3	$2u_2$	$2\sigma_d$			A_1	A_2	B_1	B_2	E
E_1'	2	0	$\sqrt{2}$	$-\sqrt{2}$	0	0	α, β	$(x+\mathrm{i}y)\alpha$, $(x-\mathrm{i}y)\beta$	E_1'	E_1'	E_2'	E_2'	$E_1'+E_2'$
E_2'	2	0	$-\sqrt{2}$	$\sqrt{2}$	0	0	$(x+\mathrm{i}y)\alpha$, $(x-\mathrm{i}y)\beta$	α, β	E_2'	E_2'	E_1'	E_1'	$E_1'+E_2'$

Groups D_6, C_{6v}, D_{3h}

D_6	e	c_2	c_3	c_3^2	c_6	c_6^5	$3u_2$	$3u_2'$	Basis functions	
C_{6v}	e	c_2	c_3	c_3^2	c_6	c_6^5	$3\sigma_v$	$3\sigma_v'$	D_6, C_{6v}	D_{3h}
D_{3h}	e	σ_h	c_3	c_3^2	s_3^2	s_3	$3u_2$	$3\sigma_v$		
E_1'	2	0	1	-1	$\sqrt{3}$	$-\sqrt{3}$	0	0	α, β	α, β
E_2'	2	0	1	-1	$-\sqrt{3}$	$\sqrt{3}$	0	0	$(x+\mathrm{i}y)^3\alpha$, $(x-\mathrm{i}y)^3\beta$	$(x-\mathrm{i}y)\alpha$, $(x+\mathrm{i}y)\beta$
E_3'	2	0	-2	2	0	0	0	0	$(x+\mathrm{i}y)\alpha$, $(x-\mathrm{i}y)\beta$	$(x+\mathrm{i}y)\alpha$, $(x-\mathrm{i}y)\beta$

Multiplication table

	A_1	A_3	A_2	A_4	E_1	E_2
	A_1^+	A_2^+	A_1^-	A_2^-	E^+	E^-
E_1'	E_1'	E_1'	E_2'	E_2'	$E_2'+E_3'$	$E_1'+E_3'$
E_2'	E_2'	E_2'	E_1'	E_1'	$E_1'+E_3'$	$E_2'+E_3'$
E_3'	E_3'	E_3'	E_3'	E_3'	$E_1'+E_2'$	$E_1'+E_2'$

Group T

T	e	$4c_3$	$4c_3^2$	$3c_2$	Basis functions
E_1'	2	1	-1	0	α, β
E_2'	2	ε_3	$-\varepsilon_3^2$	0	$\frac{1}{\sqrt{2}}\left(Y_{-1/2}^{3/2}+\mathrm{i}Y_{3/2}^{3/2}\right), \frac{1}{\sqrt{2}}\left(Y_{1/2}^{3/2}+\mathrm{i}Y_{-3/2}^{3/2}\right)$
E_3'	2	ε_3^2	$-\varepsilon_3$	0	$\frac{1}{\sqrt{2}}\left(Y_{1/2}^{3/2}-\mathrm{i}Y_{-3/2}^{3/2}\right), \frac{1}{\sqrt{2}}\left(Y_{-1/2}^{3/2}-\mathrm{i}Y_{3/2}^{3/2}\right)$

Multiplication table

	A	B_1	B_2	F
E_1'	E_1'	E_2'	E_3'	$E_1'+E_2'+E_3'$
E_2'	E_2'	E_3'	E_1'	$E_1'+E_2'+E_3'$
E_3'	E_3'	E_1'	E_2'	$E_1'+E_2'+E_3'$

Groups T_d, O

T_d	e	$4c_3$	$4c_3^2$	$3c_2$	$3s_4$	$3s_4^3$	$6\sigma_d$	Basis functions	
O	e	$4c_3$	$4c_3^2$	$3c_2$	$3c_4$	$3c_4^3$	$6u_2$	T_d	O
E_1'	2	1	-1	0	$\sqrt{2}$	$-\sqrt{2}$	0	α, β	α, β; $Y_{1/2}^{1/2}, Y_{-1/2}^{1/2}$
E_2'	2	1	-1	0	$-\sqrt{2}$	$\sqrt{2}$	0	$Y_{1/2}^{1/2}, Y_{-1/2}^{1/2}$	$\frac{1}{\sqrt{3}}[(x-iy)z\beta - ixy\alpha]$, $\frac{1}{\sqrt{3}}[(x+iy)z\alpha - ixy\beta]$
G	4	-1	1	0	0	0	0	$Y_{\pm 3/2}^{3/2}, Y_{\pm 1/2}^{3/2}$	$Y_{\pm 3/2}^{3/2}, Y_{\pm 1/2}^{3/2}$

Multiplication table

	A_1	A_2	E	F_1	F_2
E_1'	E_1'	E_2'	G'	$E_1' + G'$	$E_2' + G'$
E_2'	E_2'	E_1'	G'	$E_2' + G'$	$E_1' + G'$
G'	G'	G'	$E_1' + E_2' + G'$	$E_1' + E_2' + 2G'$	$E_1' + E_2' + 2G'$

For groups $D \times C_i \chi(gi) = \pm\chi(g)$

Using (2.10 and 8), one can represent the other functions as products of spinors by coordinate functions. Table 2.4 also presents the results of multiplication of the representation $D_{1/2}$ by the vector representations D_μ. Using these rules, one can readily establish the cases in which spin-orbit interaction results in a splitting of the term μ, namely, for this to happen, the product $D_{1/2} \times D_\mu$ should contain more than one irreducible representation.

2.7 Representations of Space Groups

The principal feature of the space groups is the presence of translations as symmetry elements. As a result, the state of an electron in a crystal is characterized by the value of the wave vector **k**, and, in accordance with Bloch's theorem, its wave function can be written in the form

$$\psi_{\mathbf{k}}(\mathbf{x}) = e^{i\mathbf{k}\cdot\mathbf{x}} u_{\mathbf{k}}(\mathbf{x}), \tag{2.20}$$

where $u_{\mathbf{k}}(\mathbf{x})$ is a periodic function of coordinates, $u_{\mathbf{k}}(\mathbf{x}+\mathbf{t}) = u_{\mathbf{k}}(\mathbf{x})$. For the lattice vibrations (phonons), the function $\psi_{\mathbf{k}}(\mathbf{x})$ is actually the displacement of crystal atoms of a species residing at point **x**, and the Bloch factors are products of the vibration amplitudes a_ν of the given branch ν by the polarization vectors $e_\kappa^\nu(\mathbf{k})$, which determine the displacement of an atom located in the basic parallelepiped under excitation of the mode.

Being a periodic function, $u_{\mathbf{k}}(\mathbf{x})$ can be expanded in a Fourier series

$$u_{\mathbf{k}}(\mathbf{x}) = \sum_{\mathbf{b}} C(\mathbf{b}) e^{i\mathbf{b}\cdot\mathbf{x}}. \tag{2.21}$$

The vectors $\mathbf{b}$ should satisfy the condition

$$\exp(i\mathbf{b}\cdot\mathbf{t}) = 1 \tag{2.22}$$

for all $\mathbf{t}$. One can readily see that this condition is met for all vectors $\mathbf{b}_n$:

$$\mathbf{b}_n = n_1\mathbf{b}_1 + n_2\mathbf{b}_2 + n_3\mathbf{b}_3, \tag{2.23}$$

where

$$\mathbf{b}_1 = \frac{2\pi}{\Omega_0}[\mathbf{a}_2\mathbf{a}_3], \quad \mathbf{b}_2 = \frac{2\pi}{\Omega_0}[\mathbf{a}_3\mathbf{a}_1], \quad \mathbf{b}_3 = \frac{2\pi}{\Omega_0}[\mathbf{a}_1\mathbf{a}_2] \tag{2.24}$$

and n_i are arbitrary integers, $-\infty < n_i < +\infty$. Note that

$$\mathbf{b}_i \cdot \mathbf{a}_j = 2\pi\delta_{ij}.$$

A lattice whose sites are given by (2.23) is called reciprocal. The point symmetry of the reciprocal lattice coincides with the symmetry of the corresponding Bravais lattice; however, the types of the reciprocal lattice coincide with that of the direct one only for simple and basal face-centered lattices. For the body-centered direct lattice, the reciprocal lattice is face-centered; for the face-centered direct lattice it is body-centered. According to (2.20), for the wave functions $\psi_{\mathbf{k}i}$, the matrix of transformation under translation is diagonal: $D^{\mathbf{k}}_{ij}(e|\mathbf{t}) = \delta_{ij}\exp(-i\mathbf{k}\cdot\mathbf{t})$. Substitution of $\mathbf{k}+\mathbf{b}$ for $\mathbf{k}$ does not change $D^{\mathbf{k}}(e|\mathbf{t})$. The vectors $\mathbf{k}$ differing by a reciprocal lattice vector are called equivalent, the Bloch functions $u_{\mathbf{k}}(\mathbf{x})$ and energies $E_{\mathbf{k}}$ being considered periodic functions in the reciprocal lattice. All inequivalent vectors $\mathbf{k}$ lie within a primitive cell of reciprocal space of volume $\Omega_1 = (\mathbf{b}_1[\mathbf{b}_2\mathbf{b}_3]) = 8\pi^3/\Omega_0$. As a rule, the symmetry of a primitive cell (basic parallelepiped) in the direct and reciprocal lattices is lower than that of the infinite lattice; indeed, some of the symmetry operations send the primitive cell not into itself but rather into one of the neighboring cells. Therefore, as the unit cell of the reciprocal lattice, one chooses usually the Brillouin zone whose volume is equal to that of Ω_1, and whose symmetry coincides with that of the reciprocal (and direct) lattices. To construct the Brillouin zone, the chosen site of the reciprocal lattice is connected by straight lines with other sites, and through the centers of these lines and perpendicular to them, planes are drawn. The polygon cut out by these planes is called the (first) Brillouin zone. The unit cell of the direct lattice constructed in a similar way is called the *Wigner-Seitz cell.* Figure 2.2 shows Brillouin zones for some types of reciprocal lattices to be discussed in the subsequent chapters. One can also find there the accepted notations of the specific points and lines.

Consider now in more detail representations of space groups. For a given point $\mathbf{k}$, all transformations of symmetry r contained in the direction group can be divided into two types. The first of them, r_1, does not change $\mathbf{k}$ or send it into an equivalent point differing from $\mathbf{k}$ by the reciprocal lattice vector. The

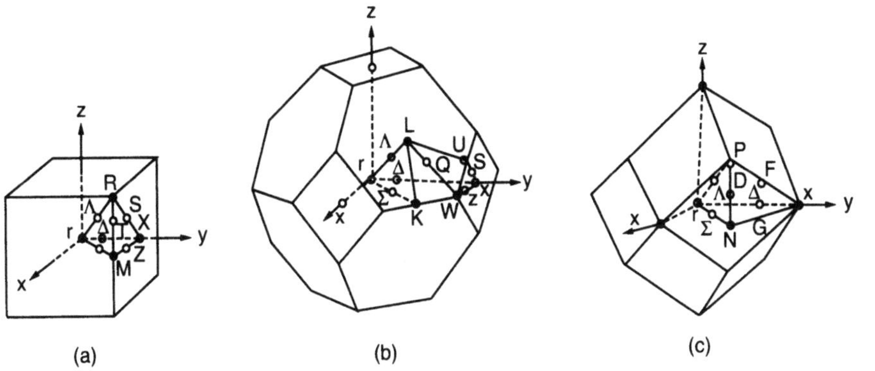

Fig. 2.2a-c. Brillouin zones for cubic lattices **a** simple cubic lattice, **b** body-centered reciprocal latice, **c** face-centered reciprocal lattice

corresponding elements of the space group $(r_1|\tau)$, together with translations (and their products $(r_1|\boldsymbol{\tau}+\mathbf{t})$), form a wave vector group. The other elements of the direction group r_1 send $\mathbf{k}$ into other inequivalent points of the Brillouin zone, $\mathbf{k}_i$. All these points $\mathbf{k}_i$, together with $\mathbf{k}$ make up a star of the wave vector $\mathbf{k}$, denoted by $\{\mathbf{k}\}$. At all these points, the energy $E(\mathbf{k}_i)$ (or phonon frequency $\omega_\nu(\mathbf{k}_i)$) for the given branch of the spectrum is the same, as is also the degeneracy, i.e., the number of the linearly independent wave functions $\psi^\nu_{\mathbf{k}_i}$ possessing the same wave vector $\mathbf{k}_i$ and the same energy. Under all operations g contained in the wave vector group, that is, those which do not change $\mathbf{k}$, these functions are transformed into one another and form the basis of some wave vector group representation, also called the small representation. The dimension of the corresponding representation of a space group is equal to the product of the dimension of the small representation by the number of different rays in the star $\{\mathbf{k}\}$. The matrices $D^{\mathbf{k}}(g)$ can be represented in the form

$$D^{\mathbf{k}}(r|\tau+\mathbf{t}) = \mathrm{e}^{-i\mathbf{k}\cdot(\tau+\mathbf{t})} D(r). \tag{2.25}$$

The matrices $D(r)$ introduced in this way form a projective representation of the point group of directions, satisfying (2.18) with

$$\omega(r_2, r_1) = \exp\left[\mathrm{i}(\mathbf{k} - r_2^{-1}\mathbf{k}) \cdot \tau_1\right]. \tag{2.26}$$

For the elements g of the wave vector group, $\mathbf{k} - r_2^{-1}\mathbf{k}$ is equal to zero or to one of the reciprocal lattice vectors. For points inside the Brillouin zone, $\mathbf{k} - r^{-1}\mathbf{k} = 0$. For symmorphic groups, $\tau_i = 0$. In these cases, as well as for a number of points on the surface of the Brillouin zone for nonsymmorphic groups, all $\omega(r_1, r_2) = 1$. Here the corresponding projective representations $D(r)$ coincide with the usual representations of point groups, i.e., with the vector or spinor representations. Note that the Bloch amplitudes $\mathbf{u}(\mathbf{x})$ transform according to vector representations, and $D(g)$, as seen from (2.25), differs from $D(r)$ only in the phase factor.

Various classes of the factor systems differ in the values of the ratios $R_{ab} = \omega(a, b)/\omega(b, a)$ for various pairs of the commuting elements a, b. The representations $D(r)$ and $D'(r)$ belonging to factor systems of the same class are related through $D(r) = D'(r)u(r)$, where $|u(r)| = 1$. Such representations are called projectively equivalent. For a class K_1 which is projectively equivalent to a vector class, $R_{ab} = 1$ for all a, b. If the functions $u(r)$ themselves form a basis of a conventional one-dimensional group representation, then the projectively equivalent representations D and D' belong to the same factor system. As an example of such p-equivalent representations, the representations F_1 and F_2 of the groups T_d or O are useful, since $F_1 = A_2 \times F_2$, or four three-dimensional representations $F_{1,2}^{\pm}$ of the group O_h.

The number of the various classes of factor systems for all finite point groups is limited. In [2.5] one can find the classes of all possible factor systems for 32 point groups, and the characters of the irreducible projective representations, as well as the corresponding matrices $D(r)$ for the generators. They can be used to determine the characters and construct the matrices $D(g)$ for any space group.

For spinor representations of space groups, the factor system $\omega(r_1, r_2)$ is a product of the factor systems (2.18 and 26). Tables of characters of the irreducible representations for all 230 space groups can be found in the literature [2.6].

2.8 Invariance Under Time Inversion

The Hamiltonian $\hat{\mathcal{H}}$ of Schrödinger's equation is invariant not only under the corresponding symmetry transformations, but under time inversion as well, which corresponds to a replacement of $\mathcal{H}$ by $\mathcal{H}^*$. In the presence of a magnetic field, invariance is preserved only under the condition of a simultaneous reversal of the direction of the magnetic field $\mathbf{B}$, and, if the spin is included, of a replacement of the Pauli matrices σ_i by $-\sigma_i^*$, i.e., σ_x by $-\sigma_x$, σ_y by σ_y, σ_z by $-\sigma_z$, since

$$\mathcal{H}^*(-\mathbf{B}, -\sigma_i) = \mathcal{H}(\mathbf{B}, \sigma_i). \tag{2.27}$$

The time inversion operation is denoted by the symbol K. This operation commutes with all symmetry elements: $Kg = gK$. For conventional representations, this operation reduces to the operation of complex conjugation K_0, i.e.,

$$K\psi = K_0\psi = \psi^*; \quad \text{and} \quad K^2 = 1. \tag{2.28}$$

For spinors, $\psi = \begin{bmatrix} \psi_1 \\ \psi_2 \end{bmatrix}$, $K = \sigma_y K_0$, and, accordingly,

$$K \begin{bmatrix} \psi_1 \\ \psi_2 \end{bmatrix} = \sigma_y \begin{bmatrix} \psi_1^* \\ \psi_2^* \end{bmatrix} = \begin{bmatrix} -i\psi_2^* \\ i\psi_1^* \end{bmatrix}. \tag{2.29}$$

Note that condition (2.27) is satisfied here, since

$$K^{-1}\mathcal{H}(\sigma_i)K = \mathcal{H}(\sigma_i).$$

In contrast to (2.28), for spinor representations (irrespective of their dimension), $K^2 = -1$.

The functions ψ_i and $K\psi_i$ correspond to the same energy and transform according to the complex conjugate representations $D(g)$ and $D^*(g)$. It should be pointed out that ψ_i and $K\psi_i$ may be linearly dependent, i.e., coupled by the relations $K\psi_i = \Sigma_j T_{ji}\psi_j$, or linearly independent. In the first case (case a), the invariance under time inversion does not lead to additional degeneracy while imposing additional restrictions on the representation $D(g)$. Indeed, for point groups (and complete representations of space groups), we have in case a:

$$D^*(g) = T^{-1}D(g)T, \quad \text{and} \quad \chi^*(g) = \chi(g). \tag{2.30}$$

For conventional representations, in case a the basis itself may be chosen real. With such a choice, all matrices $D(g)$ are also real. For spinor representations, in case a the matrices $D(g)$ are essentially complex, i.e., no unitary transformation can make them real, although the characters $\chi(g)$ are real.

If the functions ψ_i and $K\psi_i$ are linearly independent, then invariance under time inversion combines the representations $D(g)$ and D^*(g), i.e., it doubles the degeneracy. Note that the representations $D(g)$ and $D^*(g)$ may be inequivalent (case b), i.e., $\chi^*(g) \neq \chi(g)$, or equivalent (case c). In case c, just as in case a, the relations (2.30) hold. It is therefore essential to know how to distinguish between these two cases. This can be done using the criterion of Frobenius-Shur:

$$\sum = \frac{1}{h}\sum_g \chi(g^2) = \begin{cases} K^2 & \text{– case a,} \\ 0 & \text{– case b,} \\ -K^2 & \text{– case c.} \end{cases} \tag{2.31}$$

For conventional representations of point groups, all representations with real characters relate to case a, whereas representations with complex characters, to case b, and are combined pairwise. For spinor representations, summation in (2.31) is performed over all elements g and Qg. Since, however, $(gQ)^2 = g^2$, it is sufficient to carry out the summation only over the elements g, while keeping in mind that the element g^2 may include Q. For spinor representations of point groups, all one-dimensional representations of point groups relate to case c and, hence, double, whereas all the other representations with real characters belong to case a. Representations with complex characters relate to case b and are combined.

For space groups, the distribution of representations among these cases is determined by Herring's criterion: [2.7]

$$\sum = \frac{1}{h'}\sum_{g\in G'} \chi_{\mathbf{k}}(g^2)\delta_{g\mathbf{k},-\mathbf{k}} = \begin{cases} K^2 & \text{– case a,} \\ 0 & \text{– case b,} \\ -K^2 & \text{– case c.} \end{cases} \tag{2.32}$$

The summation in (2.32) is carried out over the set G' including the principal elements of the space group $g = (r|\tau) \in G'$, that is, elements which do not contain trivial translations, for which $r\mathbf{k} = -\mathbf{k}$, and h' is the number of principal elements in the wave vector group $G_\mathbf{k}$. Note that the elements $(r|\tau)$ proper may or may not be contained in $G_\mathbf{k}$, depending on whether the vectors $\mathbf{k}$ and $-\mathbf{k}$ are equivalent or not, whereas the element $g^2 = (r|\tau)^2$ is always contained in $G_\mathbf{k}$ for $r\mathbf{k} = -\mathbf{k}$. This element $(r|\tau)^2 = (r^2|\tau + r\tau)$ may contain also a trivial translation $\mathbf{t}$. For space groups, then, the role of the time inversion operation depends essentially on whether the vectors $\mathbf{k}$ and $-\mathbf{k}$ are equivalent, i.e., equal, which is possible only for $k = 0$, or differ by the reciprocal lattice vector $\mathbf{b}$ (case 1), or are inequivalent. In the latter case, the situation may again be different depending on whether there is among the symmetry elements an element R sending $\mathbf{k}$ in $-\mathbf{k}$ (case 2) or not (case 3). For different rays of the star, this element R may be different, but if it exists for one pair $(\mathbf{k}, -\mathbf{k})$, then it does as well for the other pairs of the star. If such an element does not exist, then, in accordance with (2.32), we have case b_3. Here $\mathbf{k}$ and $-\mathbf{k}$ belong to different stars, the invariance under time inversion combines these stars, and imposes the condition of equal energy at these points:

$$E(\mathbf{k}) = E(-\mathbf{k}). \tag{2.33}$$

Obviously, condition (2.33) holds in all other cases.

In case a, the functions ψ_i and $K\psi_i$ are linearly dependent, the invariance under time inversion imposing on $\chi(g)$ additional conditions:

$$\begin{aligned} &\chi^{\mathbf{k}}(g) = \chi^{\mathbf{k}*}(g)\ (a_1),\\ &\chi^{\mathbf{k}}(g) = \chi^{\mathbf{k}*}(R^{-1}gR)\ (a_2). \end{aligned} \tag{2.34}$$

In case b, the functions ψ_i and $K\psi_i$ are linearly independent and transform according to inequivalent representations. This results in a combination of the representations μ and λ with the characters

$$\begin{aligned} &\chi_\lambda^{\mathbf{k}}(g) = \chi_\mu^{\mathbf{k}*}(g)\ (b_1),\\ &\chi_\lambda^{\mathbf{k}}(g) = \chi_\mu^{\mathbf{k}*}(R^{-1}gR)\ (b_2). \end{aligned} \tag{2.35}$$

In case c, the functions ψ_i and $K\psi_i$ are also linearly independent, but transform according to inequivalent representations that should be combined. The characters of these representations in cases c_1 and c_2 satisfy the same conditions (2.34) as in cases a_1 and a_2, respectively.

The operation of time inversion, in contrast to those of space symmetry, is not unitary. Indeed, unlike (2.4), the matrices $D(gK)$ and $D(Kg)$ are related to the matrices $D(k)$ and $D(g)$ through

$$D(gK) = D(g)D(K), \quad \text{but} \quad D(Kg) = D(K)D^*(g). \tag{2.36}$$

The representations satisfying (2.36) are not conventional representations of a group. They are called corepresentations. Actually, the introduction of corepresentations is equivalent to combining in cases b and c the corresponding pairs of representations. The inclusion of invariance under time inversion turns out to be essential when defining selection rules.

2.9 Selection Rules

The solution of many problems in quantum mechanics, for instance, in the calculation of a spectrum, in the theory of quantum transitions, and in perturbation theory requires the knowledge of the matrix elements of a given operator V between the wave functions ψ:

$$V_{ij} = \langle \psi_i | V | \psi_j \rangle = \int \psi_i^* V \psi_j \, \mathrm{d}x. \tag{2.37}$$

The theory of symmetry provides the possibility of establishing which of the V_{ij} components vanish, determining the number of linearly independent components, and finding the relation between the linearly dependent components. In many cases, the linearly independent components can be derived from a comparison with experiment, after which practically all the required data are obtained based only on symmetry considerations.

To solve this problem, one has to know the irreducible representations, according to which the wave functions ψ_i and ψ_j and the operator components V_l transform. If the functions ψ_i^μ and ψ_j^ν transform according to the representations D_μ and D_ν, respectively, and the components V_l^κ, to the representation D_κ, then the integrand in (2.37), in accordance with (2.10), transforms according to the direct product of the representations D_μ^*, D_ν and D_κ. Obviously, the integral $J_i = \int F_i(x) \, \mathrm{d}x$ is nonzero only if the function $F_i(x)$ transforms according to the identity representation, i.e., if it does not change under all symmetry operations of the group, or contains a function transforming according to this representation. By (2.9), this means that the corresponding integrals are nonzero if the direct product of these representations contains the identity representation, and the number of linearly independent components of the matrix elements V_{ij}^l is

$$N_0 = \frac{1}{h} \sum_g \chi_\mu^*(g) \chi_\nu(g) \chi_\kappa(g). \tag{2.38}$$

The invariance under time inversion imposes additional conditions on the matrix elements V_{ij}^l. The operators of physical quantities possess a parity with respect to time inversion. For instance, the operators of energy and of coordinates are even, and those of velocity, momentum and current are odd.

From a mathematical standpoint, the operation of time inversion on an operator V is defined as

$$\left(KVK^{-1}\right)^+ = KV^+K^{-1} = fV, \tag{2.39}$$

where $f = +1$ for even, and $f = -1$ for odd operators. Accordingly, an additional condition is imposed on the matrix elements V_{ij}^l

$$V_{ij}^l = \langle \psi_i^\mu \left| V_l^\kappa \right| \psi_j^\nu \rangle = f \langle K\psi_j^\nu \left| V_l^\kappa \right| K\psi_i^\mu \rangle. \tag{2.40}$$

If the functions $K\psi^\mu$ and ψ^ν are linearly dependent, then this condition reduces the number of linearly independent components V_{ij}^l.

Such linear relation exists in case a for $\mu = \nu$. The number of linearly independent matrix elements will be determined here by the expression

$$N_0 = \frac{1}{2h} \sum_g \chi_\kappa(g) \left[\chi_\nu^2(g) + f K^2 \chi_\nu(g^2)\right] . \tag{2.41}$$

In cases b and c the representations D_μ and $D_\nu = D_\mu^*$ are linearly independent and are combined. There are two types of matrix elements between these states: *diagonal* $V_{ij} = \langle\psi_i|V|\psi_j\rangle$, $V_{Ki,Kj} = \langle K\psi_i|V|K\psi_j\rangle$ and *nondiagonal* $V_{Ki,j} = \langle K\psi_i|V|\psi_j\rangle$, $V_{i,Kj} = \langle\psi_i|V|K\psi_j\rangle$. The number of linearly independent *diagonal* matrix elements is determined by (2.38), these elements being related through

$$V_{Ki,Kj} = f V_{ji} . \tag{2.42}$$

The number of linearly independent *nondiagonal* matrix elements is determined by (2.41), the elements themselves being related through

$$V_{Ki,j} = f K^2 V_{Kj,i} . \tag{2.43}$$

Expressions (2.38, 41) are directly applicable to the representations of point groups. When determining selection rules for space groups, one has to consider three possible types of matrix elements. The first of them occurs, for instance, in calculations of the optical transition probabilities, or of spectra by the **k**-**p** method. These are the matrix elements between the states which correspond to the same point of a star, **k**, and different energies, and transform according to the representations D_μ^k and D_ν^k. Equation (2.38) holds in this case, but the summation should be carried out over the principal elements of the wave vector group G_k which do not contain trivial translations, with h specifying the number of these elements.

Matrix elements of the second type relate states with the same energy and of the same point of the star {**k**}. Such matrix elements are required when analyzing free carrier spectra close to the extrema. The corresponding expressions for the number of linearly independent matrix elements N_0 for this case are given in Table 2.5. If the invariance under time inversion does not impose any additional conditions on the wave functions, then the number N_0 is defined by an expression similar to (2.38).

This expression holds also in case b_3, as well as for *diagonal* matrix elements in cases b_1, b_2, c_1 and c_2. In case a_1, as well as in cases b_1 and c_1, N_0 for nondiagonal matrix elements is defined by an expression similar to (2.41). In cases a_2, b_2, and c_2, time inversion at point **k** relates $\psi_\mathbf{k}$ not with the state $K\psi_\mathbf{k}$ corresponding to point $-\mathbf{k}$, but rather to the state $KR\psi_\mathbf{k}$, where R is one of the operations sending **k** into $-\mathbf{k}$.

Matrix elements of the third type relating states with different wave vectors **k** and **k**′ represent transitions involving phonons with a wave vector **q** which preserve the total momentum

$$\mathbf{k}' - \mathbf{k} - \mathbf{q} = 0. \tag{2.44}$$

Table 2.5. The number of linearly independent matrix elements taking into account invariance under time inversion

Case	No	
b_3, diagonal components in cases b_1, b_2, c_1, c_2	$N_0 = \frac{1}{h'} \sum_{g \in G'_k} \chi_\kappa(g) \vert\chi_\nu^k(g)\vert^2$	(1)
a_1 and nondiagonal components in cases b_1, c_1	$N_0 = \frac{1}{2h'} \sum_{g \in G'_k} \chi_\kappa(g) \left[(\chi_\nu^k(g))^2 + K^2 f \chi_\nu^k(g^2)\right]$	(2)
a_2	$N_0 = \frac{1}{2h'} \sum_{g \in G'_k} \chi_\kappa(g) \vert\chi_\nu^k(g)\vert^2 + K^2 f \chi_\kappa(Rg) \chi_\nu^k((gR)^2)$	(3)
nondiagonal components in cases b_2, c_2	$N_0 = \frac{1}{2h'} \sum_{g \in G'_k} \chi_\kappa(g) \chi_\nu^k(g) \chi_\nu^k(R^{-1}gR) + K^2 f \chi_\kappa(Rg) \chi_\nu^k((gR)^2)$	(4)

For ordinary representations $K^2 = 1$, for spinor representations $K^2 = -1$, $f = +1$ for operators even under time inversion and $f = -1$ for odd ones. Summation carried out over elements of the wave vector group that do not contain trivial translations.

If normal vibrations transform according to the representation $D^{\mathbf{q}}$ of the wave vector group $G_{\mathbf{q}}$, then for transitions between the states $\mathbf{k}$ and $\mathbf{k}'$ corresponding to different energies, i.e., to different stars, we will have

$$N_0 = \frac{1}{h_0} \sum_{g_0 \in G_0} \chi_\mu^{\mathbf{k}'*}(g_0) \chi_\nu^{\mathbf{k}}(g_0) \chi_\kappa^{\mathbf{q}}(g_0). \tag{2.45}$$

The summation here is performed over the principal elements of the group G_0, which is an intersection of the groups $G_{\mathbf{k}}$, $G_{\mathbf{k}'}$, and $G_{\mathbf{q}}$ (denoted by $G_0 = G_{\mathbf{k}} \cap G_{\mathbf{k}'} \cap G_{\mathbf{q}}$), and, hence, includes only common elements of these three groups.

If transitions occur between the states $\mathbf{k}$ and $\mathbf{k}'$ belonging to the same star, and the functions ψ^ν and $K\psi^\nu$ are linearly independent, then (2.45) holds, too. Note that $\chi_\mu^{\mathbf{k}*}(g_0) = \chi_\nu^{\mathbf{k}*}(g_s^{-1} g_0 g_s)$, where g_s is an operation sending $\mathbf{k}$ into $\mathbf{k}'$, i.e., $g_s \mathbf{k} = \mathbf{k}'$. In this form, this expression relates to case b, for transitions between the points of the same star, as well as to cases $\mathrm{b}_{1,2}$ and $\mathrm{c}_{1,2}$ for diagonal components, provided that at least one of the operations $g_s^2 = g_{\mathbf{k}} g_{\mathbf{k}'}$, where $g_{\mathbf{k}}$ (and $g_{\mathbf{k}'}$) is one of the elements $G_{\mathbf{k}}$ (and $G_{\mathbf{k}'}$). If none of the elements g_s^2 is equal to $g_{\mathbf{k}} g_{\mathbf{k}'}$, then N_0 in these cases is defined by the more involved formula

$$N_0 = \frac{1}{h_0} \sum_{g_0} \chi_\kappa^{\mathbf{q}}(g_0) \left[\chi_\nu^{\mathbf{k}*}(g_s^{-1} g_0 g_s) \chi_\nu^{\mathbf{k}}(g_0) \right.$$

$$+ \chi_\nu^{\mathbf{k}*}(R^{-1} g_0 R) \chi_\nu^{\mathbf{k}}(R g_s^{-1} g_0 g_s R^{-1}) \Big] . \tag{2.46}$$

Just as before, R is here an operation sending $\mathbf{k}$ into $-\mathbf{k}$. For cases b_1 and c_1, $R = e$. For $g_s^2 \neq g_\mathbf{k} g_{\mathbf{k}'}$, the number of linearly independent components in case a, as well as of nondiagonal components in cases b and c, including transitions between points belonging to different stars $\mathbf{k}$ and $-\mathbf{k}$ in case b_3, is defined by (2.45). Note that $\chi_\mu(g_0) = \chi_\nu(g_s g_0 g_s^{-1})$. However, if at least for one element g_s, $g_s^2 = g_\mathbf{k} g_{\mathbf{k}'}$, then N_0 is given by a more complicated expression

$$N_0 = \frac{1}{2h_0} \Bigg[\sum_{g_0} \chi_\kappa^{\mathbf{q}}(g_0) \chi_\nu^{\mathbf{k}}(g_0) \chi_\nu^{\mathbf{k}}(g_s^{-1} g_0 g_s)$$

$$+ K^2 f \sum_{g'_\mathbf{k}} \chi_\kappa^{\mathbf{q}}(g_s^{-1} g'_\mathbf{k}) \chi_\nu^{\mathbf{k}} \left((g_s^{-1} g'_\mathbf{k})^2 \right) \Bigg] . \tag{2.47}$$

Just as before, the g_0 are here elements of the group G_0 representing an intersection of the groups $G_\mathbf{k} \cap G_{g_s \mathbf{k}'} \cap G_q$, and the $g'_\mathbf{k}$ are those of the elements of $G_\mathbf{k}$ for which $(g_s^{-1} g_\mathbf{k})^2$ also belong to $G_\mathbf{k}$, and $g_s^{-1} g_\mathbf{k}$ are contained in $G_\mathbf{q}$.

2.10 Determination of Linearly Independent Components of Material Tensors

By a material tensor S one understands the tensor connecting the field tensors A and B which determine the external force acting on the crystal and the change of its properties induced by this force:

$$A_i = \sum_k S_{ik} B_k . \tag{2.48}$$

These tensors A, B may have either the same or different ranks; for instance, the dielectric constant tensor κ relates two first-rank tensors, i.e., the vectors $\mathbf{D}$ and $\mathcal{I}$, the elastic constant tensor S, or its inverse, the stiffness tensor C connects two second-rank tensors, namely, the strain ε and the stress tensor X. Accordingly, the rank of the tensor S is a product of the ranks of the tensors A and B.

In general, a tensor of rank n transforms according to the representation D_n, which is actually a product of n vector or pseudovector representations of the complete spherical group, i.e., of the representations $D_1^{\mp}$ according to which transform the spherical functions $Y_m^{j\pm}$ ($m = 0, \pm 1 \ldots \pm j$) with $j = 1$, which may be even or odd under the inversion operation. The products of the spherical functions Y_m^j and Y_l^i transform according to the representation $D_i \times D_j$, which in principle is reducible and can be decomposed into irreducible representations of the complete spherical group by the formula

$$D_i \times D_j = \sum_{k=|i-j|}^{k=i+j} D_k. \tag{2.49}$$

A product of the representations $D_i \times D_j$ of one parity yields even representations D_k^+, and that of different parities, odd D_k^-. The characters of the corresponding representations $D_j^\pm$ are given by the expressions

$$\begin{aligned}
\chi^\pm(c_\varphi) &= \frac{\sin(j+1/2)\varphi}{\sin\varphi/2},\\
\chi^\pm(s_\varphi) &= \pm\chi(c_\varphi+\pi) = \pm(-1)^j\frac{\cos(j+1/2)\varphi}{\cos\varphi/2},\\
\chi(c_i) &= \pm(2j+1), \chi(\sigma) = \pm\chi(c_2) = \pm(-1)^j.
\end{aligned} \tag{2.50}$$

As the symmetry is lowered, the representations D_k in (2.49) become in turn reducible representations of the point group of directions, and can be expanded in irreducible representations of this group by (2.8).

If the components of the tensors A and B transform according to the representations D_A and D_B, respectively, then the components of the tensor S transform according to the representation

$$D_S = D_A \times D_B^*. \tag{2.51}$$

Therefore the number of linearly independent components of the tensor S is equal to that of identity representations contained in $D_A \times D_B^*$ and, in accordance with (2.9), is given by the expression

$$N_0 = \frac{1}{h}\sum_r \chi_A(r)\chi_B^*(r). \tag{2.52}$$

The summation in (2.52) is performed over all elements of the direction group F. If $D_A = \Sigma_\kappa N_\kappa^A D_\kappa$, i.e., the representation D_A contains irreducible representations D_κ, N_κ^A times each, and, similarly, $D_B = \Sigma_\kappa N_\kappa^B D_\kappa$, then

$$N_0 = \sum_\kappa N_\kappa^A N_\kappa^B. \tag{2.53}$$

Note that only the components of the tensor S which relate the components A_m and B_m^* transforming according to the same representations are nonzero, all components S_{mm} being the same, i.e.,

$$S_{mm'}^{\kappa\kappa'} = \delta_{mm'}\delta_{\kappa\kappa'}S_{mm}^{\kappa\kappa'} (m = 1, \ldots, n_\kappa; m' = 1, \ldots, n_\kappa), \tag{2.54}$$

where n_κ is the dimension of the representation D_κ.

In many cases, additional conditions are imposed on the components of the tensor S. Such a condition may be the principle of symmetry of Onsager's transport coefficients, the symmetry of the field tensors A and B, etc. As an illustration, we mention Onsager's relations for the dielectric constant tensor:

$$\kappa_{ij}(\mathbf{B}, \mathbf{q}) = \kappa_{ji}(-B, -\mathbf{q}), \tag{2.55}$$

where q is the wave vector; similarly, for the tensor of elastic constants or rigidity, we have

$$S_{iklm} = S_{lmik}, \quad T_{iklm} = T_{lmik}. \tag{2.56}$$

The symmetry of the strain tensor $\varepsilon_{ik} = \varepsilon_{ki}$ and of the stress tensor $X_{ik} = X_{ki}$ imposes on the tensors S and T additional conditions of invariance under permutation of indices within the first and second pairs. If the tensor S_{ik} is symmetrical or antisymmetrical under permutation of the indices i and k, possibilities which can occur only when the representations D_A and D_B^* coincide, then representation D_S is, respectively, equal to the symmetrized or antisymmetrized product, $[D_A^2]_{\text{symm}}$ or $\{D_A^2\}_{\text{asymm}}$.

In accordance with (2.12, 13), in these cases N_0 is given by

$$N_{0\left\{\substack{\text{symm} \\ \text{asymm}}\right.} = \frac{1}{2h} \sum_r \left[|\chi_A(r)|^2 \pm \chi_A(r^2)\right]. \tag{2.57}$$

It should be pointed out that for the representations of the complete spherical group

$$\left[D_j\right]_{\text{symm}} = \sum_{k=0}^{k=j} D_{2k}, \quad \left\{D_j\right\}_{\text{asymm}} = \sum_{k=0}^{k=j-1} D_{2k+1}. \tag{2.58}$$

As follows from these relations, the components of the tensors ε, X, κ, symmetric under index permutation, transform according to $D_0^+ + D_2^+$, the antisymmetric tensor of the second rank, according to D_1, and the tensors C and S transform according to the representations $D_4^+ + 2D_2^+ + 2D_0^+$ and, respectively, have two linearly independent components in an isotropic medium:

$$C_{xxxx}, C_{xyxy} = C_{xyyx} = \frac{1}{2}\left(C_{xxxx} - C_{xxyy}\right).$$

In cubic crystals of T_d, O, and O_h symmetries, $D_2 = E + F_2$, (or $D_2^\pm = E^\pm + F_2^\pm$) and $D_4 = A_1 + E + F_1 + F_2$ (or $D_4^\pm = A_1^\pm + E^\pm + F_1^\pm + F_2^\pm$), and the number of linearly independent components increases to three, thus implying no relation between C_{xyxy} and C_{xxxx} - C_{xxyy}.

3 Electron Spectrum in Crystals, Quantum Wells and Superlattices

We shall briefly digress from group theory to get acquainted with the so-called **k·p** *method.* It is the simplest method for calculating carrier spectra near extreme points, i.e., the conduction-band minima and valence-band maxima; it essentially represents a variant of perturbation theory. The effective-mass approximation and the theory of deformation potential, which permit description of the effect of external, magnetic and electric, fields as well as of the interactions of carriers with lattice vibrations, may be considered a natural development of the **k·p** method. All these concepts enjoy widespread use in the theory of semiconductors and of materials where the carrier concentration is usually much lower than the number of lattice atoms, and therefore the electrons and holes cluster near the extrema.

We shall acquaint ourselves with the method of invariants permitting a straightforward account of symmetry requirements without invoking perturbation theory in an explicit form.

To learn how to apply these methods in practical cases, we have to take refuge in the theory of representations, and identify the irreducible representations according to which the wave functions near a given extremum transform, as well as to understand how to use the selection rules. We shall focus our attention primarily on the cubic crystals Ge and Si, and on A_3B_5 compounds of the elements of the 3rd and 5th groups. Thereafter, we shall demonstrate how the effective-mass and the deformation-potential methods can be employed to calculate the carrier spectra and wave functions in quantum wells and superlattices, and shall consider the effect of external fields on these spectra. Obviously enough, neither the **k·p** method nor the method of invariants are able to replace more complex numerical calculations in all cases. However, these fairly simple techniques, on the one hand, provide a quite satisfactory accuracy and, on the other hand, are easily tractable and do not involve large amounts of machine computation.

3.1 The k·p Method

In semiconductors, where the number of free carriers, electrons in the conduction band, and holes in the valence band, is usually small, occupying only the lowest states of the corresponding band, most of the problems can be solved if one knows the carrier spectrum only within a comparatively narrow region near the extremum. The carrier spectrum near the extremum can be calculated using two approaches, namely, the **k**-**p** method and the method of invariants. Actually, both methods are based only on symmetry considerations, the values of the constants, e.g., of the effective masses, being determined, as a rule, from experimental data.

In the **k**-**p** method, the wave function representing the solution of Schrödinger's equation for an electron in a crystal

$$(\mathcal{H}_0 - E)\psi = 0,$$

where

$$\mathcal{H}_0 = \frac{\hat{p}^2}{2m} + U_0(\mathbf{x}), \quad \hat{\mathbf{p}} = -i\hbar\nabla \tag{3.1}$$

can be expanded near the extremum point $\mathbf{K}_0$ in the wave functions

$$\varphi_{n\mathbf{k}} = \psi_{n\mathbf{K}_0} e^{i\mathbf{k}\cdot\mathbf{x}}. \tag{3.2}$$

In contrast to the exact Bloch functions $\psi_{n\mathbf{k}}$, the $\varphi_{n\mathbf{k}}$, called *pseudo-Bloch functions*, do not represent a solution of (3.1) at the point $\mathbf{K} = \mathbf{K}_0 + \mathbf{k}$; however, they correspond to the same $\mathbf{K}$ and, therefore, the functions $\psi_{n\mathbf{k}}$ can be expanded in functions $\varphi_{n\mathbf{k}}$:

$$\psi_{n\mathbf{k}} = \sum_{n'} C_{n'} \varphi_{n'\mathbf{k}}. \tag{3.3}$$

Substituting (3.3) into (3.1), multiplying this equation on the left by $\varphi_{n\mathbf{k}}$ and integrating in $\mathbf{x}$, we come to coupled equations determining the coefficients C_n

$$\sum_{n'} \left\{ \left[E_n(\mathbf{K}_0) + \frac{\hbar^2 k^2}{2m} - E \right] \delta_{nn'} + \frac{\hbar}{m}\mathbf{k}\cdot\mathbf{P}_{nn'} \right\} C_{n'} = 0. \tag{3.4}$$

Here $P_{nn'}$ is the matrix element of the momentum operator

$$\mathbf{P}_{nn'} = \int \psi^*_{n\mathbf{K}_0} \hat{\mathbf{p}} \psi_{n'\mathbf{K}_0} \, \mathrm{d}V. \tag{3.5}$$

The term $\mathbf{k}\cdot\mathbf{P}_{nn'}/m$ in (3.4) may be considered a perturbation.

Solving (3.4) by perturbation theory yields a system of equations that determine the spectrum and the wave functions, i.e., the coefficients C_n near the extremum point $E_n(\mathbf{K}_0)$. Indeed, in the second order of perturbation theory we obtain the system

$$\sum_{n'} (\mathcal{H}_{nn'} - E\delta_{nn'}) C_{n'} = 0, \tag{3.6}$$

where

$$\mathcal{H}_{nn'} = E_n(\mathbf{K}_0)\delta_{nn'} + \frac{\hbar^2 k^2}{2m}\delta_{nn'} + \frac{\hbar}{m}\mathbf{k}\cdot\mathbf{P}_{nn'}$$

$$+\frac{\hbar^2}{4m^2}\sum_{\substack{n''\neq n,n'\\ \alpha,\beta}} k_\alpha k_\beta \left(P^\alpha_{nn''}P^\beta_{n''n'} + P^\beta_{nn''}P^\alpha_{n''n'}\right)$$

$$\times\left\{[E_n(\mathbf{K}_0) - E_{n''}(\mathbf{K}_0)]^{-1} + [E_{n'}(\mathbf{K}_0) - E_{n''}(\mathbf{K}_0)]^{-1}\right\}. \qquad (3.7)$$

Limiting ourselves to the single-band model, the number of the independent coefficients C_n is determined by the band degeneracy multiplicity at the point $\mathbf{K}_0$. In this case all $E_{n'} = E_n$. For nondegenerate bands, the coefficient $C_n = 1$, the spectrum in the quadratic-in-k approximation being determined by the expression

$$E_n(\mathbf{k}) = E_n(\mathbf{K}_0) + \frac{\hbar}{m}\sum_\alpha k_\alpha P^\alpha_{nn} + \sum_{\alpha\beta}\frac{\hbar^2}{2m_{\alpha\beta}}k_\alpha k_\beta, \qquad (3.8)$$

where

$$\frac{1}{m_{\alpha\beta}} = \frac{1}{m}\delta_{\alpha\beta} + \frac{1}{m^2}\sum_{n'}\frac{P^\alpha_{nn'}P^\beta_{n'n} + P^\beta_{nn'}P^\alpha_{n'n}}{E_n(\mathbf{K}_0) - E_{n'}(\mathbf{K}_0)}.$$

In the system of the principal axes, tensor m^{-1} has, in general, three linearly independent components $m^{-1}_{xx}, m^{-1}_{yy}, m^{-1}_{zz}$. The position of the principal axes is determined by the symmetry of the wave vector group at the point $\mathbf{K}_0$.

For degenerate bands, the energy is derived from the solution of the secular equation

$$\det\{\mathcal{H} - EI\} = 0. \qquad (3.9)$$

In semiconductors with several closely lying bands, the solution in the form of an expansion in powers of k is frequently insufficient. In these cases, one may take simultaneously into account in (3.4) several closest bands with the $E_n(\mathbf{K}_0)$ equal to $E_1(\mathbf{K}_0)$, $E_2(\mathbf{K}_0)$ ... The corresponding determinant (3.9) includes both the *diagonal* intraband matrix elements $\mathcal{H}_{nn'}$ with $E_{n'}(\mathbf{K}_0) = E_n(\mathbf{K}_0)$, and *nondiagonal* ones between different branches, i.e., with $E_{n'}(\mathbf{K}_0) \neq E_n(\mathbf{K}_0)$. It should be pointed out that the summation over n'' in the diagonal and nondiagonal terms is performed over the states of other bands, $n'' \neq n, n' \ldots$, the linear-in-k interband terms in $\mathcal{H}_{nn'}$ being taken into account exactly.

The character of the carrier spectrum in semiconductors is affected substantially by the spin-orbit interaction, which is determined by the operator

$$\mathcal{H}_{\mathrm{so}} = \frac{\hbar}{4m^2c^2}\left(\left[\nabla U_0\mathbf{p}\right]\sigma\right). \qquad (3.10)$$

Spin-orbit interaction can result in a partial lifting of degeneracy at the extremum point $\mathbf{K}_0$ and the corresponding splitting of terms. In noncentrosymmetric crystals, spin-orbit interaction results in a total lifting of degeneracy, as one moves away from the highly symmetric points $\mathbf{K}_0$ due to the appearance in $\mathcal{H}(\mathbf{k})$ of spin-dependent, linear- or cubic-in-k terms.

The spin-orbit interaction can be included in the framework of the **k-p** method in two ways. If it is small compared to the distance to other bands not included in the Hamiltonian $\mathcal{H}$ in (3.9), then one may consider $\mathcal{H}_{so}$ as a perturbation and take as an initial basis the product of the coordinate functions $\psi_{n\mathbf{K}_0}$ by the spinors α, β corresponding to the spin $\pm 1/2$. Note that besides the intraband matrix elements of the operator $\mathcal{H}_{so}$ which determine the spin-orbit splitting, one should include in (3.7) the mixed terms as well:

$$\frac{\hbar}{4m} \sum_{\substack{n'' \neq n, n' \\ \gamma}} k_\gamma \left[P^\gamma_{nn''} (\mathcal{H}_{so})_{n''n'} + (\mathcal{H}_{so})_{nn''} P^\gamma_{n''n'} \right] \\ \times [(E_n - E_{n''})^{-1} + (E_{n'} - E_{n''})^{-1}]. \tag{3.11}$$

Apart from this, one should, in principle, take into account the spin-orbit splitting in the energy denominators in (3.7) and (3.11) as well. When including the operator $\mathcal{H}_{so}$ in $\mathcal{H}$, the momentum operator $\mathbf{P}$ should be replaced by

$$\pi = \mathbf{p} + \frac{\hbar}{4mc^2} [\sigma \times \nabla U_0]. \tag{3.12a}$$

Since

$$\nabla U_0 = \{\nabla, \mathcal{H}_0\} = \frac{\mathrm{i}}{\hbar} (\mathbf{p}\mathcal{H}_0 - \mathcal{H}_0 \mathbf{p}),$$

we have

$$\pi_{nn'} = P_{nn'} - \mathrm{i} \frac{E_n - E_{n'}}{4mc^2} [\sigma \mathbf{p}]_{nn'}. \tag{3.12b}$$

We clearly see that the last term does not contribute to the intra-band matrix elements and, hence, cannot account for the appearance of the linear-in-k terms. Within the approach considered here, such terms can appear only when including the term (3.11). As for the cubic-in-k terms, they can appear in the third order of **k-p** theory when spin-orbit splitting is included in the energy denominators, and in the fourth order, i.e., third-in-k, and first in $\mathcal{H}_{so}$. The first contribution is usually predominant.

In the second method of including the spin-orbit interaction, one chooses from the outset as basis functions $\psi_{n\mathbf{K}_0}$ the functions that transform according to the corresponding spinor representations, that is, those which diagonalize the Hamiltonian $\mathcal{H}_0$ with $\mathcal{H}_{so}$ included. However, one cannot establish here from the outset which of the terms in $\mathcal{H}(\mathbf{k})$ are relativistically small, which is essential in searching for the zero-slope points, i.e., the points where $\mathrm{d}E/\mathrm{d}k_i = 0$. Finding the points where $\mathrm{d}E/\mathrm{d}k_i$ vanishes by virtue of the symmetry considerations is the first step in the **k-p** method. The relativistic, linear-in-k contributions result only in an insignificant displacement of these points and, hence, are disregarded when determining the position of the zero slope points. For a point at a zone's center, or a high-symmetry point at the boundary of the Brillouin zone to be a zero-slope point, all three components $P^i_{nn'} (i = x, y, z)$ should vanish by symmetry considerations. On the symmetry axes, identical vanishing of two components is sufficient, since the third component will vanish with a high probability at one of the points on the axis, whereas on a fixed plane, identical vanishing of one of the components will suffice.

If the basis functions $\psi_{n\mathbf{K}_0}$ transform according to the representation $D_\nu^{\mathbf{K}_0}$, then the matrix elements $P_{nn'}^{\gamma}$ will form the basis for the representation $D = D_1^- |D_\nu^{\mathbf{K}_0}|^2$. Taking into account the requirements imposed by time-reversal invariance, the number of linearly independent components will be defined by the expressions of Table 2.5, with $D_\kappa = D_1^-$, and $f = -1$. In accordance with (2.5.1), for nondegenerate bands, N_0 will be

$$N_0 = \frac{1}{h} \sum_{g \in G_{\mathbf{K}_0}} \chi_1^-(g) \tag{3.13}$$

and does not depend on the representation $D_\nu^{\mathbf{K}_0}$. In case a_1, in accordance with (2.5.2), all points in nondegenerate bands for which $\mathbf{k}$ and $-\mathbf{k}$ are equivalent, i.e., $\mathbf{k} = -\mathbf{k}$, are actually zero slope points. When spin degeneracy is included, linear-in-k terms may appear at these points in $\mathcal{H}(\mathbf{k})$ in the case of noncentrosymmetric crystals.

When determining the number of linearly independent coefficients of the quadratic-in-k terms in (3.7 and 8), one should bear in mind that the sums $\Sigma_{n''} P_{nn''} P_{n''n'}$, where the summation is done over all states with energy $E_{n''}$, form the basis of the representation $D = [D_1^2]_{\text{sym}} D_n^{\mathbf{K}_0} D_{n'}^{\mathbf{K}_0^*}$, irrespective of the representations $D_{n''}^{\mathbf{K}_0}$ according to which the corresponding functions transform. Respectively, the number of the linearly independent nonzero components is defined by the expressions in Table 2.5 with

$$f = +1, \text{ and } D_k = [D_1^2]_{\text{sym}} = D_0^+ + D_2^+.$$

The above expressions are written for the electrons. If the energy reaches a maximum at the extremum point $\mathbf{K}_0$, then the spectrum of the holes near the extremum will be defined by the expressions differing from (3.6–9,11) in $\mathcal{H}$ being replaced by $-\mathcal{H}$, and $\mathbf{k}$ by $-\mathbf{k}$. It should be pointed out that the wave functions of holes ψ_n^{h} are related to the corresponding electron wave functions ψ_n^{e}, through the expression $\psi_n^{\text{h}} = K\psi_n^{\text{e}}$, where K denotes the time inversion operation.

3.2 The Effective-Mass Method; Deformation Potential

To calculate the spectrum and wave functions of free carriers in an external (electric or magnetic) field, one has, in principle, to solve Schrödinger's equation for the electron in a crystal, whose Hamiltonian

$$\mathcal{H} = \frac{\hbar^2}{2m}\left[\mathbf{p} + \frac{e}{c}\mathbf{A}(\mathbf{x}, t)\right]^2 + U_0(\mathbf{x}) - e\varphi(\mathbf{x}, t) + \frac{1}{2}g_0\mu_{\text{B}}(\boldsymbol{\sigma}\mathbf{B}) + \mathcal{H}_{\text{so}} \tag{3.14}$$

includes both the external fields defined by the scalar potential φ and vector potential $\mathbf{A}$, and the periodic crystal potential $U_0(\mathbf{x})$. In (3.14), g_0 is the g-factor of the free electron and μ_B, the Bohr magneton.

If the external fields are smooth enough, i.e., the potentials φ and $\mathbf{A}$ change little over distances of the order of the lattice constant, and the frequency of their variations in time is small compared to $\Delta E/\hbar$, where ΔE is the distance to other bands at the point of extremum $\mathbf{K}_0$, then the solution of Schrödinger's equation with Hamiltonian (3.14) can be presented in the form of products of smoothly varying functions $F_n(\mathbf{x}, t)$ by Bloch functions at the extremum point

$$\psi(\mathbf{x}, t) = \sum_n F_n(\mathbf{x}, t)\psi_{n\mathbf{K}_0}. \tag{3.15}$$

Using perturbation theory similar to that discussed in sect. 3.2, it can be shown that the system of equations for the functions $F_n(\mathbf{x}, t)$ can be written as

$$\sum_{n'} \left(\mathcal{H}_{nn'} - \mathrm{i}\hbar\delta_{nn'}\frac{\partial}{\partial t}\right) F_{n'}(\mathbf{x}, t) = 0. \tag{3.16}$$

The Hamiltonian $\mathcal{H}_{nn'}$ can be calculated in any order of perturbation theory and differs from the operator $\mathcal{H}_{nn'}$ in (3.7) in the following features:

1) $\mathcal{H}_{nn'}$ includes the potential energy $-e\varphi(\mathbf{x}, t)\delta_{nn'}$.

2) The wave vector $\mathbf{k}$ is replaced by the operator

$$\mathbf{K} = \mathbf{k} + \frac{e}{\hbar c}\mathbf{A}, \tag{3.17}$$

where $\mathbf{k} = -\mathrm{i}\nabla$.

The electric field $\mathcal{I}$ and magnetic field $\mathbf{B}$ are related to φ and $\mathbf{A}$ through

$$\mathcal{I} = -\nabla\varphi + \frac{1}{c}\frac{\partial \mathbf{A}}{\partial t}, \qquad \mathbf{B} = \operatorname{curl}\mathbf{A}. \tag{3.18}$$

3) Since with a magnetic field present, the operators $K_i = k_i + (e/\hbar c)A_i$ no longer commute with one another, the Hamiltonian (3.16) includes, besides the symmetrized terms $[K_\alpha K_\beta]_{\mathrm{sym}}(P^\alpha_{nn''}P^\beta_{n''n'} + P^\beta_{nn''}P^\alpha_{n''n})$, also the antisymmetrized ones $\{K_\alpha K_\beta\}_{\mathrm{asym}}(P^\alpha_{nn''}P^\beta_{n''n'} - P^\alpha_{nn''}P^\beta_{n''n'})$. According to (3.17) $\{K_\alpha K_\beta\}_{\mathrm{asym}} = -\mathrm{i}e/cB_\gamma\delta_{\alpha\beta\gamma}$, where $\delta_{\alpha\beta\gamma} = 1$ if all three indices are not the same and run in direct order, i.e., as xyz, while for the reverse order (e.g., xzy) $\delta_{\alpha\beta\gamma} = -1$. If one of the pairs of the indices coincides, then $\delta_{\alpha\beta\gamma} = 0$. As a result, the Hamiltonian $\mathcal{H}_{nn'}$ in (3.16) can be written in the quadratic-in-k approximation as

$$\begin{aligned}\mathcal{H}_{nn'} &= (E_n - e\varphi)\delta_{nn'} + \frac{\hbar}{m}\mathbf{K}\cdot\mathbf{P}_{nn'} \\ &+ \frac{\hbar^2}{2m}\sum_{\alpha,\beta}\left[K_\alpha K_\beta\right]_{\mathrm{sym}}\left\{\delta_{\alpha,\beta}\delta_{nn'} + \frac{1}{2m}\sum_{n''}\left(P^\alpha_{nn''}P^\beta_{n''n'}\right.\right. \\ &\left.\left.+P^\beta_{nn''}P^\alpha_{n''n'}\right)\left[(E_n - E_{n''})^{-1} + (E_{n'} - E_{n''})^{-1}\right]\right\}\end{aligned}$$

$$+\frac{1}{2}\mu_{\mathrm{B}}\sum_{\gamma} B_{\gamma}\left\{g_0\sigma^{\gamma}_{nn'} - \frac{\mathrm{i}}{m}\sum_{\alpha,\beta,n''}\delta_{\alpha\beta\gamma}P^{\alpha}_{nn''}P^{\beta}_{n''n'}\right.$$

$$\left.\times\left[(E_n - E_{n''})^{-1} + (E_{n'} - E_{n''})^{-1}\right]\right\}. \tag{3.19}$$

It is assumed here that the eigenfunctions $\psi_{n\mathbf{K}_0}$ are actually the functions that diagonalize $\mathcal{H}_{\mathrm{so}}$ and, hence, that the spin-orbit splitting is already included in E_n. These functions can be presented in the form of a product of a coordinate function, φ_m, and spin functions χ_l (α or β), or as a superposition of such products

$$\psi_n = \sum_{ml} C^n_{ml}\varphi_m\chi_l. \tag{3.20}$$

Accordingly, in (3.19)

$$\sigma^{\gamma}_{nn'} = \sum_{mll'} C^{n*}_{ml}C^{n'}_{ml'}\sigma^{\gamma}_{ll'}. \tag{3.21}$$

We recall that $\langle\varphi_m|\varphi_{m'}\rangle = \delta_{mm'}$.

For the bands degenerate only in spin, for which the basis functions are $\psi_l = \varphi_0\chi_l(\chi_l = \alpha$ or $\beta)$, we have, in accordance with (3.8,19)

$$\mathcal{H} = E_0 - e\varphi(\mathbf{x},t) + \sum_{\alpha\beta}\frac{\hbar^2}{2m_{\alpha\beta}}[K_{\alpha}K_{\beta}]_{\mathrm{sym}} + \mu_{\mathrm{B}}\left(\frac{1}{2}g_0\sigma + \mathbf{L},\mathbf{B}\right), \tag{3.22}$$

where

$$L^{\gamma}_{ll'} = -\frac{\mathrm{i}}{\mathrm{m}}\sum_{\substack{n'\\ \alpha\beta}}\delta_{\alpha\beta\gamma}P^{\alpha}_{ln'}P^{\beta}_{n'l'}(E_0 - E_{n'})^{-1}. \tag{3.22a}$$

Note that

$$\sigma^{\gamma}_{ll'} = \langle\chi_l|\sigma^{\gamma}|\chi_{l'}\rangle,$$

$$P^{\alpha}_{ln'}P^{\beta}_{n'l'} = \sum_{m_1m_2} C^{n'}_{m_1l}C^{n'*}_{m_2l'}P^{\alpha}_{0m_1}P^{\beta}_{m_20}.$$

We readily see that the term (3.22a) contributes to the effective g-factor of the electrons due to the spin-orbit interaction which mixes the various coordinate and spin states in (3.20). The states of nondegenerate bands with $\psi_{n'} = \varphi_{m'}\chi_{l'}$ do not contribute to L. Similarly, L vanishes if one does not include spin-orbit splitting in the energy denominators in (3.22a). Consider now the applicability of the effective mass method to the calculation of spectra in quantum wells and superlattices. If the modulating potential is produced by an impurity charge, then the only criterion of applicability of this method is the smoothness of the potential. If, however, the position of the bottom of the corresponding band $E(K_0)$ is modulated by a variation in composition, then at sharp boundaries the change in the potential will no longer be smooth. Therefore, near the boundaries, a substantial contribution to the total wave function may come from states of higher bands decaying at distances from the boundary of about the lattice constant. Note that the *boundary conditions* on the envelope functions should relate these functions and their derivatives not at the boundary proper but at a certain distance to the right and left of it. In the general case, F and $\mathrm{d}F/\mathrm{d}z$ for a nondegenerate band are coupled at the boundaries of the regions A and B $(z = \pm 0)$ by linear relations

$$F_A = t_{11} F_B + t_{12} \nabla_B F_B,$$
$$\nabla_A F_A = t_{21} F_B + t_{22} \nabla_B F_B, \tag{3.23}$$

where

$$\nabla_i F_i = \frac{m}{m_i} a_0 \frac{dF_i}{dz} (i = A, B),\ F_A = F_A(-0),\ F_B = F_B(+0),$$

m_A and m_B are the effective masses in the regions A and B, m is the free electron mass, and a_0 is the lattice constant.

The condition of current continuity

$$J_z = \mathrm{i}\frac{e\hbar}{2ma_0}\left[F_i^* \nabla_i F_i - (\nabla F_i)^* F_i\right] = \text{const}$$

yields the following relation

$$t_{11}t_{22} - t_{12}t_{21} = 1. \tag{3.24a}$$

Here the components t_{ij} may depend on the boundary's structure, barrier height, and electron energy. Numerical calculations performed by different methods and different researchers [3.1] show that at the GaAs-Al_xGa_{1-x} As boundary the components t_{21}, t_{12} are small. In this case, one can write

$$t_{11}t_{22} = 1. \tag{3.24b}$$

This condition can be rewritten in the form

$$m_A^\alpha F_A = m_B^\alpha F_B,$$
$$m_A^{-(\alpha+1)} \frac{\partial}{\partial z} F_A = m_B^{-(\alpha+1)} \frac{\partial}{\partial z} F_B. \tag{3.24c}$$

Numerical calculations yield for α values ranging from $-1/2$ to 0 [3.1–4]. It should be stressed that (3.24c) do not imply that t_{11} and t_{22} depend only on the mass ratio; indeed, (3.24c) is only a convenient form of presenting relation (3.24b).

One can write boundary conditions similar to (3.23) also for the envelope functions in the case of Bloch functions corresponding to different extrema, for instance, to the two states Γ and X in AlGaAs [3.5]. For degenerate bands, one usually takes as boundary conditions

$$F_A = F_B,$$
$$-\frac{\mathrm{i}}{\hbar}[\mathbf{x}, \mathcal{H}_A]_Z F_A = -\frac{\mathrm{i}}{\hbar}[\mathbf{x}, \mathcal{H}_B]_Z F_B, \tag{3.25}$$

where F is a multicomponent function representing a solution of (3.16), (3.19). Here $-\mathrm{i}/\hbar[\mathbf{x}\mathcal{H}]$ is the velocity operator. In the **k**-representation, $\mathbf{x} = \mathrm{i}\partial/\partial\mathbf{k}$. This boundary condition ensures current continuity and corresponds to conditions (3.24b) with $\alpha = 0$.[1]

In the case of a smooth variation of composition, one also uses (3.22), where the effective masses and the g-factors are assumed to depend on coordinates. For nondegenerate bands, the kinetic energy operator is written in the form

$$H_K = -\frac{1}{2}\hbar^2 m^{\alpha}\nabla m^{-(1+2\alpha)}\nabla m^{\alpha}, \tag{3.26}$$

where, just as in (3.24c), α can take on values ranging from $-1/2$ to zero. As shown by *Morrow* and *Bronstein* [3.6], Schrödinger's equation yields, for an abrupt heterojunction at point $z = 0$, according to (3.26),

$$\lim_{\varepsilon\to 0}\int_{-\varepsilon}^{\varepsilon} m^{\alpha}\frac{\mathrm{d}}{\mathrm{d}z}m^{-(1+2\alpha)}\frac{\mathrm{d}}{\mathrm{d}z}m^{\alpha}F\,\mathrm{d}z = \lim_{\varepsilon\to 0}\frac{2}{\hbar^2}\int_{-\varepsilon}^{\varepsilon}[V(z)-E]\,F\,\mathrm{d}z = 0,$$

which is possible only if $m^{\alpha}F$ and $m^{-(1+\alpha)}\mathrm{d}F/\mathrm{d}z$ do not undergo a discontinuity at the heterojunction. This means that in this case the values of α in (3.24c and 26) coincide. One should not, however, overestimate the validity of this conclusion since, as already pointed out, near an abrupt heterojunction a substantial contribution to the total ψ function may come from states of higher lying bands.

The deformation-potential method provides the possibility of including, within the effective-mass approximation, the effect of deformations, both static, which result in a change of the spectrum, and time-variable ones produced by long-wavelength lattice vibrations, both natural and excited by an external perturbation.

In a deformed crystal, the potential $U_0(\mathbf{x})$ in the Hamiltonian (3.1) is replaced by $U_\varepsilon(\mathbf{x})$. In order to retain under these conditions the possibility of representing the solution of Schrödinger's equation in the form of an expansion (3.3 or 15) in wave functions $\psi_{n\mathbf{K}_0}$ of the unperturbed Hamiltonian, one should transform the coordinates in the Hamiltonian $\mathcal{H}_\varepsilon = P^2/2m + U_\varepsilon(\mathbf{x})$ by replacing $\mathbf{x}$ with $\mathbf{x}' = (1+\varepsilon)\mathbf{x}$, i.e., $x_i' = \Sigma_j(\delta_{ij}+\varepsilon_{ij})x_j$. As a result, the potential $U_\varepsilon[(1+\varepsilon)\mathbf{x}]$ will have the same period as $U_0(\mathbf{x})$. Under this transformation, $\mathbf{p}$ will be replaced by $(1-\varepsilon)\mathbf{p}$. Thus the Hamiltonian $\delta\mathcal{H}_\varepsilon = \mathcal{H}_0[(1+\varepsilon)\mathbf{x}] - \mathcal{H}_0(\mathbf{x})$ acting as a perturbation can be written as

$$\delta\mathcal{H}_\varepsilon = -\frac{\mathbf{P}\varepsilon\mathbf{P}}{m} + (U\varepsilon) = \sum_{ij} D_{ij}\varepsilon_{ji}. \tag{3.27}$$

[1] These boundary conditions are valid if $\mathcal{H}(\mathbf{k})$ in (3.19) contains quadratic-in-k terms. The boundary conditions for Kane's model where (3.19) retains only linear-in-k terms, are given by (3.71).

Here

$$\mathbf{p}\varepsilon\mathbf{p} = \sum_{ij} p_i \varepsilon_{ij} p_j,$$

$$(U\varepsilon) = U\varepsilon[(1+\varepsilon)\mathbf{x}] - U_0(\mathbf{x}) = \sum_{ij} U_{ij}(\mathbf{x})\varepsilon_{ji}.$$

Accordingly, one should include in the Hamiltonian (3.7, 19) linear-in-ε terms $\mathcal{H}_\varepsilon$ obtained in the first order of perturbation theory, and terms $\mathcal{H}_{\varepsilon k}$ linear in ε and k:

$$\mathcal{H}^{\varepsilon}_{nn'} = \sum_{ij} D^{ij}_{nn'} \varepsilon_{ji}, \tag{3.28a}$$

$$\mathcal{H}^{\varepsilon k}_{nn'} = -\frac{2\hbar}{m}(\mathbf{P}_{nn'}\varepsilon\mathbf{k}) + \frac{\hbar}{2m}\sum_{n''}\left[(\mathbf{k}\cdot\mathbf{P}_{nn''})\mathcal{H}^{\varepsilon}_{n''n'} + \mathcal{H}^{\varepsilon}_{nn''}(\mathbf{k}\cdot\mathbf{P}_{n''n'})\right]\left[(E_n - E_{n''})^{-1} + (E_{n'} - E_{n''})^{-1}\right]. \tag{3.28b}$$

Here

$$D^{ij}_{nn'} = -\frac{(p_i \cdot p_j)_{nn'}}{m} + U^{ij}_{nn'}. \tag{3.28c}$$

At the point of zero slope, the first term in (3.28b) vanishes. If the elements of the wave vector group include the $(c_i|\tau)$ operation, the second term also vanishes. The components $D^{ij}_{nn'}$ are called the deformation potential constants. They transform according to the representation $D = [D_1^2]_{\text{sym}} D^\mu D^\upsilon$, the number of the linearly independent components D^{ij} being defined by the expressions in Table 2.5 with $f = 1$ and $D^\kappa = D_0^+ + D_2^+$. For nondegenerate bands

$$\mathcal{H}_\varepsilon = \sum_{ij} D_{ij}\varepsilon_{ij}. \tag{3.29}$$

When transformed to the principal axes, the tensor D, just as the tensor $1/m^*$, has in general three linearly independent components D_{xx}, D_{yy}, and D_{zz}. At the low-symmetry points $\mathbf{K}_0$, the principal axes of the tensor $1/m^*$ and D may not coincide.

In higher orders of perturbation theory, the Hamiltonian $\mathcal{H}$ includes terms quadratic-in-k, and linear-in-ε, which determine the deformation-induced variation of the effective masses, and similar terms determining the variation of the g-factor under deformation, as well as relativistic terms linear-in-k and ε, which are essential when $\mathcal{H}_{\varepsilon k}$ in (3.28) vanishes in the nonrelativistic approximation. These terms appear in the third order of perturbation theory, as well as when spin-orbit splitting is included in the energy denominators in (3.29).

3.3 Method of Invariants

The effective Hamiltonian $\mathcal{H}(\mathbf{K})$ describing the spectrum of electrons and their behavior in external fields in the effective mass method is a matrix of dimension $n_s \times n_s$, where n_s is the dimension of the corresponding representation D of the wave vector group. By components K_i in $\mathcal{H}(\mathbf{K})$ we understand here both the components of the wave vector k_i or of the generalized wave vector K_i, and those of the deformation tensor ε_{ij}, of the vectors $\mathcal{I}$ and $\mathbf{B}$, and of all their appropriate products. The method of invariants permits constructing the matrix $\mathcal{H}(\mathbf{K})$ without explicitly using perturbation theory. Since the Hamiltonian $\mathcal{H}(\mathbf{K})$ is invariant under symmetry transformations, it should satisfy the condition

$$D(g)\mathcal{H}(g^{-1}\mathbf{K})D^{-1}(g) = \mathcal{H}(\mathbf{K}) \tag{3.30}$$

for all symmetry operations g. It can be shown that if the relations (3.30) hold for the generators of the group $G_{\mathbf{K}_0}$, they will hold also for all the other elements g. Equations (3.30) represent actually n_s^2 equations for the components of the matrix $\mathcal{H}(\mathbf{K})$, so that if the matrices $D(g)$ are known, one can, in principle, construct $\mathcal{H}(\mathbf{K})$ using these relations; however, in most cases, this can be done knowing only the characters of the representations $\chi(g)$.

To do this, we write the matrix $\mathcal{H}(\mathbf{K})$ as a sum of the products of components K_i and n_s^2 of linearly independent matrices X_i. These matrices X_i can be chosen in the form of matrices transforming according to irreducible representations D^κ of the group of directions $F_{\mathbf{K}_0}$. This means that the matrices X_i^κ satisfy the equations

$$D(g)X_i^\kappa D(g^{-1}) = \sum_j D_{ji}^\kappa X_j^\kappa . \tag{3.31}$$

Note that $H(\mathbf{K})$ will include the matrices X_i^κ transforming according to the representations contained in the product $D(g) \times D^*(g)$, so that the number of different sets of matrices X_i^κ forming a basis for the given representation D^κ will be determined, respectively, by expression (1) in Table 2.5. Note that $\mathcal{H}(\mathbf{K})$ can include only the components K_i^κ transforming according to the representations D^κ and can be written as

$$\mathcal{H}(\mathbf{K}) = \sum_\kappa a_\kappa \sum_l X_l^\kappa K_l^{\kappa *} . \tag{3.32}$$

Depending on the actual choice of the components X_l^κ and K_l^κ in (3.32), the constants a_κ will be real or imaginary in order for the Hamiltonian to be Hermitian, $\mathcal{H}_{ij}(\mathbf{K}) = \mathcal{H}_{ji}^*(\mathbf{K})$. Taking into account the requirements imposed by time-inversion invariance, $\mathcal{H}(\mathbf{K})$ will retain only those components K_l^κ that are even under time inversion, for which the numbers N_0 defined by the corresponding expressions in Table 2.5 are nonzero for $f = +1$ and, similarly, odd components for which $N_0 \neq 0$ for $f = -1$. As already pointed out, the k_i and B_i components are odd under time inversion as are the products containing an odd number of these components. In these cases, the matrices X_l^* can also be divided into two groups, namely, those even and odd under time inversion. Thus in cases a_1 or a_2, when the corresponding functions $\hat{K}R\psi_i$ are coupled with ψ_i by a linear relation (for case a_1, $R = e$)

$$\hat{K}R\psi_i = \sum_j T_{ji}\psi_j \tag{3.33}$$

the odd and even components $X_l^{\kappa\pm}$ transform as

$$T^{-1}X_l^{\kappa}T = fX_l^{\kappa *} = f\tilde{X}_l^{\kappa}, \tag{3.34}$$

where $f = +1$ for the even, and $f = -1$, for the odd components. In (3.34) it is assumed that all components K_l are chosen to be real, and the matrices X_l, Hermitian. Under these conditions all constants in (3.32) are real.

Obviously, in the case of one-dimensional representations, $\mathcal{H}(\mathbf{K})$ contains only the components K_l transforming according to the identity representation. For doubly degenerate representations D^{κ}, the four linearly independent 2×2 matrices can be chosen in the form of an identity matrix I, which transforms according to an identity representation, and of three Pauli matrices $\sigma_x, \sigma_y, \sigma_z$, which form a basis for the remaining representations contained in $|D^{\kappa}|^2$. Note that all the components K_l are assumed to be real. The distribution of the corresponding matrices σ_i in these irreducible representations is arbitrary, each choice corresponding to a different unitary transformation of the basis.

If $|D^{\kappa}|^2$ includes complex conjugate representations according to which the components K_2 and $K_3 = K_2^*$ transform, then in place of the matrices σ_x and σ_y one has to use the matrices

$$\sigma_+ = \frac{1}{2}\left(\sigma_x + i\sigma_y\right) = \begin{bmatrix} 0 & 1 \\ 0 & 0 \end{bmatrix} \text{ and } \sigma_- = \frac{1}{2}\left(\sigma_x - i\sigma_y\right) = \begin{bmatrix} 0 & 0 \\ 1 & 0 \end{bmatrix} \tag{3.35}$$

the corresponding constants a_2 and a_3 in (3.32) being also complex conjugate, $a_3 = a_2^*$. If one of the representations in $|D^{\kappa}|^2$ is two- or three-dimensional, then the number of independent constants in (3.32) will decrease, respectively, to three or two.

If D^{κ} is a vector or spinor representation, i.e., the basis functions $\psi^{\kappa}_{\mathbf{K}_0}$ transform as spherical functions Y_m^j with an integer or half-integer j, then as basis matrices X_l one may choose the matrices of the axial vector components J_i^j constructed using the corresponding functions Y_m^j, and their products. Under appropriate transformations of the basis functions given by the matrices $D(g)$, these matrices J_i^j will transform as the components of the pseudovector J_i:

$$D(g)J_i^j D(g^{-1}) = \sum_{i'} D_1^+(g)_{i'i} J_{i'}^j. \tag{3.36}$$

In accordance with (3.32), the Hamiltonian $\mathcal{H}(\mathbf{K})$ can be written as a product of the components J_i and of their products, and of the components K_l which form the basis of conjugate representations of the wave vector group, with a subsequent replacement of J_i, $[J_iJ_k]_{\text{sym}} \ldots$ by the corresponding matrices in the basis Y_m^j. The total number of linearly independent matrices J_i, $[J_i J_k]_s$ is equal to the square of the dimension of the representation D_j, i.e., $(2j+1)^2$, namely, to nine for $j = 1$, and to 16 for $j = 3/2$. If the representation D_j in a given group $G_{\mathbf{K}_0}$ is reducible, then the matrices J_i, $[J_i J_k]_s$ for the corresponding irreducible representation of dimension n_s are constructed using the appropriate *incomplete* basis functions, the number of the linearly independent matrices being n_s^2, and their dimension, $n_s \times n_s$.

As pointed out in Sect. 2.7, all three-dimensional representations in the groups T_h, T_d, O and O_h, and the corresponding representations of the wave vector group at point $\mathbf{K}_0 = 0$ for crystals of these classes are projectively equivalent. This means that the matrices $D^\mu(g)$ for each of them can be presented as a product of the matrices of one of these representations according to which the functions Y_m^1 transform, by the character $\chi^v(g)$ of one of the one-dimensional representations of the same group. Therefore for all these representations one can write $|D^\mu(g)|^2 = |D^1(g)|^2 |\chi^v(g)|^2 = |D^1(g)|^2$. This implies that the Hamiltonians $\mathcal{H}(\mathbf{K})$ for all projectively equivalent representations of one group $G_{\mathbf{K}_0}$ also coincide. The same applies also to two four-dimensional spinor representations of the group O_h. Therefore in practice there is no need to use matrices J_m^j with $j > 3/2$ although, for instance, the basis functions proper of the four three-dimensional representations of the group O_h correspond to $j = 1, 2, 3, 4$. The same rule applies also to symmorphic groups whose projective representations coincide with the vector representations of point groups, representations $D(g)$ differing from them only in the phase factors that cancel out in (3.30). For nonsymmorphic groups whose projective representations are inequivalent to the vector ones for the four- and six-dimensional representations available among them, the basis matrices should be determined directly from (3.38). *Bir* and *Pikus* [3.7] presented basis matrices transforming according to the vector components x, y, z for all P-inequivalent representations of point groups. The other basis matrices can be obtained by constructing their products.

To take into account the invariance under time inversion, one has to determine which components of the matrices $J_i, [J_i J_j]_s \dots$ are even or odd under this operation. In case a_1, i.e., for $\mathbf{K} = -\mathbf{K}$, the matrices J_i and the symmetrized products containing an odd number of the components J_i are odd, and those containing an even number of the components, even. The antisymmetrized products $\{J_i J_k\}_{\text{asym}}$ are not linearly independent, since $J_i J_k - J_k J_i = \mathrm{i}\delta_{ikl} J_l$. In case a_2, if only one of the sets of matrices $J_i, [J_i J_l]_s$ transforms according to one representation D, their parity can be determined from expression (2) of Table 2.5: if $N_0 \neq 0$ for $f = 1$, then these matrices are even, and they are odd if $N_0 \neq 0$ for $f = -1$. If several sets of matrices $J_i, [J_i J_l]_s \dots$ transform according to representation D^κ, then their parity can be derived from (3.34). In cases $b_{1,2}$ and $c_{1,2}$, where representations D_I and D_{II} combine by force of the requirements imposed by invariance under time inversion, no additional constraints are imposed on the diagonal submatrices $\mathcal{H}^{I,I}(\mathbf{K})$ and $\mathcal{H}^{II,II}(\mathbf{K})$, so that the number of different sets of matrices X_i^κ, just as in case b_3, will be defined by expression (1) of Table 2.5. Note that the submatrices $\mathcal{H}^{I,I}$ and $\mathcal{H}^{II,II}$ are related through:

$$\begin{aligned} &\text{case } b_1, c_1 \mathcal{H}^{II,II}(\mathbf{K}) = \mathcal{H}^{I,I}(f\mathbf{K}), \\ &\text{case } b_2, c_2 \mathcal{H}^{II,II}(\mathbf{K}) = \mathcal{H}^{I,I}(fR\mathbf{K}). \end{aligned} \tag{3.37}$$

While for spinor representations the Hamiltonian $\mathcal{H}(\mathbf{K})$ is constructed in the same way, one has to take into account that in the expressions of Table 2.5 we have to use $K^2 = -1$. This method, however, is not always capable of providing information on which of the coefficients in the Hamiltonian are relativistically small. Therefore one can conveniently construct $\mathcal{H}(\mathbf{K})$ first for the conventional, that is, the nonspinor representation; after this, spin-orbit splitting and other relativistic contributions can be taken into account by including together with the components K_i also the operators σ_i acting on the spinor functions α, β and, if required, their products by other components as well, for instance, terms linear, quadratic and cubic in k, etc. Subsequently, basis functions are chosen, in the form of products of the coordinate functions $\psi(\mathbf{x})$ by the spinor functions α, β (or directly by the sums of such products which diagonalize the spin-orbit interaction operator), with the operators X_i acting only on the coordinate functions, and the σ_l, on the spinor ones. The terms containing the products of X_i and σ_l describe spin-orbit splitting which can be included without any need of finding the matrix elements of the operator $\mathcal{H}_{\text{so}}$. When taking into account the requirements imposed by symmetry and time inversion invariance, one has to bear in mind that the components σ_i transform similar to those of a pseudovector, and reverse sign under time inversion. The use of the method of invariants for the construction of $\mathcal{H}(\mathbf{k})$ for electrons and holes in cubic crystals will be exemplified in the next section.

The method of invariants can also be employed in constructing the spectra of other elementary excitations: excitons, phonons, etc. near the extremum, as well as the spectra of bound single- and multiparticle states, for instance, of multiexciton complexes in external fields. One has, however, to keep in mind that the Hamiltonians thus obtained are valid only for energies small compared with the distance to other terms not included in this Hamiltonian. This method is also applicable to the construction of matrix elements of transitions, a point particularly convenient when describing transitions which involve several quasiparticles, for example, phonon-assisted optical transitions. In contrast to perturbation theory, there is no need here for considering any particular intermediate states. If a transition occurs between states $\psi^{\upsilon}_{\mathbf{k}_1}$ and $\psi^{\mu}_{\mathbf{k}_2}$, which transform according to the irreducible representations $D^{\upsilon}_{\mathbf{k}_1}$ and $D^{\mu}_{\mathbf{k}_2}$ of the groups $G_{\mathbf{k}_1}$ and $G_{\mathbf{k}_2}$, then the basis matrices X_l should transform according to the irreducible representations $D^{\kappa}_{\mathbf{q}}$ of the group $G_{\mathbf{q}}$, which is an intersection of the groups $G_{\mathbf{k}_1}$ and $G_{\mathbf{k}_2}$ and corresponds to $\mathbf{q} = \mathbf{k}_1 - \mathbf{k}_2$. The number of linearly independent matrices X^{κ}_l is defined by (2.45). The matrices X^{κ}_l are constructed by the methods discussed above. If the representations $D^{\mu}_{\mathbf{k}_1}$ and $D^{\upsilon}_{\mathbf{k}_2}$ are projectively equivalent to the vector representations according to which transform spherical functions Y_m^l with l equal or differing by unity, then for X_{lik}^{κ} one may choose matrix elements of the components of the polar or, accordingly, axial vector $\mathbf{R}_i$.

By the above rules, $\mathcal{H}(\mathbf{K})$ includes, as components of $\mathbf{K}$, products of the light polarization vectors $\mathbf{e}_i$ and amplitudes of normal lattice vibrations a_{vq}, which transform according to the same irreducible representations $D_{\mathbf{q}}^{\kappa}$ of the group $G_{\mathbf{q}}$ as the components of $X_l^{\kappa *}$. If a transition occurs with scattering not by phonons but by impurity centers, then one has first to determine the irreducible representations D^s of the group of directions $G_{\mathbf{q}}$ according to which transforms the impurity potential V_0 that depends on the actual location of the centers. $\mathcal{H}(\mathbf{K})$ includes the product of the components X_i^{κ} and $e_i^{\kappa *}$ transforming according to these representations D^s.

The Hamiltonian $\mathcal{H}(\mathbf{K})$ for quantum wells and superlattices can be constructed based on three-dimensional Hamiltonians and considering the potential creating a well or superlattice within the framework of the effective mass method. This method offers the possibility of expressing the constants of the corresponding Hamiltonian in terms of the parameters of the original crystal.

On the other hand, one can readily construct the two-dimensional Hamiltonian $\mathcal{H}(\mathbf{K})$ for quantum wells by one of the techniques discussed above for the basis functions which transform according to the corresponding representations of surface or doubly periodic groups specified in Sect. 2.2.

It should be pointed out that the components K_i form the basis for the representations of the corresponding groups of directions, i.e., of the point groups C_s, C_n, C_{nv} for surface groups, and of the other, except for cubic, point groups for doubly periodic space groups.

3.4 Electron and Hole Spectrum in Diamond- and Zincblende-Type Cubic Crystals

The most widely used materials for heterostructures, such as quantum wells and superlattices, are A_3B_5 compounds with zincblende-type structure, as well as A_2B_6 compounds, and diamond-type crystals, namely, germanium, silicon, and their alloys. In this section, we are going to present relevant information on the band structure of these materials. Figure 3.1 shows the band structure of Ge, Si and GaAs. For the points $\Gamma(0,0,0)$ and $L((\pi/a_0)(111))$, and Λ and Δ, lying, respectively, on the [111] and [100] axes inside the Brillouin zone, and the point $X((2\pi/a_0)(100))$ in the A_3B_5 lattice, the representations of the wave vector group are projectively equivalent to those of the corresponding point groups. The relations between these representations are given in Tables 3.1, 2. Note that the characters of representations for the elements $(r|\tau)$ of the wave vector group at these points differ from those of the corresponding representation for the elements r of the point group in the factor $e^{-i\mathbf{k}\cdot\tau}$. At the point X in the diamond lattice, the factor system $\omega(g_1, g_2) \neq 1$, the representations of this group being projectively inequivalent to the vector representations of the group. This group has four conventional doubly degenerate representations $X_1 - X_4$, and one spinor, four-fold-degenerate representation X_5. The characters of these representations are presented in Table 3.3.[2] The top of the valence band in these crystals, just as in practically all A_3B_5 and A_2B_6 cubic crystals, lies at the point $\Gamma(k=0)$, and is triply degenerate if spin is not included. In O_h class crystals, the basis functions X, Y, Z transform according to the representation F_2^+, i.e., as yz, xz, xy, respectively. This representation is denoted Γ'_{25}. In crystals of the T_d class, both x, y, z and yz, zx, xy functions form the basis for the representation F_2 denoted as Γ_{15} (or Γ_5). Spin-orbit interaction splits the representation Γ'_{25} in a four-fold degenerate $G^+(\Gamma_8^+)$ and a two-fold degenerate $E_2'^+$ (Γ_7^+), and in crystals of the class T_d, Γ_{15} splits in $G(\Gamma_8)$ and $E_2(\Gamma_7)$.

In Ge the lowest extrema of the conduction band lie at the point Γ (representation Γ'_2 corresponding to A_1^- of the group O_h), at the point L at the boundary of the Brillouin zone (representation L_1 corresponding to A_1^+ of the group D_{3d}), and at the point Δ on the [100] axes (representation Δ_1 corresponding to A_1 of the group C_{4v}). The lowest of them is the point L. All these representations are nondegenerate if spin is not included, and transform into the corresponding two-fold degenerate spinor representations: $\Gamma'_2 \rightarrow \Gamma_7^-$, $L_1 \rightarrow L_6$, $\Delta_1 \rightarrow \Delta_6$.

In contrast to Ge, in Si in the lowest extremum at the point Γ, the conduction band is three-fold degenerate (representation Γ_{15}), the lowest being the point Δ. The point X on the [100] axes at the boundaries of the Brillouin zone in Ge

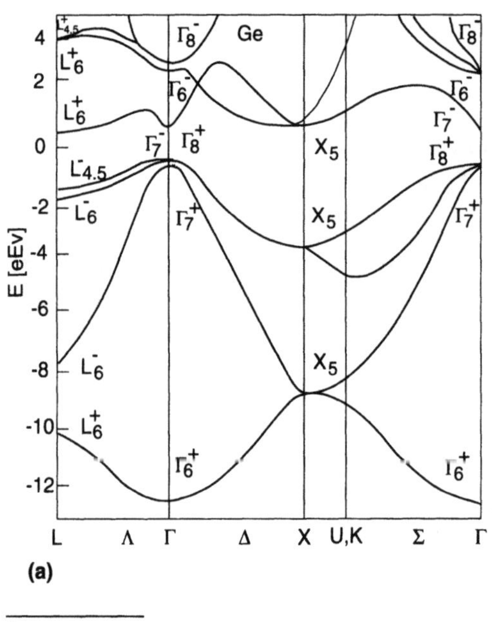

Fig. 3.1a-c. Band structure of cubic crystals: **a** Ge, **b** Si, **c** GaAs

[2] It should be kept in mind that different authors use different notations for the irreducible representations of the wave vector group. The tables of the characters provide the possibility of establishing the correspondence between these notations.

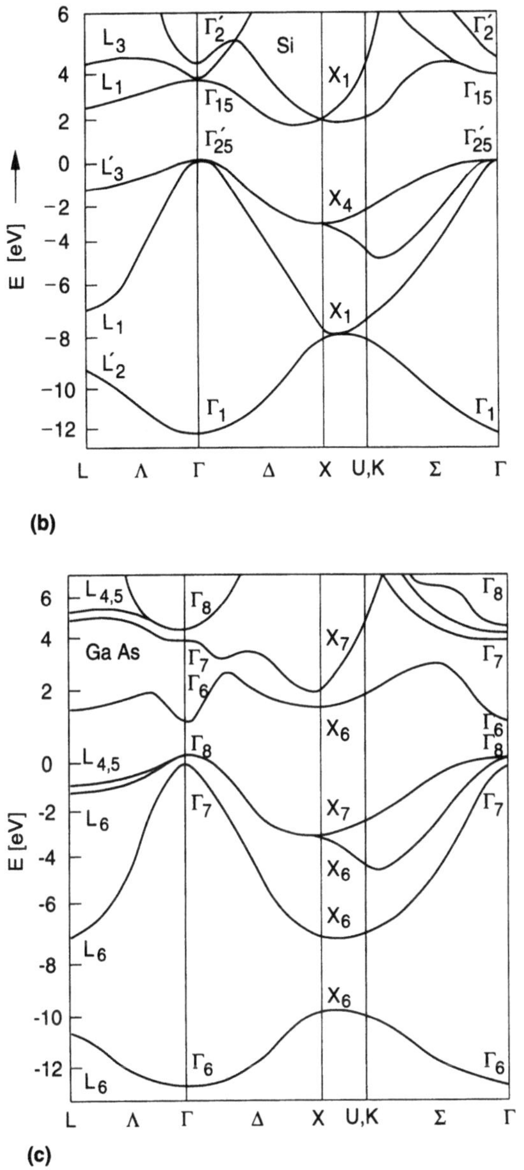

Fig. 3.1b,c

and Si is not the zero slope point since all the representations in which spin is not included are two-fold degenerate here, and with spin included, they are four-fold degenerate. In crystals of the class T_d, the degeneracy is lifted at the point X with the formation of two closely spaced branches X_1 and X_3, this becoming a zero slope point.[3] The wave functions in the lowest extrema of the

[3] With **k-p** interaction between the X_1 and X_3 branches included, the conduction band minimum lies not at the point X_1 proper but near it on the [100] axis, whereas X is the saddle point.

Table 3.1. Representations of the wave-vector groups at points Γ, Δ, L for Ge and Si and the corresponding representations of the point groups

Representations of $O_h = T_d \times C_i$	A_1^+	A_1^-	A_2^+	A_2^-	E^+	E^-	F_1^+	F_1^-	F_2^+	F_2^-	$E_1^{'+}$	$E_1^{'-}$	$E_2^{'+}$	$E_2^{'-}$	$G^{'\pm}$
Representations at Γ	Γ_1	Γ_2'	Γ_2	Γ_1'	Γ_{12}	Γ_{12}'	Γ_{15}'	Γ_{25}	Γ_{25}'	Γ_{15}	Γ_7^+	Γ_6^-	Γ_6^+	Γ_7^-	$\Gamma_8^\pm$

Representations of C_{4v}	A_1	A_2	B_1	B_2	E	E_1'	E_2'
Representations at Δ	Δ_1	Δ_2	Δ_3	Δ_4	Δ_5	Δ_6	Δ_7

Representations of $D_{3d} = C_3 \times C_i$	A_1^+	A_1^-	A_2^+	A_2^-	E^+	E^-	$A_1^{'+}$	$A_1^{'-}$	$A_2^{'+}$	$A_2^{'-}$	$E^{'+}$	$E^{'-}$
Representations at L	L_1	L_2'	L_2	L_1'	L_3	L_3'	L_4	L_5'	L_5	L_4'	L_6	L_6'

Table 3.2. Representations of the wave-vector groups at points Γ, X, Δ, L for A_3B_5 cubic crystals and the corresponding representations of the point groups

Representation of T_d	A_1	A_2	E	F_1	F_2	E'_1	E'_2	G'
Representation at Γ	Γ_1	Γ_2	Γ_{12}	Γ_{25}	Γ_{15}	Γ_6	Γ_7	Γ_8

Representations of D_{2d}	A_1	A_2	B_1	B_2	E	E'_1	E'_2
Representations at X	X_1	X_4	X_2	X_3	X_5	X_6	X_7

Representations of C_{2v}	A_1	A_2	B_1	B_2	E'
Representations at Δ	Δ_1	Δ_4	Δ_2	Δ_3	Δ_5

Representations of C_{3v}	A_1	A_2	E	A'_1	A'_2	E'
Representations at L	L_1	L_2	L_3	L_4	L_5	L_6

Table 3.3. Characters of representations at point X for Ge, Si

	X_1	X_2	X_3	X_4	X_5
$(e\|0)$	2	2	2	2	4
$(c_{2x}\|0)$	2	2	−2	−2	0
$(c_{2y}\|0)$	0	0	0	0	0
$(c_{2z}\|0)$	0	0	0	0	0
$(s_{4x}\|0)$	0	0	0	0	0
$(s^3_{4x}\|0)$	0	0	0	0	0
$(\sigma_{yz}\|0)$	2	−2	0	0	0
$(\sigma_{y\bar{z}}\|0)$	2	−2	0	0	0
$(\mathrm{i}\|\tau)$	0	0	0	0	0
$(\sigma_x\|\tau)$	0	0	0	0	0
$(\sigma_y\|\tau)$	0	0	0	0	0
$(\sigma_z\|\tau)$	0	0	0	0	0
$(c^3_{4x}\|\tau)$	0	0	0	0	0
$(c_{4x}\|\tau)$	0	0	0	0	0
$(c_{2y\bar{z}}\|\tau)$	0	0	−2	2	0
$(c_{2yz}\|\tau)$	0	0	2	−2	0

conduction band at the points Γ, X and L in these crystals transform according to the representations Γ_1 (corresponding to the representation A_1 of the group T_d), L_1 (representation A_1 of the group C_{3v}), X_1 (representation A_1 of the group D_{2d}). When spin is included, they transform to spinor representations $\Gamma_1 \to \Gamma_6$, $L_1 \to L_6$, $X_1 \to X_6$. The lowest extremum of the conduction band in most A_3B_5 crystals is the point Γ. The total degeneracy of the conduction band at the extremum is equal to the product of that of the corresponding representation and the number of the star points which is one for the point Γ, four for L, three for X, and six for Δ. For the bands degenerate in spin only the spectrum is defined by (3.22); note that at the point Γ all three components of the tensor m^{*-1} are equal, while at the points L, X and Δ we have $m^{*-1}_{xx} = m^{*-1}_{yy} = m^{*-1}_{\perp}, m^{*-1}_{zz} = m^{-1}_{\|}$. The z axis is assumed to be aligned along the principal axis [111] or [100] of the corresponding extremum. In crystals of the class T_d, as one moves away from the extremum points Γ, L, or X, the spin degeneracy is lifted (except for the points lying on the [100] and [111] axes). At Γ this splitting is proportional to k^3 and is described by the Hamiltonian

$$\mathcal{H}_{c3} = \gamma_c \sum_i \sigma_i \kappa_i, \tag{3.38}$$

where $\kappa_z = k_z(k_x^2 - k_y^2)$, $\kappa_x = k_x(k_y^2 - k_z^2)$, $\kappa_y = k_y(k_z^2 - k_x^2)$.

At the points X and L the splitting is linear-in-k:

$$\begin{aligned} \mathcal{H}_X &= \beta_X(\sigma_x k_x - \sigma_y k_y) (\text{at } X \text{ points}), \\ \mathcal{H}_L &= \beta_L[\boldsymbol{\sigma}\mathbf{k}]_z (\text{at } L \text{ points}). \end{aligned} \tag{3.39}$$

Here σ_i are the Pauli matrices, and $S_i = \sigma_i/2$ the matrices of the operator J_i in the basis $Y_m^{1/2}(m = \pm 1/2)$ with the quantization axis z directed along [001] for the points Γ or X, and along [111] (or equivalent directions) for the point L.

For the degenerate band Γ_{25}' or Γ_{15}, one chooses as basis matrices X_i in the method of invariants and the matrices of the operators J_i in the basis $Y_m^1(m = 0, \pm 1)$, or their products given in Table 3.4. The Hamiltonian $\mathcal{H}(\mathbf{k})$ constructed for holes by the rules specified in Sect. 3.3 will have the form

$$\mathcal{H}(\mathbf{k}) = -Ak^2 I + 3B \sum_i J_i^2 \left(k_i^2 - \frac{k^2}{3}\right) + 2\sqrt{3}D \sum_{i>j} [J_i J_j]_s k_i k_j. \tag{3.40}$$

Using the **k-p** method, one can express the constants A, B and D in terms of the corresponding matrix elements of the momentum operator

$$\begin{aligned} A &= \frac{1}{3}(L + 2M),\ B = \frac{1}{3}(L - M),\ D = \frac{N}{\sqrt{3}}, \quad \text{where} \\ L &= \frac{\hbar^2}{2m} + \frac{\hbar^2}{m^2} \sum_n \frac{|\langle X|p_x|n\rangle|^2}{E_0 - E_n}, \\ M &= \frac{\hbar^2}{m^2} \sum_n \frac{|\langle X|p_y|n\rangle|^2}{E_0 - E_n}, \\ N &= \frac{\hbar^2}{m^2} \sum_n \frac{\langle X|p_x|n\rangle\langle n|p_y|Y\rangle + \langle X|p_y|n\rangle\langle n|P_x|Y\rangle}{E_0 - E_n}. \end{aligned} \tag{3.41}$$

Table 3.4. Matrices J_i and their products for the D_1 representation in the basis Y^1_m; $m = 1, 0, -1$ (canonical basis)

$$J_x = \begin{bmatrix} 0 & 1/\sqrt{2} & 0 \\ 1/\sqrt{2} & 0 & 1/\sqrt{2} \\ 0 & 1/\sqrt{2} & 0 \end{bmatrix}, \quad J_y = \begin{bmatrix} 0 & -i/\sqrt{2} & 0 \\ i/\sqrt{2} & 0 & -i/\sqrt{2} \\ 0 & i/\sqrt{2} & 0 \end{bmatrix}, \quad J_z = \begin{bmatrix} 1 & 0 & 0 \\ 0 & 0 & 0 \\ 0 & 0 & -1 \end{bmatrix},$$

$$J_x^2 = \begin{bmatrix} 1/2 & 0 & 1/2 \\ 0 & 1 & 0 \\ 1/2 & 0 & 1/2 \end{bmatrix}, \quad J_y^2 = \begin{bmatrix} 1/2 & 0 & -1/2 \\ 0 & 1 & 0 \\ -1/2 & 0 & 1/2 \end{bmatrix}, \quad J_z^2 = \begin{bmatrix} 1 & 0 & 0 \\ 0 & 0 & 0 \\ 0 & 0 & 1 \end{bmatrix},$$

$$2[J_x J_y]_s = \begin{bmatrix} 0 & 0 & -i \\ 0 & 0 & 0 \\ i & 0 & 0 \end{bmatrix}, \quad 2[J_x J_z]_s = \begin{bmatrix} 0 & 1/\sqrt{2} & 0 \\ 1/\sqrt{2} & 0 & -1/\sqrt{2} \\ 0 & -1/\sqrt{2} & 0 \end{bmatrix}, \quad 2[J_y J_z]_s = \begin{bmatrix} 0 & -i/\sqrt{2} & 0 \\ i/\sqrt{2} & 0 & i/\sqrt{2} \\ 0 & -i/\sqrt{2} & 0 \end{bmatrix}.$$

For the spinor representations Γ_8^+ or Γ_7^+ (or Γ_8 and Γ_7 of the class T_d), the $\mathcal{H}(k)$ can be obtained by including in (3.41) the term

$$\mathcal{H}_{so} = \frac{\Delta}{3}\sum_i J_i\sigma_i \tag{3.42}$$

describing spin-orbit interaction. The basis is formed by six functions: $X\alpha$, $X\beta$, $Y\alpha$, $Y\beta$, $Z\alpha$ and $Z\beta$. Here Δ is the separation between the bands Γ_8^+ (Γ_8) and Γ_7^+ (Γ_7).

In the canonic basis, the functions diagonalizing $\mathcal{H}_{so}$ are

$$\Gamma_8\begin{cases}\psi_{3/2}^{3/2} &= \psi_1^1\alpha\\ \psi_{1/2}^{3/2} &= \frac{1}{\sqrt{3}}\left(\psi_1^1\beta+\sqrt{2}\psi_0^1\alpha\right)\\ \psi_{-1/2}^{3/2} &= \frac{1}{\sqrt{3}}\left(\psi_{-1}^1\alpha+\sqrt{2}\psi_0^1\beta\right)\\ \psi_{-3/2}^{3/2} &= \psi_{-1}^1\beta\end{cases}$$
$$\Gamma_7\begin{cases}\psi_{1/2}^{1/2} &= \frac{1}{\sqrt{3}}\left(\sqrt{2}\psi_1^1\beta-\psi_0^1\alpha\right)\\ \psi_{-1/2}^{1/2} &= \frac{1}{\sqrt{3}}\left(-\sqrt{2}\psi_{-1}^1\alpha+\psi_0^1\beta\right)\end{cases} \tag{3.43}$$

where $\psi_1^1 = -\frac{1}{\sqrt{2}}(X+\mathrm{i}Y)$, $\psi_0^1 = Z$, $\psi_{-1}^1 = \frac{1}{\sqrt{2}}(X-\mathrm{i}Y)$.
Under O_h or T_d symmetry operations, the functions ψ_m^l transform as the eigenfunctions of the operator J with $l = 3/2$ and $l = 1/2$.

One can use the method of invariants to construct $\mathcal{H}(\mathbf{k})$ directly for the representation $\Gamma_8^+(\Gamma_8)$ by choosing as basis matrices X_i the matrices of the operators J_i in the basis $Y_m^{3/2}$ and their products presented in Table 3.5. The corresponding Hamiltonian for the holes can be written

$$\mathcal{H}(\mathbf{k}) = \frac{\hbar^2}{2m}\left\{\gamma_1 k^2 I - 2\gamma_2\sum_i J_i^2\left(k_i^2-\frac{k^2}{3}\right) - 4\gamma_3\sum_{i>j}[J_iJ_j]k_ik_j\right\}. \tag{3.44}$$

The constants γ_i are connected with the constants A, B and D in (3.40) through the relations

$$\frac{\hbar^2}{2m}\gamma_1 = -A,\ \frac{\hbar^2}{m}\gamma_2 = -B,\ \frac{\hbar^2}{m}\gamma_3 = -\frac{D}{\sqrt{3}}. \tag{3.45}$$

In the matrix form, the Hamiltonian (3.44) can be rewritten as

$$\mathcal{H}(\mathbf{k}) = \begin{bmatrix} F & H & I & O\\ H^* & G & O & I\\ I^* & O & G & -H\\ O & I^* & -H^* & F\end{bmatrix}, \tag{3.46}$$

Table 3.5. Matrices J_i and their products for the representation $D_{3/2}$ in the basis $Y_m^{3/2}$; $m = 3/2, 1/2, -1/2, -3/2$ (canonical basis)

$$J_x = \begin{bmatrix} 0 & \sqrt{3}/2 & 0 & 0 \\ \sqrt{3}/2 & 0 & 1 & 0 \\ 0 & 1 & 0 & \sqrt{3}/2 \\ 0 & 0 & \sqrt{3}/2 & 0 \end{bmatrix}, \quad J_y = \begin{bmatrix} 0 & -\mathrm{i}\sqrt{3}/2 & 0 & 0 \\ \mathrm{i}\sqrt{3}/2 & 0 & -\mathrm{i} & 0 \\ 0 & \mathrm{i} & 0 & -\mathrm{i}\sqrt{3}/2 \\ 0 & 0 & \mathrm{i}\sqrt{3}/2 & 0 \end{bmatrix}, \quad J_z = \begin{bmatrix} 3/2 & 0 & 0 & 0 \\ 0 & 1/2 & 0 & 0 \\ 0 & 0 & -1/2 & 0 \\ 0 & 0 & 0 & -3/2 \end{bmatrix},$$

$$J_x^2 = \begin{bmatrix} 3/4 & 0 & \sqrt{3}/2 & 0 \\ 0 & 7/4 & 0 & \sqrt{3}/2 \\ \sqrt{3}/2 & 0 & 7/4 & 0 \\ 0 & \sqrt{3}/2 & 0 & 3/4 \end{bmatrix}, \quad J_y^2 = \begin{bmatrix} 3/4 & 0 & -\sqrt{3}/2 & 0 \\ 0 & 7/4 & 0 & -\sqrt{3}/2 \\ -\sqrt{3}/2 & 0 & 7/4 & 0 \\ 0 & -\sqrt{3}/2 & 0 & 3/4 \end{bmatrix}, \quad J_z^2 = \begin{bmatrix} 9/4 & 0 & 0 & 0 \\ 0 & 1/4 & 0 & 0 \\ 0 & 0 & 1/4 & 0 \\ 0 & 0 & 0 & 9/4 \end{bmatrix}$$

$$2[J_x J_y]_s = \begin{bmatrix} 0 & 0 & -\mathrm{i}\sqrt{3} & 0 \\ 0 & 0 & 0 & -\mathrm{i}\sqrt{3} \\ \mathrm{i}\sqrt{3} & 0 & 0 & 0 \\ 0 & \mathrm{i}\sqrt{3} & 0 & 0 \end{bmatrix}, \quad 2[J_x J_z]_s = \begin{bmatrix} 0 & \sqrt{3} & 0 & 0 \\ \sqrt{3} & 0 & 0 & 0 \\ 0 & 0 & 0 & -\sqrt{3} \\ 0 & 0 & -\sqrt{3} & 0 \end{bmatrix}, \quad 2[J_y J_z]_s = \begin{bmatrix} 0 & -\mathrm{i}\sqrt{3} & 0 & 0 \\ \mathrm{i}\sqrt{3} & 0 & 0 & 0 \\ 0 & 0 & 0 & \mathrm{i}\sqrt{3} \\ 0 & 0 & -\mathrm{i}\sqrt{3} & 0 \end{bmatrix},$$

$$J_x^3 = \begin{bmatrix} 0 & 7\sqrt{3}/8 & 0 & 3/4 \\ 7\sqrt{3}/8 & 0 & 5/2 & 0 \\ 0 & 5/2 & 0 & 7\sqrt{3}/8 \\ 3/4 & 0 & 7\sqrt{3}/8 & 0 \end{bmatrix}, \quad J_y^3 = \begin{bmatrix} 0 & -\mathrm{i}7\sqrt{3}/8 & 0 & \mathrm{i}3/4 \\ \mathrm{i}7\sqrt{3}/8 & 0 & -\mathrm{i}5/2 & 0 \\ 0 & \mathrm{i}5/2 & 0 & -\mathrm{i}7\sqrt{3}/8 \\ -\mathrm{i}3/4 & 0 & \mathrm{i}7\sqrt{3}/8 & 0 \end{bmatrix}, \quad J_z^3 = \begin{bmatrix} 27/8 & 0 & 0 & 0 \\ 0 & 1/8 & 0 & 0 \\ 0 & 0 & -1/8 & 0 \\ 0 & 0 & 0 & -27/8 \end{bmatrix},$$

$$V_x = \begin{bmatrix} 0 & -\sqrt{3}/4 & 0 & -3/4 \\ -\sqrt{3}/4 & 0 & 3/4 & 0 \\ 0 & 3/4 & 0 & -\sqrt{3}/4 \\ -3/4 & 0 & -\sqrt{3}/4 & 0 \end{bmatrix}, \quad V_y = \begin{bmatrix} 0 & -\mathrm{i}\sqrt{3}/4 & 0 & \mathrm{i}3/4 \\ \mathrm{i}\sqrt{3}/4 & 0 & \mathrm{i}3/4 & 0 \\ 0 & -\mathrm{i}3/4 & 0 & -\mathrm{i}\sqrt{3}/4 \\ -\mathrm{i}3/4 & 0 & \mathrm{i}\sqrt{3}/4 & 0 \end{bmatrix}, \quad V_z = \begin{bmatrix} 0 & 0 & \sqrt{3}/2 & 0 \\ 0 & 0 & 0 & -\sqrt{3}/2 \\ \sqrt{3}/2 & 0 & 0 & 0 \\ 0 & -\sqrt{3}/2 & 0 & 0 \end{bmatrix},$$

$$2[J_x J_y J_z]_s = \begin{bmatrix} 0 & 0 & -\mathrm{i}\sqrt{3}/2 & 0 \\ 0 & 0 & 0 & \mathrm{i}\sqrt{3}/2 \\ \mathrm{i}\sqrt{3}/2 & 0 & 0 & 0 \\ 0 & -\mathrm{i}\sqrt{3}/2 & 0 & 0 \end{bmatrix}$$

$$V_i = \left[J_i \left(J_{i+1}^2 - J_{i+2}^2\right)\right]_s$$

where $F = -Ak^2 - \frac{B}{2}(k^2 - 3k_z^2)$, $G = -Ak^2 + \frac{B}{2}(k^2 - 3k_z^2)$,

$$H = Dk_z(k_x - \mathrm{i}k_y),\ I = \frac{\sqrt{3}}{2}B(k_x^2 - k_y^2) - \mathrm{i}Dk_xk_y.$$

The corresponding determinant $\|\mathcal{H}(\mathbf{k}) - E\| = 0$ will be

$$(E - F)(E - G) - |H|^2 - |I|^2. \tag{3.47}$$

The roots of the equation $\|\mathcal{H}(\mathbf{k}) - E\| = 0$ are

$$E_{1,2} = -Ak^2 \pm \left[B^2k^4 + C^2\sum_{i>j} k_i^2k_j^2\right]^{1/2}, \tag{3.48}$$

where $C^2 = D^2 - 3B^2$. The plus sign corresponds to the light and the minus, to the heavy holes.

The eigenfunctions diagonalizing $\mathcal{H}$ (k) have the form

$$F_{1j} = \left[(E_i - F)(E_i - E_j)\right]^{-1/2}\begin{bmatrix} H \\ E_i - F \\ 0 \\ I^* \end{bmatrix}$$

$$F_{2j} = \left[(E_i - F)(E_i - E_j)\right]^{-1/2}\begin{bmatrix} -I \\ 0 \\ -(E_i - F) \\ H^* \end{bmatrix} \tag{3.49}$$

$i = 1, 2;\ j \neq i$.

The Hamiltonian (3.40) in the matrix form for the band Γ'_{25} was first constructed by *Shockley* (1950) [3.8], for the band Γ_8^+ in the form (3.46) by *Elliot* (1954) [3.9,10], *Dresselhaus* et al. (1955) [3.11], and in the invariant form (3.44) by *Luttinger* (1956) [3.12]. The constants γ_i in (3.44) are called Luttinger's constants.

In many A_3B_5 crystals, the constants γ_2 and γ_3 are close in magnitude thus permitting one to use the spherical approximation, i.e.,

$$\gamma_2 = \gamma_3 = \bar{\gamma} = \frac{1}{5}(2\gamma_2 + 3\gamma_3).$$

In the spherical approximation

$$\mathcal{H}(\mathbf{k}) = \frac{\hbar^2}{2m}\left[(\gamma_1 + \frac{5}{2}\bar{\gamma})k^2 I - 2\bar{\gamma}(\mathbf{J}\cdot\mathbf{k})^2\right] \tag{3.50}$$

the hole energy

$$E_{1,2} = \frac{\hbar^2}{2m}(\gamma_1 \pm 2\bar{\gamma})k^2. \tag{3.51}$$

Taking for the quantization axis for the $\psi_{\mathbf{k}}^m$ state the $\mathbf{k}$ direction, the heavy holes of energy $E_2(\mathbf{k})$ will be characterized by the states $\psi_{\mathbf{k}}^{\pm 3/2}$, and the light ones of energy $E_1(\mathbf{k})$, by the states $\psi_{\mathbf{k}}^{\pm 1/2}$. The basis functions $\psi_{\mathbf{k}}^{m'}$ are related to the state $\psi_{\mathbf{k}_0}^{m}$ in the fixed coordinate frame $X_0, Y_0, Z_0 (Z_0 \| \mathbf{k}_0)$ through

$$\psi_{\mathbf{k}}^{m'} = \sum_m d_{mm'}^{j}(\beta) e^{-im\alpha} e^{-im'\gamma} \psi_{\mathbf{k}_0}^{m}. \tag{3.52}$$

Here $\beta = \cos\theta$, where θ, α and γ are Euler's angles defining the position of the new frame $X, Y, Z(Z\|\mathbf{k})$ with respect to the system X_0, Y_0, Z_0.[4] The matrices $d_{mm'}^{j}$, for $j = 1/2$ and $j = 3/2$ are given in Table 3.6.

In crystals of the class T_d, the Hamiltonian $\mathcal{H}(\mathbf{k})$ for the bands Γ_8 and Γ_7 includes terms odd in k which remove the two-fold degeneracy. For the band $\Gamma_7, \mathcal{H}_{k3}$ is described by an expression differing from (3.38) in γ_c being replaced by γ_{v2}. For the band Γ_8 the main contribution is described by a similar expression

$$\mathcal{H}_{v3} = \gamma_v \sum_i J_i \kappa_i. \tag{3.53}$$

In contrast to the conduction band, in this band the term (3.53) is of nonrelativistic nature, i.e., it is not connected with spin-orbit interaction; indeed, the Hamiltonian $\mathcal{H}(\mathbf{k})$ for the representation Γ_{15} also includes a term similar to (3.53) (with γ_v replaced by $\gamma_v' = 3/2\gamma_v$). Besides (3.50), the Hamiltonian $\mathcal{H}(\mathbf{k})$ for the band Γ_8 includes also relativistic terms, linear and cubic in k:

$$\mathcal{H}_{v1} = \frac{4}{3} k_0 \sum_i V_i k_i, \tag{3.54}$$

where

$$V_z = [J_z(J_x^2 - J_j^2)]_{\rm sym} \text{ etc.}$$

$$\delta\mathcal{H}_{v3} = \gamma_v \left[a_2 \sum_i J_i^3 \kappa_i + a_3 \sum_i V_i k_i \left(k_i^2 - \frac{k^2}{3} \right) + \alpha_4 k^2 \sum_i V_i k_i \right]. \tag{3.55}$$

Besides the quadratic-in-k terms, those linear in k include also *nondiagonal* components $\mathcal{H}(\mathbf{k})$, which determine the mixing of the states Γ_8 and Γ_7 for $k \neq 0$. In a magnetic field $\mathbf{B}$ the Hamiltonian $\mathcal{H}$ includes linear-in-B terms. In accordance with (3.22), for nondegenerate bands we have

$$\mathcal{H}_B = \frac{1}{2}\mu_{\rm B}(\sigma g \mathbf{B}) = \frac{1}{2}\mu_{\rm B} \sum_{ij} \sigma_i g_{ij} B_j. \tag{3.56}$$

[4] Euler's transformation includes the following operations:

1. rotation through angle α about Z_0, with $Y_0 \to Y_1, X_0 \to X_1$;
2. rotation through angle θ about Y_1, with $Z_0 \to Z, X_1 \to X_2$;
3. rotation through angle γ about Z, with $X_2 \to X, Y_1 \to Y$.

Table 3.6. Matrices $d^j_{mm'}(\beta)$ in the basis $j = 1/2$ and $j = 3/2$

$j = 1/2$

m' / m	1/2	−1/2
1/2	$\frac{1}{\sqrt{2}}(1+\beta)^{1/2}$	$-\frac{1}{\sqrt{2}}(1-\beta)^{1/2}$
−1/2	$\frac{1}{\sqrt{2}}(1-\beta)^{1/2}$	$\frac{1}{\sqrt{2}}(1+\beta)^{1/2}$

$j = 3/2$

m' / m	3/2	1/2	−1/2	−3/2
3/2	$\frac{1}{2\sqrt{2}}(1+\beta)^{3/2}$	$-\frac{\sqrt{3}}{2\sqrt{2}}(1+\beta)(1-\beta)^{1/2}$	$\frac{\sqrt{3}}{2\sqrt{2}}(1-\beta)(1+\beta)^{1/2}$	$-\frac{1}{2\sqrt{2}}(1-\beta)^{3/2}$
1/2	$\frac{\sqrt{3}}{2\sqrt{2}}(1+\beta)(1-\beta)^{1/2}$	$\frac{1}{2\sqrt{2}}(3\beta-1)(1+\beta)^{1/2}$	$-\frac{1}{2\sqrt{2}}(3\beta+1)(1-\beta)^{1/2}$	$\frac{\sqrt{3}}{2\sqrt{2}}(1-\beta)(1+\beta)^{1/2}$
−1/2	$\frac{\sqrt{3}}{2\sqrt{2}}(1-\beta)(1+\beta)^{1/2}$	$\frac{1}{2\sqrt{2}}(3\beta+1)(1-\beta)^{1/2}$	$\frac{1}{2\sqrt{2}}(3\beta-1)(1+\beta)^{1/2}$	$-\frac{\sqrt{3}}{2\sqrt{2}}(1+\beta)(1-\beta)^{1/2}$
−3/2	$\frac{1}{2\sqrt{2}}(1-\beta)^{3/2}$	$\frac{\sqrt{3}}{2\sqrt{2}}(1-\beta)(1+\beta)^{1/2}$	$\frac{\sqrt{3}}{2\sqrt{2}}(1+\beta)(1-\beta)^{1/2}$	$\frac{1}{2\sqrt{2}}(1+\beta)^{3/2}$

The tensor g has three components g_{xx}, g_{yy}, g_{zz} along the principal axes. At the point Γ, all three components are equal. At the points L, Δ, X, we have $g_{xx} = g_{yy} = g_{\perp}, g_{zz} = g_{\parallel}$; just as before, the Z axis is aligned along the principal axis of the corresponding extremum. For the Γ_8^+ or Γ_8 bands

$$\mathcal{H}_B = \mu_B g_o \left[\mathcal{K}(\mathbf{J} \cdot \mathbf{B}) + q \sum_i J_i^3 B_i \right] . \tag{3.57}$$

3.4.1 Kane's Model

As already pointed out, in direct band-gap A_3B_5 semiconductors with a narrow gap the inclusion of only quadratic-in-k terms often turns out to be insufficient [3.13, 14]. On the other hand, the main contribution to m^* or the constants L, M, N in (3.41) comes in these semiconductors from **k-p** interaction between the closest bands Γ_1 and Γ_{15}, or, if spin is included, between the bands Γ_6 and Γ_8, Γ_7. In this case, one can construct within the framework of the **k-p** method a Hamiltonian for both these bands by including only the interaction between them.

A convenient form of writing Kane's equation was proposed by *Suris* [3.15]. The total wave function is written in the form of a product of s smooth functions u_l, $v_{il}(l = \pm 1/2, i = x, y, z)$ by the corresponding Bloch functions SX_l, $R_i X_l (R_x = X, R_y = Y, R_z = Z)$:

$$\psi = uS + \mathbf{v} \cdot \mathbf{R}, \quad u = \begin{bmatrix} u_{1/2} \\ u_{-1/2} \end{bmatrix}, \quad \mathbf{v} = \begin{bmatrix} \mathbf{v}_{1/2} \\ \mathbf{v}_{-1/2} \end{bmatrix}, \tag{3.58}$$

The Hamiltonian $\mathcal{H}$, according to (3.4, 42), can be written as

$$\mathcal{H} = \frac{\hbar}{m}(\mathbf{k} \cdot \mathbf{p}) + \frac{\Delta}{3}(\mathbf{J} \cdot \sigma).$$

The operator $\mathbf{p}$ acts on the coordinate functions and has one linearly independent component $\langle S|p_z|Z\rangle$. In accordance with Table 3.4, the nonzero matrix elements of the operators J_k are $\langle R_j|J_k|R_l\rangle = \mathrm{i}\delta_{jkl}$. Multiplying $\hat{\mathcal{H}}\psi$ on the left by $\langle S|$ or $\langle R|$ and integrating over the coordinates, we obtain a system of equations for u and $\mathbf{v}$:

$$-\mathrm{i}\hbar \frac{\partial u}{\partial t} = -\mathrm{i}P(\mathbf{k}\cdot\mathbf{v}), \tag{3.59a}$$

$$-\mathrm{i}\hbar \frac{\partial \mathbf{v}}{\partial t} = -E_g'\mathbf{v} + \mathrm{i}P\mathbf{k}u + \mathrm{i}\frac{\Delta}{3}[\sigma \mathbf{v}]. \tag{3.59b}$$

Here $E_g' = E_g + \Delta/3$, $P = \mathrm{i}(\hbar/m)\langle S|p_z|Z\rangle$.

For eigenstates

$$Eu + \mathrm{i}P(\mathbf{k}\cdot\mathbf{v}) = 0, \tag{3.60a}$$

$$(E + E_g')\mathbf{v} + \frac{\Delta}{3}\hat{\lambda}\mathbf{v} - \mathrm{i}P\mathbf{k}u = 0. \tag{3.60b}$$

Here

$$\hat{\lambda}\mathbf{f} = -\mathrm{i}[\sigma\mathbf{f}].$$

Taking a vector product of (3.60b) by $\mathbf{v}$ and taking into account that $[\sigma[\sigma\mathbf{v}]] = \mathrm{i}[\sigma\mathbf{v}] + 2\mathbf{v}$, we come to the equation

$$\frac{2}{3}\Delta\mathbf{v} + \left(E + E_g' + \frac{\Delta}{3}\right)\hat{\lambda}\mathbf{v} - \mathrm{i}P\hat{\lambda}\mathbf{k}u = 0. \qquad (3.61)$$

Excluding $\hat{\lambda}\mathbf{v}$ from (3.60b, 61), we obtain an equation relating $\mathbf{v}$ and u

$$\mathbf{v} = \frac{\mathrm{i}P}{3}\left(\frac{2-\hat{\lambda}}{E+E_g} + \frac{1+\hat{\lambda}}{E+E_g+\Delta}\right)\mathbf{k}u. \qquad (3.62)$$

Substituting (3.62) in (3.60b) yields an equation for u:

$$Eu + \frac{1}{3}P^2\left(\frac{2}{E+E_g} + \frac{1}{E+E_g+\Delta}\right)\nabla^2 u = 0. \qquad (3.63)$$

For $u = u_0 e^{\mathrm{i}\mathbf{k}\cdot\mathbf{r}}$ one can obtain from (3.63) a secular equation for $E(\mathbf{k})$:

$$E(E+E_g)(E+E_g+\Delta) - P^2k^2\left(E + E_g + \frac{2}{3}\Delta\right) = 0. \qquad (3.64)$$

Kane's Hamiltonian is given in Table 3.7 in the basis of the valence band functions $\psi_m^{3/2}$ and $\psi_m^{1/2}$, which diagonalize H_{so} and are determined by (3.43), and of the conduction band functions $S\alpha$ and $S\beta$. Just as in (3.59–64), the term $\hbar^2k^2/2m$, which is comparable with the contribution of higher lying bands, is dropped here. The secular equation $\|\mathcal{H} - E\| = O$ splits into two, one of them being $(E + E_g) = 0$, and the other coinciding with (3.64). This equation describes the spectrum of electrons for the band Γ_6, of the light hole branch of the bands Γ_8, and of the bands Γ_7, and the first equation the spectrum of the heavy hole branch of Γ_8 for which in the Kane model $E = -E_g$, and the effective mass is infinite, since it is determined by $\mathbf{k}$-$\mathbf{p}$ interaction with the higher lying bands. Writing (3.64) in the form $k^2 = F(E)$, one immediately finds the three roots of the equation by choosing E in the intervals:

$$\text{band } \Gamma_6 : E > 0; \ \text{band } \Gamma_8 : E_g + \frac{2}{3}\Delta > -E > E_g;$$

$$\text{band } \Gamma_7 : -E > E_g + \Delta.$$

Knowing $k^2 = F(E)$, it becomes possible to determine the effective mass $m(E) = \hbar^2(\mathrm{d}^2E/\mathrm{d}k^2)^{-1}$, the density-of-states mass $m_\mathrm{d} = \hbar^2/(2E)^{1/3} \times [\pi^2\rho(E)]^{2/3}$, where $\rho(E) = 2/(2\pi)^{-3}\int\delta(E(k)-E)\mathrm{d}^3k$, as well as the expansion $E(k)$ with any desired accuracy:

$$m^*(E) = \frac{\hbar^2}{2}\left(\frac{\mathrm{d}F}{\mathrm{d}E}\right)^3\left[\left(\frac{\mathrm{d}F}{\mathrm{d}E}\right)^2 - 2F\frac{\mathrm{d}^2F}{\mathrm{d}E^2}\right]^{-1}, \qquad (3.65a)$$

Table 3.7. Kane's Hamiltonian

	$\lvert S\alpha\rangle$	$\lvert S\beta\rangle$	$\lvert 3/2, 3/2\rangle$	$\lvert 3/2, 1/2\rangle$	$\lvert 3/2, -1/2\rangle$	$\lvert 3/2, -3/2\rangle$	$\lvert 1/2, 1/2\rangle$	$\lvert 1/2, -1/2\rangle$
$\langle S\alpha\rvert$	0	0	$-K_+$	$\sqrt{\frac{2}{3}}K_z$	$\frac{1}{\sqrt{3}}K_-$	0	$-\frac{1}{\sqrt{3}}K_z$	$-\sqrt{\frac{2}{3}}K_-$
$\langle S\beta\rvert$	0	0	0	$-\frac{1}{\sqrt{3}}K_+$	$\sqrt{\frac{2}{3}}K_z$	K_-	$-\sqrt{\frac{2}{3}}K_+$	$\frac{1}{\sqrt{3}}K_z$
$\langle 3/2, 3/2\rvert$	$-K_+^*$	0	$-E_g$	0	0	0	0	0
$\langle 3/2, 1/2\rvert$	$\sqrt{\frac{2}{3}}K_z^*$	$-\frac{1}{\sqrt{3}}K_-^*$	0	$-E_g$	0	0	0	0
$\langle 3/2, -1/2\rvert$	$\frac{1}{\sqrt{3}}K_-^*$	$\sqrt{\frac{2}{3}}K_z^*$	0	0	$-E_g$	0	0	0
$\langle 3/2, -3/2\rvert$	0	K_-^*	0	0	0	$-E_g$	0	0
$\langle 1/2, 1/2\rvert$	$-\frac{1}{\sqrt{3}}K_z^*$	$\sqrt{\frac{2}{3}}K_+^*$	0	0	0	0	$-E_g-\Delta$	0
$\langle 1/2, -1/2\rvert$	$-\sqrt{\frac{2}{3}}K_-^*$	$\frac{1}{\sqrt{3}}K_z^*$	0	0	0	0	0	$-E_g-\Delta$

Here $K_i = Pk_i$, $P = \mathrm{i}\frac{\hbar}{m}\langle S\lvert p_z\rvert Z\rangle$, $K_\pm = \frac{1}{\sqrt{2}}(K_x \pm \mathrm{i}K_y)$.

$$m_{\mathrm{d}}(E) = \frac{\hbar^2}{2}\left(\frac{F(E)}{E}\right)^{1/3}\left(\frac{\mathrm{d}F}{\mathrm{d}E}\right)^{2/3}. \tag{3.65b}$$

To within terms with k^4 we have

$$E(k) = \frac{\hbar^2 k^2}{2m^*}\left(1 - \frac{\hbar^2 k^2}{2m^*}\frac{1}{E_{\mathrm{g}}}\frac{3-2\eta+\eta^2}{3-\eta}\right), \tag{3.66}$$

where $\eta = \Delta/(E_{\mathrm{g}}+\Delta)$.

In Kane's model, the effective mass of electrons at the bottom of Γ_6, $m^* \equiv m_{\mathrm{c}}$, that of the holes at the top of Γ_7, m_{h}, Luttinger's constants γ_i, and the g factors of the electron g_{c}, of the holes in Γ_8, $\mathcal{K}$, and of Γ_7, g_{v}, are related through

$$\frac{m}{m_{\mathrm{c}}} = \frac{2}{3}\frac{m}{\hbar^2}P^2\left(\frac{2}{E_{\mathrm{g}}} + \frac{1}{E_{\mathrm{g}}+\Delta}\right), \quad \frac{m}{m_{\mathrm{h}}} = \frac{2}{3}\frac{m}{\hbar^2}\frac{P^2}{E_{\mathrm{g}}+\Delta},$$

$$\frac{1}{2}\gamma_1 = \gamma_2 = \gamma_3 = \frac{1}{3}\frac{m}{\hbar^2}\frac{P^2}{E_{\mathrm{g}}}, \quad g_{\mathrm{c}} = 2 - \frac{4}{3}\frac{m}{\hbar^2}P^2\frac{\Delta}{E_{\mathrm{g}}(E_{\mathrm{g}}+\Delta)},$$

$$g_{\mathrm{v}} = -\frac{2}{3}\left(1 + \frac{2m}{\hbar^2}\frac{P^2}{E_{\mathrm{g}}+\Delta}\right), \quad \mathcal{K} = \frac{1}{3}\left(1 - \frac{m}{\hbar^2}\frac{P^2}{E_{\mathrm{g}}}\right). \tag{3.67}$$

Quantitatively, the Kane model describes satisfactorily only the spectrum for narrow-gap semiconductors, e.g., InSb. In most of the other $A_3\,B_5$ compounds, the separation between the conduction band Γ_1^{c} and the valence band Γ_{15}^{v} is not much smaller than that between Γ_1^{c} and the higher-lying conduction band Γ_{15}^{c}. One can generalize this model by including in Kane's Hamiltonian the quadratic-in-k contribution of the other bands by the **k-p** technique. This approach, however, deprives the model of its main asset, namely, the possibility of expressing all spectral parameters in terms of a limited number of constants of the given model. This accounts for the current recognition of a five-band model which, similar to Kane's model, takes exactly into account the **k-p** interaction of the above mentioned three bands, Γ_1^{c}, Γ_{15}^{v}, and Γ_{15}^{c}, or, if the spin is included, of five bands: Γ_6^{c}, Γ_8^{v}, Γ_7^{v}, and of the bands Γ_8^{c}, Γ_7^{c} with the basis functions $X^{\mathrm{c}}, Y^{\mathrm{c}}, Z^{\mathrm{c}}$. Besides P, this model includes also as parameters two matrix elements:

$$P' = \mathrm{i}\frac{\hbar}{m}\langle S|p_x|X^{\mathrm{c}}\rangle, \quad Q = \mathrm{i}\frac{\hbar}{m}\langle X^{\mathrm{c}}|p_y|Z^{\mathrm{v}}\rangle \tag{3.68}$$

the distance E_{g}' between the bands Γ_6^{c} and Γ_8^{c}, the spin-orbit splitting $\Delta' = E^{\mathrm{c}}_{\Gamma_8} - E^{\mathrm{c}}_{\Gamma_7}$ of the band Γ_{15}^{c}, and the constant of spin-orbit mixing of the bands Γ_{15}^{v} and Γ_{15}^{c}:

$$\Delta^- = 3\langle\psi_m^{3/2\mathrm{c}}|\mathcal{H}_{\mathrm{so}}|\psi_m^{3/2\mathrm{v}}\rangle = \frac{3}{2}\langle\psi_n^{1/2\mathrm{c}}|\mathcal{H}_{\mathrm{so}}|\psi_n^{1/2\mathrm{v}}\rangle;$$

here $\psi_m^{3/2}$ and $\psi_n^{1/2}$ are, respectively, the basis functions for the bands Γ_8, Γ_7.

It is the **k**-**p** interaction of these three bands, $\Gamma_1^c, \Gamma_{15}^v, \Gamma_{15}^c$, that makes a nonzero contribution to the constants γ_i determining the cubic-in-k splitting of the bands $\Gamma_6^c, \Gamma_8^v, \Gamma_7^v$. For $\Delta' \ll \Delta, \Delta^- \ll \Delta$

$$\gamma_v = -\frac{4}{3}pp'Q\frac{1}{E_g(E_g+E_g')}, \quad \gamma_{v2} = \gamma_v\frac{E_g}{E_g+\Delta},$$

$$\gamma_c = \gamma_{v2}\left(\frac{\Delta}{E_g}+\frac{\Delta}{E_g'}\right), \quad a_2 = -a_3 = \frac{1}{2}\frac{p}{p'}\frac{\Delta^-}{\Delta}, a_r = 0. \tag{3.69}$$

The inclusion of **k**-**p** interaction with the band Γ_{15}^c also results in the appearance in the Hamiltonian in Table 3.5 of quadratic-in-k terms, namely, in the replacement of Pk_l by $Pk_l - \mathrm{i}/m_{vc}k_{l+1}k_{l+2}(l = x, y, z)$, where $m_{vc}^{-1} = (P'Q/m)$ $[E_g^{-1} + (E_g + E_g')^{-1}]$; as for the linear-in-k term (3.54), it is zero within the framework of the five-band model and originates from the **k**-**p** and $\mathcal{H}_{so}$ interaction with higher-lying bands, and this is what accounts for the smallness of the corresponding constant k_0.

To derive the boundary conditions in the Kane model we will, following *Suris* [3.15], multiply (3.59a) by u^+, and (3.59b) by v^+. Similarly, we will multiply the complex conjugate equations, accordingly, by $-u$ and $-v$, add the right- and left-hand parts of all four expressions thus obtained, and carry out summation over the spin indices. We come to

$$\frac{\partial}{\partial t}(|u|^2 + |v|^2) + P\,\mathrm{div}(u^+v + uv^+) = 0. \tag{3.70a}$$

Equation (3.69) is the continuity equation $\frac{\partial P}{\partial t} + \mathrm{div}\, I = 0$, from which follows the condition of flux conservation

$$I = P(u^+v + uv^+) = \mathrm{const} \tag{3.70b}$$

at the boundaries.

This condition is met if at the boundaries $P^\alpha u$ and $P^{1-\alpha}v_\mathbf{n}$ are continuous (here $v_\mathbf{n}$ is the component of **v** normal to the surface, i.e. for $\mathbf{n}\|\tilde{z}$) at the boundary

$$P_A^\alpha u_A = P_B^\alpha u_B, P_A^{1-\alpha}v_{zA} = P_B^{1-\alpha}v_{zB}. \tag{3.71}$$

From the continuity of $P^{1-\alpha}v_z$, according to (3.62), follows the continuity of the quantity

$$P^{2-\alpha}\left[\left(\frac{2}{E+E_g}+\frac{1}{E+E_g+\Delta}\right)\frac{\partial u}{\partial z} - \frac{\Delta[\sigma\mathbf{k}]_z u}{(E+E_g)(E+E_g+\Delta)}\right]. \tag{3.72}$$

As seen from (3.67), for $E << E_g$ these boundary conditions for u reduce to (3.23) with $t_{11} = \left(\frac{P_B}{P_A}\right)^\alpha$, $t_{22} = \left(\frac{P_B}{P_A}\right)^{-\alpha}$, $t_{12} = t_{21} = 0$. These conditions differ from (3.24c) in $m_{A,B}^\alpha$ being replaced by $P_{A,B}^\alpha$, and $m_{A,B}^{-(1+\alpha)}$, by $m_{A,B}^{-1}P_{A,B}^{-\alpha}$.

3.4.2 Gapless Semiconductors

Crystals of grey tin and of some A_2B_6 compounds, for instance, HgTe, HgSe, exhibit band inversion, i.e., the extremum of the band $\Gamma_8^+(\Gamma_8)$ lies above the band $\Gamma_7^-(\Gamma_6)$. As a result, the light hole branch of the band $\Gamma_8^+(\Gamma_8)$ acts as a conduction band, the heavy hole branch remains the valence band, and the band Γ_7^- (Γ_6) becomes the lower valence subband and lies between Γ_8^- (Γ_8) and Γ_7^+ (Γ_7) originating from $\Gamma_{25}'(\Gamma_{15})$. Therefore in such crystals the forbidden band, i.e., the gap between the conduction and valence bands, does not actually exist. The spectrum in the merging bands is described by the expressions (3.40–48). Note that here, in contrast to Ge, Si, and A_3B_5 crystals, the constants γ_i are negative (i.e., A, B and D are positive), and $|\gamma_1| < 2|\gamma_2|, 2|\gamma|_3$. The spectrum in the three bands Γ_6, Γ_8 and Γ_7 for these crystals is also described by the Kane model if the **k**-**p** interaction with the other bands is disregarded, with $E_g < 0$ and $|E_g| < \frac{2}{3}\Delta$ in (3.61–64). To the conduction band correspond the energies $E \geqslant |E_g|$, for the heavy hole branch of Γ_8, $E = |E_g|$, the band Γ_6 lies in the interval $|E_g| - \frac{2}{3}\Delta < E < 0$, and to Γ_7 correspond the energies $E < -\Delta + E_g$. It should naturally be kept in mind that the Kane model describes the behavior of $E(\mathbf{k})$ not throughout the Brillouin zone but only in the vicinity of the point $k = 0$.

3.4.3 Deformation-Induced Change of Spectrum

The band-edge shift at the point of extremum for nondegenerate bands is described by (3.29). At the point Γ, it is determined by one constant of the deformation potential $C = D_{xx} = D_{yy} = D_{zz}$, and at the points L, X and Δ, - by two constants, $\Xi_d = D_{xx} = D_{yy}$ and $\Xi_u = D_{zz} - D_{xx}$. Just as before, the Z axis is directed along the principal axis of the extremum. In deformed crystals of the class T_d, the Hamiltonian $\mathcal{H}(\mathbf{k})$ for the band Γ_6 includes, besides the cubic-in-k terms (3.38), also a spin-dependent term which is linear in ε and k:

$$\mathcal{H}_{\varepsilon k} = \frac{1}{2} C_3 \sum_i \sigma_i \varphi_i, \tag{3.73}$$

where $\varphi_z = \varepsilon_{zx} k_x - \varepsilon_{zy} k_y$, etc.

The spin-dependent Hamiltonian for the valence band Γ_7 has a similar form. Denote the corresponding constant by C_4. These terms appear due to the presence in T_d symmetry crystals of an interband constant of deformation potential C_2 determining the mixing of the states of the bands Γ_1 and Γ_{15} under deformation. The corresponding matrix elements can be written as

$$\langle S|\mathcal{H}_\varepsilon|Z\rangle = -\mathrm{i} C_2 \varepsilon_{xy} \qquad \text{etc.}, \tag{3.74}$$

i.e., $C_2 = \mathrm{i}(\langle S|D_{xy}|Z\rangle + \langle S|D_{yx}|Z\rangle)$. The constants C_3 and C_4 are related to C_2 by the expressions

$$C_4 = -\frac{4}{3}\frac{C_2}{E_g + \Delta}P, \quad C_3 = -C_2\frac{\Delta}{E_g}. \tag{3.75}$$

The Hamiltonian $\mathcal{H}$ of the band Γ_7 also contains a term

$$\mathcal{H} = \frac{1}{2}C_4'\sum_i \sigma_i\psi_i, \tag{3.76a}$$

where $\psi_z = k_z(\varepsilon_{xx} - \varepsilon_{yy})$ and so on.
In the five-band model

$$C_4' = -\frac{4Q\Delta^-}{(E_g + E_g')\Delta}b. \tag{3.76b}$$

For the degenerate valence band Γ_8, $\mathcal{H}_\varepsilon$ is described by the *Bir-Pikus Hamiltonian* [3.7]. For the holes it has the form:

$$\mathcal{H}_\varepsilon = -a\varepsilon + b\sum_i J_i^2\left(\varepsilon_{ii} - \frac{1}{3}\varepsilon\right) + \frac{2}{\sqrt{3}}d\sum_{i>j}[J_iJ_j]_s\varepsilon_{ij}. \tag{3.77}$$

Here $a = (l + 2m)/3, b = (l - m)/3, d = n/\sqrt{3}, l = \langle X|D_{xx}|X\rangle, m = \langle X|D_{yy}|X\rangle, n = \langle X|D_{xy}|Y\rangle + \langle X|D_{yx}|Y\rangle$.
In the matrix form $\mathcal{H}_\varepsilon$ differs from (3.46) in the replacement of

$$F \text{ by } f = -a\varepsilon - \frac{b}{2}(\varepsilon - 3\varepsilon_{zz}), \quad G \text{ by } g = -a\varepsilon + \frac{b}{2}(\varepsilon - 3\varepsilon_{zz}),$$

$$H \text{ by } h = d(\varepsilon_{zx} - \mathrm{i}\varepsilon_{zy}) \text{ and } I \text{ by } i = \frac{\sqrt{3}}{2}b(\varepsilon_{xx} - \varepsilon_{yy}) - \mathrm{i}d\varepsilon_{xy}. \tag{3.77a}$$

Taking into account in $\mathcal{H}$ both linear-in-ε and quadratic-in-k terms yields the following expression for the hole energy:

$$E_{1,2} = -a\varepsilon - Ak^2 \pm \left\{\frac{3}{2}\sum_i\left[b\left(\varepsilon_{ii} - \frac{1}{3}\varepsilon\right) + B\left(k_i^2 - \frac{1}{3}k^2\right)\right]^2 + \sum_{i>j}\left(d\varepsilon_{ij} + Dk_ik_j\right)^2\right\}^{1/2}. \tag{3.78}$$

At the point k = 0, deformation results in a branch splitting by

$$\delta E = 2\left[\frac{3}{2}\sum_i b^2\left(\varepsilon_{ii} - \frac{1}{3}\varepsilon\right)^2 + \sum_{i>j} d^2\varepsilon_{ij}^2\right]^{1/2}. \tag{3.79}$$

In Ge, Si and A_3B_5 compounds the constants b and d, just as A and B, are negative. Therefore compression shifts upward the band of the light holes and tension, that of the heavy ones.

In gapless semiconductors, uniaxial deformation results in the appearance of a gap at $k = 0$. As seen from (3.78), under tension, when the heavy hole band at $k = 0$ rises above the light-hole branch, deformation along the [100] or [111] principal axes results in merging of the bands at the points $k_{100}^2 = -b\varepsilon_{xx}/B$ or $k_{111}^2 = -d\varepsilon_{111}/D$. (For gapless semiconductors $A, B, D > 0$, so that the merging occurs for $\varepsilon_{xx}, \varepsilon_{111} > 0$). Tension in other directions makes the bands approach without merging.

In the case of spherical isotropy, i.e., for $C_{xyxy} = (C_{xxxx} - C_{xxyy})/2$ and $b/B = d/D$, the bands can touch under any uniaxial tensile deformation. In crystals of the class T_d, the Hamiltonian $\mathcal{H}$ for the band Γ_8 also includes terms linear in k and ε:

$$\mathcal{H}_{\varepsilon k} = C_5(J\varphi) + C_6(J\psi) + C_7 \sum_i J_i^3 \varphi_i$$

$$+ C_8 \sum_8 \sum_i J_i^3 \psi_i + C_9(V_\chi) + C_{10}\varepsilon(Vk), \tag{3.80a}$$

where $\chi_z = k_z(\varepsilon_{zz} - \frac{1}{3}\varepsilon)$.

In the five-band model

$$C_5 = -\frac{C_2 P}{E_g}, \quad C_6 = -\frac{1}{24}C_4', \quad C_7 = -\frac{2}{3\sqrt{3}}C_4' \frac{d}{b};$$

$$C_8 = \frac{1}{6}C_4', \quad C_9 = -\frac{1}{2}C_4', \quad C_{10} = 0, \tag{3.80b}$$

where C_4' is defined by (3.76b).

As pointed out by *Trebin* et al. [3.16], second-order terms containing products of interband matrix elements of the operators P and D defined by (3.5 and 28c) can as well contribute to the constants C_5–C_9.

In the five-band model, the matrix of operator $\mathcal{H}_\varepsilon$ between the wave functions of the Γ_{15}^{c} and Γ_{15}^{v} bands differs from the corresponding matrix for the Γ_{15}^{v} band only in the replacement of the intraband deformation-potential constants a, b, and d by the interband constants a', b', and d', the latter being related through expressions similar to (3.77) to the constants l', m', and n', which differ from l, m, and n in one of the X^{v} functions in the corresponding matrix elements being replaced by X^{c}.

The inclusion of these terms yields a contribution to the C_4' and C_6 constants

$$\delta C_4' = \frac{4b'Q}{E_g + E_g'}, \quad \delta C_6 = \frac{1}{2}\delta C_4'. \tag{3.80c}$$

The corresponding contributions to the relativistic constants C_7–C_9 are nonzero only if the spin-orbit splitting Δ' of the Γ_{15}^{g} band is included

$$\delta C_7 = -\frac{2}{3\sqrt{3}} \frac{\Delta'}{E_g + E_g'} \frac{d'}{b'} \delta C_4',$$

$$\delta C_8 = \frac{1}{6} \frac{\Delta'}{E_g + E_g'} \delta C_4', \quad \delta C_9 = 3 \delta C_8 . \tag{3.80d}$$

The constant C_5 also contains similar additional terms

$$\delta C_5 = \frac{2}{\sqrt{3}} \frac{Q}{E_g + E_g'} \left(-d' + \frac{10}{3} \frac{\Delta^-}{\Delta} d \right) . \tag{3.80e}$$

The main term in C_5 is determined by (3.80b), the contributions (3.80b and c) to the nonrelativistic constants C_4' and C_6 may be comparable; the principal terms in the relativistic constants C_7–C_9 are described by (3.80b), and it is seen that they can be compared to C_4' and C_6.

3.5 Electron Spectra of Quantum Wells and Superlattices

There are two radically different methods to calculate the carrier spectrum in quantum wells and superlattices. In the first, the superlattice is considered as a specific crystalline structure whose spectra can be obtained by the techniques used for conventional crystals, e.g., by the tight-binding, pseudopotential, orthogonalized plane-wave methods, etc. These methods are indispensable when calculating the spectra of short-period superlattices or thin quantum wells, the computer time involved increasing with the number of atomic layers in the wells and barriers.

The second is the envelope-function method. As already pointed out, this method presupposes that the effective masses and other parameters in each well and barrier coincide with those of the corresponding bulk materials, and that the equations for the envelopes are the conventional equations used in the effective-mass technique. The parameters of the bulk materials are either taken from experiment or calculated. The accuracy of the envelope function method is higher, the larger are the well and barrier sizes. Typically, a satisfactory accuracy is obtained for well and barrier sizes in excess of eight to ten lattice periods. In the present section, we will discuss only the second method, the other techniques being considered in detail, e.g., in the review by *Smith* and *Mailhiot* [3.17].

3.5.1 Nondegenerate Bands

If the carrier spectrum is isotropic, or, in the case of its being anisotropic, if the normal to the interface **n** is directed along one of the principal axes of the effective-mass tensor, then the solution of Schrödinger's equation

$$\left[\sum_i \frac{\hbar^2 k^2}{2m_{ii}} + U(z) - E(k_\perp) \right] F(x) = 0, \tag{3.81}$$

where $V(z)$ is the quantum well or superlattice potential, has the form

$$F(\mathbf{x}) = e^{i\mathbf{k}_\perp \cdot \boldsymbol{\rho}} \varphi(z). \tag{3.82}$$

Here the z axis is directed along the normal $\mathbf{n}$, and ρ is the component of $\mathbf{x}$ perpendicular to $\mathbf{n}$. The electron spectrum in a quantum well consists of a series of subbands corresponding to different values of $E(O)$.

The superstructure potential in (3.81) determines the electron energy at the band minimum, i.e., for $k_\perp = O$. It includes the potentials both resulting from the change in the band minimum position caused by changes in the composition and structure of the crystal, as well as by external or internal deformations, and created by the impurity charge, by free carriers and external fields. For rectangular wells of size a the functions $\varphi(z)$ have a certain parity with respect to reflection in the plane passing through the well center and, in the region $|z| < a/2$, they can be written

$$\begin{aligned} \varphi &= c_1 \cos kz \qquad \text{for even states,} \\ \varphi &= c_1 \sin kz \qquad \text{for odd states.} \end{aligned} \tag{3.83}$$

Here the point $z = O$ corresponds to the well center,

$$k^2 = \frac{2m_{zz}^A}{\hbar^2}\left(E - E_\perp^A\right), \quad E_\perp^A = \frac{\hbar^2 k_x^2}{2m_{xx}^A} + \frac{\hbar^2 k_y^2}{2m_{yy}^A}, \tag{3.84}$$

the energy E being reckoned from the well bottom where $V = 0$.

For a well of width a with infinitely high barriers, i.e., for

$$\begin{aligned} V(z) &= 0 \quad \text{for } |z| < a/2, \\ V(z) &= \infty \quad \text{for } |z| > a/2, \end{aligned} \tag{3.85}$$

we have

$$\begin{aligned} k &= \frac{\pi}{a}(2n+1) \quad \text{for even states,} \\ k &= \frac{\pi}{a}2n \quad \text{for odd states,} \end{aligned} \tag{3.86a}$$

the coefficient c_1 in (3.83) being equal to $(2/a)^{1/2}$.

For wells with barriers of finite height

$$V(z) = V_0 \quad \text{for } |z| > a/2,$$

we have

$$\begin{aligned} \varphi &= c_2 \exp[\lambda(z + a/2)] \quad \text{for } z < -a/2 \\ &= c_3 \exp[-\lambda(z - a/2)] \quad \text{for } z > a/2, \end{aligned} \tag{3.86b}$$

where

$$\lambda^2 = \frac{2m_{zz}^B}{\hbar^2}\left(V_0 - E - E_\perp^B\right), \quad E_\perp^B = \frac{\hbar^2 k_x^2}{2m_{xx}^B} + \frac{\hbar^2 k_y^2}{2m_{yy}^B}. \tag{3.87}$$

For the boundary conditions (3.24c)

$$\begin{aligned} m_A^\alpha \varphi_A &= m_B^\alpha \varphi_B, \\ m_A^{-(1+\alpha)} \frac{d\varphi_A}{dz} &= m_B^{-(1+\alpha)} \frac{d\varphi_B}{dz} \quad \left(z = \pm\frac{a}{2}\right), \end{aligned} \tag{3.88}$$

where $m_A = m_{zz}^A, m_B = m_{zz}^B$, the transcendental equation defining the positions of the levels can be written in the form

$$\tan\frac{ka}{2} = \frac{\lambda}{k}\left(\frac{m_A}{m_B}\right)^{1+2\alpha} \quad \text{for even states,} \tag{3.89a}$$

$$\cot\frac{ka}{2} = -\frac{\lambda}{k}\left(\frac{m_A}{m_B}\right)^{1+2\alpha} \quad \text{for odd states.} \tag{3.89b}$$

The electron spectrum in a lattice consisting of square wells and barriers was calculated already at the dawn of quantum mechanics, the corresponding Kronig-Penney model being considered a one-dimensional model of the crystal. In such a periodic lattice, the wave function $\varphi(z)$ in the neighboring wells has the form

$$\begin{aligned} &\varphi = c_1\cos(kz) + c_2\sin(kz) \quad 0 < z < a, \\ &\varphi = c_3\cos[k(z-d)] + c_4\sin[k(z-d)] \quad d < z < d+a \end{aligned} \tag{3.90a}$$

and in the barrier separating these wells,

$$\varphi = c_5\cosh\lambda(z-a) + c_6\sinh\lambda(z-a) a < z < d. \tag{3.90b}$$

Here a is the well thickness, b is the barrier width, $d = a+b$ is the superlattice period, k and λ being defined by (3.84, 87). Besides the boundary conditions (3.88) for $z = a$ and $z = d$, the condition of periodicity $\varphi(z+d) = \varphi(z)\mathrm{e}^{\mathrm{i}qd}$ is imposed on the function (3.90) in accordance with Bloch's theorem, whence it follows that $c_3 = c_1\mathrm{e}^{\mathrm{i}qd}, c_4 = c_2 e^{\mathrm{i}qd}$. As a result, we obtain four coupled equations for the coefficients c_1, c_2, c_5, and c_6. The condition for the determinant of this system to be zero is

$$\cos(qd) = F(E, k_x, k_y), \tag{3.91}$$

where

$$F = \cos(ka)\cosh(\lambda b) + \frac{1}{2}\sin(ka)\sinh(\lambda b)(R - R^{-1}), \tag{3.92}$$

$$R = \frac{\lambda}{k}\left(\frac{m_A}{m_B}\right)^{1+2\alpha}. \tag{3.92a}$$

This equation differs from the equation of Kronig and Penney only in that λ/k is replaced by R.

The regions of allowed energies called minibands correspond to the solution of (3.91) with real q. The position of the minimum of the even minibands is determined by the condition $F(E_n^0) = 1$ for $k_x = k_y = 0$, which at the miniband maximum, i.e., for $q = \pm\pi/d$, $F(E_n^0) = -1$. For odd minibands, the point $q = 0$ corresponds to maximum energy, and $q = \pm\pi/d$, to the lowest energy, i.e., to the miniband bottom. The effective masses m_{zz}^*, m_{xx}^* and m_{yy}^* determining the miniband bottom spectrum can be found from the equations

$$m_{zz}^* = -\frac{\hbar^2}{d^2}\left(\frac{\partial F(E)}{\partial E}\right)_0,$$

$$m_{ii}^* = -\frac{\hbar^2}{2}\left(\frac{\partial F(E)}{\partial E}\right)_0 \Big/ \left(\frac{\partial F}{\partial k_i^2}\right)_0 \quad (i = x, y). \tag{3.93}$$

All the derivatives are taken at $E = E_0, k_x = k_y = 0$.

It has been shown [3.18] that (3.92) retains its general form for arbitrary periodic lattices as well. Indeed, if the transmission of a single barrier $V(z)$ extending over $|z| < d/2$ ($V(z) = 0$ for $|z| > d/2$) is $t = |t|e^{i\delta}$, then for a lattice made up of such barriers in (3.91)

$$F = \frac{1}{|t|}\left[\cos(kd)\cos\delta + \sin(kd)\sin\delta\right]. \tag{3.94}$$

For the coefficient of reflection from a single barrier we have

$$\tau = \mp(1 - |t|^2)^{1/2} e^{i\delta}.$$

In the Kronig-Penney model

$$\frac{1}{t} = e^{-ikb}\left[\cosh(\lambda b) + \frac{i}{2}\left(R - R^{-1}\right)\sinh(\lambda b)\right]. \tag{3.94a}$$

It should also be stressed that (3.94) describes the spectrum of any (electronic, phonon, etc.) excitation in a one-dimensional periodic lattice. If the wave functions in regions A and B are defined by (3.90a, b), and the boundary conditions at the interface between them have the form

$$\varphi_A = \varphi_B,\; c_A\frac{\partial\varphi_A}{\partial z} = c_B\frac{\partial\varphi_B}{\partial z},$$

then the spectrum will be described by (3.92) with

$$R = \frac{\lambda}{k}\frac{c_A}{c_B}. \tag{3.94b}$$

3.5.2 Many-Valley Semiconductors

If the carrier spectrum in the vicinity of the extrema is anisotropic, and the extrema lie in nonequivalent positions with respect to the normal, then the positions of levels in the quantum wells and superlattices will also be different for different valleys, namely, the lowest will be the levels of the valleys with the largest mass m_{zz} along the normal. For the Si(001) lattice, these are the Δ extrema $(0, 0, k_0)$ and $(0, 0, -k_0)$ and, for Ge(111), the L(111) extremum.

The value of the transverse component of the wave vector $\mathbf{k}_0$ determining the extremum position is conserved in a quantum well or superlattice. In a superlattice, the period t_z increases from a_0 to d, the Brillouin subband decreasing in size in this direction, accordingly, from $\pm\pi/a_0$ to $\pm\pi/d$. Note that the point k_{0z} transfers to the point k'_{0z} of the miniband, $k'_{0z} = k_{0z} - \nu 2\pi/d$, where $\nu = [k_{0z}/(2\pi/d)]$, that is, it is the largest integer which does not exceed the ratio $k_{0z}/(2\pi/d)$. If the point k_{0z} lies at the boundary of the Brillouin zone, i.e., $k_{0z} = \pi/a_0$, then $k'_{0z} = 0$ for $d/a_0 = 2n$, and $k'_{0z} = \pi/d$ for $d/a_0 = (2n + 1)$.

This means that the point k_{0z} will be at the miniband center if the superlattice period contains an even number of primitive cells, and at its boundary, if the number of primitive cells fitting into the period is odd. If the carrier spectrum is anisotropic, and the normal to the surface is directed in an arbitrary way with respect to the principal axes of the effective mass tensor for the given extremum, then in the coordinate frame x, y, z with the z axis along the normal **n**

$$\mathcal{H} = \sum_{ij} \frac{\hbar^2}{2} \frac{k_i k_j}{m_{ij}} + V(z), \tag{3.95}$$

the tensor m_{ij}^{-1} containing nondiagonal components m_{xz}^{-1}, m_{yz}^{-1}. In this case, the solution of Schrödinger's equation can be also represented in the form (3.82), however, here

$$\varphi(z) = \exp\left[-\mathrm{i}\left(\frac{m_{zz}}{m_{xz}}k_x + \frac{m_{zz}}{m_{yz}}k_y\right)z\right]\left(c_1\mathrm{e}^{\mathrm{i}kz} + c_2\mathrm{e}^{-\mathrm{i}kz}\right) \tag{3.96}$$

and the energy

$$E = \frac{\hbar^2 k^2}{2m_{zz}} + E_\perp,$$

where

$$E_\perp = \frac{\hbar^2 k_x^2}{2m_{xx}}\left(1 - \frac{m_{xx}m_{zz}}{m_{xz}^2}\right) + \frac{\hbar^2 k_y^2}{2m_{yy}}\left(1 - \frac{m_{yy}m_{zz}}{m_{yz}^2}\right) + \frac{\hbar^2 k_x k_y}{m_{xy}}\left(1 - \frac{m_{xy}m_{zz}}{m_{xz}m_{yz}}\right). \tag{3.97}$$

Just as in (3.86a), for infinitely high barriers, $k = \frac{\pi}{a}n$.

In accordance with (3.25, 95), for the velocity we obtain

$$v_z\varphi = \frac{-\mathrm{i}}{\hbar}[z\mathcal{H}]\varphi = \frac{\hbar k}{m_{zz}}\left(c_1 e^{ikz} - c_2 e^{-ikz}\right) \times \exp\left[-\mathrm{i}\left(\frac{m_{zz}}{m_{xz}}k_x + \frac{m_{zz}}{m_{yz}}k_y\right)z\right].$$

The conditions of flux continuity at the boundary are met if $\varphi(z)$ and $\hat{v}_z\varphi_z$ are continuous, which corresponds to the boundary conditions (3.88) with $\alpha = 0$ and $m_A = m_{zz}^A, m_B = m_{zz}^B$ (in this case $t_{11} = t_{22} = 1$ in (3.23)). Therefore if the masses m_{xz} and m_{yz} in the well and in the barriers are equal, the secular equations (3.89, 91, 92) remain valid for wells of finite depth or superlattices with $\alpha = 0$ (taking into account the corresponding contribution to $E_\perp$ in (3.81)). If the values of the energy in two extrema corresponding to the same value of $\mathbf{K}_0$ turn out to be close, then in quantum wells or superlattices the states of these valleys will mix. Similar mixing occurs also in the cases where the lowest extrema in wells and barriers correspond to different $\mathbf{K}_0$ points. Such a situation can be realized, for instance, in GaAl-Al_xGa_{1-x}As superlattices for $x > 0.4$, when the lowest extremum in AlGaAs is one of the X points, and in GaAs, the Γ point. Here the splitting of the extrema X_1 and X_3 is small, and the mixing of the envelope functions ξ_u and ξ_v corresponding to these extrema results in a shift of the minimum to a certain distance away from the point X. The spectrum in the vicinity of the point X is given by the equations

$$\left[E_u - \frac{\hbar^2}{2m_{zz}}\frac{d^2}{dz^2} + V(z) - E\right]\xi_u - \mathrm{i}\frac{\hbar}{m}p\frac{d}{dz}\xi_v = 0,$$
$$\left[E_v - \frac{\hbar^2}{2m_{zz}}\frac{d^2}{dz^2} + V(z) - E\right]\xi_v - \mathrm{i}\frac{\hbar}{m}p^*\frac{d}{dz}\xi_u = 0. \tag{3.98}$$

Here, $p = \mathrm{i}\langle u|p_z|v\rangle$ where u and v are the Bloch functions at the points X_1 and X_3. For bulk materials $\mathrm{i}\dfrac{d\xi_{v,u}}{dz} = k_z\xi_{v,u}$ and

$$E = \frac{E_u + E_v}{2} \pm \left[\left(\frac{E_u - E_v}{2}\right)^2 + |p|^2\frac{\hbar^2k_z^2}{m^2}\right]^{1/2} + \frac{\hbar^2k}{2m_{zz}}. \tag{3.99}$$

According to *Ando* and *Akera* [3.5], the envelope functions of the band Γ, ξ_Γ, and of the bands $X_{1,3}$ and their derivatives

$$\nabla_\Gamma\xi_\Gamma = \frac{m}{m_\Gamma}a_0\frac{d\xi_\Gamma}{dz},\ \nabla_u\xi_u = \frac{m}{m_{zz}}a_0\frac{d\xi_u}{dz},\ \nabla_v\xi_v = \frac{m}{m_{zz}}a_0\frac{d\xi_v}{dz}$$

are related at the boundaries through

$$\xi_i^A = \sum_j\left(t_{11}^{ij}\xi_j^B + t_{12}^{ij}\nabla_j\xi_j^B\right),$$
$$\nabla_i\xi_i^A = \sum_j\left(t_{21}^{ij}\xi_j^B + t_{22}^{ij}\nabla_j\xi_j^B\right), \tag{3.100}$$

where the indices i, j run through the values Γ, u, v.

Table 3.8 lists the values of some components of t calculated by *Ando* and *Akera* for the GaAs/$Al_xGa_{1-x}As$ lattice for $x = 0.3$ and $x = 0.6$. The diagonal components t_{11}^{ii} and t_{22}^{ii} differ from unity by not more than 0.1. The other components not given in Table 3.8 are not larger than 0.05. (Table 3.9 presents the band parameters for these crystals).

As follows from calculations [3.5], the components $t_{\alpha\alpha}^{ij}$ relating the functions of opposite parity with respect to the operation S_{4z} (As), i.e., even Γ, u and odd v, are zero. The coefficients $t_{\alpha\beta}^{ij}$ with $\alpha \neq \beta$ relating the i, j functions of the same parity, in particular with $i = j$, are likewise zero. While the vanishing of the $t_{\alpha\beta}^{ij}$ components does not follow from symmetry considerations, apparently, it is a specific feature of the model chosen in [3.5]. One may expect them to be always small. One may also expect that, in accordance with Table 3.8, t_{21}^{vr} and t_{21}^{vr} should be the largest nondiagonal components. If we retain in (3.100)

Table 3.8. Parameters t_{21}^{ij} for GaAs/Al_xGa_{1-x}As lattices

	$x = 0.3$	$x = 0.6$
$t_{21}^{\Gamma v}$	−0.360	−0.686
t_{21}^{uv}	0.061	0.125
$t_{21}^{v\Gamma}$	0.244	0.554
t_{21}^{vu}	−0.099	−0.203

only these nondiagonal components, then the condition of current conservation at the boundary implies that they should be related through $t_{21}^{rv} = t_{21}^{vr*}$. The phase of the constants t^{ij} depends on the choice of the wave function and, for a unit barrier, these constants may be considered real. As shown by *Aleiner* and *Ivchenko* [3.19] the phases of the components t^{ij} at the adjacent well and barrier boundaries may differ, and should depend on the actual number of monolayers in the barrier, M, and the well, N.

If the coefficients $t^{ij}(0)$ at the boundary of the well (A-GaAs layer) and of the barrier (B-AlAs layer) are given, the boundary conditions at the opposite side of the barrier can be derived from (3.100) by applying the operation S_{4z} (As) and translation T by $M\mathbf{a}_2$ or $M\mathbf{a}_3$, where $\mathbf{a}_2 = (a_0/2)(1,0,1)$, $\mathbf{a}_3 = (a_0/2)(0,1,1)$ are elementary translation vectors. Similarly, the boundary conditions on the opposite side of the well can be obtained from (3.100) by applying the operation $T(-Na_2)S_{4z}$.

Translation by $(M+N)a_2$ yields $t_{\alpha\beta}^{ij}$ at the boundary of the adjacent wells.

We finally obtain the following relations connecting the coefficients $t_{\alpha\beta}^{ij}(0)$ and $t_{\alpha\beta}^{ij}(R)$ with $R = M, -N, M+N$:

$$t_{\alpha\beta}^{ij}(R) = S(R)t_{\alpha\beta}^{ij}(0),$$

for $R = M$ and $R = -N$:

$$S = 1 \quad \text{for} \quad \alpha = \beta, \quad i = j \quad \text{or} \quad \alpha \neq \beta, \quad i, j = u, v$$

$$S = -1 \quad \text{for} \quad \alpha \neq \beta, \quad i = j \quad \text{or} \quad \alpha = \beta, \quad i, j = u, v$$

$$S = (-1)^R \quad \text{for} \quad \alpha = \beta, \quad i, j = \Gamma, u \quad \text{or} \quad \alpha \neq \beta, \quad i, j = \Gamma, v$$

$$S = -(-1)^R \quad \text{for} \quad \alpha \neq \beta, \quad i, j = \Gamma, u \quad \text{or} \quad \alpha = \beta, \quad i, j = \Gamma, v$$

for $R = M + N$:

$$S = 1 \quad \text{for} \quad i = j \quad \text{and} \quad i, j = u, v; \quad \Gamma, u$$

$$S = (-1)^{M+N} \quad \text{for} \quad i, j = \Gamma, v.$$

As pointed out in [3.19], similar mixing of the functions Γ, u, and v may occur in the case where the lower extrema of the conduction band in AlAs lie at the points X_x and X_y corresponding to $K_{0x} = (2\pi/a_0)(1,0,0)$ and $K_{0y} = (2\pi/a_0)(0,1,0)$. Under these conditions it may be expected that the

Table 3.9. Band parameters for Ge, Si, cubic crystal A_3B_5

	Indirect-gap semiconductors					
	Ge	Si	AlP	AlSb	AlAs	GaP
Indirect gap E_g [eV] (4.2 K)	0.744	1.17	2.53	1.70	2.23	2.35
E_g [eV] (300 K)	0.664	1.11	2.45	1.63	2.15	2.27
Direct gap Γ Point E'_g [eV] (4.2 K)	0.898	3.4	3.6	2.3	3.0	2.8÷ 2.9
Conduction band extremum point	L	Δ	X	X	X	X
$m_\parallel/m$ (4.2 K)	1.58÷ 1.64	0.916	-	1.0–1.6	1.56	2.2
$m_\perp/m$ (4.2 K)	0.082	0.191	-	0.23÷ 0.26	0.19	0.2÷ 0.4
g_c	1.8	1.999	-	-	1.52	2.0
Ξ_u [eV]	18.7÷ 19.3	8.1÷ 9.2	-	-	-	-
Ξ_d [eV]	−10.5÷ −12.3	5	-	-	-	-
Valence band Δ [eV] (4.2 K)	0.296	0.044	-	0.7	0.275	0.08
γ_1	13.25÷ 13.38	4.26÷ 4.28	3.47	4.15	4.04	4.2
γ_2	4.20÷ 4.24	0.34÷ 0.38	0.06	1.01	0.78	0.98
γ_3	5.56÷ 5.69	1.45÷ 1.56	1.15	1.75	1.57	1.66
$\mathcal{K}$	3.27÷ 3.41	−0.26	−0.54	0.31	0.12	0.34
q	0.06	0.01	0.01	0.07	0.03	0.01
a [eV]	−12.7	−5	-	−5.9	-	−9.3÷ −9.9
b [eV]	−2.2÷ −2.6	−1.92÷ −2.27	-	−1.35	-	−1.5÷ −1.8
d [eV]	−4.7÷ −5.3	−4.84÷ −5.1	-	−4.3	-	−4.5÷ −4.6
d_o/a_o [eV]	33÷ 36	-	-	37	-	44
κ_o	15.8÷ 16.5	11.7– 12.1	9.8	12	10.06	11.0
κ_∞	$=\kappa_o$	$=\kappa_o$	7.5	10	8.16	9.0

	Direct-gap semiconductors				
	GaAs	GaSb	InP	InAs	InSb
E_g [eV] (4.2 K)	1.518	0.812	1.423	0.418	0.235
E_g [eV] (300 K)	1.428	0.70	1.35	0.36	0.180
Conduction band m^*/m (4.2 K)	0.065	0.041	0.079	0.023	0.014

Table 3.9. (*Contd.*)

	Direct-gap semiconductors				
	GaAs	GaSb	InP	InAs	InSb
m^*/m (300 K)	0.067	-	0.077÷ 0.073	0.027	0.013
g_c	1.96÷ 1.99	−7.8÷ −9.2	1.26÷ 1.48	−14.7÷ −15.4	−51.3÷ −47.8
Valence band Δ [eV]	0.34	0.75	0.108	0.39	0.81÷ 0.98
γ_1	6.8	13.1	5.0	19.7	33.5÷ 40.1
γ_2	2.1÷ 2.4	4.5	1.6	8.4	14.5÷ 18.1
γ_3	2.9	6.0	1.7÷ 2.3	9.3	15.7÷ 19.2
$\mathcal{K}$	1.2	3.5	0.97	7.68	13.5÷ 17.0
q	0.04	-	0.02	0.04	0.5
a [eV]	−6.7÷ −9.8	−8.3	−6.4÷ 6.6	−6	−7.7
b [eV]	−1.7÷ −2.0	−1.8÷ −2.0	−1.55 −2.0	−1.8	−2.0÷ −2.05
d [eV]	−4.55÷ −5.4	−4.6÷ −4.8	−4.2÷ −5.0	−3.6	−4.8÷ −5.0
d_0/a_0 [eV]	41	32 ÷ 37	35 ÷ 42	42	39
κ_0	12.5÷ 12.9	15.7	12.6	15.2	16.8÷ 18.0
κ_∞	10.9	14.4	9.6	12.3	15.7
γ_c [eVÅ^3]	24.5	187	8	-	220
C_3 [eVÅ]	5.2	19.7	2.6	-	45

For AlP-GaP, the values of $m_{\|}/m$ above the saddle point. After Landolt-Börnstein Band 17a Halbleiter (Springer, Berlin, Heidelberg 1982) γ_c and C_3 after G.E. Pikus, B.A. Maruschak, A.N. Titkov: Sov. Phys. Semicond. *22*, 185 (1988)

largest of the non-diagonal components $t^{ij}_{\alpha\beta}$ with $i \neq j$ or $\alpha \neq \beta$ are the coefficients $t_{21}^{v_x v_y} = t_{21}^{v_y v_x}$. Since the phase of the functions u_x, u_y, v_x, v_y changes in translation by a_2 or a_3, and $(K_{0y} - K_{0x})a_2 = \pi$, the components $t^{ij}_{\alpha\beta}$ with $i, j = v_x, v_y$ or u_x, u_y at the adjacent boundaries are related through

$$t^{ij}_{\alpha\beta}(R) = S(R)t^{ij}_{\alpha\beta}(0) \quad (i, j = u_x, u_y;\, v_x, v_y).$$

For $\alpha \neq \beta, i, j = v_x, v_y$, $S = -(-1)^R$ if $R = M$ or $R = -N$, and $S = (-1)^R$ if $R = M + N$ or if $\alpha = \beta, i, j = u_x, u_y$. The explicit dependence of the coefficients $t^{ij}_{\alpha\beta}$ on the number of the monolayers M and N in the wells and barriers results in that both the superlattice spectra and the coefficients of the structures containing several wells and barriers may differ qualitatively for even and odd values of N and M.

3.5.3 Spin Splitting

As already pointed out, the spin splitting of the conduction band near the point Γ in crystals of T_d class is proportional to k^3. In quantum wells and superlattices, the spin-dependent Hamiltonian, $\mathcal{H}_s$, contains the two contributions linear and cubic in the in-plane electron wavevector, $\mathbf{k}$, and can be written as

$$\mathcal{H}_s = (\boldsymbol{\sigma}\boldsymbol{\Omega}). \tag{3.101}$$

The contribution to $\boldsymbol{\Omega} \equiv \boldsymbol{\Omega}(k_x, k_y)$ arising from the cubic term in the bulk semiconductor depends on the interface crystallographic orientation. In accordance with (3.3), for the [001] structures, $\Omega_x = -\Omega_1 \cos\varphi - \Omega_3 \cos 3\varphi$, $\Omega_y = \Omega_1 \sin\varphi - \Omega_3 \sin 3\varphi$, $\Omega_z = 0$, where

$$\Omega_1 = \gamma k \left(\langle k_z^2 \rangle - \frac{1}{4} k^2 \right), \qquad \Omega_3 = \frac{1}{4} \gamma k^3, \qquad k^2 = k_x^2 + k_y^2,$$

$$\tan\varphi = \frac{k_y}{k_x}, \qquad \langle k_z^2 \rangle = \int \varphi(z) \hat{k}_z^2 \varphi(z) \mathrm{d}z.$$

For the [111] structures, $\Omega_x = \Omega_1 \sin\varphi$, $\Omega_y = -\Omega_1 \cos\varphi$, $\Omega_z = -\Omega_3 \sin 3\varphi$, where

$$z \| [111], \qquad y \| [1\bar{1}0], \qquad x \| [11\bar{2}],$$

$$\Omega_1 = \frac{2}{\sqrt{3}} \gamma k \left(\langle k_z^2 \rangle - \frac{1}{4} k^2 \right), \qquad \Omega_3 = \frac{1}{\sqrt{6}} \gamma k^3.$$

For the [110] structures, $\Omega_x = \Omega_y = 0$, $\Omega_z = -\Omega_1 \cos\varphi + \Omega_3 \cos 3\varphi$, where

$$z \| [110], \qquad x \| [1\bar{1}0], \qquad y \| [001],$$

$$\Omega_1 = \frac{1}{2} \gamma k \left(\langle k_z^2 \rangle - \frac{1}{4} k^2 \right), \qquad \Omega_3 = \frac{3}{8} \gamma k^3.$$

As follows from general symmetry considerations, in structures with a noncentrosymmetric potential $V(z)$, $\mathcal{H}_s$ may also include the so-called Rashba term

$$\mathcal{H}_R = \delta [\boldsymbol{\sigma} \mathbf{k}]_n, \tag{3.102}$$

where n is the normal to the interface.

In a homogeneous electric field, $\mathcal{I}$, the constant δ calculated in Kane's model is

$$\delta = e\mathcal{I} \frac{P^2}{3} \left[\frac{1}{E_g^2} - \frac{1}{(E_g + \Delta)^2} \right]. \tag{3.102a}$$

An additional contribution to δ can arise due to different properties of the opposite interfaces.

3.5.4 Inclusion of Band Nonparabolicity

The above expressions were derived under the assumption that the energy of the states measured from the band edge in the wells and barriers is small compared to the width of the corresponding band gap. If this condition is not met, one has to take into account the band non-parabolicity. Kane's model provides the simplest solution of this problem. The wave functions of the conduction-band electrons can be written in the forms (3.83, 86b, 90), where the C_i are now spinors with components $C_{i\alpha}$ and $C_{i\beta}$. For an infinitely deep well, k is defined by (3.86) as before, the relation between **k** and E being given by (3.64)

$$P^2(k^2+k_\perp^2) = E(E+E_g)(E+E_g+\Delta)(E+E_g+\frac{2}{3}\Delta)^{-1}. \qquad (3.103)$$

Here and in what follows, the energy E is reckoned from the minimum of the conduction band in the well. In principle, (3.103) describes both the spectrum of the light-hole levels (for an infinite heavy-hole mass), and that of the split-off valence band Γ_7^+ (or Γ_6). Note that E should be chosen in the corresponding interval, $-E_g > E > -(E_g+2\Delta/3)$, or $E < -(E_g+\Delta)$. However, for a finite heavy-hole mass, the mixing of the light and heavy-hole states usually cannot be neglected.

When including the band nonparabolicity in the case of wells of finite depth and square-well lattices, one should take into account also the mixing of states with different spin at the interfaces, which is described by the second term in (3.72). Thus the position of states for a well with barriers of equal height is described by the equation

$$2A_1A_2k\lambda\cos ka + \left[A_2^2\lambda^2 - A_1^2k^2 - (B_1-B_2)^2k_\perp^2\right]\sin ka = 0. \qquad (3.104)$$

Here

$$A_i = P_i^2\left(\frac{2}{E+E_{gi}-V_i} - \frac{1}{E+E_{gi}+\Delta_i-V_i}\right),$$

$$B_i = P_i^2\frac{\Delta}{(E+E_{gi}-V_i)(E+E_{gi}+\Delta_i-V_i)},$$

the index $i=1$ refers to the well parameters ($V_1=0$), $i=2$ refers to barriers, k is related to E through (3.103), and λ, through an expression differing from (3.103) in the replacement of k^2 by - λ^2, and of E by $E-V_2$, where V_2 is the barrier height. Kane's model does not take into account band anisotropy, while, for instance, for the point Γ of a cubic crystal and for nondegenerate bands $\mathcal{H}(k)$, it contains, besides the term k^4 included in (3.104), also the term $k_x^4+k_y^4+k_z^4$.

Band anisotropy can be taken into account in Kane's model by including in $\mathcal{H}(k)$ quadratic-in-k terms caused by **k**·**p** interaction with higher bands. The corresponding matrix for the valence bands differs from (3.46) in the absence of the contribution to the constants γ_i coming from the **k**·**p** interaction with the conduction band and defined by (3.41). Likewise, the quantity $1/m^*$ for the conduction band does not include the contribution (3.6) coming from interaction with the valence band. A similar system of equations for superlattices was considered, for instance, in refs. [3.20–27], a more complete list of publications on this subject being contained in the review by *Smith* and *Mailhiot* [3.17]. This system of equations can be solved only by numerical methods which will not be dealt with here.

3.6 Hole Spectrum in Quantum Wells and Superlattices for Degenerate Bands

If we limit ourselves in the Γ_8 band to hole energies small compared to the separation from the conduction band and the split-off band $\Gamma_7(\Gamma_7^+)$, then the secular equation determining the spectrum of holes can be obtained analytically. For an infinitely deep (001) well, such an equation was obtained in 1970 by *Nedorezov* [3.28]. In accordance with (3.46, 49) the solution of Schrödinger's equation for the envelope function F can be presented in the form

$$F = \sum_{i=1,2} \left[c_1^i F_1(+k_i)\mathrm{e}^{\mathrm{i}k_i z} + c_2^i F_1(-k_i)\mathrm{e}^{-\mathrm{i}k_i z}\right.$$
$$\left. + c_3^i F_2(+k_i)\mathrm{e}^{\mathrm{i}k_i z} + c_4^i F_2(-k_i)\mathrm{e}^{-\mathrm{i}k_i z}\right], \qquad (3.105)$$

where, by (3.49),

$$F_1 = \begin{bmatrix} H(k) \\ E - F(k) \\ O \\ I^*(k) \end{bmatrix}, \quad F_2 = \begin{bmatrix} -I(k) \\ O \\ -(E - F(k)) \\ H^*(k) \end{bmatrix}. \qquad (3.106)$$

Here the z axis is directed along the normal to the boundary, $k_i \equiv k_{iz} (i = 1, 2)$ are the two roots of (3.48) for fixed values of k_x, k_y and of energy E, F, H and I in (3.106) which are defined by relations (3.46). Note that for the (001) well considered here, $H(-k_z) = -H(k_z)$, and I does not depend on k_z. For an infinitely deep well, at both its boundaries, $z = 0$ and $z = a$, each row in (3.105) should vanish. This yields eight equations for eight coefficients c_m^i. By equating to zero the determinant of this system of equations one can obtain a secular equation defining the light and heavy-hole states.

In 1985, *Broido* and *Sham* [3.29] proposed a transformation T permitting one to diagonalize partially the Hamiltonian (3.46) by transforming it into two 2×2 Hamiltonians:

$$T = \frac{1}{\sqrt{2}} \begin{bmatrix} \mathrm{e}^{-\mathrm{i}\varphi} & O & O & -\mathrm{e}^{\mathrm{i}\varphi} \\ O & \mathrm{e}^{-\mathrm{i}\eta} & \mathrm{e}^{\mathrm{i}\eta} & O \\ O & -\mathrm{e}^{-\mathrm{i}\eta} & \mathrm{e}^{\mathrm{i}\eta} & O \\ \mathrm{e}^{-\mathrm{i}\varphi} & O & O & \mathrm{e}^{\mathrm{i}\varphi} \end{bmatrix}, \qquad (3.107)$$

where

$$\mathrm{e}^{\mathrm{i}(\varphi-\eta)} = \frac{H}{|H|}, \quad \mathrm{e}^{\mathrm{i}(\varphi+\eta)} = -\mathrm{i}\frac{I}{|I|}.$$

This transformation converts the Hamiltonian $\mathcal{H}$ (3.46) in the basis set $|3/2\rangle, |-1/2\rangle, |1/2\rangle, |-3/2\rangle$ to

$$\tilde{\mathcal{H}} = T\mathcal{H}T^{-1} = \begin{bmatrix} F & R & O & O \\ R^* & G & O & O \\ O & O & G & R \\ O & O & R^* & F \end{bmatrix}, \qquad (3.108)$$

where $R = |H| + \mathrm{i}|I|$,

The basis of Hamiltonian (3.108) is formed by the functions

$$\tilde{\varphi}_i = \sum_j T_{ji}^{-1}\varphi_j = \sum_j T_{ij}^{*}\varphi_j .$$

Hence

$$\tilde{\varphi}_{1,4} = \frac{1}{\sqrt{2}}\left(e^{i\varphi}|3/2\rangle \mp e^{-i\varphi}|-3/2\rangle\right),$$

$$\tilde{\varphi}_{2,3} = \frac{1}{\sqrt{2}}\left(e^{i\eta}|1/2\rangle \pm e^{-i\eta}|-1/2\rangle\right). \tag{3.109}$$

Later, similar transformations, which partially diagonalize the 6×6 Hamiltonian for both valence bands, and the 8×8 Hamiltonian including the conduction band as well, were proposed [3.30, 31]. However, since the functions $F_i(k_z)$ in (3.105) correspond to different values of k_z, they can both simultaneously be reduced to the form (3.109) only if φ and η in (3.107) do not depend on k_z; this is possible only for the (001) wells, where I does not depend on k_z at all, and H can be presented as $H = hk_z$. By choosing $e^{i(\varphi-\eta)} = h/|h|$, one can reduce in this case the Hamiltonian $\tilde{\mathcal{H}}$ and the basis functions $\tilde{F}$ to the form (3.108, 109) with $R = i|I| + k_z|h|$. Note that k_z in (3.108) may be considered as an operator, and the envelope function presented in the form

$$F = \sum_{i=1,2}\left[c_1^i F_l(+k_i)e^{ik_iz} + c_2^i F_l(-k_i)e^{-ik_iz}\right], \tag{3.110}$$

where $F_l\,(l = 1, 2)$ is one of the functions (3.109).

By equating to zero each of the rows in (3.110) at $z = 0$ and $z = d$, we obtain four equations for the coefficients c_m^i $(m = 1, 2; i = 1, 2)$ in place of eight equations for the function (3.105). The condition for the determinant of this system to be zero can be written as

$$\begin{aligned}&\sin(k_1a)\sin(k_2a)\left[|H_1|^2(E-F_2)^2 + |H_2|^2(E-F_1)^2\right.\\&\quad\left.+(F_1-F_2)^2|I|^2\right] - [1-\cos(k_1a)\cos(k_2a)]\\&\quad\times(H_1H_2^* + H_1^*H_2)(E-F_1)(E-F_2) = 0.\end{aligned} \tag{3.111}$$

Here $F_i = F(\mathbf{k}_i)$, $H_i = H(\mathbf{k}_i)(i = 1, 2)$.

Equation (3.111) was first derived by *Nedorezov* [3.28] and later reproduced in many publications. For $k_\perp \to 0$ from (3.111) it follows that for the light holes $\sin k_1a = 0$, i.e. $k_1 = \pi n/a$, for the heavy ones $\sin k_2a = 0$, i.e. $k_2 = \pi n/a$, and, accordingly, the spectrum of the light holes can be written as

$$E_n^{(1)} = \frac{\hbar^2}{2m}(\gamma_1 + 2\gamma_2)\left(\frac{\pi n}{a}\right)^2 \tag{3.112a}$$

and that of the heavy holes,

$$E_n^{(2)} = \frac{\hbar^2}{2m}(\gamma_1 - 2\gamma_2)\left(\frac{\pi n}{a}\right)^2, \tag{3.112b}$$

which is seen directly from the original equation for F, since for $k_\perp \to 0$ the Hamiltonian $\mathcal{H}(\mathbf{k})$ determined by (3.46, 108) retains only diagonal components. For odd n, the wave function $\varphi(z) = \sin \pi z/a$ does not reverse its sign with a c_{2x} rotation about the X axis passing through the center of the well, i.e., with z replaced by $a - z$, and transforms according to the representation A_1 of the group D_{2d}. For even n, $\varphi(z) = \sin \pi z/a$ reverses the sign for a c_{2x} rotation and transforms according to the representation B_2. Since the Bloch functions of the heavy holes transform according to the spinor representation $\Gamma_6(E_2')$ of the group D_{2d}, and those of the light ones according to $\Gamma_7(E_1')$, the heavy-hole wavefunctions transform according to Γ_6 for odd n, and according to Γ_7, for even n. Conversely, the light-hole wave functions transform according to Γ_7 for odd n, i.e., for even $\varphi(z)$, and according to Γ_6 for even n, i.e., for odd $\varphi(z)$.

By (3.111), the transverse masses for the light and heavy-hole states are determined by the expressions [3.28]

$$\begin{aligned}
\frac{m}{m_{1n}} &= \frac{\gamma_2(\gamma_1 - \gamma_2) + 3\gamma_3^2}{\gamma_2} + \frac{3\gamma_3^2\left(\gamma_1^2 - 4\gamma_2^2\right)^{1/2}}{\gamma_2^2}\,\frac{(-1)^{n+1} + \cos\varphi_n^{-1}}{\pi n \sin\varphi_n^{-1}},\\
\frac{m}{m_{2n}} &= \frac{\gamma_2(\gamma_1 + \gamma_2) - 3\gamma_3^2}{\gamma_2} + \frac{3\gamma_3^2\left(\gamma_1^2 - 4\gamma_2^2\right)^{1/2}}{\gamma_2^2}\,\frac{(-1)^{n+1} + \cos\varphi_n}{\pi n \sin\varphi_n},
\end{aligned} \tag{3.113}$$

where $\varphi_n = \left(\dfrac{\gamma_1 - 2\gamma_2}{\gamma_1 + 2\gamma_2}\right)^{1/2} \pi n.$

To calculate the spectrum for the [111] quantum wells, one can conveniently cross over to the coordinate frame $z \| [111], x \| [1\bar{1}0], y \| [11\bar{2}]$. In this frame, [3.7],

$$\begin{aligned}
\mathcal{H} = \frac{\hbar^2}{2m}\Bigg\{&\gamma_1 k^2 + \gamma_3\left(J_z^2 - \frac{5}{4}\right)\left(k^2 - 3k_z^2\right)\\
&- \frac{1}{3}(\gamma_2 + 2\gamma_3)\left(J_+^2 k_-^2 + J_-^2 k_+^2\right) + \frac{2}{3}(\gamma_2 - \gamma_3)\Big[\left(J_+^2 k_+ k_z\right.\\
&\left.+ J_-^2 k_- k_z\right) + \sqrt{2}\left(\left[J_z J_+\right]_s k_+^2 + \left[J_z J_-\right]_s k_-^2\right)\Big]\\
&\frac{2\sqrt{2}}{3}(2\gamma_2 + \gamma_3)\left(\left[J_z J_+\right]_s k_z k_- + \left[J_z J_-\right] k_z k_+\right)\Bigg\}.
\end{aligned} \tag{3.114}$$

Here, $J_\pm = J_x \pm \mathrm{i} J_y$ and $k_\pm = \dfrac{1}{\sqrt{2}}(k_x \pm \mathrm{i} k_y)$.

When presented in the matrix form, this Hamiltonian resembles closely (3.46). In units of $\hbar^2/2m$, we can write

$$F = \gamma_1 k^2 + \gamma_3\left(k^2 - 3k_z^2\right),$$

$$G = \gamma_1 k^2 - \gamma_3 \left(k^2 - 3k_z^2\right),$$

$$H = -2\sqrt{\frac{2}{3}}\left[(2\gamma_2 + \gamma_3) k_z k_- - (\gamma_2 - \gamma_3) k_+^2\right],$$

$$I = -\frac{2}{\sqrt{3}}\left[(\gamma_2 + 2\gamma_3) k_-^2 - 2(\gamma_2 - \gamma_3) k_+ k_z\right]. \tag{3.115}$$

While the wave function can also be written in the form (3.104), in contrast to the [001] wells, H and I now contain terms both depending on, and independent of, k_z. Therefore one cannot lower the order of the secular equation by means of a transformation similar to (3.10). If we set $\gamma_2 - \gamma_3 = 0$ in H and I in (3.115), such a transformation becomes possible, and the spectrum will in this case be determined by an equation differing from (3.111) only in different expressions for F, I and H. At $k_\perp = 0$ the position of the states is determined by expressions differing from (3.112) in γ_2 being replaced by γ_3. In the group C_{3V}, even and odd $\varphi(z)$ functions transform according to A_1. Therefore the heavy-hole states transform according to $\Gamma_5 + \Gamma_6(A_1' + A_2')$, and the light-hole states, according to $\Gamma_4(E')$, irrespective of the parity of $\varphi(z)$.

In the isotropic approximation, i.e., for $\gamma_2 = \gamma_3$, the expression for the spectrum can be obtained most conveniently with the technique first used by *Matulis* and *Piragas* [3.32]. In contrast to the generally accepted choice of the axes, they directed the X axis along the normal to the surface, and the Y axis, along $\mathbf{k}$, and, accordingly, set $k_z = 0$. With such a choice of the axes, $H = 0$, $I(-k_x) = I(k_x)$, the columns F in (3.106) split into two independent systems of functions, similar to (3.110). Accordingly, one obtains two independent systems of equations for the coefficients $c^i_{1,2}$ and $c^i_{3,4}$. As a result, the secular equation takes the form [3.32]:

$$\sin k_1 a \sin k_2 a \left[k_1^2 k_2^2 + \frac{1}{4} k_\perp^2 \left(k_1^2 + k_2^2\right) + k_\perp^4\right]$$
$$= \frac{3}{2} k_\perp^2 k_1 k_2 \left(\cos k_1 a \cos k_2 a - 1\right), \tag{3.116}$$

where $k_1^2 = 2m_l\hbar^{-2}E - k_\perp^2$, $k_2^2 = 2m_h\hbar^{-2}E - k_\perp^2$, m_l, m_h are the light and heavy-hole effective masses, $\dfrac{m}{m_l} = \gamma_1 + 2\bar{\gamma}$, $m/m_h = \gamma_1 - 2\bar{\gamma}$.

In this approximation, the position of states for $k_\perp = 0$ is determined by an expression differing from (3.112) in γ_2 being replaced by $\bar{\gamma}$, the transverse effective masses being

$$\frac{1}{m_{1n}} = \frac{1}{m_l} + 3\left(m_l m_h\right)^{-1/2} \frac{(-1)^{n+1} + \cos \pi n (m_h/m_l)^{1/2}}{\pi n \sin \pi n (m_h/m_l)^{1/2}},$$

$$\frac{1}{m_{2n}} = \frac{1}{m_h} + 3\left(m_l m_h\right)^{-1/2} \frac{(-1)^{n+1} + \cos \pi n (m_l/m_h)^{1/2}}{\pi n \sin \pi n (m_l/m_h)^{1/2}}. \tag{3.117}$$

For $m_1 \ll m_h$, for the first heavy-hole level, we obtain

$$\frac{1}{m_{21}} = \frac{1}{2}\frac{1}{m_h} + \frac{6}{\pi^2}\frac{1}{m_l}. \tag{3.118}$$

We see this mass to be close to the transverse mass of heavy holes in deformed crystals, which, by (3.78), can be written as

$$\frac{1}{m_{h\perp}} = \frac{1}{4}\left(\frac{1}{m_h} + \frac{3}{m_l}\right).$$

As for the light-hole transverse mass, it depends substantially on the ratio m_l/m_h and differs from that in deformed crystals

$$\frac{1}{m_{l\perp}} = \frac{1}{4}\left(\frac{3}{m_h} + \frac{1}{m_l}\right).$$

The longitudinal masses determining the position of levels for $k_\perp = 0$ coincide with those in deformed crystals.

For quantum wells of finite depth and superlattices with square wells, one can write the wave function in the barriers in the form similar to (3.104), and for (001) lattices or wells, in the form (3.110) (with imaginary k_i, i.e., $ik_i = \lambda_i$) and, using boundary conditions, obtain the corresponding secular equations. The equations thus derived being very cumbersome, one uses, as a rule, numerical methods. This is done ordinarily by expanding the exact solution in the wave functions φ_{h}^0 and φ_{e}^0 representing the solution of the corresponding Schrödinger's equation for $k_\perp = 0$. These solutions differ from (3.83, 90) in the effective masses $m_{A,B}$ being replaced, respectively, by the longitudinal effective masses of heavy and light holes in the wells and barriers. After this one limits oneself to a desired number of terms in this expansion and solves numerically the coupled equations for the expansion coefficients thus obtained. An analytical solution with inclusion of mixing of the valence band Γ_8 and conduction band obtained for a well of finite depth in the isotropic approximation is given by *Gerchikov* and *Subashiev* [3.33].

Symmetry considerations permit one to draw general conclusions on the character of the $E(k_z)$ spectrum in the (001) and (111) superlattices. In a (001) superlattice, states with $k_z \neq 0$ do not have a definite parity and transform at $k_\perp = 0$ according to the representation $\Gamma_5(E')$ of the group G_{2v} for the light and heavy holes. Therefore the crossing of the light and heavy hole branches results in their repulsion (anticrossing). In a (111)-growth-axis superlattice for $k_z \neq 0$, the symmetry does not change and, just as in the case of quantum wells, for $k_\perp = 0$ the light-and heavy-hole states transform according to the different representations Γ_6 and $\Gamma_4 + \Gamma_5$; note that these branches can cross. For $\mathbf{k}_\perp \neq 0$, both in quantum wells and superlattices, the light- and heavy-hole states transform according to the same representation, the branches always anticrossing.

3.6.1 Spin Splitting

In the T_d class [001]- and [111]-oriented superlattices, just as in quantum wells, $\mathcal{H}(\mathbf{k})$ for the valence bands includes linear-in-k terms. In this geometry, $\mathcal{H}(\mathbf{k})$ does not contain terms linear in k_z. The corresponding Hamiltonians for the light-and heavy-hole branches have the same form, (3.101) or (3.102), as those for the conduction band electrons. The corresponding constants β for the light holes are connected with γ in (3.53) through the same relations as the ones valid for the electrons. For the heavy holes, $\mathcal{H}_{v3}$ defined by (3.53) does not contribute to the linear-in-k splitting and, therefore, the constants β are proportional to the small coefficients a_i in $\mathcal{H}_{v3}$ in (3.53).

3.7 Deformed and Strained Superlattices

The load used in studies of superlattices or quantum wells is ordinarily applied along the normal to the interface. In this case F in the Hamiltonian (3.46) is replaced, in accordance with (3.77a), by $\tilde{F} = F + f$, and G, by $\tilde{G} = G + g$. For the (001)-oriented lattices

$$f = a\varepsilon + \frac{b}{2}(\varepsilon - 3\varepsilon_{zz}), \quad g = a\varepsilon - \frac{b}{2}(\varepsilon - 3\varepsilon_{zz}) \tag{3.119a}$$

and for the (111)–growth-axis lattices

$$f = a\varepsilon + \frac{d}{2\sqrt{3}}(\varepsilon - 3\varepsilon_{z'z'}), \quad g = a\varepsilon - \frac{d}{2\sqrt{3}}(\varepsilon - 3\varepsilon_{z'z'}), \tag{3.119b}$$

where the z'-axis is directed along [111]. The spectrum in a deformed lattice is given by (3.78), the secular equation for the [001]- or [111]-oriented quantum wells differing from (3.111) only in $F_{1,2}$ being replaced by $\tilde{F}_{1,2} = F_{1,2} + f$.

Under tensile stress, which shifts the heavy-hole states down in energy, the separation between the light- and heavy-hole states increases, so that for not too thin wells and for large enough deformations only the heavy hole contribution has to be included in ground-state calculations. In this case, the spectrum will be defined by the relations derived for the conduction band, the expression for the effective masses in (001)-growth axis structures being

$$\frac{m}{m_{\parallel}} = \gamma_1 - 2\gamma_2, \quad \frac{m}{m_{\perp}} = \gamma_1 + \gamma_2 \tag{3.120a}$$

and for the (111)–oriented structures

$$\frac{m}{m_{\parallel}} = \gamma_1 - 2\gamma_3, \quad \frac{m}{m_{\perp}} = \gamma_1 + \gamma_3. \tag{3.120b}$$

In contrast, compressive deformation brings down the light-hole state making it the lowest-lying at large enough deformations, i.e., producing inversion of states. In the lattices of many-valley semiconductors, where confinement results in the splitting of states belonging to different valleys, deformation induces an additional shift of these states and can also result in their inversion.

In structures with a noticeable mismatch between the lattice period constants a_0 in the wells and barriers, stress appears, namely, the crystal undergoes tensile stress in the wells and compression in the barriers (or vice versa); for example, in the (001)-oriented lattices

$$\varepsilon^a_{xx} - \varepsilon^b_{xx} = \varepsilon^a_{yy} - \varepsilon^b_{yy} = \delta = \frac{\Delta a_0}{\bar{a}_0}, \tag{3.121}$$

where $\Delta a_0 = a_b - a_a$; $\bar{a}_0 = (a_a + a_b)/2$. Here a_a and a_b are the lattice constants in the A and B regions. The condition of the stress-tensor component X_{zz} being zero implies that

$$c^l_{11}\varepsilon^l_{zz} + c^l_{12}(\varepsilon^l_{xx} + \varepsilon^l_{yy}) = 0 (l = a, b) \tag{3.122}$$

and the condition for the mean force acting upon the lattice to be zero can be written as

$$\sum_l h_l \left[\left(c^l_{11} + c^l_{12} \right) \varepsilon_{ll} + c^l_{12}\varepsilon^l_{zz} \right] = 0. \tag{3.123}$$

Here h_a, h_b are the sizes of the A and B regions. Equation (3.119) takes into account that $\varepsilon^l_{xx} = \varepsilon^l_{yy}$.

Disregarding the difference between the elastic moduli c^a_{ij} and c^b_{ij}, it follows from (3.121–123) that

$$\begin{aligned} \varepsilon^a_{xx} &= \varepsilon^a_{yy} = \delta \frac{h_b}{h_a + h_b}, \quad \varepsilon^b_{xx} = \varepsilon^b_{yy} = -\delta \frac{h_a}{h_a + h_b} \\ \varepsilon^l_{zz} &= -\frac{2C_{12}}{C_{11}} \varepsilon^l_{xx}, \quad \varepsilon_l = \mathrm{Tr}\varepsilon^l = 2\varepsilon^l_{xx} \frac{C_{11} - C_{12}}{C_{11}}. \end{aligned} \tag{3.124a}$$

Whence one can write for the quantity $3\varepsilon_{zz} - \varepsilon$ which, by (3.119), determines the shift of the light- and heavy-hole states

$$3\varepsilon^l_{zz} - \varepsilon_l = -\frac{2(C_{11} + 2C_{12})}{C_{11}} \varepsilon^l_{xx}. \tag{3.124b}$$

In accordance with the above symmetry considerations, in the case of (111)–oriented lattices the difference $c_{11} - c_{12}$ is replaced by $2c_{44}$, the quantity $c_{11} + 2c_{12}$, which determines the deformation ε under hydrostatic pressure, remaining unchanged. Then in the coordinate system $z \| [111], x, y \perp [111]$, we will obtain

$$3\varepsilon^l_{zz} - \varepsilon_l = -\frac{6(C_{11} + 2C_{12})}{C_{11} + 2C_{12} + 4C_{44}} \varepsilon^l_{xx}, \tag{3.125}$$

where ε^l_{xx}, just as is the case with (001)-oriented lattices, is defined by (3.124a). In the isotropic approximation $2c_{44} = c_{11} - c_{12}$ and the expressions (3.124b, 125) coincide. For a single quantum well $h_b \gg h_a$, and $\varepsilon^a_{xx} = \delta$. A specific feature of strained (111)–oriented lattices in T_d class crystals is the appearance of a polarization $\mathcal{P}$ induced by the piezoelectric effect. The quantity $\mathcal{P}$ is defined by

$$\mathcal{P}_k = 2e_{14}\varepsilon_{ij}|\delta_{ijk}|.$$

The quantity e_{14} is called the piezoelectric modulus. For the (111)-oriented lattices, $\varepsilon_{xy} = \varepsilon_{xz} = \varepsilon_{yz} = \varepsilon_{z'z'} - \frac{1}{3}\varepsilon$, and, respectively

$$\mathcal{P}^l_{z'} = \frac{2}{\sqrt{3}}e^l_{14}\left(3\varepsilon^l_{z'z'} - \varepsilon_l\right). \tag{3.126}$$

In the absence of free carriers screening the electric field $\mathcal{I}$, the field normal to the surface can be written as

$$\mathcal{I}^l_{111} = \frac{4\pi\mathcal{P}_{111}}{\kappa_l}, \tag{3.127}$$

where κ_l is the dielectric constant.

For equal elastic and piezoelectric moduli in the a and b materials, the mean field $\bar{\mathcal{I}} = (\mathcal{I}_a h_a + \mathcal{I}_b h_b)/(h_a + h_b)$ is zero. If these moduli are not equal, a mean field will appear, and, as is usually the case with pyro- and piezoelectrics, it will be compensated by the surface charge creating a field equal to the mean one but of opposite sign. In (001)-oriented lattices there is no polarization. For other orientations there is also a component $\mathcal{P}_\perp$ in the well plane. Here the electric field $\mathcal{I}_\perp$ appears only at the **side faces,** in a region where $h_a + h_b$ is thick. In the main region $\mathcal{I}_\perp = 0$, and the displacement field $\mathbf{D}_\perp = 4\pi\mathcal{P}_\perp$. This polarization can be revealed from the associated optical birefringence.

Later on we will discuss the specific features of the carrier spectrum in quantum wells and superlattices in the presence of a built-in or external electric field.

3.8 Quantum Wells and Superlattices in a Magnetic Field

The electron spectrum in quantum wells and superlattices in the presence of a magnetic field is obtained from the solution of Schrödinger's equation with the Hamiltonian $\mathcal{H}$ given by (3.19) and for nondegenerate bands by (3.22). If the magnetic field B is directed along the normal of the interface z, coinciding in the case of an anisotropic spectrum with one of the principal axes of the effective-mass tensor, then the variables in Schrödinger's equation (3.16) can be separated, and its solution written in the form

$$F(\mathbf{x}) = \varphi(z)\Phi(\rho). \tag{3.128}$$

Here ρ is the component of $\mathbf{x}$ perpendicular to $\mathbf{B}$.

The function $\varphi(z)$ retains the form it had without the magnetic field, the only difference being that states with different spin S_z will have different energies because of the Zeeman splitting described by the last term in (3.22). For an isotropic spectrum, the solutions of Schrödinger's equation for the function $\Phi(\rho)$

$$\left[-\frac{1}{2m^*}\left(-i\hbar\nabla_\rho + \frac{e}{c}A\right)^2 - E_\perp\right]\Phi = 0 \tag{3.129}$$

in the Landau gauge $A_x = -By$, $A_y = A_z = 0$ have the form

$$\Phi = e^{ik_x x}\chi(y), \tag{3.130}$$

where

$$\chi(y) = \frac{1}{\pi^{1/4}(2^n n! r_B)^{1/2}} e^{-\frac{(y-y_0)^2}{2r_B^2}} H_n\left(\frac{y-y_0}{r_B}\right).$$

Here $y_0 = -\hbar c k_x/eB = -r_B^2 k_x$, $r_B = (\hbar c/eB)^{1/2}$ is the magnetic length, $H_n(\xi)$ is the Hermite polynomial

$$H_n(\xi) = (-1)^n e^{\xi^2}\frac{d^n}{d\xi^n}e^{-\xi^2}. \tag{3.131}$$

For the energy one can write

$$E_{\perp n} = (n + 1/2)\hbar\omega_c, \tag{3.132}$$

where $\omega_c = eB/m^*c$ is the cyclotron frequency.
In a number of problems, for instance, when calculating the spectra of impurity centers or excitons in a magnetic field, the Landau gauge is less convenient than the gauge:

$$\mathbf{A} = \frac{1}{2}[\mathbf{B}\rho]. \tag{3.133}$$

In the latter, the solutions of Schrödinger's equation (3.128) have the form

$$\Phi(\rho) = \frac{e^{im\varphi}}{\sqrt{2\pi}}\Phi_{n_\rho,m}(|\rho - \rho_0|)e^{i\mathbf{k}\cdot\rho}. \tag{3.134}$$

Here φ is the angular coordinate, $\rho_0 = r_B^2[\mathbf{B}\cdot\mathbf{k}]/|\mathbf{B}|$,

$$\Phi_{n_\rho,m} = \left(\frac{n_\rho}{(n_\rho + |m|)!}\right)^{1/2}\frac{1}{r_B}\left(\frac{\rho^2}{2r_B^2}\right)^{\frac{|m|}{2}} e^{-\frac{\rho^2}{4r_B^2}} \times L^{|m|}_{n_\rho+|m|}\left(\frac{\rho^2}{2r_B^2}\right), \tag{3.135}$$

$L^m_{n+m}(\xi)$ is the associated Laguerre polynomial:

$$L^m_{n+m}(\xi) = \frac{(n+m)!}{n!}e^{\xi}\xi^{-m}\frac{d^n}{d\xi^n}\left(e^{-\xi}\xi^{n+m}\right). \tag{3.136}$$

For the energy $E_{n_\rho,m}$ we have

$$E_{\perp n_\rho,m} = \left[n_\rho + \frac{1}{2}(1 + |m| + m)\right]\hbar\omega_c. \tag{3.137}$$

If the transverse electron masses in the A and B layers are unequal, the electron wave function $\Phi(\rho)$ will be defined by (3.130, 134) as before; however, the Landau levels will not be equidistant any more, since in this case the roots of (3.89, 92) depend on the quantity $E_{\perp B} - E_{\perp A} = \hbar(n+1/2)(\omega_B - \omega_A)$, i.e., on the level number. In an electric field $\mathcal{I}$ perpendicular to $\mathbf{B}$, the equivalence of the directions X and Y in the Landau gauge breaks down, and the Y axis in (3.134) should preferably be directed along the field $\mathcal{I}$. Under these conditions, $\mathbf{k}$ in (3.130, 134) is replaced by $\mathbf{k} + m^*\mathbf{v}_0/\hbar$, where $\mathbf{v}_0 = c[\mathcal{I}\mathbf{B}]/B^2$, and the energy (3.132, 137) includes an additional term

$$\delta E = -\hbar(\mathbf{k}\cdot\mathbf{v}_0) - \frac{m^*}{2}v_0^2. \tag{3.138}$$

In the Landau gauge $\mathbf{k}\|\mathbf{v}_0$.

If the carrier spectrum is anisotropic, one can introduce the variables

$$x' = x\left(\frac{m_{xx}}{\bar{m}}\right)^{1/2}, \quad y' = y\left(\frac{m_{yy}}{\bar{m}}\right)^{1/2} \quad \left(\bar{m} = (m_{xx}m_{yy})^{1/2}\right) \tag{3.139a}$$

and, accordingly,

$$k_i' = k_i\left(\frac{\bar{m}}{m_{ii}}\right)^{1/2}, \quad A_i' = A_i\left(\frac{\bar{m}}{m_{ii}}\right)^{1/2}, \quad \mathcal{I}_i' = \mathcal{I}_i\left(\frac{\bar{m}}{m_{ii}}\right)^{1/2}, \tag{3.139b}$$

after which Schrödinger's equation reduces to the form obtained for an isotropic spectrum with $m^* = m$, and with the same value of B_z. Note that $\mathcal{I}' = (\mathcal{I}_x^2 m_{yy} + \mathcal{I}_y^2 m_{xx})^{1/2}/\bar{m}^{1/2}$, and the direction of $\mathcal{I}'$ can differ from that of $\mathcal{I}$. If the magnetic field $\mathbf{B}$ is not directed along the normal to the surface or, for an anisotropic spectrum, along one of the principal axes of the effective-mass tensor, then the electron energy and wave function for square wells can be calculated only numerically.

For a parabolic well, however, the spectrum and wave functions can be found analytically for an arbitrary orientation of the magnetic field [3.34]. Indeed, for $V(z) = az^2$, one can conveniently introduce the vector potential $A_y = xB_z - zB_x$, $A_x = A_z = 0$, after which Schrödinger's equation will acquire the form

$$\left\{-\frac{\hbar^2}{2m^*}\left(\frac{\partial^2}{\partial x^2} + \frac{\partial^2}{\partial z^2}\right) + \frac{1}{2m^*}\left[-i\hbar\frac{\partial}{\partial y} + \frac{e}{c}(xB_z - zB_x)\right]^2 \right.$$
$$\left. + az^2 - E\right\}\psi = 0. \tag{3.140}$$

Its solution can be written as

$$\psi(x, y, z) = \mathrm{e}^{ik_y y}\Phi(x, z). \tag{3.141}$$

Introducing $x' = x - x_0$, where $x_0 = \hbar k_y/eB_z$, one can eliminate in (3.140) the terms linear in x' and z and the term containing k_y^2, the equation for $\Phi(x, z)$

acquiring the form

$$\left\{-\frac{\hbar^2}{2m^*}\left(\frac{\partial^2}{\partial x'^2}+\frac{\partial^2}{\partial z^2}\right)+\frac{m^*}{2}\left[\omega_z^2 x'^2+\left(\omega_x^2+\omega_0^2\right)z^2\right.\right.$$
$$\left.\left.+2\omega_x\omega_z x'z\right]-E\right\}\psi=0, \tag{3.142}$$

where $\omega_z = eB_z/m^*c$, $\omega_x = eB_x/m^*c$, $\omega_0^2 = 2a/m^*$. Crossing over to the coordinates ξ, η

$$x' = \xi\cos\varphi + \eta\sin\varphi,\ z = -\xi\sin\varphi + \eta\cos\varphi,$$

where

$$\tan 2\varphi = \frac{2\omega_x\omega_z}{\omega_z^2-\omega_x^2-\omega_0^2},$$

we will have in the new variables in (3.142) only terms quadratic in ξ and η, and the wave function can be written in the form of a product $\Phi(\xi,\eta) = \Phi_1(\xi)\Phi_2(\eta)$, where each of the functions $\Phi_{1,2}$ is a solution of the harmonic-oscillator equation and has the form (3.130) with $y_0 = 0$, $r_{B1,2} = (\hbar/m^*\omega_{1,2})^{1/2}$, where

$$\omega_{1,2} = \left\{\frac{1}{2}\left(\omega_c^2+\omega_0^2\right)\pm\frac{1}{2}\left[\left(\omega_c^4+\omega_0^4\right)-2\omega_0^2\left(\omega_x^2-\omega_z^2\right)\right]^{1/2}\right\}^{1/2}. \tag{3.143}$$

Here $\omega_c = (\omega_z^2+\omega_x^2)^{1/2} = eB/m^*c$, and the energy

$$E = \hbar\omega_1(n_1+1/2)+\hbar\omega_2(n_2+1/2). \tag{3.144}$$

For $B_x = 0$ we have $\omega_1 = \omega_z$, $\omega_2 = \omega_0$, i.e., the frequency of one of the oscillators does not depend on B. For $B_z \to 0$, $\omega_2 \to 0$, $\omega_1 = (\omega_0^2+\omega_x^2)^{1/2}$. However for $B_z \to 0$, $x_0 = \hbar k_y/eB_z \to \infty$, and therefore (3.142) is no longer valid. In this case, the solution of (3.142) can be chosen in the form $\psi = \exp[\mathrm{i}(k_x x + k_y y)]\chi(z)$. Introducing now $z' = z - z_0$, where $z_0 = (\hbar k_y/m^*)\omega_c/(\omega_c^2+\omega_0^2)$, we come to the harmonic-oscillator equation whose solution is the function (3.130) with $y_0 = 0$, $r_B = (\hbar/m\omega_1)^{1/2}$, $\omega_c = eB_x/m^*c$. The total energy can be written as

$$E = \hbar\omega_1(n+1/2)+\frac{\hbar^2}{2m^*}\left(k_x^2+k_y^2\frac{\omega_0^2}{\omega_0^2-\omega_c^2}\right). \tag{3.145}$$

One can ask the question at what values of B_z (3.144) transforms into (3.145). The condition of validity of the relations obtained is $\omega_{1,2}\tau \gg 1$, where τ is the momentum relaxation time. Obviously, $\omega_2 \gg 1/\tau$ for $\omega_z\tau \gg (1+\omega_c^2/\omega_0^2)^{1/2}$. Note that the densities of states $g(E)$ calculated from (3.144,145) coincide for $\omega_z \to 0$ and are equal to

$$g(E) = \frac{m^*}{\pi\hbar^2}\left(1+\frac{\omega_c^2}{\omega_0^2}\right)^{1/2}.$$

We note in conclusion that, in weak magnetic fields where the spacing between the Landau levels is small compared to the miniband width, the spectrum in superlattices can be calculated from the above expressions using the values of the effective masses defined by (3.93).

3.9 Spectrum of Quantum Wells and Superlattices in an Electric Field

The solution of Schrödinger's equation in a constant electric field $\mathcal{I} = \mathcal{I}_x$

$$\left(-\frac{\hbar^2}{2m^*}\frac{d^2}{dx^2} + e\mathcal{I}x - E\right)\Phi = 0 \tag{3.146}$$

is represented by Airy's functions Ai(z) and Bi(z), where

$$z = -\left(x + \frac{E}{e\mathcal{I}}\right)\left(\frac{2m^*e\mathcal{I}}{\hbar^2}\right)^{1/3} = -\kappa^{2/3}(x + x_0),\, x_0 = \frac{E}{e\mathcal{I}},$$

$$\kappa = \left(\frac{2m^*e\mathcal{I}}{\hbar^2}\right)^{1/2}.$$

For an infinitely deep quantum well of thickness α, the energy E is determined from the condition of vanishing of the wave function

$$\varphi(z) = c_1\mathrm{Ai}(z) + c_2\mathrm{Bi}(z) \tag{3.147}$$

for $x = \pm a/2$ or, accordingly, for $z = z_{1,2} = -\kappa^{2/3}(x_0 \pm a/2)$. The corresponding secular equation can be written as

$$\mathrm{Ai}(z_1)\mathrm{Bi}(z_2) - \mathrm{Ai}(z_2)\mathrm{Bi}(z_1) = 0. \tag{3.148}$$

To calculate spectra in the limits of weak or strong electric fields, one can conveniently use the asymptotic expressions for Airy's functions:

for $|z| \ll 1$

$$\begin{aligned}\mathrm{Ai}(z) &= D_1 f(z) - D_2 g(z)\\ \mathrm{Bi}(z) &= \sqrt{3}[D_1 f(z) + D_2 g(z)],\end{aligned} \tag{3.149}$$

where

$$\begin{aligned} f(z) &= 1 + \frac{z^3}{3!} + \frac{4}{6!}z^6 + \dots\\ g(z) &= z + \frac{2}{4!}z^4 + \frac{2.5}{7!}z^7 + \dots\\ D_1 &= 3^{-2/3}\Gamma(2/3) \simeq 0.355\\ D_2 &= 3^{-1/3}\Gamma(1/3) \simeq 0.259,\end{aligned}$$

for $|z| \gg 1$

$$Ai(|z|) = \frac{1}{2}\pi^{-1/2}|z|^{-1/4}e^{-\xi}\left(1 - \frac{5}{72}\xi^{-1} + \ldots\right) \tag{3.150}$$

$$Bi(|z|) = \pi^{-1/2}|z|^{-1/4}e^{\xi}(1 + \frac{5}{72}\xi^{-1} + \ldots)$$

$$Ai(-|z|) = \pi^{-1/2}|z|^{-1/4}\left[\sin\left(\xi + \frac{\pi}{4}\right) - \frac{5}{72}\xi^{-1}\cos\left(\xi + \frac{\pi}{4}\right) + \ldots\right]$$

$$Bi(-|z|) = \pi^{-1/2}|z|^{-1/4}\left[\cos\left(\xi + \frac{\pi}{4}\right) + \frac{5}{72}\xi^{-1}\right.$$

$$\left.\sin\left(\xi + \frac{\pi}{4}\right) + \ldots\right], \tag{3.151}$$

where $\xi = \frac{2}{3}|z|^{3/2}$.

Thus, in weak electric fields, i.e., for $-z \gg 1$, one derives from (3.148, 151)

$$\sin(\xi_2 - \xi_1) = \frac{5}{72}(\xi_1^{-1} - \xi_2^{-1})\cos(\xi_2 - \xi_1), \tag{3.152}$$

where

$$\xi_{1,2} = \frac{2}{3}|z_{1,2}|^{3/2} = \frac{2}{3}\kappa(x_0 \pm \frac{a}{2})^{3/2}.$$

Since in this approximation $\xi_{1,2} \gg 1$, the solution of (3.152) can be looked for in the form $\xi_2^n - \xi_1^n = \pi n + \varphi_n$. Expanding $\xi_1^{-1} - \xi_2^{-1}$ in powers of $a/2x_0$, we can now find φ_n and, thus, obtain the values of E_n

$$E_n = E_n^0\left[1 + \frac{1}{48}\left(\frac{e\mathcal{I}a}{E_n^0}\right)^2\left(1 - \frac{15}{(\pi n)^2}\right)\right]. \tag{3.153}$$

Here $E_n^0 = \hbar^2\pi^2 n^2/(2m^* a^2)$ is the energy of the nth level in the absence of electric field. A comparison with the exact solution of (3.148) shows that for $(e\mathcal{I}_a/E_1^0) \leqslant 1$, (3.153) determines the field-induced shift of a level to better than 1%. As seen from this expression, in weak fields the first level is shifted downwards, and the others, upwards. In a strong electric field, i.e., for $e\mathcal{I}_a \gg E_1^0$, the energy of a state reckoned from the bottom of the well is determined by the condition

$$\mathrm{Ai}(-\kappa^{2/3}x_0) = 0 \tag{3.154}$$

and can be written as

$$E_n = a_n\frac{(e\mathcal{I}\hbar)^{2/3}}{(2m^*)^{1/3}}, \tag{3.155}$$

where $-a_n$ are the roots of the equation $\mathrm{Ai}(z) = 0$. The numerical values of the first roots are:

$$a_1 = 2.338;\ a_2 = 4.088;\ a_3 = 5.520;\ a_4 = 6.787;\ a_5 = 7.944.$$

As seen from a comparison with the exact solution of (3.148), for $e\mathcal{I}a/E_1^0 > 10$ the expression (3.155) is accurate to better than 1%. The variational function used frequently in the case of a triangular barrier

$$\varphi = 2\alpha^{3/2} z e^{-\alpha z}, \alpha = \left(\frac{3}{2}\frac{e\mathcal{I}m^*}{\hbar^2}\right)^{1/3} \tag{3.156}$$

yields for the energy of the first level an expression coinciding with (3.155) for $a_1 = 2.47$, which differs from the exact value $a_1 = 2.34$ by 6%. Note that the quasiclassical approximation yields the values of energy coinciding with (3.155) to better than 1%. It was shown [3.35] that the variational function

$$\varphi = ce^{-\alpha z} sin\frac{\pi z}{a}$$

for a well with infinitely high walls at $z = 0$ and $z = a$ yields throughout the whole field range the energy of the first level differing from the exact solution of (3.148) by not more than 8%. In particular, for weak fields it gives for the shift of the energy of the first level a value differing from (3.153) by a factor 0.97. *Bastard* et al. [3.35] present also the variational functions and specify the energies for wells of finite width.

The electron spectrum in a superlattice in the presence of a weak electric field when $e\mathcal{I}d \ll \Delta E$, where d is the superlattice period and ΔE is the miniband width, can be readily found using the effective mass method. In the k-representation

$$e\mathcal{I}\hat{x} = \mathrm{i}e\mathcal{I}\frac{\partial}{\partial q}$$

and for the solution of the equation

$$\left[\mathrm{i}e\mathcal{I}\frac{\partial}{\partial q} + E(q) - E\right]\psi = 0, \tag{3.157}$$

one obtains

$$\psi(q) = c\exp\left\{\frac{\mathrm{i}}{e\mathcal{I}}\int_{-\pi/d}^{q}\left[E(q') - E\right]\mathrm{d}q'\right\}. \tag{3.158}$$

As follows from the condition of periodicity $\psi(q + 2\pi/d) = \psi(q)$

$$\frac{\mathrm{i}}{e\mathcal{I}}\int_{-\pi/d}^{+\pi/d}\mathrm{d}q'\left[E(q') - E_n\right] = -2\pi \mathrm{i}n,$$

since $E(q) = E_0 + \sum_N C_N \cos q\,\mathrm{d}N$ is a periodic function, we have

$$E_n = e\mathcal{I}\,\mathrm{d}n + E_0. \tag{3.159}$$

We see that electric field splits the miniband into a series of equidistant levels, the Stark ladder, the spacing between them being $\delta E = e\mathcal{I}d$. This effect is called the Wannier-Stark quantization [3.36].

The character of electron motion under the conditions of Wannier-Stark quantization can be better visualized if one uses another gauge, namely, $A = -\mathcal{I}ct$ [3.37]. The solution of Schrödinger's equation

$$i\hbar\frac{\partial\psi}{\partial t} = E\left(q + \frac{eA}{\hbar c}\right)\psi = E\left(q - \frac{e\mathcal{I}t}{\hbar}\right)\psi \tag{3.160}$$

can be written as

$$\psi(t) = c\exp\left[-\frac{i}{\hbar}\int_0^t E\left(q - \frac{e\mathcal{I}t'}{\hbar}\right)\mathrm{d}t'\right], \tag{3.161}$$

$E(q)$ being a periodic function of q with a period $2\pi/d$, the electron motion in an electric field is periodic with a period

$$T = \frac{2\pi}{d}\frac{\hbar}{e\mathcal{I}}.$$

Since the electron velocity is

$$v = \frac{1}{\hbar}\frac{\partial E}{\partial q} = \frac{1}{e\mathcal{I}}\frac{\partial E}{\partial t},$$

the distance passed in moving from the minimum to maximum of the miniband, i.e., in the time $T/2$, will be

$$L = \frac{1}{e\mathcal{I}}\int_0^{T/2}\frac{\partial E}{\partial t}\,\mathrm{d}t = \frac{\Delta}{e\mathcal{I}}, \tag{3.162}$$

where Δ is the width of the allowed miniband.

In the second half-period, the electron moves along the same path in the opposite direction, the probability of finding the electron at the point z, $\rho(z)$, being inversely proportional to the time the electron spends close to the point z, i.e., to the velocity $v(z)$. For illustration, let us calculate $\rho(z)$ for the case where the tight binding approximation holds. In this approximation

$$E_m(q) = E_{0m} - \frac{\Delta_m}{2}\cos qd. \tag{3.163}$$

The miniband width Δ_m is determined by the overlap of the wave functions of the neighboring cells

$$\Delta_m = 4\int_{d-a/2}^{d+a/2} dz\psi_m^*(z)\Delta\mathcal{H}(z)\psi_m(z+d). \tag{3.164}$$

Here $\psi_m(z)$ is the wave function of a single well centered at the point $z = 0$, and $\Delta\mathcal{H}$ is the difference between the Hamiltonians $\mathcal{H}$ for the superlattice and a single well, the integration performed over the neighboring cell.

The expression for Δ_m in the Kronig-Penney model can be readily obtained from (3.92) for the case of $\lambda b \gg 1$ corresponding to the tight-binding approximation. According to (3.91–92), the position of a level in a single well E_0 is determined by the equation

$$\cos k_0 a + R_-^0 \sin k_0 a = 0, \tag{3.165}$$

where

$$R_-^0 = \frac{1}{2}\left(R_0 - R_0^{-1}\right), k_0 = k\left(E_0\right), \lambda = \lambda(E_0), R_0 = R(k_0, \lambda_0).$$

Replacing $\cosh \lambda b$ and $\sinh \lambda b$ by $e^{\lambda b}/2$, and expanding $F(E)$ defined by (3.92) in powers of $E - E_0$ we come to

$$\Delta = -2\left(\frac{\partial F}{\partial E}\right)_{E_0}^{-1} = 4E_0 e^{-\lambda_0 b} \frac{\sin k_0 a}{k_0 a - \dfrac{V_0}{2V_0 - E} \sin 2k_0 a}. \tag{3.166}$$

In the tight-binding approximation $v = \dfrac{\Delta d}{2\hbar} \sin qd$, and

$$\rho_m(z) = \frac{4}{\pi}\frac{e\mathcal{I}}{\Delta_m}\left[1 - \left(\frac{4e\mathcal{I}(z - z_0)}{\Delta_m}\right)^2\right]^{-1/2}. \tag{3.167}$$

At z_0 the energy $E(z_0) = E_0 + e\mathcal{I}z_0$ is equal to the energy of the given Stark level E_n. We see that $\rho(z)$ reaches a maximum near the turning points $|z - z_0| = \Delta_m/4e\mathcal{I}$ corresponding to electron reflection from the band edge. In accordance with (3.158, 163), $\rho(z)$ decays in the region of forbidden energies exponentially as $\exp\{-2/d[(z-z_0)^2 - (\Delta_m/4e\mathcal{I})^2]^{1/2}\}$. This penetration of the mini-band wave functions into the forbidden region results in interband tunneling. Knowing the wave functions in the forbidden region, i.e., for imaginary values of q, one can calculate the interband tunneling probability in terms of a model taking simultaneously into consideration two or more mini-bands in a manner which is similar to what was done for bulk crystals [3.38, 39]. In a strong electric field, where the condition of smallness of the subband tunneling probability fails, (3.157) becomes invalid. The condition of validity of this equation, just as of the conventional effective-mass approximation, is $e\mathcal{I}d \ll \Delta$, or $T \gg \hbar/\Delta$, $L \gg d$; on the other hand, quantization occurs only if the period T is small compared to the momentum relaxation time, i.e., $T \ll \tau_p$, or $e\mathcal{I}d \gg 2\pi\hbar/\tau_p$. For bulk crystals, these conditions cannot be met simultaneously in practice. Therefore it is in superlattices that the Stark ladder was reliably observed for the first time.

In the opposite limiting case, $e\mathcal{I}d > \Delta$, the electron wave functions are localized at separate wells. The position of a level in each well is practically determined by an equation similar to (3.148), the spacing between levels in the neighboring wells, just as that between the Stark ladder levels, being $e\mathcal{I}d$. In intermediate cases, such as for $e\mathcal{I}d \simeq \Delta$, the wave function of a given well penetrates partially into the neighboring ones.

Numerical calculations offer the possibility of determining the wave-function amplitudes in neighboring wells. In strained superlattices, the mean field is zero so that the potential $V(z)$ remains periodic. In this case, the spectrum can be calculated by setting the wave functions in the wells and barriers in the form (3.147) and using the periodicity condition. To do this, it is sufficient, in accordance with (3.94), to calculate the single-barrier transmission coefficient.

4 Vibrational Spectra of Crystals and Superlattices Electron-Phonon Interaction

The present monograph primarily deals with electronic properties. Information on the vibrational spectra and electron-phonon interaction will mainly be needed in Chap. 8 which is devoted to light scattering. Nevertheless, application of group theory to the classification of vibrational spectra of bulk crystals and superlattices illustrates the power of these methods. Their use permits one, without recourse to concrete calculations, to establish the general pattern of phonon spectra, to find the irreducible representations according to which normal vibrations transform near singular points, to determine how long-range dipole-dipole interaction affects the spectrum, and how the spectrum changes when superlattices form. All these aspects are considered in the first part of the present chapter.

In its second part we shall carry out some simple calculations of superlattice spectra. For the vibrations originating from short-range forces the presence of a superstructure only modifies the bulk crystal spectrum; it can be included by introducing the corresponding boundary conditions. For long-range forces, they account for the appearance of new types of long-wavelength vibrations due to polarization charges forming either in one of the layers or at their boundaries.

Rather than discussing here more precise numerical methods for calculating the phonon spectra of superlattices, we shall refer the reader to some original papers. For the electron-phonon interaction, the theory of symmetry and, in particular, the method of invariants, represent the simplest way of constructing effective Hamiltonians for describing the acoustic and optical vibrations when both short- and long-range forces are to be taken into account.

4.1 Normal Vibrations: Distribution in Irreducible Representations

Each atom in a periodic lattice can be characterized by two indices, namely, the number of the primitive cell, f, and the number of the atom in the cell, κ. Accordingly, we will denote the position of an atom (f, κ) in the crystal by a vector $\mathbf{X}_f^\kappa$

$$\mathbf{X}_f^\kappa = \mathbf{X}_f + \mathbf{X}_\kappa, \tag{4.1}$$

where $\mathbf{X}_f$ are the coordinates of a given site in cell f, reckoned from a similar site in the cell chosen as the origin and $\mathbf{X}_\kappa$ the coordinates of atom (f, κ)

reckoned from the corresponding site of cell f. The displacement $\mathbf{u}_{f\kappa}$ of atom (f, κ) can be written in the form

$$\mathbf{u}_{f\kappa} = \sum_{\mathbf{q}} \mathbf{u}_{\kappa}(\mathbf{q}) e^{i\mathbf{q}\mathbf{X}_f}. \tag{4.2}$$

Since the displacements $\mathbf{u}_{f\kappa}$ are real, we can write

$$\mathbf{u}_{\kappa}(-\mathbf{q}) = \mathbf{u}_{\kappa}^{*}(\mathbf{q}).$$

In accordance with (4.2), we obtain

$$\mathbf{u}_{\kappa}(\mathbf{q}) = \frac{1}{N} \sum_{f} \mathbf{u}_{f\kappa} e^{-i\mathbf{q}\mathbf{X}_f}, \tag{4.3}$$

where N is the number of unit cells in the crystal considered to be of finite dimensions.

Since $\mathbf{X}_f$ is one of the vectors of the direct lattice, $\mathbf{u}_{\kappa}(\mathbf{q})$ is a periodic function in the reciprocal space. Therefore the summation in (4.2) is limited to the values of q within the first Brillouin zone. Note that for a finite crystal $\mathbf{q}$ takes on N discrete values. The components $u_{\kappa i}(\mathbf{q})$ make up a basis for the representation of the wave vector group $G_{\mathbf{q}}$ of dimension $3n_a$, where n_a is the number of atoms in the unit cell. It can be expanded into irreducible representations.

Each irreducible representation corresponds to a definite branch of the vibrational spectrum with the frequency $\Omega_{\nu}(\mathbf{q})$. We will denote in what follows with the index ν each of the branches of the spectrum bearing in mind that at high symmetry points of the Brillouin zone there may be degeneracy, that is, the same frequency may correspond to several branches. Just as in the case of an electron spectrum, the degree of degeneracy is determined by the dimension of the corresponding irreducible representation of the group $G_{\mathbf{q}}$, frequencies $\Omega_{\nu}(\mathbf{q})$ at different points of the star $\{\mathbf{q}\}$ being the same. The total number of the branches ν is $3n$. Accordingly, the displacements $\mathbf{u}_{f\kappa}$, by (4.2), can be written

$$\mathbf{u}_{f\kappa} = \sum_{\nu,q} \left[a_{\mathbf{q}\nu} \mathbf{e}_{\kappa}^{\nu}(\mathbf{q}) e^{i\mathbf{q}\mathbf{X}_f} + a_{\mathbf{q}\nu}^{*} e_{\kappa}^{\nu *}(\mathbf{q}) e^{-i\mathbf{q}\mathbf{X}_f} \right]. \tag{4.4}$$

The quantities $a_{\mathbf{q}\nu}$ are called normal vibrations and form the basis of the corresponding irreducible representations of the group $G_{\mathbf{q}}$, the polarization vectors $\mathbf{e}_{\kappa}^{\nu}(\mathbf{q})$ determining the displacement of the atom κ in vibrations of the corresponding mode.

The components $B_{f\kappa,i}^{\nu\mathbf{q}} = (M_{\kappa}/M_0 N)^{1/2} \mathrm{e}_{\kappa,i}^{\nu}(\mathbf{q}) e^{i\mathbf{q}\mathbf{X}_f}$, where M_{κ} is the mass of the atom κ, and $M_0 = \Sigma_{\kappa} M_{\kappa}$, form a matrix of dimension $3nN \times 3nN$ whose rows are given by $3n$ indices ν and N values of $\mathbf{q}$, and the columns, by $3n$ indices κ and i and N values of f. This matrix provides a transition from $3nN$ displacements $u_{f\kappa,i}$ to $3nN$ normal vibrations $a_{\mathbf{q}\nu}$. The matrix $B_{f\kappa,i}^{\nu\mathbf{q}}$ is orthogonal in rows and columns and, accordingly, the polarization vectors relating to different branches ν or different atoms κ in a cell are orthogonal too:

$$\sum_{\kappa} M_{\kappa} \mathbf{e}_{\kappa}^{\nu}(\mathbf{q}) \mathbf{e}_{\kappa}^{\nu' *}(\mathbf{q}) = M_0 \delta_{\nu\nu'}, \tag{4.5a}$$

$$\sum_{\nu} (M_{\kappa} M_{\kappa'})^{1/2} e_{\kappa,i}^{\nu}(\mathbf{q}) e_{\kappa',i'}^{\nu *}(\mathbf{q}) = M_0 \delta_{\kappa\kappa'} \delta_{ii'}. \tag{4.5b}$$

In quantum-mechanical formalism, the amplitudes $a_{\mathbf{q}\nu}$ and $a_{\mathbf{q}\nu}^*$ are replaced by the operators $a_{\mathbf{q}\nu}$ and $a_{\mathbf{q}\nu}^+$:

$$a_{\mathbf{q}\nu} = \left(\frac{\hbar}{2\rho \Omega_{\mathbf{q}\nu} V} \right)^{1/2} b_{\mathbf{q}\nu}, \quad a_{\mathbf{q}\nu}^+ = \left(\frac{\hbar}{2\rho \Omega_{\mathbf{q}\nu} V} \right)^{1/2} b_{\mathbf{q}\nu}^+, \tag{4.6}$$

where $\rho = N M_0 / V$ is the density, V is the crystal volume, and b^+ and b are the phonon creation and annihilation operators whose matrix elements can be written

$$\langle n' | b | n \rangle = n^{1/2} \delta_{n',n-1}, \quad \langle n' | b^+ | n \rangle = (n+1)^{1/2} \delta_{n',n+1}. \tag{4.7}$$

Here n and n' are the numbers of phonons of a given mode $\mathbf{q}$, ν in the initial and final states. Accordingly

$$\mathbf{u}_{f\kappa} = \left(\frac{\hbar}{2\rho V} \right)^{1/2} \sum_{\mathbf{q},\nu} \Omega_{\mathbf{q}\nu}^{-1/2} \left[\mathbf{e}_{\kappa}^{\nu}(\mathbf{q}) b_{\mathbf{q}\nu} \mathrm{e}^{\mathrm{i}\mathbf{q}\mathbf{X}_f} + \mathbf{e}_{\kappa}^{\nu *}(\mathbf{q}) b_{\mathbf{q}\nu}^+ \mathrm{e}^{-\mathrm{i}\mathbf{q}\mathbf{X}_f} \right]. \tag{4.8}$$

In the basis of normal vibrations, the potential energy of the phonon system is diagonal

$$U = 2 M_0 N \sum_{\mathbf{q},\nu} \Omega_{\mathbf{q}\nu}^2 a_{\mathbf{q}\nu} a_{-\mathbf{q}\nu} \tag{4.9}$$

and, accordingly, we obtain for the quantum-mechanical Hamiltonian

$$\mathcal{H} = \sum_{\mathbf{q},\nu} \hbar \Omega_{\mathbf{q}\nu} \left(b_{\mathbf{q}\nu}^+ b_{\mathbf{q}\nu} + \frac{1}{2} \right) \tag{4.10}$$

and for its eigenvalues

$$E_{n,\mathbf{q}\nu} = \hbar \Omega_{\mathbf{q}\nu} \left(n + \frac{1}{2} \right). \tag{4.11}$$

To determine the irreducible representations according to which the normal vibrations $a_{\mathbf{q}\nu}$ transform, one has to find the characters of the representation of the wave vector group which forms the basis of the components $u_{\kappa,i}(\mathbf{q})$ defined by (4.3).

According to (2.3), the operation $g \in G_{\mathbf{q}}$ yields

$$g^{-1} u_{\kappa,i}(\mathbf{q}) = \sum_{\kappa',i'} D_{\kappa'\kappa,i'i}(g) u_{\kappa',i'}(\mathbf{q}).$$

The fact that g preserves $\mathbf{q}$ to within the reciprocal lattice vector is included here. The matrix $D_{\kappa'\kappa,i'i}$ represents a direct product of the matrix $R_{ii}(g)$ determining

the transformation of the vector components u_i, i.e., of components Y_m^l with $l = 1$, $m = 0, \pm 1$, and of the matrix $D_{\kappa'\kappa}$ identifying the position κ' to which the atom κ is sent by the operation g.
For the character we will have

$$\chi(g) = \chi_R(g) \sum_{\kappa\kappa'} D_{\kappa'\kappa}(g)\delta_{\kappa'\kappa} = \chi_R(g) \sum_{\kappa} D_{\kappa\kappa}(g). \tag{4.12}$$

For an element $g = (r|\tau)$, $\chi_R(g) = \chi_R(r)$. In accordance with (2.17), under rotation c_φ

$$\chi_R(c_\varphi) = 1 + 2\cos\varphi. \tag{4.13a}$$

Since $s_\varphi = c_{\varphi+\pi} c_i$ and $R_{m'm}(c_i) = -\delta_{m'm}$ we have

$$\chi_R(s_\varphi) = -\chi_R(c_{\varphi+\pi}) = -1 + 2\cos\varphi, \tag{4.13b}$$

$$\chi_R(c_i) = -3, \quad \chi_R(\sigma) = \chi_R(s_0) = 1. \tag{4.13c}$$

Let us find now the trace of the matrix $D_{\kappa'\kappa}$. By applying the operation $g = (r|\tau)$, we obtain for the displacement of the atom (f, κ):

$$(r|\tau)^{-1}(\mathbf{X}_f + \mathbf{X}_\kappa) - (\mathbf{X}_f + \mathbf{X}_\kappa) = r^{-1}(\mathbf{X}_\kappa - \tau) - \mathbf{X}_\kappa + (r^{-1}\mathbf{X}_f - \mathbf{X}_f).$$

The exponent can be written

$$\mathbf{q}(r^{-1}\mathbf{X}_f - \mathbf{X}_f) = \mathbf{X}_f(r\mathbf{q} - \mathbf{q}) = 2\pi m \quad (m = 0, \pm 1, \ldots),$$

since $r\mathbf{q} - \mathbf{q}$ for $g \in G_\mathbf{q}$ is equal to zero or one of the reciprocal lattice vectors and $\mathbf{X}_f$ one of the direct lattice vectors. It is thus sufficient to consider only the displacement

$$\delta\mathbf{X}_\kappa = r^{-1}(\mathbf{X}_\kappa - \tau) - \mathbf{X}_\kappa. \tag{4.14}$$

The following variants can be envisaged here:

1) Atom κ does not leave the site, i.e., $\delta\mathbf{X}_\kappa = 0$.
2) Atom κ transfers to an equivalent site in one of the neighboring cells, i.e., $\delta\mathbf{X}_\kappa$ is equal to one of the direct lattice vectors $\mathbf{a}_i$.
3) Atom κ transfers to an inequivalent site, i.e., $\delta\mathbf{X}_\kappa \neq \mathbf{a}_i$.

In the first two cases $D_{\kappa\kappa} = e^{i\mathbf{q}\cdot\delta\mathbf{X}_\kappa}$, with $D_{\kappa\kappa} = 1$ in the first case. In the third case $D_{\kappa\kappa} = 0$. Hence

$$\chi(g) = \chi_R(g) \sum_{\kappa}{}' \exp\left\{-i\mathbf{q} \cdot \left[r^{-1}(\mathbf{X}_\kappa - \tau) - \mathbf{X}_\kappa\right]\right\} \delta_{g\kappa,\kappa}, \tag{4.15}$$

where $\delta_{g\kappa,\kappa} = 1$ for $\delta\mathbf{X}_\kappa = \mathbf{a}_i$, and $\delta_{g\kappa,\kappa} = 0$ for $\delta\mathbf{X}_\kappa \neq \mathbf{a}_i$. For the points with $r\mathbf{q} = \mathbf{q}$, in particular for points inside the Brillouin zone, $\mathbf{q}\cdot(r^{-1}(\mathbf{X}_\kappa - \tau) - \mathbf{X}_\kappa) = -\mathbf{q}\cdot\tau$, and

$$D(g) = R(g) n_g e^{-i\mathbf{q}\cdot\tau}, \tag{4.16}$$

where n_g is the number of atoms left in an equivalent position under the operation g^{-1}. Note that the character of the corresponding projective representation (see sect. 2.7)

$$\chi(r) = \chi_R(r) n_g. \tag{4.17}$$

For lattices which do not contain identical atoms in the primitive cell, one can write for $r\mathbf{q} = \mathbf{q}$:

$$\chi(r) = \chi_R(r) n_a, \tag{4.18}$$

where n_a is the number of atoms in the primitive cell. This implies that the projective representation in this case is p – equivalent to the vector one and contains only irreducible representations contained in $R(r)$ n_a times each.

Just as with the electronic states, the condition of invariance under time inversion can result in combination of the representations, i.e., in the doubling of the degeneracy. Since the displacements $u_{f\kappa,i}$ are real, only cases a and b can be realized for the irreducible representations contained in (4.16) (sect. 2.8). In cases a_1 and a_2 the functions ψ and $K\psi$ are linearly dependent, the characters of the corresponding representations satisfying condition (2.34). In cases b_1 and b_2 the functions ψ and $K\psi$ are linearly independent, and the representations μ and λ, whose characters are connected through relation (2.35), are combined. In case b_3 the points $\mathbf{q}$ and $-\mathbf{q}$ correspond to different stars, and the invariance under time inversion combines these stars and imposes the condition of equality of the frequencies $\Omega_\nu(\mathbf{q})$ and $\Omega_\nu(-\mathbf{q})$, which naturally is met in the other cases $a_{1,2}$ and $b_{1,2}$ as well.

We will use now the above relations to determine the irreducible representations according to which normal vibrations transform in the high symmetry points of $\mathbf{q}$ space in diamond (Ge, Si) and zincblende (GaAs and other A_3B_5 compounds) lattices, as well as in the Ge/Si-and GaAs/AlAs-type superlattices. The primitive cell of these crystals contains two atoms. Choosing one of them as the origin, the coordinate of the second will be $(a_0/4)(111)$, where a_0 is the lattice constant, the vectors of elementary translations being $\mathbf{a}_1 = (a_0/2)(110)$, $\mathbf{a}_2 = (a_0/2)(101)$, $\mathbf{a}_3 = (a_0/2)(011)$. Accordingly, the phonon spectrum has six branches. Just as in any crystal, three of them are acoustic, which implies that for them $\Omega_\nu(\mathbf{q}) \to 0$ as $\mathbf{q} \to 0$, and, correspondingly, for $\mathbf{q} = 0$ the components a_ν transform as components of a polar vector, so that $\chi(g) = \chi_R(g)$. In a diamond-type lattice both atoms are identical, and out of 48 elements of the factor group $(r|\boldsymbol{\tau})$ which are isomorphic with the group O_h, for 24 elements $\boldsymbol{\tau} = 0$, and for the others, $\boldsymbol{\tau} = \boldsymbol{\tau}_0 = (a_0/4)(111)$. These 24 elements send both atoms into inequivalent positions and, for them, according to (4.14), $\chi(r|\boldsymbol{\tau}_0) = 0$. The elements $(r|0)$ leave the first atom in its place and displace the second one by one of the vectors $\mathbf{a}_i$. Table 4.1 specifies these displacements and lists the values of $\chi_R(g)$ and $\chi(g)$ for arbitrary $\mathbf{q}$, as well as for the points $\Gamma(0,0,0)$ and X, i.e., for $\mathbf{q} = (2\pi/a_0(100))$ for all elements contained in the corresponding wave vector group. For the points Δ and Λ lying on the axes [100] and [111], as well as for the point L, in accordance with (4.16), $\chi(r|0) = 2\chi_R(r)$. The zincblende lattice contains two different atoms, and its factor group isomorphic with the group T_d contains the same 24 elements $(r|0)$ given in Table 4.1. Using the tables of characters of the point groups (Table 2.3) and of the space group $G_{\mathbf{q}}$ at point X (Table 3.3), one can expand the corresponding representations into irreducible ones. The results are shown in Table 4.2. The relation between the representations of the corresponding point groups and of the wave vector groups at the points Γ, X, L, Λ and Δ is presented in Tables 3.1, 2. By using the compatibility conditions, one can readily establish the character of the spectral branch splitting as one moves away from higher symmetry points. To do this, one should expand the irreducible representations of the group G at this point into irreducible representations of the wave vector group at a low symmetry point. For example, as one crosses over from point Γ to point Δ the threefold degenerate representation Γ_{15} corresponding to the acoustic branch splits into Δ_1 and Δ_5, while the representation Γ_{25} corresponding to optical vibrations in the diamond lattice splits into Δ_4 and Δ_5. Note that the nondegenerate representations Δ_1 and Δ_4 correspond to longitudinal, and the doubly degenerate Δ_5, to transverse vibrations.

In considering the compatibility conditions at the point X one should bear in mind that the character of the representation of the element $(r|\tau)$ belonging to the wave vector group at point Δ differs from that of the element r of the

Table 4.1. Characters of representation D_u for a diamond lattice

Group element	$\chi_1(r)$	Displacement of atom	Characters: arbitrary $\mathbf{q}$	at Γ (000)	at $X\left(\frac{2\pi}{a_0}(100)\right)$	at $L\left(\frac{\pi}{a_0}(111)\right)$
$(e\|0)$	3	0	6	6	6	6
$(c_{2x}\|0)$	−1	$\mathbf{a}_3$	$-1-e^{i\mathbf{q}\cdot\mathbf{a}_3}$	−2	−2	–
$(c_{2y}\|0)$	−1	$\mathbf{a}_2$	$-1-e^{i\mathbf{q}\cdot\mathbf{a}_2}$	−2	0	–
$(c_{2z}\|0)$	−1	$\mathbf{a}_1$	$-1-e^{i\mathbf{q}\cdot\mathbf{a}_1}$	−2	0	–
$(s_{4x}\|0)$	−1	$\mathbf{a}_1$	$-1-e^{i\mathbf{q}\cdot\mathbf{a}_1}$	−2	0	–
$(s_{4y}\|0)$	−1	$\mathbf{a}_3$	$-1-e^{i\mathbf{q}\cdot\mathbf{a}_3}$	−2	–	–
$(s_{4z}\|0)$	−1	$\mathbf{a}_2$	$-1-e^{i\mathbf{q}\cdot\mathbf{a}_2}$	−2	–	–
$(s_{4x}^3\|0)$	−1	$\mathbf{a}_2$	$-1-e^{i\mathbf{q}\cdot\mathbf{a}_2}$	−2	0	–
$(s_{4y}^3\|0)$	−1	$\mathbf{a}_1$	$-1-e^{i\mathbf{q}\cdot\mathbf{a}_1}$	−2	–	–
$(s_{4z}^3\|0)$	−1	$\mathbf{a}_3$	$-1-e^{i\mathbf{q}\cdot\mathbf{a}_3}$	−2	–	–
$(\sigma_{xy}\|0)$	1	0	2	2	–	2
$(\sigma_{x\bar{y}}\|0)$	1	$\mathbf{a}_1$	$1+e^{i\mathbf{q}\mathbf{a}_1}$	2	–	–
$(\sigma_{xz}\|0)$	1	0	2	2	–	2
$(\sigma_{x\bar{z}}\|0)$	1	$\mathbf{a}_2$	$1+e^{i\mathbf{q}\mathbf{a}_2}$	2	–	–
$(\sigma_{yz}\|0)$	1	0	2	2	2	2
$(\sigma_{y\bar{z}}\|0)$	1	$\mathbf{a}_3$	$1+e^{i\mathbf{q}\mathbf{a}_3}$	2	2	–
$8(c_3\|0), c_3^2\|0$	0	–	0	0	0	0
$24(gi\|\tau)$	–	Goes to inequivalent position	0	0	0	0

Table 4.2. Distribution of vibrations according to irreducible representations of $G_{\mathbf{k}}$

Type of crystal	Γ ac.	Γ opt.	**X**	Λ	L	Δ	Σ, K
Ge, Si	Γ_{15}	Γ'_{25}	X_1, X_2, X_4	$2\Lambda_1, 2\Lambda_3$	L_1(LO), L'_2(LA), L_3(TA), L'_3(TO)	Δ_1(LA), Δ_2(LO), $2\Delta_5$(TO, TA)	$2\Sigma_1, 2\Sigma_3, \Sigma_4, \Sigma_5$
GaAs	Γ_{15}	Γ_{15}	$X_1, X_4, 2X_5$	$2\Lambda_1, 2\Lambda_3$	$2L_1, 2L_3$	$2\Delta_1, 2\Delta_3, 2\Delta_4$	$4\Sigma_1, 2\Sigma_2$

point group C_{4v}, i.e., from the character of the projective representation of the element $(r|\tau)$, in the factor $e^{-i\mathbf{q}\cdot\tau}$. Close to the point X, $\mathbf{q}\cdot\tau = \pi/2$.

One can readily check that at X the representation X_1 splits into Δ_1 and Δ_4, i.e., the acoustic and optical branches merge, while X_2 and X_4 transform to Δ_5. Therefore which of these representations corresponds to acoustic or optical vibrations can be established only by numerical calculations. For instance, for Ge the representation X_2 corresponds to acoustic, and X_4, to optical vibrations.[1]

In zincblende-type crystals, the representation Γ_{15} splits into three one-dimensional representations as one transfers to the point Δ. At the point X, Δ_1 transforms to X_1 or X_3, and Δ_3 and Δ_4 combine to form a two-dimensional representation X_5. Prior to considering the distribution of normal vibrations in irreducible representations in superlattices, let us first establish the space groups that can occur in different types of superlattices.

4.1.1 GaAs/AlAs-type superlattices [4.1–4]

Let us take a [001]-growth-axis lattice. In this case, the three above mentioned elementary translation vectors $\mathbf{a}_1, \mathbf{a}_2, \mathbf{a}_3$ can be combined into two vectors orthogonal to the [001] axis:

$$\mathbf{t}_1 = \mathbf{a}_1 = \frac{a_0}{2}(110), \mathbf{t}_2 = \mathbf{a}_2 - \mathbf{a}_3 = \frac{a_0}{2}(1\bar{1}0) \tag{4.19}$$

with $\mathbf{t}_1 \perp \mathbf{t}_2$. As for the third elementary translation $\mathbf{t}_3$ it depends on the numbers l and n of monomolecular layers in layers A and B. If $l+n$ is even, then one can choose

$$\mathbf{t}_3^{(1)} = \frac{l+n}{2}(\mathbf{a}_3 + \mathbf{a}_2 - \mathbf{a}_1) = \frac{a_0}{2}(0, 0, l+n). \tag{4.20}$$

In this case $\mathbf{t}_3 \perp \mathbf{t}_1, \mathbf{t}_2$ is the primitive cell containing at least two pairs of atoms. Under these conditions, the Bravais lattice is a simple lattice of the orthogonal system $T_0(O)$ and crystal class D_{2d}, the corresponding space group containing no nontrivial translations since it is a symmorphic group $D_{2d}^5(P\bar{4}m2)$. If, however, $l+n$ is odd, then

$$\mathbf{t}_3^{(2)} = \frac{l+n-1}{2}(\mathbf{a}_3 + \mathbf{a}_2 - \mathbf{a}_1) + \mathbf{a}_2 = \frac{a_0}{2}(1, 0, l+n). \tag{4.21}$$

In this case the primitive cell, just as in a bulk crystal, can contain one pair of atoms ($l+n=1$) and the Bravais lattice is body-centered, $\Gamma_0^v(O_v)$, the space group being the symmorphic group $D_{2d}^9(I\bar{4}m2)$.

In the [111]-growth-axis superlattice, one may conveniently convert to the coordinate frame $x' \| [1\bar{1}0]$, $y' \| [11\bar{2}]$, $z' \| [111]$ and choose

$$\mathbf{t}_1 = \mathbf{a}_1 - \mathbf{a}_2 = \frac{a_0}{2\sqrt{6}}(-\sqrt{3}, 3, 0), \mathbf{t}_2 = \mathbf{a}_1 - \mathbf{a}_3 = \frac{a_0}{2\sqrt{6}}(\sqrt{3}, 3, 0). \tag{4.22}$$

[1] As already pointed out, one has to bear in mind that different researchers use different notations for irreducible representations.

For $l+n=3m$ one can take

$$\mathbf{t}_3^{(1)} = \frac{l+n}{3}(\mathbf{a}_1+\mathbf{a}_2+\mathbf{a}_3) = \frac{a_0}{\sqrt{3}}(0,0,l+n). \tag{4.23}$$

In this case, the Bravais lattice belongs to the hexagonal system Γ_h, the corresponding space group being $C_{3v}^1(P3m1)$. If, however, $l+n \neq 3m$, then we will have, accordingly

$$\begin{aligned} l+n &= 3m+1 \\ \mathbf{t}_3^{(2)} &= \frac{l+n-1}{3}(\mathbf{a}_1+\mathbf{a}_2+\mathbf{a}_3)+\mathbf{a}_1 \\ &= \frac{a_0}{\sqrt{6}}(0,1,\sqrt{2}(l+n)), \end{aligned} \tag{4.24}$$

$$\begin{aligned} l+n &= 3m+2 \\ \mathbf{t}_3^{(3)} &= \frac{l+n-2}{3}(\mathbf{a}_1+\mathbf{a}_2+\mathbf{a}_3)+\mathbf{a}_2+\mathbf{a}_3 \\ &= \frac{a_0}{\sqrt{6}}(0,-1,\sqrt{2}(l+n)). \end{aligned} \tag{4.25}$$

In both cases, the Bravais lattices belong to the rhombohedral (trigonal) system Γ_{rh}, which can be readily checked by choosing $\tilde{\mathbf{t}}_1$, $\tilde{\mathbf{t}}_2$ and $\tilde{\mathbf{t}}_3$ such that $\tilde{t}_1=\tilde{t}_2=\tilde{t}_3$, and $(\tilde{\mathbf{t}}_1\cdot\tilde{\mathbf{t}}_2)=(\tilde{\mathbf{t}}_2\cdot\tilde{\mathbf{t}}_3)=(\tilde{\mathbf{t}}_3\cdot\tilde{\mathbf{t}}_2)$, namely:
for $l+n=3m+1$:

$$\tilde{\mathbf{t}}_1=\mathbf{t}_3^{(2)}-\mathbf{t}_1, \quad \tilde{\mathbf{t}}_2=\mathbf{t}_3^{(2)}-\mathbf{t}_2, \quad \tilde{\mathbf{t}}_3=\mathbf{t}_3^{(2)}, \tag{4.26}$$

for $l+n=3m+2$:

$$\tilde{\mathbf{t}}_1=\mathbf{t}_3^{(3)}+\mathbf{t}_1, \quad \tilde{\mathbf{t}}_2=\mathbf{t}_3^{(3)}+\mathbf{t}_2, \quad \tilde{\mathbf{t}}_3=\mathbf{t}_3^{(3)}.$$

The corresponding space group is $C_{3v}^5(R3m)$.

[110]-growth-axis superlattice: choosing $x'\|[1\bar{1}0]$, $y'\|[001]$, $z'\|[110]$, one can set for even $l+n$:

$$\mathbf{t}_1=\mathbf{a}_2-\mathbf{a}_3=\frac{a_0}{\sqrt{2}}(100), \quad \mathbf{t}_2=\mathbf{a}_2+\mathbf{a}_3-\mathbf{a}_1=a_0(010) \tag{4.27}$$

and

$$\mathbf{t}_3^{(1)}=\frac{l+n}{2}\mathbf{a}_1=\frac{a_0}{2\sqrt{2}}(0,0,l+n).$$

All three vectors $\mathbf{t}_i$ are orthogonal, and the Bravais lattice is a simple lattice of the orthogonal system $\Gamma_0(O)$, the space group being C_{2v}^1 (Pmm2). For odd $l+n$ we have

$$\mathbf{t}_3^{(2)}=\frac{l+n-1}{2}\mathbf{a}_1+\mathbf{a}_2=\frac{a_0}{\sqrt{2}}(1,\sqrt{2},l+n). \tag{4.28}$$

In this case, the Bravais lattice is body centered, of the same system $\Gamma_o^v(O_v)$, and the space group is C_{2v}^{20} (Imm2).

For the Ge/Si superlattices grown along the [001], [111] and [110] directions, the Bravais lattice systems are the same as for the A_3B_5 lattices, the space groups depending on the numbers of monomolecular layers of the atoms of Ge(l) and Si(n) (in the case of $l = n = 1$ the Ge/Si superlattice is an analog of bulk A_3B_5) [4.2, 3] (Table 4.3).

Consider now the distribution of normal vibrations in irreducible representations in GaAs/AlAs-type superlattices. The total number of atoms in the primitive cell being $2(l+n)$, the number of the spectral branches is $6(l+n)$. When acted upon by the e, c_2 and $2\sigma_d$ operations of the group D_{2d}, all atoms in the [001]-growth axis lattices remain in place or displace by one of the direct lattice vectors, $\mathbf{t}_1$ or $\mathbf{t}_2$ (or by their combinations), and under the operations $2s_4$ and $2u_2$ one of the atoms remains in place, and another shifts by $\mathbf{t}_3$, the others transferring to inequivalent positions. Therefore at the point Γ we will have for these

Table 4.3. Space symmetry of Ge/Si superlattices

(001) Ge/Si superlattices

$l+n=4$ m	l, n odd	D_{2d}^5
Γ_0 system	l,n even	D_{2h}^5
$l+n=2(m+1)$	l,n odd	D_{2d}^9
Γ_0^v system	l,n even	D_{2h}^{28}
$l+n=2m+1$ Γ_0^v system		D_{4h}^{19}

For even $l+n$, the number of atoms in the unit cell is $l+n$ and for odd $l+n$, $2(l+n)$. In the latter, two layers of Ge and Si atoms fit into the period $\mathbf{t}_3$, and there is a nontrivial translation $\tau = \frac{1}{2}\mathbf{t}_3(\mathbf{t}_3 \parallel [001])$.

(111) Ge/Si superlattices

$l+n=3m$	l and/or n even	D_{3d}^3
Γ_h system	l and n odd	C_{3v}^1
$l+n=3m+1$, $3m+2$	l and/or n even	D_{3d}^5
Γ_{rh} system	l and n odd	C_{3v}^5

In both cases these space groups are symmorphic, the number of atoms in the unit cell being $l+n$ for even $l+n$ and $2(l+n)$ for odd $l+n$.

[110] Ge/Si superlattices

$l+n$ even	l, n odd	D_{2h}^5
Γ_0 system	l, n even	D_{2h}^7
$l+n$ odd Γ_0^v system		D_{2h}^{28}

In all cases the number of atoms in the unit cell is $2(l+n)$

operations, according to (4.15, 16), $\chi_u(g) = 2\chi_R(g)$, whereas at the point Z at the boundary of the Brillouin zone, i.e., for $\mathbf{q} = \mathbf{b}_3/2 = \pi[\mathbf{t}_1\mathbf{t}_2]/\Omega_0$, $\chi_u(g) = 0$. Accordingly, at the Γ point

$$D_u = (l+n-1)\Gamma_1 + (l+n+1)\Gamma_2 + 2(l+n)\Gamma_3. \quad (4.29)$$

At the point Z

$$D_u = (l+n)(Z_1 + Z_2 + 2Z_5) \quad (4.30)$$

($\Gamma_1, \Gamma_2, \Gamma_5$ or Z_1, Z_2, Z_5 correspond to the representations A_1, B_2, and E of the group D_{2d}).

In the lattices grown along [111] and [110], all operations of the group c_{3v} or c_{2v} leave all atoms in place or shift them by one of the vectors $\mathbf{t}_1$ or $\mathbf{t}_2$ (or by their combinations). Therefore, according to (4.15, 16), $\chi_u(g) = 2(l+n)\chi_R(g)$, so that in a [111]-growth-axis lattice we will have at the Γ point

$$D_u = 2(l+n)(\Gamma_1 + \Gamma_3) \quad (4.31)$$

and at the point Z, i.e., at the boundary of the Brillouin zone, for $\mathbf{q} = \mathbf{b}_3/2$

$$D_u = 2(l+n)(Z_1 + Z_3) \quad (4.32)$$

(Γ_1, Γ_3 or Z_1, Z_3 correspond to the representations A_1 and E of the group C_{3v}). In the lattice grown along [110], we have at the point Γ:

$$D_u = 2(l+n)(\Gamma_1 + \Gamma_3 + \Gamma_4) \quad (4.33)$$

and at the point Z (for $\mathbf{q} = \mathbf{b}_3/2$):

$$D_u = 2(l+n)(Z_1 + Z_3 + Z_4) \quad (4.34)$$

($\Gamma_1, \Gamma_3, \Gamma_4$ or Z_1, Z_3, Z_4 correspond to the representations A_1, B_1 and B_2 of the group C_{2v}).

4.2 Vibrational Spectra of Superlattices

Similar to the electronic spectra, the phonon spectra of superlattices can be calculated by two radically different methods. In the first of them, the superlattice is considered as a new crystal containing a large number of atoms in the primitive cell proportional to the number $l+n$ of monolayers or monomolecular layers. Such methods are usually employed when calculating spectra of short-period superlattices. The second method assumes the vibrational spectrum of each layer to be the same as that of the bulk crystal, the effect of the neighboring layer being felt only in a narrow near-boundary region, which is taken into account by introducing the corresponding boundary conditions. Basically, this method is similar to the envelope method used with electronic spectra. We will consider here the latter. Readers interested in numerical calculations of spectra can refer to the reviews [4.5–8].

4.2.1 Acoustical Vibrations

When calculating long-wavelength acoustic vibrational spectra, the crystal may be considered a homogeneous elastic medium characterized by the elasticity moduli C_{ij}. This model was used to calculate the spectrum of superlattices by *Rytov* as far back as 1956 [4.9]. For waves propagating along the principal axes of a cubic crystal, no mixing of longitudinal and transverse vibrations occurs, the wave vector of the corresponding vibrations being $k^i_{\mathrm{l,t}} = \Omega/v^i_{\mathrm{l,t}}$ in each of the layers A and B. Here Ω is the vibration frequency, $v_{\mathrm{l,t}}$ the velocity of longitudinal (l) or transverse (t) waves:

$$v^2_{\mathrm{l,t}} = C_{\mathrm{l,t}}/\rho, \tag{4.35}$$

where ρ is the crystal density and $C_{\mathrm{l,t}}$ the corresponding combination of the elastic moduli depending on the $\mathbf{k}$ direction. The values of $C_{\mathrm{l,t}}$ for the vector $\mathbf{k}$ directed along the principal axes are listed in Table 4.4.

In each of the layers A and B, the displacements u_z or u_x, u_y can be written

$$\begin{aligned} u_A &= A_1\cos(k_1 z) + A_2\sin(k_1 z)(0 < z < a), \\ u_B &= A_3\cos[k_2(z-a)] + A_4\sin[k_2(z-a)](a < z < d). \end{aligned} \tag{4.36}$$

In a similar way one can represent the displacements in the layer A located at $d < z < d + a$. Note that

$$u(z+d) = u(z)\mathrm{e}^{\mathrm{i}q_z d}.$$

The boundary conditions require that at the layer boundaries the displacements u_A and u_B and the stresses $X^{A,B}_{zz}$ for longitudinal, and X^{AB}_{xz}, X^{AB}_{yz} for transverse vibrations be equal, where

$$X_{zz} = (C_{11} - C_{12})\sum_i \frac{\partial u_i}{\partial x_i} + (C_{\mathrm{l}} - C_{11} + C_{12})\frac{\partial u_z}{\partial z},$$

$$X_{xz} = C_{\mathrm{t}}\left(\frac{\partial u_x}{\partial z} + \frac{\partial u_z}{\partial x}\right), X_{yz} = C_{\mathrm{t}}\left(\frac{\partial u_y}{\partial z} + \frac{\partial u_z}{\partial y}\right).$$

As a result, we obtain four coupled equations for the coefficients A_i, their determinant differing from expression (3.92) only in the replacement $k \to k_1, \lambda \to \mathrm{i}k_2$. Note that, in accordance with (3.92a), $R = \mathrm{i}\eta_{\mathrm{l,t}}$, where $\eta_{\mathrm{l,t}} = (C^B_{\mathrm{l,t}}/C^A_{\mathrm{l,t}})(k_2/k_1)$ and

$$\cos(q_z d) = \cos(k_1 a)\cos(k_2 b) - F\sin(k_1 a)\sin(k_2 b), \tag{4.37}$$

Table 4.4. Effective elastic moduli C_{l} and C_{t} for cubic crystals

k	[001]	[111]	[110]
C_{l}	C_{11}	$\frac{1}{3}(C_{11}+2C_{12}+4C_{44})$	$\frac{1}{2}(C_{11}+C_{12}+2C_{44})$
C_{t}	C_{44}	$\frac{1}{3}(C_{11}-C_{12}+C_{44})$	$C_{44}(\mathbf{u}\Vert[001]), \frac{1}{2}(C_{11}-C_{12})(\mathbf{u}\Vert[110])$

where

$$F_{\mathrm{l,t}} = \frac{1}{2}\left(\eta_{\mathrm{l,t}} + \eta_{\mathrm{l,t}}^{-1}\right) = \frac{1}{2}\left(\frac{\rho_A v_{\mathrm{l,t}}^A}{\rho_B v_{\mathrm{l,t}}^B} + \frac{\rho_B v_{\mathrm{l,t}}^B}{\rho_A v_{\mathrm{l,t}}^A}\right).$$

In the isotropic approximation, i.e., for $C_{11} - C_{12} = 2C_{44}$, the velocities of the longitudinal and transverse waves do not depend on the direction of **q**, (4.37) being valid for any orientation of the layers if **q** is directed along the normal **n** to the layer boundary. For a normal transverse wave, for which the displacement $\mathbf{u} \perp \mathbf{q}, \mathbf{n}$ but **q** is not parallel to **n**, it also remains valid for any direction of **q**. Under these conditions, we have in (4.37) $k_1^2 = k_{01}^2 - k_\perp^2$, $k_2^2 = k_{02}^2 - k_\perp^2$, where $k_{0i} = \Omega/v_i$, and $\mathbf{k}_\perp$ is the component of **q** perpendicular to **n**. The remaining two modes are mixed transverse-longitudinal modes, their spectrum for arbitrary **q** given by a more complex equation.

The densities and sound velocities in the layers A and B are usually not much different, i.e., $|F-1| \ll 1$. Therefore (4.37) can be conveniently rewritten in the form

$$\cos(q_z d) = \cos(k_1 a + k_2 b) - (F - 1)\sin(k_1 a)\sin(k_2 b) \tag{4.38}$$

and solved by iterations in the parameter $F - 1$. For $F - 1 = 0$ we have $q_z = \Omega/v_{\mathrm{SL}} = (a v_A^{-1} + b v_B^{-1})/d$, which means that in the extended zone scheme one obtains the usual acoustic vibration spectrum with $v = v_{\mathrm{SL}}$. The inclusion of corrections of order $F - 1$ results in a splitting of the spectral branches at the points $q_m = \pi m/d$, i.e., at the center ($m = 2, 4, 6 \ldots$) and at the boundary ($m = 1, 3, 5 \ldots$) of the Brillouin zone. According to (4.38), at these points

$$\Omega(q_m) = v_{\mathrm{SL}} q_m \pm \delta, \tag{4.39}$$

where

$$\delta^2 = \frac{1}{2}\frac{(v_A\rho_A - v_B\rho_B)^2}{\bar{\rho}^2 d^2}\left[1 - (-1)^m \cos\frac{\pi m}{d}(a - b)\right].$$

Although (4.38) has been obtained in the elastic continuum approximation and for close elastic moduli C_A and C_B, it can be used as well for an approximate description of the total spectrum of the superlattice acoustic vibrations if one sets in (4.37) $F = 1/2(k_1/k_2 + k_2/k_1)$ and finds $k_1(\Omega)$ and $k_2(\Omega)$ from the spectra of bulk crystals A and B. In the frequency region where the acoustic branches of these crystals overlap, k_1 and k_2 are real, while if one of the branches is absent, the corresponding values of k_i become imaginary, the vibrations being practically localized in one of the layers. The corresponding spectrum may be considered as the acoustic branch of an *averaged* bulk crystal folded within the Brillouin zone of the superlattice, i.e., for $-\pi/d < q_z \leqslant \pi/d$, and split at the center and boundaries of the zone (Fig. 4.1). Note that, strictly speaking, only the first of these branches, for which $\Omega(q) \to 0$ as $q \to 0$, is the acoustic branch for a superlattice, whereas the other branches are optical.

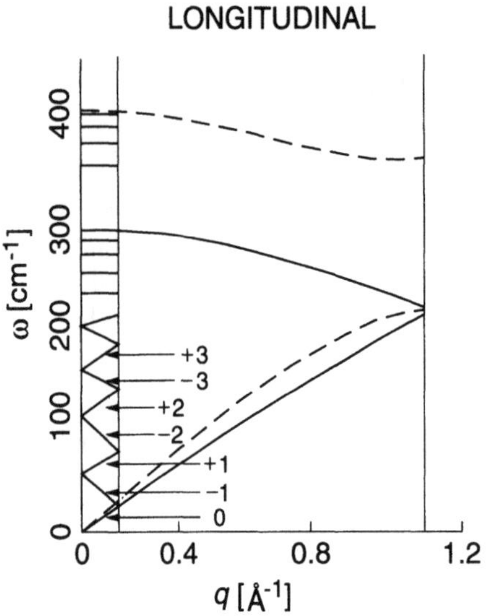

Fig. 4.1. Dispersion curves for longitudinal acoustic and optical phonons in the $(GaAs)_5$ $(AlAs)_4$ superlattice. Solid line: spectrum in bulk GaAs crystals, dotted line: same for AlAs (after ref. [4.10])

4.2.2 Optical Vibrations

One can apply a similar analysis to nonpolar optical vibrations; however, in order to derive the boundary conditions, one will have to use here a certain lattice model. Thus, the GaAs/AlAs [001] superlattice can be considered [4.10] as consisting of isolated chains of atoms GaAsGaAsAlAsAlAs ... with masses M_{11}(Ga), M_{12}(Al), M_2(As). If only interaction with the nearest neighbors is included, and the constants k of interaction between the pairs GaAs and AlAs are assumed to be equal, then one can write the following dispersion relation for individual links of each chain containing only GaAs or only AlAs atoms:

$$1 - \cos k_i a_0 = \Omega^2(\Omega_{0i}^2 - \Omega^2)\frac{M_{1i}M_2}{2\tilde{k}^2}, \tag{4.40}$$

where Ω_{0i} is the optical vibration frequency for a bulk crystal, a_0 is the distance between the nearest pairs of like atoms, GaGa, AlAl, or AsAs. For the relative displacement of the nearest atoms As(u_2) and Ga(u_{11}) or Al(u_{12}), we can write

$$\gamma_i^{\pm} = \frac{u_2}{u_{1i}} = -\frac{k(1 + e^{\pm i k_i a_0})}{M_2\Omega^2 - 2\tilde{k}}. \tag{4.41}$$

The plus sign corresponds to the wave $u \sim e^{ik_i z}$ and the minus, to $u \sim e^{-ik_i z}$. If we assume that the displacements of the atoms Ga or Al, $u_{1A,B}(z)$, in the layers A and B are described by expressions similar to (4.36), that the displacement of the central As atom contained in both sequences is determined by (4.41), and that those of the nearest Ga or Al atoms are equal, i.e.,

$$u_1^A(a_0) - u_1^A(0) = u_1^B(a_0) - u_1^B(0), \tag{4.42}$$

then the determinant of the corresponding system of equations for the coefficients A_i will differ from (4.37) only in the value of η:

$$\eta = \tan(k_2 a_0/2)/\tan(k_1 a_0/2). \tag{4.43}$$

In principle, the corresponding equation, just as (4.40, 43), describes both the acoustic and the optical spectral branches. If the optical branches of the layers A and B overlap, then k_1 and k_2 in (4.37, 43) are real, and if there is no overlap, then one of the values of k_i in the corresponding frequency region is imaginary, the vibrations localizing in one of the regions. In the first case, the corresponding spectrum may be considered a folded optical branch of an averaged crystal, and, in the second, for $\Omega = \Omega^A$, $u_A(z) \sim \sin \pi m z/a$, $u_B = 0$, while for $\Omega = \Omega^B$, $u_B(z) \sim \sin \pi m z/b$; accordingly, the spectrum represents a series of discrete lines with $\Omega_m^{A,B} = \Omega^{A,B}(k_m)$, where $k_m^A = \pi m/a, k_m^B = \pi m/b$ $(m = 1, 2 \ldots)$, the $\Omega_i(k_i)$ dependence determined, for instance, by (4.40). In contrast to the electronic spectra, here the frequency Ω_m decreases with increasing m. These modes are called confined optical phonons.

4.2.3 Dipole–Dipole Interaction

The above expressions do not take into account the variation of the spectrum due to long-range interaction between the dipole moments produced by polar optical vibrations. If normal optical vibrations $a_{\mathbf{q}\nu}$ form for $q \to 0$ a basis for the same representation D_R according to which the polar vector compounds transform, then such vibrations cause a polarization of the medium

$$P_{i\mathbf{q}} = \frac{2}{\Omega_0} \sum_\nu Q_{\nu i} a_{\mathbf{q}\nu}. \tag{4.44}$$

The constants $Q_{\nu i}$ have the dimensionality of charge, Ω_0 being the primitive-cell volume. Such polarization is produced in crystals containing in the primitive cell two or more unlike atoms, or not less than three like ones. Since the induced charge $\rho_{\mathbf{q}} = \operatorname{div} \mathbf{P}_{\mathbf{q}}$, it follows from Poisson's equation that the potential $\varphi_{\mathbf{q}}$ created by polar vibrations is

$$\varphi_{\mathbf{q}} = -4\pi \mathrm{i}(\mathbf{q} \cdot \mathbf{P}_{\mathbf{q}}) / \sum_{ij} \kappa_{ij}^\infty q_i q_j. \tag{4.45}$$

Here, κ^∞ is the high-frequency dielectric constant originating from electronic-shell deformation. The interaction of the field $\varphi_{\mathbf{q}}$ with the charge $\rho_{\mathbf{q}}$ leads to an additional contribution to the potential energy U of the phonon system:

$$\Delta U = \frac{1}{2} N \Omega_0 \sum_q \varphi_{\mathbf{q}} \rho_{-\mathbf{q}} = 2 M_0 N \sum_{\nu,\nu'} \Theta_\nu(-\mathbf{q}) \Theta_\nu(\mathbf{q}) a_{\mathbf{q}\nu} a_{-\mathbf{q}\nu}, \tag{4.46}$$

where

$$\Theta_\nu(\mathbf{q}) = \left(\frac{4\pi}{M\Omega_0 \sum_{ij} \kappa_{ij}^{\infty} q_i q_j}\right)^{1/2} \sum_i Q_{\nu i} q_i . \quad (4.47)$$

In accordance with (4.9, 46), we can thus write the equation of motion in the form

$$\Omega^2 a_{\mathbf{q}\nu} = \frac{1}{4M_0 N} \frac{\partial(\mathbf{U} + \Delta\mathbf{U})}{\partial a_{-\mathbf{q}\nu}} = \Omega_{\mathbf{q}\nu}^{02} a_{\mathbf{q}\nu} + \sum_{\nu'} \Theta_\nu(-\mathbf{q})\Theta_{\nu'}(\mathbf{q}) a_{\mathbf{q}\nu'} . \quad (4.48)$$

Here $\Omega_{\mathbf{q}\nu}^0$ are the eigenfrequencies in the absence of long-range interaction. When this interaction is included, the eigenfrequencies $\Omega_{\mathbf{q}\nu}$ can be found from the condition of a vanishing determinant of the coupled equations (4.48):

$$\left[\prod_\nu \left(\Omega_{\mathbf{q}\nu}^{02} - \Omega^2\right)\right] \left(1 + \sum_\nu \frac{\Theta_\nu(\mathbf{q})\Theta_\nu(-\mathbf{q})}{\Omega_{\mathbf{q}\nu}^{02} - \Omega^2}\right) = 0. \quad (4.49)$$

If for a given n_ν -fold degenerate term corresponding to the frequency $\Omega_{\mathbf{q}\nu}^0$ all constants $Q_{\nu i} = 0$, i.e. this term is optically inactive, then for it $\Omega_{\mathbf{q}\nu} = \Omega_{\mathbf{q}\nu}^0$. For optically active terms (4.49) has $n_\nu - 1$ unshifted roots $\Omega_{\mathbf{q}\nu} = \Omega_{\mathbf{q}\nu}^0$. From each of these terms one vibration splits, their frequencies, according to (4.47, 49), determined by the equation

$$\sum_{ij} \left(\kappa_{ij}^{\infty} q_i q_j + \frac{4\pi}{M_0\Omega_0} \sum_\nu \frac{Q_{\nu i}(\mathbf{q}) Q_{\nu j}(-\mathbf{q}) q_i q_j}{\Omega_{\mathbf{q}\nu}^{02} - \Omega^2}\right) = 0. \quad (4.50)$$

The order of this equation is equal to the number of the optically active terms. The latter equation can be derived from Maxwell's equations if one neglects retardation, that is, mixing of the phonon and photon branches of the spectrum. In this approximation $\mathrm{curl}\mathcal{I} = 0$ and $\mathrm{div}\,\mathbf{D} = 0$, whence it follows that $\mathcal{I} = -\nabla\varphi$ and

$$\sum_{ij} \kappa_{ij} \frac{\partial^2\varphi}{\partial x_i \partial x_j} = 0. \quad (4.51)$$

In the $\mathbf{q}$-representation the last expression can be written as $\Sigma_{ij} q_i q_j \kappa_{ij} \varphi_{\mathbf{q}} = 0$, so that for $\varphi_{\mathbf{q}} \neq 0$ the vibrational spectrum will be determined by the equation

$$\sum_{ij} \kappa_{ij} q_i q_j = 0. \quad (4.52)$$

The expression for the dielectric tensor

$$\kappa_{ij}(\Omega) = \kappa_{ij}^{\infty} + \frac{4\pi}{M\Omega_0} \sum_\nu \frac{Q_{\nu i}(\mathbf{q}) Q_{\nu j}(-\mathbf{q})}{\Omega_{\mathbf{q}\nu}^{02} - \Omega^2} \quad (4.53)$$

can be derived from the relation

$$D_i = \kappa_{ij}^{\infty} \mathcal{I}_j + 4\pi P_i = \kappa_{ij} \mathcal{I}_j$$

by substituting P_i from (4.44) and using the relation between $a_{\nu\mathbf{q}}$ and $\mathcal{I}_{\mathbf{q}} = -\mathrm{i}\mathbf{q}\varphi_{\mathbf{q}}$ following from (4.44, 48):

$$a_{\mathbf{q}\nu} = \frac{1}{\Omega_{\mathbf{q}\nu}^{02} - \Omega^2} \frac{1}{2M_0} \sum_j Q_{\nu j}(-\mathbf{q}) \mathcal{I}_{j\mathbf{q}}. \tag{4.54}$$

According to (4.48), to the root of (4.50) and $\Omega^2 = \Omega_{\mathbf{q}\mu}^2$, correspond the displacements

$$a_{\mathbf{q}\nu}^{\mu} = A \frac{\Theta_\nu(-\mathbf{q})}{\Omega_{\mathbf{q}\mu}^2 - \Omega_{\mathbf{q}\nu}^{02}} \tilde{a}_{\mathbf{q}\mu}, \tag{4.55}$$

where $A = \sum_\nu \Theta_\nu(\mathbf{q}) a_{\mathbf{q}\nu}$ is a constant determined from the condition of normalization:

$$|A|^2 \sum_\nu \frac{|\Theta_\nu(-\mathbf{q})|^2}{(\Omega_{\mathbf{q}\mu}^2 - \Omega_{\mathbf{q}\nu}^{02})^2} = 1.$$

As follows from (4.44, 55), the polarization $\mathbf{P}_\mu$ induced by the displacement $\tilde{a}_{\mathbf{q}\mu}$ corresponding to the frequency $\Omega_{\mathbf{q}\mu}$ is

$$P_{\mu j} = \frac{2}{\Omega_0} \sum_\nu Q_{\nu j} \frac{\Theta_\nu(-\mathbf{q})}{\Omega_{\mathbf{q}\mu}^2 - \Omega_{\mathbf{q}\nu}^{02}} A \tilde{a}_{\mathbf{q}\mu}. \tag{4.56}$$

In zincblende or CsCl-type crystals, the primitive cell contains two atoms whose masses we denote by M_1 and M_2. The polarization vectors for the three optical vibration branches can be chosen in the form

$$e_{1i}^j = \left(\frac{M_2}{M_1}\right)^{1/2} \delta_{ij}, \; e_{2i}^j = -\left(\frac{M_1}{M_2}\right)^{1/2} \delta_{ij}, \tag{4.57}$$

where $i = 1, 2, 3$ refer to x, y, z.

The normal coordinates

$$a_i^{\mathrm{opt}} = \frac{\sqrt{M_1 M_2}}{2(M_1 + M_2)} (u_{1i} - u_{2i}), \tag{4.58}$$

where u_{li} are the displacements of the l-th atom transforming as polar vector components. The effective charge Q has only diagonal components

$$Q_{\nu j} = Q \delta_{\nu j}. \tag{4.59}$$

Therefore, according to (4.45)

$$\varphi_{\mathbf{q}} = -\frac{8\pi \mathrm{i}(\mathbf{q} \cdot \mathbf{a}^{\mathrm{opt}})}{\Omega_0 q^2 \kappa^\infty} Q, \tag{4.60}$$

i.e., the field $\varphi_{\mathbf{q}}$ is generated only by longitudinal components with $\mathbf{a}^{\mathrm{opt}} \| \mathbf{q}$. In accordance with (4.50), the frequency of these components

$$\Omega_{\mathrm{l}}^2 = \Omega_{\mathrm{t}}^2 + \frac{4\pi Q^2}{M_0 \Omega_0 \kappa^\infty}. \tag{4.61}$$

The frequency of the two transverse vibration branches is Ω_t. According to (4.53), the dielectric constant

$$\kappa(\Omega) = \kappa^{\infty} + \frac{4\pi Q^2}{M_0\Omega_0(\Omega_t^2 - \Omega^2)}. \tag{4.62}$$

One can use (4.62) to express the effective charge in terms of the difference between the low frequency dielectric constant κ^0 for frequencies $\Omega \ll \Omega_t$ and the high frequency one κ^{∞}:

$$Q^2 = \frac{1}{4\pi}(\kappa^0 - \kappa^{\infty})M_0\Omega_0\Omega_t^2. \tag{4.63}$$

Equations (4.62, 63) yield the well-known Lyddane, Sachs and Teller relation:

$$\frac{\Omega_l^2}{\Omega_t^2} = \frac{\kappa^0}{\kappa^{\infty}} \tag{4.64}$$

and the relation (4.62) for $\kappa(\Omega)$ can be written in the form

$$\kappa(\Omega) = \kappa^{\infty}\frac{\Omega_l^2 - \Omega^2}{\Omega_t^2 - \Omega^2}. \tag{4.65}$$

4.2.4 Coulomb Modes in Superlattices

The dielectric model considers optically active vibrations in superlattices as oscillations of the electric field $\mathcal{I}$ or potential φ [4.11–14]. If each of the layers is isotropic, i.e. $\kappa_{ij}^{A,B} = \kappa^{A,B}\delta_{ij}$, then, according to (4.51), for each layer we can write

$$\kappa\nabla^2\varphi = 0 \tag{4.66}$$

and at the layer boundary

$$\mathcal{I}_{\perp}^A = \mathcal{I}_{\perp}^B, \quad D_z^A = D_z^B, \quad \text{i.e.} \quad \kappa_A\mathcal{I}_z^A = \kappa_B\mathcal{I}_z^B. \tag{4.67}$$

Two types of Coulomb modes are distinguished in GaAs/AlAs-type lattices. The first of them, bulk-like modes, occur at the longitudinal optical vibration frequencies of GaAs or AlAs, Ω_{LO}^A or Ω_{LO}^B. Since for $\Omega = \Omega_{LO}^A$ according to (4.65), $\kappa_A = 0$, we obtain $\kappa_B \neq 0$. The solution of (4.66) for the layer B has the form

$$\varphi_2 = e^{i\mathbf{k}\cdot\rho}(A_1e^{\kappa z} + A_2e^{-\kappa z}), \tag{4.68}$$

where $\mathbf{k}$ and ρ are the components of $\mathbf{q}$ and $\mathbf{r}$ normal to z.

As follows from the boundary condition (4.67) for $\kappa_A = 0$, we have $\mathcal{I}_z^B = 0$ for $z = a, d$, which is possible only for $A_1 = A_2 = 0$, i.e., $\varphi_2 = 0$. In layer A

$$\varphi_1 = e^{i\mathbf{k}\cdot\rho}\Phi(z). \tag{4.69}$$

For $\kappa_A = 0$, (4.66) does not impose any conditions on $\Phi(z)$, while the boundary conditions (4.67) for $\mathcal{I}_{\perp}$ imply that $\Phi(z) = 0$ at the boundary of layer A. One can conveniently place the origin $z = 0$ at the layer's center and write this condition in the form

$$\Phi(z) = 0 \quad \text{for} \quad z = \pm a/2. \tag{4.70}$$

In early publications, $\Phi(z)$ was chosen in the form of the so-called zero-node modes

$$\Phi_m(z) = \begin{cases} A_m \cos\dfrac{\pi m z}{a} & m = 1, 3, \dots \\ A_m \sin\dfrac{\pi m z}{a} & m = 2, 4, \dots \end{cases} \quad (-a/2 < z < a/2). \tag{4.71}$$

However, with such a choice of $\Phi(z)$, the field $\mathcal{I}_z$ becomes discontinuous at the boundary between the layers A and B and, accordingly, also the displacements $a_z \sim (u_{1z} - u_{2z})$ become discontinuous, since, according to (4.54, 55)

$$a_z = \frac{Q}{\Omega_\mathrm{T}^2 - \Omega_\mathrm{L}^2}\frac{\mathcal{I}_z}{2M_0}. \tag{4.72}$$

Therefore *Huang* and *Zhu* [4.13] proposed the following relations for $\Phi(z)$, which make $\Phi(z)$ and $\mathcal{I}_z$ vanish at the layer boundary ($z = \pm a/2$):

$$\Phi_m(z) = \begin{cases} A_m \left[\cos\dfrac{\pi m z}{a} - (-1)^m\right] & m = 2, 4, 6, \dots \\ A_m \left[\sin\dfrac{\pi \mu_m z}{a} + \dfrac{c_m Z}{a}\right] & m = 3, 5, \dots \end{cases} \tag{4.73}$$

The condition of periodicity implies that

$$\Phi_m(z + nd) = \mathrm{e}^{\mathrm{i}q_z nd}\Phi_m(z)$$

the constants μ_m and c_m in (4.73) being derived from the boundary conditions, i.e.,

$$\tan\frac{\pi\mu_m}{2} = \frac{\pi\mu_m}{2}, \quad c_m = -2\sin\frac{\pi\mu_m}{2}.$$

For $m > 7$, μ differs from m by not more than 0.05, and c_m from $2(-1)^{(m+1)/2}$ by not more than 0.01. Similarly, at $\Omega = \Omega_\mathrm{LO}^B$ when $\kappa_B = 0$, the field is localized in region B and is defined by expressions differing from (4.71, 73) in that a is replaced by b.

The second type of the Coulomb modes is the interface modes appearing at frequencies $\Omega \neq \Omega_\mathrm{LO}^A, \Omega_\mathrm{LO}^B$. If $\kappa_A, \kappa_B \neq 0$, then it follows from (4.66) that, both in layer A and B, $\varphi(\rho, z)$ is determined by an expression similar to (4.68); note that the boundary condition (4.67) for $\mathcal{I}_\perp$ implies that the value of $\mathbf{k}_\perp$ in both layers is the same. Then, from these boundary conditions, one can derive four equations for the coefficients A_i in layers A and B, the corresponding secular equation differing from (4.37) in that k_1 and k_2 are replaced by ik, and η, by $\eta_0 = \kappa_A/\kappa_B$, since, by (4.66), $k_1^2 + k^2 = k_2 + k^2 = 0$ [4.16, 19],

$$\cos(qd) = \cosh(ka)\cosh(kb) + R_0\sinh(ka)\sinh(kb), \tag{4.74}$$

where

$$R_0 = \frac{1}{2}\left(\frac{\kappa_A}{\kappa_B} + \frac{\kappa_B}{\kappa_A}\right).$$

While the bulk-like modes are due to the vibrations of the charge generated in the bulk of the layers, the interface modes originate from the vibration of the charges appearing at the layer boundary and their amplitude decays as one moves away from it into the bulk. In the long-wavelength limit, i.e., for $q \to 0$ and $k \to 0$, the frequencies of these vibrations, according to (4.74), are defined by the relation

$$\tan^2\theta = \frac{q^2}{k^2} = -\frac{(a\kappa_A + b\kappa_B)(a\kappa_B + b\kappa_A)}{\kappa_A\kappa_B d^2}. \tag{4.75}$$

This expression can be readily derived from (4.52) if one considers the superlattice as a homogeneous medium for which, in accordance with the conventional rules of parallel and series connection of capacitance, one can write

$$\kappa_\perp = (\kappa_A a + \kappa_B b)/d, \quad \kappa_{zz}^{-1} = (\kappa_A^{-1} a + \kappa_B^{-1} b)/d.$$

According to (4.52),

$$\kappa_\perp k^2 + \kappa_{zz} q^2 = 0, \tag{4.75a}$$

which coincides with (4.75). For $\theta \to \pi/2$, i.e., $\mathbf{q} \parallel z$, (4.75) has four roots

$$\kappa_A \to \infty, \quad \text{i.e.,} \Omega \to \Omega_\mathrm{T}^A; \kappa_A \to 0, \quad \text{i.e.,} \Omega \to \Omega_\mathrm{L}^A;$$

$$\kappa_B \to \infty, \quad \text{i.e.,} \Omega \to \Omega_\mathrm{T}^B; \kappa_B \to 0, \quad \text{i.e.,} \Omega \to \Omega_\mathrm{L}^B. \tag{4.76}$$

In the reverse limiting case, $\theta \to 0$, i.e., $\mathbf{q} \perp z$, the four roots of (4.75) are given by the expressions

$$a\kappa_A + b\kappa_B = 0, \tag{4.77a}$$

$$a\kappa_A^{-1} + b\kappa_B^{-1} = 0. \tag{4.77b}$$

The corresponding frequencies lie in the interval between Ω_L and Ω_T for each of the layers (Fig. 4.2). Note that to the roots of (4.77a) correspond the vibrations with $A_1 \approx -A_2$, i.e., with $\mathcal{I} \perp z$ (LO mode) and to the roots of (4.77b), the vibrations with $A_1 \approx A_2$, i.e., with $\mathcal{I} \| z$ (TO mode).

We conclude by summing up the above discussion. In GaAs/AlAs-type superlattices the primitive cell contains $2(l+n)$ atoms, where $l = a/a_0, n = b/a_0$. Accordingly, the number of modes for fixed k and q_z is $6(l+n)$ (the values of q_z differing by $2\pi/d$ being considered equal). Out of these $6(l+n)$ modes, $3(l+n)$ originate from the acoustic branches of the bulk crystal spectrum. The spectrum $\Omega(\mathbf{k}, q_z)$ of these modes for $k = 0$ or $\mathbf{u} \perp \mathbf{k}$ and $\mathbf{q}$ is defined by (4.37) (for $\eta \simeq k_2(\Omega)/k_1(\Omega)$). Out of the $3(l+n)$ modes originating from the optical branches, the transverse mode with $\mathbf{u} \perp \mathbf{q}$ is optically inactive and its spectrum $\Omega(\mathbf{k}, q_z)$ determined by an equation of the type (4.37), with the value of η depending on the actual lattice model chosen. Four out of the remaining $2(l+n)$ modes are the interface modes, their spectrum being described by (4.74). In the dielectric model, $l-1$ modes have the frequency Ω_L^A and $n-1$ modes, Ω_L^B. The frequency of the remaining $l+n-2$ modes is either Ω_T^A, or Ω_T^B, and in terms of this model they are optically inactive. These $2(l+n-2)$ modes are mixed longitudinal-transverse, with the vectors $\mathbf{u}$ lying in the $\mathbf{k}$, $\mathbf{q}$

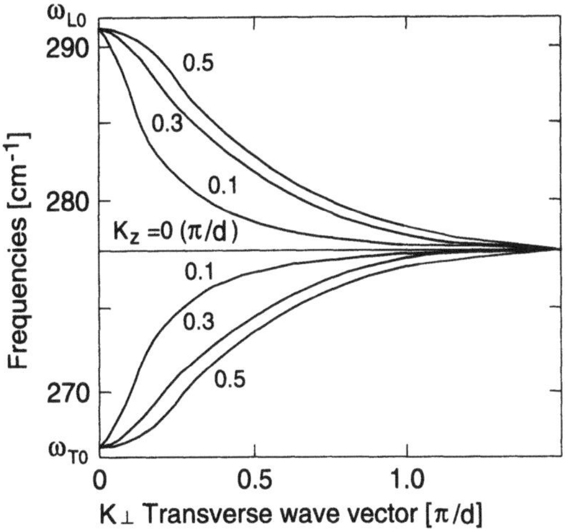

Fig. 4.2. Dependence of the frequency of GaAs-like interface modes on $k_\perp d/\pi$ in the dielectric model (after ref. [4.13])

plane. For $k = 0$, these modes are actually the result of splitting of the doubly degenerate representation E and of mixing of one of the split-off branches with the representation A_1 (or A_1 and B_2 for $q_z = 0$ and $q_z = \pi/d$), caused by dipole-dipole interaction. If spatial dispersion is included, the frequency of these modes will depend on the number m and be approximately equal to that of the corresponding bulk-crystal modes, Ω_L^{AB} or Ω_T^{AB}, for a given $\mathbf{q}$, and for q_z equal to $\pi m/a$ or $\pi m/b$.

4.3 Electron-Phonon Interaction

The electron mass being much smaller than the mass of the atoms, the electron potential energy may be considered to depend on the instantaneous position of the atoms rather than on their velocity. In this approximation called adiabatic, the field acting upon the electron, $V(\mathbf{x})$, may be represented as a sum of the potential of a perfect lattice, $V_0(\mathbf{x})$, determining the electron spectrum and wave functions, and a perturbation $\Delta V(\mathbf{x}) = V(\mathbf{x}) - V_0(\mathbf{x})$ which depends on the atomic displacements from the equilibrium positions. We will consider here only the displacements caused by the lattice vibrations. The forces responsible for the perturbation may be divided into short-range ones determined by the displacement of the neighboring atoms, and long-range forces created by the lattice polarization due to the atomic displacements.

4.3.1 Short-Range Forces

In the deformation-potential theory, the interaction with acoustic phonons is described by the same expressions that determine the deformation-induced change of the spectrum, i.e., (3.29) for nondegenerate bands and (3.77) for the degenerate valence band Γ_8. In accordance with (4.4), the deformation produced by acoustic vibrations may be represented in the form

$$\varepsilon_{ij}(\mathbf{x}) = \sum_{\nu,q} \left(\varepsilon_{ij,\mathbf{q}\nu} a_{\mathbf{q}\nu} \mathrm{e}^{\mathrm{i}\mathbf{q}\cdot\mathbf{x}} + \varepsilon^*_{ij,\mathbf{q}\nu} a^*_{\mathbf{q}\nu} \mathrm{e}^{-\mathrm{i}\mathbf{q}\cdot\mathbf{x}} \right), \tag{4.78}$$

where

$$\varepsilon_{ij,\mathbf{q}\nu} = \frac{1}{2}[\mathrm{e}^{\nu}_i(\mathbf{q})q_j + \mathrm{e}^{\nu}_j(\mathbf{q})q_i].$$

Here we have assumed that for long-wavelength acoustic vibrations the polarization vectors of all atoms in the primitive cell coincide to within a small parameter qa_0; for the same reason, the atomic coordinate $\mathbf{X}_f$ in (4.4) is replaced by the running coordinate $\mathbf{x}$.

The Hamiltonian describing the short-range interaction with optical phonons contains usually only terms linear in the displacement $\mathbf{u}$. For nondegenerate bands, these terms do not vanish if there is a corresponding normal coordinate transforming according to the identity representation of the wave vector group, for example, at the point L for Ge, Si, and cubic A_3B_5 crystals. For the degenerate valence band Γ_8, the corresponding Hamiltonian is

$$\mathcal{H}^{\mathrm{opt}}_u = \frac{2}{\sqrt{3}} d_0 \sum_{ikl} a^{\mathrm{opt}}_i [J_k J_l] |\delta_{ikl}|, \tag{4.79}$$

where a^{opt}_i is defined by (4.58)

It should be pointed out that, for the acoustic vibrations, the Hamiltonian $\mathcal{H}_{\mathrm{eff}}$ also contains terms proportional to the displacement u:

$$\mathcal{H}_u = \frac{\mathrm{i}}{\hbar}(\{\mathbf{p}H_0\}\mathbf{u})$$

and

$$\mathcal{H}_L = -\mathrm{i}(\{\mathbf{L}H_0\}\boldsymbol{\omega}). \tag{4.80}$$

Here $\mathbf{L} = -[\mathbf{pr}]/\hbar$ is the angular momentum operator, $\boldsymbol{\omega} = 1/2[\nabla \times \mathbf{u}]$ the rotation angle, $\mathbf{p}$ the momentum operator, and $\{AB\} = AB - BA$.

In contrast to the optical vibrations, the diagonal matrix elements of the operators (4.80) are zero, and the matrix elements for transitions between the eigenstates of the operator H_0 with energies E_n and $E_{n'}$, are proportional to $(E_n - E_{n'})$. Therefore such contributions are usually neglected, although in some cases they are comparable with the deformation terms.

4.3.2 Long-Range Forces

Long-range forces are associated with the electric fields created by the lattice polarization. In cubic crystals containing two atoms in the primitive cell these fields are produced only by longitudinal vibrations, the corresponding potential, according to (4.60, 63), being

$$\mathcal{H} = -e\varphi^{\mathrm{opt}} = \sum_{\mathbf{q}} \frac{ie}{q}\left(\frac{2\pi\hbar\Omega_{\mathrm{LO}}}{N\Omega_0\kappa^*}\right)^{1/2}(b_{\mathbf{q}}\mathrm{e}^{\mathrm{i}\mathbf{q}\cdot\mathbf{x}} - b_{\mathbf{q}}^{+}\mathrm{e}^{-\mathrm{i}\mathbf{q}\cdot\mathbf{x}}) \tag{4.81}$$

where $\kappa^{*-1} = \kappa_\infty^{-1} - \kappa_0^{-1}$. Equation (4.81) was derived using (4.60, 63), expressing the amplitudes of the normal vibrations $a_{\mathbf{q}\nu}$ and $a_{\mathbf{q}\nu}^{+}$ in terms of the operators $b_{\mathbf{q}\nu}$ and $b_{\mathbf{q}\nu}^{+}$ in accordance with (4.6). Hamiltonian (4.81) is called the Fröhlich Hamiltonian.

For transverse optical vibrations, as well as in nonpolar crystals, the polarization is proportional to the derivative of the displacements with respect to the coordinate. For instance, in diamond-type lattices,

$$\varphi_{\mathbf{q}}^{\mathrm{opt}} = -\frac{8\pi\mathrm{i}}{\kappa q^2\Omega_0} Q_0 \sum_{ikl} q_i q_k a_l^{\mathrm{opt}} |\delta_{ikl}|. \tag{4.82}$$

In acoustic vibrations, the polarization in piezoelectric crystals (3.126) is proportional to deformation, so that

$$\varphi_{\mathbf{q}}^{\mathrm{ac}} = -\frac{8\pi\mathrm{i}e_{14}}{\kappa_0 q^2\Omega_0} \sum_{ikl} q_i \varepsilon_{kl} |\delta_{ikl}|. \tag{4.83}$$

In nonpiezoelectric crystals, polarization is proportional to the gradient of deformation and usually disregarded.

In superlattices, the spectrum of folded phonons is usually close to that of bulk crystal vibrations, and the corresponding vibrations may be considered bulk-type, so that the electron – phonon interaction Hamiltonian may be written in the form

$$\mathcal{H}_{\mathrm{e-ph}} = \sum_{\mathbf{q},\nu} a_{\mathbf{q}}^{\nu}\varphi_{\mathbf{q}}^{\nu}\mathrm{e}^{\mathrm{i}\mathbf{q}\cdot\mathbf{x}}. \tag{4.84}$$

In accordance with (3.82, 90), for the electron wave function in a superlattice we obtain

$$F_l(\mathbf{x}) = \mathrm{e}^{\mathrm{i}\mathbf{k}_\perp\cdot\rho}\sum_n \mathrm{e}^{\mathrm{i}k_z z_n}\varphi_l(z - z_n), \tag{4.85}$$

where $z_n = dn$, and $\varphi(z - z_n)$ is nonzero within the nth cell: $z_n < z < z_{n+1}$. The electron and phonon wave functions are assumed to be normalized to unit length, and $\varphi(z)$, to the length d.

The wave vector k_z in (4.85) lies within the Brillouin zone of the superlattice, $|k_z| < \pi/d$. Similarly, the phonon wave vector component q_z can be written in the form $q_z = q_m + q$, where $q_m = 2\pi m/d$, and $|q| \leqslant \pi/d$. Then, according to (4.84 85), for the matrix element of an electron transition from state $(l, \mathbf{k}_\perp, k_z)$ to state $(l', \mathbf{k}'_\perp, k'_z)$ involving absorption (emission) of a phonon of the branch ν with $\mathbf{q}_\perp = \mathbf{k}$, $q_z = q_m + q$, one can write

$$\langle l', \mathbf{k}'_\perp, k'_z|\mathcal{H}_{\mathrm{e-ph}}(\nu, \mathbf{k}, q_m + q)|l, \mathbf{k}_\perp, k_z\rangle$$
$$= a_{\mathbf{k},q_m+q}^{\nu}\varphi_{\mathbf{k},q_m+q}^{\nu}\delta_{\mathbf{k}_\perp,\pm\mathbf{k}+\mathbf{k}'_\perp}\delta_{k_z,\pm q+k'_z}$$

$$\times \int_o^d e^{\pm iq_m z} \varphi_{l'}^*(z)\varphi_l(z)dz. \tag{4.86}$$

For the zero-node and interface modes, according to (4.43, 68, 73), we obtain

$$\mathcal{H}_{\text{e-ph}}(\nu, \mathbf{k}, q) = a_{\mathbf{k},q}^{\nu} e^{i\mathbf{k}\rho} \sum_n e^{iqz_n} V_\nu(z - z_n). \tag{4.87}$$

For the Coulomb modes, $V_\nu(z)$ is the potential (4.68, 73) generated by vibrations, the index ν including the mode number. In this case, the matrix element of transition with emission (absorption) of a phonon (ν, $\mathbf{k}$, q) is

$$\begin{aligned}&\langle l', \mathbf{k}'_\perp, k'_z | \mathcal{H}_{\text{e-ph}}(\nu, \mathbf{k}, q) | l, \mathbf{k}_\perp, k_z \rangle \\ &= a_{\mathbf{k},q}^{\nu} \delta_{\mathbf{k}_\perp, \pm\mathbf{k}+\mathbf{k}'_\perp} \delta_{k_z, \pm q + k'_z} \\ &\times \int_o^d \varphi_{l'}^*(z)\varphi_l(z) V_\nu(z) dz.\end{aligned} \tag{4.88}$$

If the vibrations are localized within one of the layers, as is the case with the zero-node modes, integration in (4.88) is carried out only over this layer, the principal contribution coming from vibrations in the layers that form the well for the electrons. If the barrier is high enough, the electron wave functions, just as in the case of a single well, have a definite parity, i.e., they either do or do not reverse the sign upon reflection in the plane passing through the well center. Therefore transitions between the states l and l' of the same parity will occur under these conditions only if they involve even phonon modes, and transitions between the states l and l' of opposite parity, only if odd phonon modes are involved.

We note in conclusion that for sufficiently high concentrations of free electrons ($n \gtrsim 10^{16}$ cm^{-3}), the change in the optical phonon frequency caused by the plasmon-phonon interaction becomes essential. Under these conditions, the electric field generated by lattice vibrations is partially screened by the free electrons [4.15,16].

5 Localized Electron States and Excitons in Heterostructures

One may suggest a variety of direct and convincing proofs for the validity of the effective mass approximation in the physics of semiconductors. Particularly elegant among them is the argumentation based on the prediction and observation of hydrogen-like states of Coulomb center-bound carriers and of the Wannier-Mott excitons. One can say without overstating the case that the physics of shallow impurity centers and excitons has had a rebirth with the development of high-quality heterostructures with microscopically thin layers of compositional materials. A need has been found for calculating Coulomb states with account for the superstructure potential, for the difference between the band parameters in adjoining layers, etc.

In this chapter we formulate principles for the application of the effective-mass approximation to calculate the binding energies and wave functions in such quantum-mechanical problems. We shall employ reasonably simple analytical methods to explain a number of properties of the Coulomb states in multilayer structures, for example, the dependence of the donor-bound electron binding energy on the spatial position of the impurity center, and the behavior of the exciton states in periodic structures with variation of the layer width, namely, the transition from three-dimensional excitons quantum-confined as a whole to two-dimensional excitons as the well width decreases, and from quasi two- to three-dimensional excitons with decreasing barrier thickness.

In degenerate-band semiconductors, the role of the operator of single-particle kinetic energy in Schrödinger's equation for the Coulomb problem is played by the matrix effective Hamiltonian. Symmetry imposes a number of constraints on the diversity of eigenstates in this complex case as well, thus permitting substantial simplification of the procedure for their calculation. The concluding section outlines a general symmetry-based approach to the analysis of the fine structure of exciton states with account of the electron-hole exchange interaction. We shall consider the exchange-induced splitting of exciton levels in more detail in Chap. 9.

5.1 Shallow Impurity Centers

To determine the wave functions and energy E of a shallow impurity center in a three-dimensional semiconductor in the effective-mass approximation, one has to solve a system of equations

$$\sum_{n'} \mathcal{H}_{nn'}(\hat{\mathbf{k}}, \mathbf{x})\varphi_{n'}(\mathbf{x}) = E\varphi_n(\mathbf{x}), \tag{5.1}$$

where $\hat{\mathbf{k}} = -\mathrm{i}\nabla$,

$$\mathcal{H}(\mathbf{k}, \mathbf{x}) = \mathcal{H}_0(\mathbf{k}) - \frac{\mathrm{e}^2}{\kappa|\mathbf{x} - \mathbf{x}_\mathrm{i}|}. \tag{5.2}$$

$\mathcal{H}_0(\mathbf{k})$ is the effective Hamiltonian of the electron (or hole) in a homogeneous semiconductor representing a matrix of dimension $N \times N$, where N is the number of branches in the conduction (or valence) band included in the effective-mass approximation, κ is the static dielectric constant, $\mathbf{x}_\mathrm{i}$ is the position of the Coulomb impurity center. Equation (5.1) for stationary states is obtained from (3.16) by the replacement

$$F_n(\mathbf{x}, t) \to \mathrm{e}^{-\mathrm{i}Et/\hbar}\varphi_n(\mathbf{x}).$$

The wave function of a stationary state is written as

$$\Psi(\mathbf{x}) = \sum_{n=1}^{N} \varphi_n(\mathbf{x})\psi_{n\mathbf{K}_0}(\mathbf{x}). \tag{5.3}$$

In the problem of a shallow impurity center in a heterostructure consisting of layers of different semiconductors whose thickness is in excess of the lattice constant, one can also use the effective-mass method. For matrix (5.2) one employs in this case the matrix

$$\mathcal{H}(\mathbf{k}, \mathbf{x}) = \mathcal{H}_0(\mathbf{k}, z) - \frac{\mathrm{e}^2}{\kappa|\mathbf{x} - \mathbf{x}_\mathrm{i}|} + V(z), \tag{5.4}$$

where the z axis is perpendicular to the layers, $V(z)$ is a potential describing the variation of the band edge along the axis. The effective Hamiltonian $\mathcal{H}_0$ does not depend on z within a homogeneous layer; however, its parameters may change stepwise when one goes over to another layer. For the sake of simplicity, we disregard here the difference between the dielectric constants in the layers. Therefore (5.4) does not contain the contribution from the image forces. The electron wave function in a heterostructure is likewise written in the form of an expansion of (5.3) keeping in mind that the Bloch function $\psi_{n\mathbf{K}_0}$ changes stepwise when crossing an interface. The envelopes $\varphi_n(\mathbf{x})$ are matched at the boundary between the layers by properly defining the boundary conditions. The boundary conditions for an N-component function $\varphi(\mathbf{x})$ take on the form

$$\varphi_A = \varphi_B, v_z(\hat{\mathbf{k}}, A)\varphi_A = v_z(\hat{\mathbf{k}}, B)\varphi_B, \tag{5.5}$$

where $v_z(\hat{\mathbf{k}}, z)$ is the projection on the z axis of the velocity operator

$$\mathbf{v}(\mathbf{k}, z) = \hbar^{-1}\nabla_\mathbf{k}\mathcal{H}_0(\mathbf{k}, z). \tag{5.6}$$

It is these conditions that we will be using in what follows unless otherwise specified.

Operator (5.2) satisfies the invariance conditions (3.30) under operations of the point group of the wave vector $\mathbf{K}_0$. In some particular cases, the symmetry of this operator can be still higher. For instance, for the band Γ_6 in a semiconductor of the class T_d we have in the parabolic approximation

$$\mathcal{H}_0(\mathbf{k}) = \hbar^2 k^2/2m^* \tag{5.7}$$

and Hamiltonian (5.2) is invariant under the point group transformations of the orthogonal group K_h. The envelope of the wave function $\varphi(\mathbf{x})$ in this case can be represented as a product of the spinor

$$\Phi = C_{1/2}\alpha + C_{-1/2}\beta \tag{5.8}$$

and a scalar function $\varphi(\mathbf{x})$, for which the second of the boundary conditions can be simplified to

$$\frac{1}{m_A}\left(\frac{\partial\varphi}{\partial z}\right)_A = \frac{1}{m_B}\left(\frac{\partial\varphi}{\partial z}\right)_B . \tag{5.9}$$

The Hamiltonian (3.46) for holes in the band Γ_8 of a zincblende semiconductor (class T_d) is characterized by the symmetry O_h, since the terms quadratic in wave vector are invariant under the inversion $\mathbf{k} \to -\mathbf{k}$. The presence of the heterostructure potential $V(z)$ and the dependence of the matrix $\mathcal{H}_0(\mathbf{k})$ on z lowers the symmetry of the Hamiltonian. For a centrosymmetric structure with the principal axis $z \| [001]$, the group K_h goes into $D_{\infty h}$, and group O_h, into D_{4h}. Placing an impurity atom at the well center does not lower the symmetry any further. With an off-center impurity atom the symmetry of the Hamiltonian (5.4) is lowered to $C_{\infty v}$ and C_{4v}, respectively, in place of $D_{\infty h}$ and D_{4v}.

5.1.1 Nondegenerate Band

We first consider the states of a donor-bound electron within the simplest model of a semiconductor quantum well with a nondegenerate conduction band characterized by an isotropic effective mass m^*. The axial symmetry of operator (5.4) implies that the states $\varphi(\mathbf{x})$ can be classified according to the projection of the angular momentum onto the axis z

$$L_z\varphi(\mathbf{x}) = l\varphi(\mathbf{x}),$$
$$\mathbf{L} = -\mathrm{i}\,[\mathbf{x} \times \nabla]\,, \quad l = 0, \pm 1, \pm 2, \ldots \tag{5.10}$$

For the function $\varphi_l(\mathbf{x})$, one can use the representation

$$\varphi_l(\mathbf{x}) = \rho_{\pm}^{|l|} R_{|l|}(\rho, z), \tag{5.11}$$

where $\rho_{\pm} = \rho_x \pm \mathrm{i}\rho_y$, $\rho = |\boldsymbol{\rho}|$, $\boldsymbol{\rho}$ is the in-plane component of the vector $\mathbf{x} - \mathbf{x}_{\mathrm{i}}$. This permits one to preserve the nomenclature $1s$, $2s$, $2p_{\pm 1}$, $2p_0$ for the neutral donor states. If the impurity atom is at the center of a symmetric well, then the Hamiltonian (5.4) is invariant under space inversion, the states (5.3) having a definite parity under the replacement of z by $-z$. The invariance of $\mathcal{H}(\mathbf{k}, \mathbf{x})$

under time reversal implies that the states $\varphi_l(\mathbf{x})$ and $\varphi_{-l}(\mathbf{x}) = \varphi_l^*(\mathbf{x})$ have the same energy.

Schrödinger's equation (5.1) for an impurity center in a heterostructure is usually solved by variational techniques with the trial function of the type

$$\varphi(\mathbf{x}) = f_\nu(z)G(\rho, z), \tag{5.12}$$

where $f_\nu(z)$ is the eigenfunction of the operator $\mathcal{H}_0(\mathbf{k}, z) + V(z)$ in the subband ν for $k_\perp = 0$. In the infinite potential barrier approximation, the even and odd solutions f_ν are defined by (3.83, 86a). For $G(\rho, z)$ one uses functions acceptable to the variational procedure and vanishing for $\rho \to \infty$. In the simplest case of the $1s$ state of a neutral donor, one can take this function in the form

$$G(\rho, z) = C\exp(-|\mathbf{x} - \mathbf{x}_i|/\bar{\lambda}), \tag{5.13}$$

where $\bar{\lambda}$ is the variational parameter. Hence for the ground state of a localized electron in the $\nu = 1$ subband we will have in this case

$$\varphi(\mathbf{x}) = \begin{cases} C\sqrt{\dfrac{2}{\pi}}\cos\dfrac{\pi z}{a}\exp\left\{-\dfrac{1}{\bar{\lambda}}\left[\rho^2 + (z - z_i)^2\right]^{1/2}\right\} & \text{if } |z| < a/2, \\ 0 & \text{if } |z| > a/2, \end{cases} \tag{5.14}$$

where z and z_i are measured from the center of the quantum well. The condition of normalization of $\varphi(\mathbf{x})$ to unity yields

$$\begin{aligned} C^2 = \frac{1}{\bar{\lambda}^3}\Bigg[& 1 - \cosh\frac{2z_i}{\bar{\lambda}}\mathrm{e}^{-a/\bar{\lambda}} + \frac{\cos 2k_1 z_i}{(1 + k_1^2\bar{\lambda}^2)^2} \\ & + \left(\frac{1}{(1 + k_1^2\bar{\lambda}^2)^2} - \frac{a}{2\bar{\lambda}}\frac{k_1^2\bar{\lambda}^2}{1 + k_1^2\bar{\lambda}^2}\right)\mathrm{e}^{-a/\bar{\lambda}}\cosh\frac{2z_i}{\bar{\lambda}} \\ & + \frac{k_1^2\bar{\lambda}^2}{1 + k_1^2\bar{\lambda}^2}\frac{z_i}{\bar{\lambda}}\mathrm{e}^{-a/\bar{\lambda}}\sinh\frac{2z_i}{\bar{\lambda}}\Bigg]^{-1}, \end{aligned} \tag{5.15}$$

where $k_1 = \pi/a$. The electron energy for the state (5.14) is [5.1]

$$\begin{aligned} E(\bar{\lambda}, a, z_i) = \langle\varphi|\mathcal{H}|\varphi\rangle = & \frac{\hbar^2 k_1^2}{2m^*} + \frac{\hbar^2}{2m^*\bar{\lambda}^2} \\ & - \frac{e^2 C^2\bar{\lambda}^2}{\kappa}\left(1 + \frac{\cos 2k_1 z_i}{1 + k_1^2\bar{\lambda}^2} - \frac{k_1^2\bar{\lambda}^2}{1 + k_1^2\bar{\lambda}^2}\mathrm{e}^{-a/\bar{\lambda}}\cosh\frac{2z_i}{\bar{\lambda}}\right). \end{aligned} \tag{5.16}$$

The equation for $\bar{\lambda}$ is found from the condition of the minimum energy

$$\frac{\partial}{\partial\bar{\lambda}}E(\bar{\lambda}, a, z_i) = 0. \tag{5.17}$$

Consider now some limiting cases:

A. $a \gg \bar{\lambda}$, $|z_i| \ll a$. In this case, the functions $\varphi(\mathbf{x})$ reduce to the hydrogenlike functions of a neutral donor in a three-dimensional crystal. The choice of $\varphi(\mathbf{x})$ in the form (5.14) is in accordance with this limit. Indeed, setting in (5.14–16)

$\cos k_1 z$, $\cos 2k_1 z \to 1$, and $\exp(-a/\bar{\lambda}) \to 0$, we obtain

$$\varphi(\mathbf{x}) \to \varphi_{1s}(\mathbf{x}) = \frac{1}{\sqrt{\pi \bar{\lambda}^3}} \mathrm{e}^{-|\mathbf{x}-\mathbf{x}_i|/\bar{\lambda}}, \tag{5.18}$$

where

$$\bar{\lambda} = a_B \equiv \frac{\kappa \hbar^2}{m^* e^2}, \; E = -E_B \equiv -\frac{m^* \mathrm{e}^4}{2\kappa^2 \hbar^2}. \tag{5.19}$$

For $m^* = 0.067$ m, $\kappa = 13$ the Bohr radius is $a_B = 102.7$ Å, and the Bohr energy (Rydberg) $E_B = 5.4$ meV.

B. $a \gg a_B$, $|z_i| = a/2$. If the impurity center is at the interface, then for an infinitely high barrier V_0 the eigenfunctions inside a wide quantum well will coincide with those $\varphi(\mathbf{x} - \mathbf{x}_i)$ of the homogeneous problem which vanish at the boundary $z = z_i$. To the ground state (of energy $-E_B/4$) corresponds the bulk $2p_0$ state, and to the first excited state (of energy $-E_B/9$), the $3d_{\pm 1}$ states. The trial function (5.14) of the ground state goes into the function (for $|z| < a/2$):

$$\varphi_{2p_0}(\mathbf{x}) = \frac{z - z_i}{4\sqrt{2\pi a_B^5}} \exp\left(-\frac{|\mathbf{x} - \mathbf{x}_i|}{2a_B}\right), \tag{5.20}$$

since for $z_i/a = \pm 1/2$ and $|z - z_i|/a \to 0$ we have

$$\cos\frac{\pi z}{a} = \cos\left(\pi \frac{z - z_i}{a} \pm \frac{\pi}{2}\right) = \mp \sin \pi \frac{z - z_i}{a} \to \mp \pi \frac{z - z_i}{a}.$$

C. $a \ll a_B$, $|z_i| < a/2$. In this limit, the eigenfunctions can be represented in the multiplicative form (5.12), where G does not depend on z and satisfies the two-dimensional Schrödinger's equation with the Hamiltonian

$$\mathcal{H}^{2D} = -\frac{\hbar^2}{2m^*}\left(\frac{\partial^2}{\partial \rho_x^2} + \frac{\partial^2}{\partial \rho_y^2}\right) - \frac{\mathrm{e}^2}{\kappa \rho}. \tag{5.21}$$

In this limiting case, the trial function also reduces to the exact solution

$$\varphi^{2D}(\boldsymbol{x}) = \begin{cases} \dfrac{2}{\sqrt{\pi \bar{\lambda}^2 a}} \cos\dfrac{\pi z}{a} \mathrm{e}^{-\rho/\bar{\lambda}} & \text{for} \quad |z| < a/2, \\ 0 & \text{for} \quad |z| > a/2, \end{cases} \tag{5.22}$$

where the variational parameter $\bar{\lambda}$ is equal to the Bohr radius of the two-dimensional electron $a_B^{(2D)}$, which is one half the three-dimensional Bohr radius a_B defined in (5.19). For the ground-state energy, we obtain $E = -E_B^{(2D)} = -4E_B$. The binding energy $\varepsilon_n^{(2D)}$ of the excited electron localized states decreases with increasing principal quantum number $n = 1, 2, \ldots$ as

$$\varepsilon_n^{(2D)} = \frac{E_B^{(2D)}}{(2n-1)^2} = \frac{E_B}{\left(n - \frac{1}{2}\right)^2}. \tag{5.23}$$

We recall that in the three-dimensional case

$$\varepsilon_n^{(3D)} = E_B/n^2 \quad (n = 1, 2, \ldots). \tag{5.24}$$

An essential difference between these states and the conventional 3D ones consists in the dependence of the binding energy on the position of the impurity center in space. For states below the lower $\nu = 1$ subband, the binding energy is defined as the difference

$$\varepsilon = E_1 - \langle \varphi | \mathcal{H} | \varphi \rangle , \tag{5.25}$$

where E_1 is the electron energy at subband bottom which for infinitely high barriers is $\hbar^2 k_1^2/2m^*$. Figure 5.1 presents the dependence of the binding energy for the ground state of a neutral donor (a) on the quantum-well thickness for two impurity center positions, $z_i = 0$ and $z_i = a/2$, and (b) on position z_i for the well thickness $a = a_B$. In accordance with the above analysis of the limiting cases A, B, and C, in a thick well $\varepsilon(0)$ and $\varepsilon(a/2)$ are close to E_B and $E_B/4$; these values increase with decreasing thickness and tend to $4E_B$ for $a \to 0$. It should be stressed that the absolute value of electron energy $E = E_1 - \varepsilon$ grows without limit with decreasing a, since $E_1 \propto a^{-2}$. The electron localized state in a quantum well is formed by an impurity center's Coulomb field also in the case where this center resides in the barrier. The binding energy decreases as the impurity moves away from the interface (to zero in the limit as $|z_i| \to \infty$), the electron cloud spreading in the (x, y) plane.

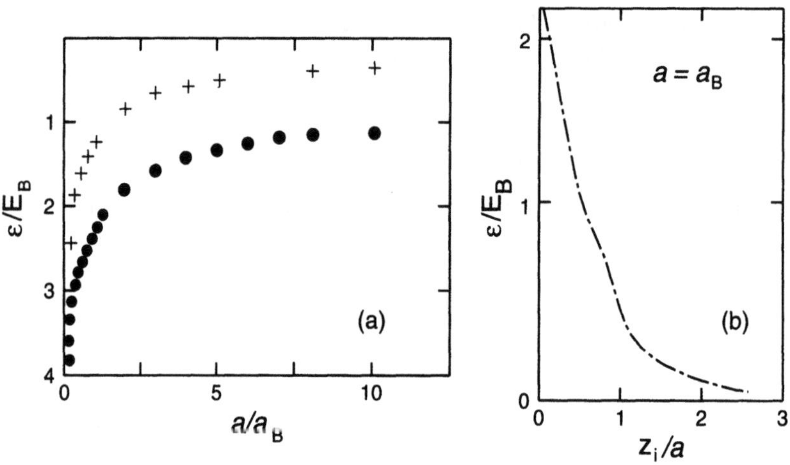

Fig. 5.1. **a** Dependence of the binding energy of donor-bound electron on quantum well thickness for an impurity atom at well center (dots) and at interface (crosses), **b** dependence of ε on donor position with respect to well center. Well thickness chosen equal to a_B. Calculation was made in terms of the infinite barrier model [5.1]

Taking into account the symmetry $V(z) = V(-z)$, the binding energy is an even function of z_i and has a maximum at $z_i = 0$. Therefore, for the density of states (per unit area), we obtain

$$g(\varepsilon) = \int dz n_i(z) \delta[\varepsilon - \varepsilon(z)] = \frac{2n_i(z_\varepsilon)}{|d\varepsilon/dz|}. \tag{5.26}$$

Here, $n_i(z)$ is the three-dimensional impurity concentration at the point z and z_ε the center position corresponding to the ground state binding energy ε. Since the function $\varepsilon(z_i)$ reaches a maximum at $z_i = 0$, i.e., in the vicinity of this point

$$\varepsilon(z_i) = \varepsilon_o - Bz_i^2, \tag{5.27}$$

the density of states has a root singularity

$$g(\varepsilon)|_{\varepsilon \to \varepsilon_o} \to \frac{n(0)}{\sqrt{B(\varepsilon_o - \varepsilon)}}. \tag{5.28}$$

In the limit as $a \gg a_B$, when the terms proportional to $\exp(-a/\bar{\lambda})$ in (5.15, 16) can be neglected, the coefficient $B = E_B k_1^4 a_B^2$.

In real conditions, the barrier height V_0 being finite, one has to take into account electron tunneling out of the quantum well, and the function $f(z)$ is nonzero both in the well and in the barrier (3.86b). Figure 5.2 presents the dependence on a of the binding energy of the ground ($1s$) and first excited even ($2s$) states of a neutral donor residing at the well's center. The calculation was performed for four different barrier heights V_0. According to (5.25), the binding energy ε is the difference between two quantities depending differently on the well thickness. For $a \gg a_B$, λ^{-1}, and for the donor in the central position, the main contribution to the $\varepsilon(a)$ dependence comes from the shift of the bottom of the lower subband E_1, since in this case one can neglect the effect of barriers on the bound state of the electron, its energy (to within an exponential accuracy) does not change, and the edge of the $\nu = 1$ subband shifts with decreasing a following a power law $E_1 \propto a^{-2}$. For a well thickness comparable to the Bohr radius a_B, the binding energy ε depends on a both because of the subband edge shift E_1 and because of compression along the z axis of the electron envelope wave function, with the result that the average separation between the electron and the impurity center decreases. As a decreases still further until the well thickness and the penetration length λ^{-1} become comparable, the probability for the electron to be inside the barrier increases noticeably, and the binding energy, on passing through a maximum, decreases as well. In the limit as $a/a_B \to 0$, the functions $\varphi(\mathbf{x})$ transform into hydrogen-like states of the electron localized at an impurity center in the homogeneous material of the barrier, the 1s electron binding energy approaching $(e^4 m_B / 2\hbar^2 \kappa^2)$, where m_B is the electron effective mass in the barrier layer.

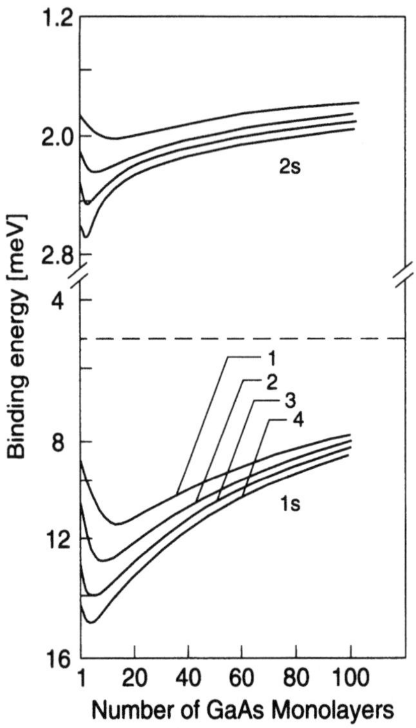

Fig. 5.2. Dependence of the binding energy for the ground and first excited states of donor-bound electron at the center of quantum well on well thickness in GaAs/Al$_x$Ga$_{1-x}$ As heterostructure for $x = 0.1$ (curve 1), 0.2 (2), 0.3 (3) and 0.4 (4) [5.2]. The dashed line identifies the ground state binding energy for neutral donor in bulk GaAs. The thickness of one monomolecular GaAs layer is 2.8 Å

5.1.2 Degenerate Band

Consider now the structure of a shallow impurity center for the band Γ_8 with an effective Hamiltonian (3.46). Since (5.1) does not render an explicit solution, one also uses here the variational method to find the wave functions and the energy. Rather than being arbitrary, the choice of the trial functions should satisfy certain conditions imposed by symmetry considerations. We are going to show how this choice is made in the method of invariants used in Chap. 3 in the construction of effective Hamiltonians.

It is well known that the ground state of a hole bound to an acceptor in a 3D zincblende semiconductor is four-fold degenerate, the four-component envelopes of the wave function $\varphi^{(m)}(\mathbf{x})$ in this state also transforming according to the representation Γ_8. Here the index m identifies degenerate states. The functions $\varphi^{(m)}$ can be conveniently written in the matrix form

$$\boldsymbol{\varphi}^{(m)}(\mathbf{x}) = \hat{R}(\mathbf{x})\mathbf{F}^{(m)}, \tag{5.29}$$

where $\mathbf{F}^{(m)}$ is the column with components $F_{m'}^{(m)} = \delta_{mm'}$ and, hence, $R_{m',m} \equiv \varphi_{m'}^{(m)}$. Since the columns $\boldsymbol{\varphi}^{(m)}$ and $\mathbf{F}^{(m)}$ transform under symmetry operation according to equivalent representations, the matrix $\hat{R}$ satisfies the condition of invariance

$$D(g)\hat{R}(g^{-1}\mathbf{x})D^{-1}(g) = \hat{R}(\mathbf{x}) \tag{5.30}$$

similar to condition (3.30) for the matrix $\mathcal{H}(\mathbf{k})$. This permits presenting the matrix $\hat{R}$ in the following general form

$$\begin{aligned}\hat{R} = C &+ P(\chi_1 M_1 + \chi_2 M_2) \\ &+ Q\left(\varepsilon_1 \left[J_y J_z\right]_s + \varepsilon_2 \left[J_z J_x\right]_s + \varepsilon_3 \left[J_x J_y\right]_s\right) \\ &+ \mathrm{i}S \sum_l \varepsilon_l V_l ,\end{aligned} \tag{5.31}$$

where

$$\begin{aligned} M_1 &= \frac{\sqrt{3}}{2}\left(J_x^2 - J_y^2\right), \\ M_2 &= J_z^2 - \frac{1}{2}\left(J_x^2 + J_y^2\right) = \frac{3}{2}\left(J_z^2 - \frac{5}{4}\right), \\ V_x &= \left[J_x, J_y^2 - J_z^2\right]_s \text{ etc.,}\end{aligned}$$

the even functions χ_1, χ_2 transform according to the representation E of group T_d as

$$\chi_1^0 = \frac{\sqrt{3}}{2}\left(x^2 - y^2\right), \chi_2^0 = z^2 - \frac{1}{2}\left(x^2 + y^2\right) \tag{5.32}$$

and the even functions ε_l, according to the representation F_2, as yz, zx, xy; C, P, Q and S are functions invariant under transformation of the group $T_d \times \mathrm{i} = O_h$. In the expansion (5.31) we have restricted ourselves to harmonics with $l \leqslant 2$, i.e., to s- and d-type functions. Taking into account the time-inversion symmetry, the functions introduced in (5.31) may be considered real. In the isotropic approximation $D = \sqrt{3}B$, the matrix $\hat{R}$ becomes a spherical invariant, the functions Q and P being related through $Q = 3P$, $S \equiv O$ and

$$\hat{R} = C + \frac{1}{2}Q\left[(\mathbf{J}\cdot\mathbf{x})^2 - \frac{5}{4}|\mathbf{x}|^2\right]. \tag{5.33}$$

The acceptor ground state in a quantum well splits into a pair of doubly degenerate levels. If the acceptor is at the center of a well with the principal axis $z \| [001]$, then the states Γ_8 go into the states Γ_6 and Γ_7 of the group D_{2d}. For the levels Γ_6 (or Γ_7) of a neutral acceptor the envelopes $\varphi_{3/2}^{(m)}, \varphi_{-3/2}^{(m)}$ and $\varphi_{1/2}^{(m)}, \varphi_{-1/2}^{(m)}$ should transform, respectively, according to the representations $\Gamma_6 \times \Gamma_6$ and $\Gamma_6 \times \Gamma_7$ (or $\Gamma_7 \times \Gamma_6$ and $\Gamma_7 \times \Gamma_7$). If the Hamiltonian for free holes includes only quadratic-in-k terms, the Hamiltonian H ($\mathbf{k}$, $\mathbf{x}$) in (5.4) has the symmetry D_{4h}, and in a calculation of the ground state Γ_6 or Γ_7 one should choose for the trial functions $\varphi^{(m)}$ only even functions of $\mathbf{x} - \mathbf{x}_\mathrm{i}$. Taking into account that as the symmetry is lowered, $T_d \to D_{2d}$, the representations A_1, E, F_2 of group T_d transform, respectively, according to the representations $A_1, A_1 + B_1, E + B_2$ of the group D_{2d}, we will have in place of (5.29), (5.31)

$$\boldsymbol{\varphi}^{(\Gamma_6, m)}(\mathbf{x}) = \hat{R}^{(6)}\mathbf{F}^{(m)}\ (m = \pm 3/2),$$

$$\varphi^{(\Gamma_7,m)}(\mathbf{x}) = \hat{R}^{(7)}\mathbf{F}^{(m)}\ (m = \pm 1/2)\,, \tag{5.34}$$

where

$$\begin{aligned}\hat{R}^{(t)} &= C^{(t)} + P_2^{(t)}\chi_2 + P_1^{(t)}\chi_1 M_1 \\ &+ Q_1^{(t)}(\varepsilon_1[J_yJ_z]_s + \varepsilon_2[J_zJ_x]_s) + Q_2^{(t)}\varepsilon_3[J_xJ_y]_s \\ &+ \mathrm{i}S_1^{(t)}(\varepsilon_1 V_x + \varepsilon_2 V_y) + \mathrm{i}S_2^{(t)} V_z\varepsilon_3.\end{aligned} \tag{5.35}$$

For $a \to \infty$ we have to set

$$\begin{aligned}&C^{(t)} \to C, \quad P_2^{(6)} \to \frac{3}{2}P, \quad P_2^{(7)} \to -\frac{3}{2}P, \\ &P_1^{(t)} \to P; \quad Q_1^{(t)}, \quad Q_2^{(t)} \to Q; \quad S_1^{(t)}, \quad S_2^{(t)} \to S.\end{aligned} \tag{5.36}$$

Equation (5.35) includes the matrices contained in the expansion (5.31). Strictly speaking, to these matrices one should add the matrices J_x, J_y and J_x^3, J_y^3 forming the basis for the representation E of the group D_{2d}, and the matrix $[J_z[J_xJ_y]_s]_s$ (representation B_1). There is no need to introduce in (5.35) also the contribution proportional to M_2 since the operation of the matrix M_2 on a linear combination of columns $F^{(m)}$ with $m = \pm 3/2$ (or with $m = \pm 1/2$) reduces to multiplying them by 3/2 (or by $-3/2$). For $D = \sqrt{3}B$, the Hamiltonian (5.4) has cylindrical symmetry, and one has to set in (5.35) $Q_2^{(t)} = 3P_1^{(t)}$, $S_{1,2}^{(t)} \equiv 0$. In their variational calculations of the ground state of an acceptor located at the well center, *Masselink* et al. [5.3] limited themselves to linear combinations of s- and d-type functions by choosing $\chi_1, \chi_2, \varepsilon_1, \varepsilon_2, \varepsilon_3$ in the form $\chi_1^0, \chi_2^0, yz, zx, xy$, respectively, and $C^{(t)}$, $P_{1,2}^{(t)}$, $Q_{1,2}^{(t)}$, in the form of linear combinations of exponentials $\exp\left[-\alpha_i(\rho^2 + \mu z^2)\right]$ with prescribed values of α_i. No contributions to $\hat{R}^{(t)}$ proportional to V_α were included. For the off-center impurity position when the Hamiltonian (5.4) is of symmetry C_{4v}, *Masselink* et al. added to the functions $\varphi^{(m)}$ $(\mathbf{x})$ an additional term $zD(\rho^2, z^2)\mathbf{F}^{(m)}$, which takes roughly into account the wave function asymmetry with respect to the replacement of $\mathbf{x}$ by $-\mathbf{x}$. Instead of solving the problem with the boundary conditions (5.5), the difference between the band parameters in the well and in the barrier was taken into account by a simple although somewhat artificial method.

Calculations show that the position of a level of a hole bound to an on-center acceptor in the GaAs/Al_x Ga_{1-x}As structure remains fixed to $a \simeq 150$ Å, while the binding energy measured from the heavy hole subband edge increases reaching ≈ 30 meV for x = 0.1–0.3 and a thickness $a = 150$ Å(the 3D value of the binding energy of an acceptor-bound hole in GaAs is 27 meV). For $a < 100$ Å the ground-state energy begins to increase with a simultaneous growth of splitting $\Delta E = E(\Gamma_7) - E(\Gamma_6)$ (Fig. 5.3). This splitting, however, is much smaller than that of the band states Γ_7 and Γ_6. It is only natural to relate this with the fact that the curvature of the lowest hole subband hh1 exceeds that of the lh1.

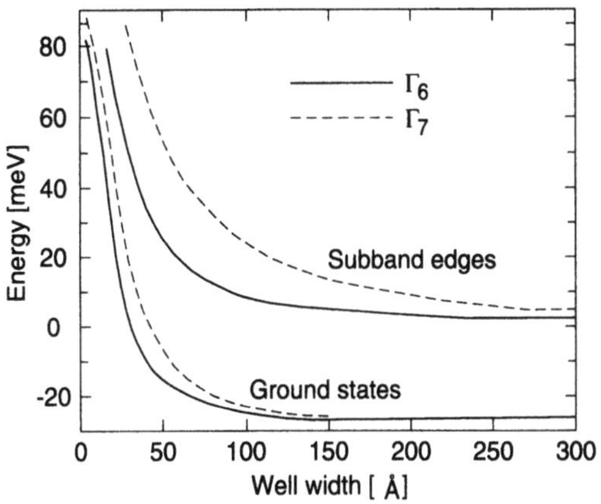

Fig. 5.3. Hole energy at the bottom of hh1 (Γ_6 symmetry) or lh1 (Γ_7 symmetry) subband and acceptor-bound hole energy at the lowest Γ_6, Γ_7 levels vs. quantum well thickness in $GaAs/Al_{0.3}Ga_{0.7}As$ heterostructure [5.3]

5.1.3 Effect of Magnetic and Electric Field and of Uniaxial Strain

In the presence of a magnetic field **B,** one has to add to the Hamiltonian (5.4) in the case of a nondegenerate band the term

$$V_{\mathbf{B}} = \mu_{\mathrm{B}} \left(\frac{m_0}{m^*} \mathbf{L} + g\mathbf{s} \right) \mathbf{B} + \frac{e^2}{8m^* c^2} |\mathbf{B} \times \mathbf{x}|^2, \tag{5.37}$$

where g is the g-factor, **s** the electron spin, and **L** the orbital angular momentum operator (5.10). We introduce here a dimensionless parameter

$$\gamma = \frac{\hbar \omega_{\mathrm{c}}}{2E_{\mathrm{B}}}, \tag{5.38}$$

where $\omega_{\mathrm{c}} = |e|B/m^*\mathrm{c}$ is the cyclotron frequency. For $\mathbf{B} \parallel z$ this permits us to rewrite V_{B} in the form

$$V_{\mathbf{B}} = E_{\mathrm{B}} \left[\gamma L_z + \frac{1}{4} \gamma^2 (x^2 + y^2) \right] + g\mu_{\mathrm{B}} (s \cdot \mathbf{B}) \cdot \tag{5.39}$$

Figure 5.4 presents the dependence $\varepsilon(a)$ for the 1s ground state and $2p_{-1}$ excited state of a neutral donor calculated for different values of the magnetic field (or parameter γ). The binding energy ε is defined, in accordance with (5.25), by the relation

$$\varepsilon = E_1 + \Delta E_1 - \langle \varphi | \mathcal{H} | \varphi \rangle, \tag{5.40}$$

where ΔE_1 is the shift of the lowest Landau level relative to E_1 which for $\mathbf{B} \parallel z$ and the same mass m^* in the well and the barrier is $\hbar\omega_{\mathrm{c}}/2 = \gamma E_{\mathrm{B}}$. The diamagnetic shift of the 1s state in the low-field limit $\gamma \ll 1$

$$\delta E = \frac{1}{4} \gamma^2 E_{\mathrm{B}} \int (x^2 + y^2) \varphi_{1s}^2 (\rho, z) \mathrm{d}\boldsymbol{x}. \tag{5.41}$$

Substituting the functions (5.18, 22) for φ_{1s} and integrating, we find that as the dimension is lowered, 3D $\to$ 2D, the integral in (5.41) changes from $2a_{\mathrm{B}}^2$ to (3/8) a_{B}^2.

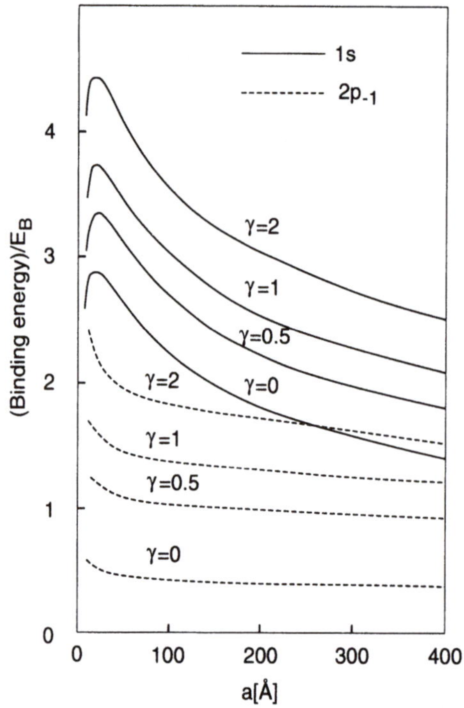

Fig. 5.4. Binding energy of donor-bound electron in the ground and $2p_{-1}$ excited states vs. quantum well thickness in $GaAs/Al_xGa_{1-x}As$ heterostructure ($x = 0.3$) for four different values of magnetic field specified by dimensionless parameter $\gamma = e\hbar B/2m^*cE_B$ [5.4]

When defining ε as in (5.25), paramagnetic interaction is not included and the spin splitting of the impurity level is taken into account separately. This splitting is

$$\Delta E = |g_{\text{eff}}|\mu_B B, \tag{5.42}$$

where

$$g_{\text{eff}} = g_A w_a + g_B w_b, \tag{5.43}$$

$g_{A,B}$ is the electron g-factor in the corresponding layer, and $w_{a,b}$ is the probability for an electron to be in the well or in the barrier. In a single quantum-well structure

$$w_a = \int\limits_{|z|<a/2} \varphi^2(\mathbf{x})\mathrm{d}\mathbf{x}, \quad w_b = 1 - w_a.$$

The g-factor for the electron in a superlattice will be calculated in Chap. 7.

The effect of magnetic field on the state of a neutral acceptor in the GaAs/AlGaAs quantum well structure was considered by *Masselink* et al. [5.3]. It was found that the paramagnetic splitting of the acceptor states Γ_6 and Γ_7 in a field $\mathbf{B} \parallel z$ in a quantum well practically coincides with that in a homogeneous material despite the fact that in the first case the levels Γ_6 and Γ_7 are split while, in the second, these levels form a fourfold degenerate level Γ_8.

An important feature of the effect of an electric field $\mathcal{I} \parallel z$ on neutral impurity centers in a quantum well is the difference of the binding energies of the ground state for centers located at z_i and $-z_i$, in particular, at the right- and left-hand edges of the well, $z_i = \pm a/2$. In the linear-in-field approximation, the difference

$$\varepsilon\left(\frac{a}{2}\right) - \varepsilon\left(-\frac{a}{2}\right) = e\mathcal{I}\int z\left[\varphi_{1s}^2\left(\mathbf{x};\frac{a}{2}\right) - \varphi_{1s}^2\left(-\mathbf{x};\frac{a}{2}\right)\right]\mathrm{d}\mathbf{x}$$

is determined by the component of the function $\varphi_{1s}^2(\mathbf{x})$, which is odd under replacement of z by $-z$. In order of magnitude, this splitting is close to $e\mathcal{I}a$.

Under uniaxial deformation along $z \parallel [001]$, we have for the strain tensor components $\varepsilon_{xy} = \varepsilon_{yz} = \varepsilon_{zx} = 0$, $\varepsilon_{xx} = \varepsilon_{yy} = S_{12}X$ and $\varepsilon_{zz} = S_{11}X$, where S_{ij} are the moduli of elasticity, and X is stress. By (3.77), the strain-induced contribution to the Hamiltonian in this case will be

$$\mathcal{H}_{\varepsilon} = \left[a(S_{11} + 2S_{12}) - \frac{2}{3}b(S_{11} - S_{12})M_2\right]X, \tag{5.44}$$

where the matrix M_2 was introduced in (5.31). In the electron representation, $b < 0$. Therefore in the case of compression of a bulk crystal ($X < 0$) the band electron states $\pm 1/2, \mathbf{k} = 0$ lie above the $\pm 3/2, \mathbf{k} = 0$ states (in the hole representation we have the reverse pattern). Whence it follows that compression of a quantum well structure (or superlattice) along the z axis should reduce the splitting between the heavy (hh1) and light (lh1) hole subbands for $\mathbf{k}_\perp = 0$, the splitting pattern of the hh1 and lh1 states becoming inverted at a certain stress. A similar inversion of splitting occurs also for the neutral acceptor states. Figure 5.5 shows the dependence of the Γ_6 and Γ_7 splitting on well thickness

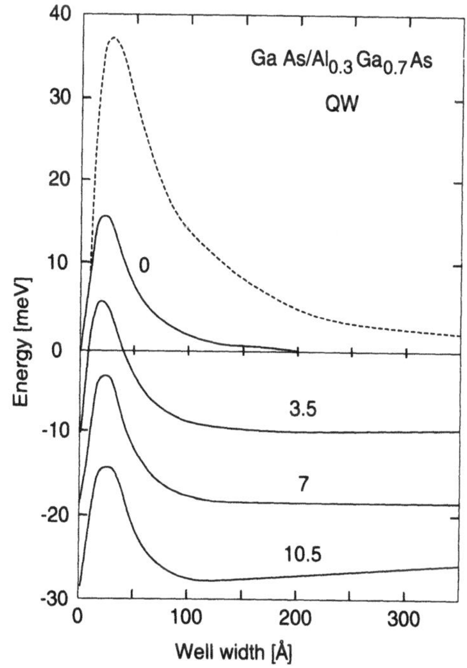

Fig. 5.5. Splitting of neutral acceptor ground state $E(\Gamma_7) - E(\Gamma_6)$ levels in GaAs/Al_xGa_{1-x}As SQW without and in the presence of uniaxial strain. The numbers specify the pressure (in kbar). For comparison, dotted line shows the splitting of hh1 and lh1 subbands [5.3]

in the absence of strain and for three different values of external pressure. At moderate pressures, $X \simeq 3.5$ kbar, there is a width $a(X)$ such that in a well with $a < a(X)$, *crystal-field* splitting dominates, and $\Delta E = E^{\rm h}(\Gamma_7) - E^{\rm h}(\Gamma_6) > 0$, while for a $> a(X)$ the strain-induced splitting is predominant, so that $\Delta E < 0$ (here $E^{\rm h}$ is the energy in the hole representation). At higher pressures (7 and 10.5 kbar) the splitting does not reverse its sign.

5.1.4 Periodic Heterostructure

Consider now how the presence of other quantum wells separated by barriers from the well containing a donor can affect the electron binding energy. In connection with this, one could study structures with two or three wells [5.5, 6]; we will, however, start immediately with an analysis of a periodic structure consisting of alternating layers of two semiconductors, say, GaAs and $Al_xGa_{1-x}As$, with well thickness a and barrier width b. For $\lambda b \gg 1$ such a structure may be considered a system of isolated quantum wells. As b decreases, the wave functions $f(z)$ in (5.12) corresponding to the neighboring wells start to overlap, the electron energy E_1 at the bottom of the lowest subband goes down and the subbands become minibands characterized by energy dispersion not only in k_x, k_y but in k_z as well. Obviously, initially the decrease of b affects E_1 more strongly than it does the energy $E = \langle\varphi|\mathcal{H}|\varphi\rangle$ of a donor-bound electron. Therefore the binding energy ε defined by (5.25) decreases with the appearance of additional quantum wells.

As already mentioned, for a single quantum well and for $a \to 0$, the effect of the well on the bound-electron state vanishes and $\varepsilon \to E_{\rm B}$ ($Al_xGa_{1-x}As$). In a short-period superlattice with $d = a + b \to 0$ the relative probability for a donor-bound electron to be found in a well and in a barrier depends on the relative magnitude of a and b, the binding energy lying between $E_{\rm B}$ (GaAs) and $E_{\rm B}$ ($Al_xGa_{1-x}As$).

To obtain the exact solution of the problem, the wave function of a donor-localized electron in a periodic structure can be expanded in super-Bloch functions (3.90)

$$\varphi(\mathbf{x}) = \sum_{\nu\mathbf{k}} G_{\nu\mathbf{k}}\chi_{\nu\mathbf{k}}(\boldsymbol{x}), \tag{5.45}$$

where

$$\chi_{\nu\mathbf{k}}(\boldsymbol{x}) = \mathrm{e}^{\mathrm{i}\mathbf{k}\cdot\boldsymbol{x}}U_{\nu\mathbf{k}}(z), \quad U_{\nu\mathbf{k}}(z+d) = U_{\nu\mathbf{k}}(z). \tag{5.46}$$

In superlattices with periods small compared to the Bohr radius a_B one may neglect in (5.45) the mixing of states from different minibands and use in the calculation of the function

$$G(\rho, z) = \sum_{\mathbf{k}} \mathrm{e}^{\mathrm{i}\mathbf{k}\cdot\mathbf{x}}G_{\mathbf{k}} \tag{5.47}$$

the effective mass approximation. In this approximation, the function $G(\boldsymbol{\rho}, z)$ satisfies Schrödinger's equation

$$\left(-\sum_j \frac{\hbar^2}{2M_{jj}}\frac{\partial^2}{\partial x_j^2} - \frac{e^2}{\kappa|\mathbf{x}-\mathbf{x}_\mathrm{i}|}\right) G = -\varepsilon G \tag{5.48}$$

with the effective masses of the electron in a SL calculated in terms of the Kronig-Penney model, see (3.93). Equation (5.48) can be solved by the variational technique just as it is done for homogeneous semiconductors with an anisotropic free-carrier effective mass. If in compositional materials the effective masses m_A and m_B are isotropic, the transverse masses M_{xx} and M_{yy} will coincide. In this case, good results for the ground-state energy can be obtained by the simplest choice of the trial function

$$G(\boldsymbol{\rho}, z) = \frac{1}{\sqrt{\pi a_\parallel a_\perp^2}} \exp\left[-\left(\frac{\rho^2}{a_\perp^2}+\frac{z^2}{a_\parallel^2}\right)^{1/2}\right], \tag{5.49}$$

where $a_\parallel$ and $a_\perp$ are the variational parameters. These parameters are determined from the condition of minimum of the functional

$$\langle G|\mathcal{H}|G\rangle$$
$$= E_\mathrm{B}^\perp\left[\frac{1}{3}\left(\frac{a_\mathrm{B}^\perp}{a_\perp}\right)^2\left(2+\frac{\gamma}{1-\beta^2}\right) - 2\frac{a_\mathrm{B}^\perp}{a_\perp}\frac{\arcsin\beta}{\beta}\right], \tag{5.50}$$

where $\beta^2 = (a_\perp^2 - a_\parallel^2)/a_\perp^2$, $\gamma = M_\perp/M_\parallel$, $M_\perp \equiv M_{xx} = M_{yy}$, $M_\parallel \equiv M_{zz}$, $E_\mathrm{B}^\perp = M_\perp e^4/2\hbar^2\kappa^2$, $a_\mathrm{B}^\perp = \hbar^2\kappa/M_\perp e^2$. By minimizing this functional we come to the following equations for $a_\perp$ and β:

$$a_\perp = a_\mathrm{B}^\perp \frac{1}{3}\left(2+\frac{\gamma}{1-\beta^2}\right)\frac{\beta}{\arcsin\beta},$$

$$\gamma = 2(1-\beta^2)^{3/2}\frac{\sqrt{1-\beta^2}\arcsin\beta}{\beta\sqrt{1-\beta^2}-\arcsin\beta}. \tag{5.51}$$

Equations (5.51) are written for the case $\gamma < 1$. For $\gamma > 1$, one should replace in them $1-\beta^2$ by $1+\beta^2$, and $\arcsin\beta$, by $\operatorname{arc\,sinh}\beta$. An analysis shows that the error of the variational solution compared with the exact value of the binding energy is $\simeq 7.5\%$ for $\gamma = 0$ and is smaller for other values $\gamma > 0$. The exact solution of (5.48) is obtained at $\gamma = 1$. Taking into account the first of (5.51), the expression for the binding energy simplifies to

$$\varepsilon = -\langle G|\mathcal{H}|G\rangle = E_\mathrm{B}^\perp \frac{a_\mathrm{B}^\perp}{a_\perp}\frac{\arcsin\beta}{\beta}. \tag{5.52}$$

If the effective mass is weakly anisotropic, the relations (5.51, 52) in the first approximation in the small parameter $1-\gamma$ take on the form

$$\beta^2 = \frac{5}{7}(1-\gamma), \quad a_\perp = a_\mathrm{B}^\perp\left[1-\frac{3}{14}(1-\gamma)\right], \tag{5.53}$$

$$\varepsilon = E_{\mathrm{B}} m_A^{-1}\left(\frac{2}{3}M_\perp + \frac{1}{3}M_\perp\right). \tag{5.54}$$

In the case of a SL with a short period satisfying the inequality

$$d \ll \left(m^* V_0/\hbar^2\right)^{-1/2} \tag{5.55}$$

the expressions for $M_\parallel$ and $M_\perp$ can be found by expanding the functions $\cos qd$, $\cos ka$, $\cosh \lambda b$, $\sin ka$, $\sinh \lambda b$ in (3.91) in powers of qd, ka, λb. For the boundary conditions (5.5, 9) when $\alpha = 0$ in (3.92a), the masses $M_\parallel$, $M_\perp$ are related to the effective masses m_A, m_B in the homogeneous materials of the well and the barrier by the following expressions

$$M_\parallel = \frac{m_A a + m_B b}{a+b}, \quad \frac{1}{M_\perp} = \frac{1}{a+b}\left(\frac{a}{m_A} + \frac{b}{m_B}\right). \tag{5.56}$$

Hence in the limiting case (5.55)

$$\gamma = \left[1 + \frac{ab}{m_A m_B}\left(\frac{m_A - m_B}{a+b}\right)^2\right]^{-1}. \tag{5.57}$$

Taking into account the linear relation of the effective mass with composition in the $Al_xGa_{1-x}As$ solid solution

$$m^*(x) = (0.067 + 0.083x)m. \tag{5.58}$$

We can see that in the $GaAs/Al_xGa_{1-x}As$ superlattice with $x \leqslant 0.4$ the electron effective mass anisotropy does not exceed 4%. Hence the expression in the parentheses in (5.54) may be replaced by $M_\perp$ or $M_\parallel$, so that

$$\varepsilon \simeq \frac{e^4}{2\kappa^2\hbar^2}\frac{m_A m_B(a+b)}{am_B + bm_A} \simeq \frac{e^4}{2\kappa^2\hbar^2}\frac{am_A + bm_B}{a+b}.$$

In our approximation the electron binding energy does not depend on donor position in the SL. This dependence appears, however, if in the calculation of the potential energy

$$V_{\mathrm{C}} = -\frac{e^2}{\kappa}\int \varphi^2(\mathbf{x})\frac{\mathrm{d}x}{|\mathbf{x} - \mathbf{x}_{\mathrm{i}}|} \tag{5.59}$$

one takes for $\varphi(\mathbf{x})$ not $G(\rho, z)$ but rather the function $G(\rho, z)U_0(z)$, where $U_0(z)$ is the periodic amplitude of the super-Bloch function (5.46) with $\mathbf{k} = 0$ and $G(\rho, z)$ a trial function which can also be chosen in the form (5.49). As a result, in (5.50), to the potential energy [5.7]

$$V_{\mathrm{C}}^0 = -2E_{\mathrm{B}}^\perp \frac{a_{\mathrm{B}}^\perp}{a_\perp}\frac{\arcsin\beta}{\beta}$$

one will have to add the term

$$\Delta V_{\mathrm{C}} = -\frac{e^2}{\kappa}\int G^2(\boldsymbol{x} + \boldsymbol{x}_{\mathrm{i}})\left[U_0^2(z + z_{\mathrm{i}}) - 1\right]\frac{\mathrm{d}\boldsymbol{x}}{|\boldsymbol{x}|}. \tag{5.60}$$

We expand now the periodic function $U_0(z)$ in a Fourier series

$$U_0(z) = \sum_l C_l e^{ib_l z}$$

where $b_l = 2\pi l/d (l = 0, \pm 1, \ldots)$. Using the integral relation

$$\int e_{iq\cdot x} \frac{dx}{|x|} = \frac{4\pi}{q^2} \tag{5.61}$$

we obtain for a short-period SL with $d < a_B$

$$\begin{aligned}
\Delta V_C &= -\frac{4e^2}{\kappa a_\parallel a_\perp^2} \sum_{l \neq l'} \frac{C_l^* C_{l'}}{(b_l - b_{l'})^2} e^{i(b_{l'} - b_l)z_i} \\
&= -\frac{4e^2}{\kappa a_\parallel a_\perp^2} \lim_{\eta \to +0} \frac{d}{d\eta} \int_0^\infty \left[U_0^2(z + z_i) - 1\right] e^{-\eta z}\, dz \\
&= -\frac{e^2}{\kappa a_\parallel a_\perp^2} \left[\frac{d^2}{3} - 2 \int_0^d U_0^2(z + z_i) \left(\frac{z^2}{d} - z\right) dz\right].
\end{aligned} \tag{5.62}$$

After some algebra, the expression for ΔV_C can be transformed to

$$\Delta V_C = \frac{e^2}{\kappa a_\parallel a_\perp} \left(\frac{d^2}{6} - R_1^2 + R_2^2 - 2z_i^2\right),$$

$$R_1^2 = \frac{4}{d} \int_0^{d/2} U^20 \left(z + \frac{d}{2}\right) z^2\, dz,\ R_2^2 = 4 \int_0^{z_i} U_0^2(z)(z_i - z)\, dz. \tag{5.63}$$

According to (3.90), the Bloch function at the bottom of the lowest miniband can be written as

$$U_0(z) = \begin{cases} C_a \cos k(z - \bar{z}_a) & \text{in wells,} \\ C_b \cosh \lambda(z - \bar{z}_b) & \text{in barriers.} \end{cases} \tag{5.64}$$

Here

$$k = \left(\frac{2m_A}{\hbar^2} E^0\right)^{1/2}, \quad \lambda = \left[\frac{2m_B}{\hbar^2}(V_0 - E^0)\right]^{1/2}, \tag{5.65}$$

E^0 is the bottom energy of the lowest miniband and $\bar{z}_{a,b}$ the position of the center of the corresponding well or barrier. For the boundary conditions (5.5, 9) and normalization

$$\int_d U_0^2(z)\, dz = d, \tag{5.66}$$

the coefficients C_a and C_b are determined by the expressions

$$C_a^2 = 2(a + b) \left[a\left(1 + \frac{\sin ka}{ka}\right) + b\left(1 + \frac{\sinh \lambda b}{\lambda b}\right) \frac{1 + \cos ka}{1 + \cosh \lambda b}\right]^{-1},$$

$$C_b^2 = \frac{1 + \cos ka}{1 + \cosh \lambda b} C_a^2. \tag{5.67}$$

Note that with the normalization (5.66)

$$R_2^2(Nd) = 2(Nd)^2,$$

where N is an integer. Therefore the function $R_2^2(z_i) - 2z_i^2$ and, hence, the function $\Delta V_C(z_i)$ are periodic with a period d.

Substituting (5.64) in (5.63) and integrating in z, we obtain

$$R_1^2 = \frac{1}{4d} \left(C_a^2 \left\{ \frac{d^3 - b^3}{3} + \frac{1}{k^3} [-2kb \cos ka \right.\right.$$
$$\left. +2kd + (k^2b^2 - 2) \sin ka] \right\} + C_b^2 \left\{ \frac{1}{3} b^3 \right.$$
$$\left.\left. + \frac{1}{\lambda^3} \left[(\lambda^2 b^2 + 2) \sinh \lambda b - 2\lambda b \cosh \lambda b \right] \right\} \right), \tag{5.68}$$

$$R_2^2 = C_a^2 \left(z_i^2 + \frac{\sin^2 kz_i}{k^2} \right) \quad \text{for} \quad |z_i| \leqslant a/2,$$

$$R_2^2 = C_a^2 \left[a \left(z_i - \frac{a}{4} \right) + \frac{1}{2k^2} (1 - \cos ka - ka \sin ka) \right.$$
$$\left. + \frac{z_i}{k} \sin ka \right] + C_b^2 \left[\left(z_i - \frac{a}{2} \right)^2 + \frac{1}{\lambda^2} (\cosh 2\lambda z_i - \cosh \lambda a) \right.$$
$$\left. - \frac{1}{\lambda} \left(z_i - \frac{a}{2} \right) \sinh \lambda a \right] \quad \text{for} \quad a/2 \leqslant |z_i| \leqslant d/2. \tag{5.69}$$

Figure 5.6 presents the behaviour of the electron binding energy to a donor in the GaAs/AlGaAs SL. With the impurity at well center (in the middle of

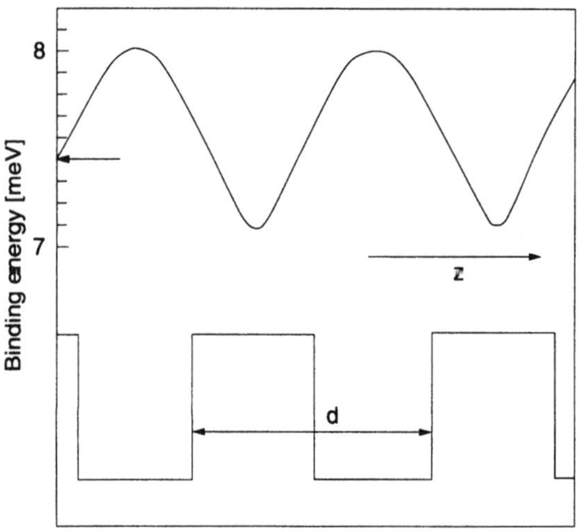

Fig. 5.6. Dependence of electron binding energy on impurity center position in GaAs/$Al_{0.35}Ga_{0.65}$As SL with $d = 50$ Å ($a = b$). Arrow specifies the binding energy neglecting the correction ΔV_C. For conveniency the figure shows, besides $\varepsilon(z_i)$, also the superstructural potential V(z) [5.7]

the barrier) ε is seen to be larger (smaller) than the energy (5.52) calculated neglecting the contribution ΔV_C.

5.2 Localized States at Superlattice Defects

A SL with defects can have states for carriers localized in one dimension. Such defects can be various distortions in the superlattice periodicity which leave the layers themselves homogeneous. For illustration, Fig. 5.7 shows several types of defects: (a) nonstandard well, (b) heteroboundary displacement, (c) symmetrical displacement of two neighboring heteroboundaries. Note that a perturbation of the electronic Hamiltonian (a change in potential energy and effective mass) compared with a perfect SL is limited along the z axis for defects b, c and not localized for type-a defects. The defects can be built into the SL on purpose and serve as elements of quantum microelectronics structures. They may, however, appear accidentally in the course of structure growth.

Consider the general procedure of calculating the binding energy and localization length of the electron bound to a defect which consists of a finite number of layers filling the region $z_- < z < z_+$. It is assumed that the regions $z < z_-$ and $z > z_+$ contain alternating layers of type A and B of thickness a and b, respectively. The equation for the energy of the localized electron E can be derived by matching at the points z_- and z_+ the solution $\varphi(\mathbf{x})$ at the defect with those in the regular regions dying out as $z \to \pm\infty$. The latter coincide with the solutions of Schrödinger's equation for a perfect SL, which satisfy,

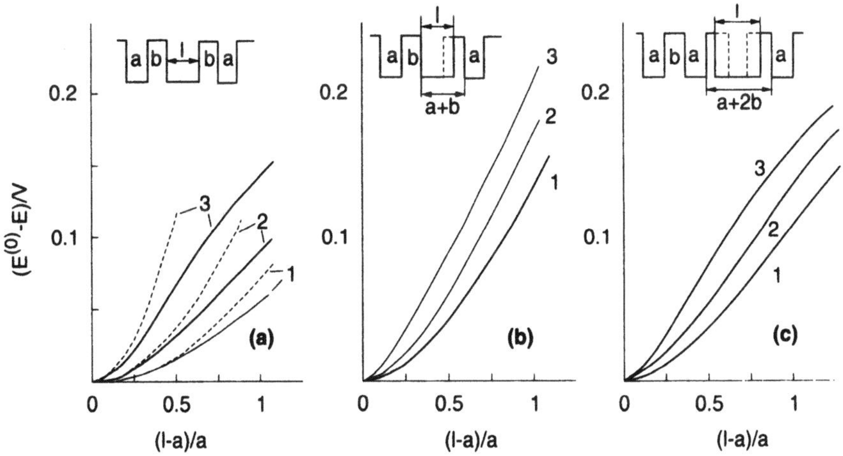

Fig. 5.7a-c. Dependence of binding energy for the electron at defects of the type **a** nonstandard well, **b** shifted heteroboundary, and **c** symmetric displacement of two neighboring heteroboundaries on l in GaAs/Al_xGa_{1-x}As superlattice for $x = 0.35$ ($V_0 = 0.25$ eV) and $a = b = 15$ Å (curves 1), 20 Å (2), and 30 Å (3) [5.8]

accordingly, the conditions

$$\varphi^{(\pm)}(z+a+b) = \mathrm{e}^{\pm\beta(a+b)}\varphi^{(\pm)}(z), \tag{5.70}$$

where Re $\{\beta\} > 0$. The dispersion $\beta(E, \mathbf{k}_\perp)$ is found from the equation

$$\cosh \beta d = F\left(E, k_x^2, k_y^2\right), \tag{5.71}$$

which differs from (3.91) in cosh βd being substituted for $\cos qd$. This is only natural since localized states should lie in the region of forbidden states where the function $F(E, k_x^2, k_y^2)$ takes on values outside the interval $[-1, 1]$. Note that the quantity $(\mathrm{Re}\ \{\beta\})^{-1}$ determines the electron localization length along the z axis. We consider next states below the bottom of the lowest miniband for which $F > 1$ and the values of $\beta(E, \mathbf{k}_\perp)$ are real. In the upper forbidden subbands where $F < -1$, the quantity β contains an imaginary contribution $\mathrm{i}\pi/\mathrm{d}$ defined unambiguously to within $2\pi\mathrm{i}N/\mathrm{d}$, where N is an integer.

We introduce now dimensionless transfer matrices $\hat{T}_i$ relating pairs of values of φ and $\tilde{\varphi} \equiv (m_A/m_i)k^{-1}(d\varphi/dz)$ at the left- and right-hand boundaries of the ith layer. For a layer of type A or B of thickness L, these matrices can be written as

$$\hat{T}_A(L) = \begin{bmatrix} \cos kL & \sin kL \\ -\sin kL & \cos kL \end{bmatrix}, \quad \hat{T}_B(L) = \begin{bmatrix} \cosh \lambda L & \eta^{-1}\sinh \lambda L \\ \eta \sinh \lambda L & \cosh \lambda L \end{bmatrix}, \tag{5.72}$$

where

$$\eta = \frac{m_A}{m_B}\frac{\lambda}{k}. \tag{5.73}$$

Taking into account the continuity of the functions φ and $\tilde{\varphi}$ at the heteroboundaries (5.5, 9), we can find the relation between the values of the functions (5.70) for $z = z_+$ and $z = z_-$

$$\begin{bmatrix} \varphi^{(+)}(z_+) \\ \tilde{\varphi}^{(+)}(z_+) \end{bmatrix} = \hat{T} \begin{bmatrix} \varphi^{(-)}(z_-) \\ \tilde{\varphi}^{(-)}(z_-) \end{bmatrix}, \tag{5.74}$$

where $\hat{T}$ is the matrix of transfer through the entire defected region:

$$\hat{T} = \prod_{i=1}^{N} \hat{T}_i(L_i). \tag{5.75}$$

This matrix satisfies the condition

$$\mathrm{Det}\{\hat{T}\} = T_{11}T_{22} - T_{12}T_{21} = 1 \tag{5.76}$$

since

$$\mathrm{Det}\{\hat{T}\} = \prod_{i=1}^{N} \mathrm{Det}\{\hat{T}_i(L_i)\}$$

and, by (5.72), $\mathrm{Det}\{\hat{T}_i(L_i)\} = 1$. From the condition of solvability of (5.74), one obtains the equation for the energy of the localized electron states

$$w^{+}(z_{+}) = \frac{T_{21} + T_{22}w^{-}(z_{-})}{T_{11} + T_{12}w^{(-)}(z_{-})} \tag{5.77}$$

where

$$w^{(\pm)} = \tilde{\varphi}^{(\pm)}/\varphi^{\pm}.$$

At the heteroboundaries AB and BA of a regular superlattice, we have

$$\begin{aligned} w_{AB}^{\pm} = -w_{BA}^{\mp} &= \frac{\cos ka - \mathrm{e}^{\pm\beta(a+b)}\cosh\lambda b}{\sin ka + \eta^{-1}\mathrm{e}^{\pm\beta(a+b)}\sinh\lambda b} \\ &= \frac{\mathrm{e}^{\mp\beta(a+b)} - \cos ka\cosh\lambda b - \eta\sin ka\sinh\lambda b}{\sin ka\cosh\lambda b + \eta^{-1}\cos ka\sinh\lambda b}. \end{aligned} \tag{5.78}$$

The first equality here follows from the symmetry of a perfect SL under replacement of z by $-z$, and the last one is another form of writing the dispersion relation (5.71). In place of $w^{(+)}(z_{+})$ in (5.77), one has to substitute $w_{AB}^{(+)}$ or $w_{BA}^{(+)}$ depending on whether the layer adjoining the defected region is type B or A. Similarly, $w^{(-)}(z_{-})$ has to be replaced by $w_{AB}^{(-)}$ or $w_{BA}^{(-)}$ if the layer on the left of the defect is type A or B, respectively.

The above matrix formalism is valid also for defects with a continuous or quasicontinuous distribution of the parameters $m^{*}(z)$ and $V_0(z)$ in the region $z_{-} < z < z_{+}$. We will have in this case

$$\hat{T} = t_z \exp\left(\int_{z_-}^{z_+} \hat{M}(z)\,\mathrm{d}z\right),$$

where

$$\hat{M}(z) = \begin{bmatrix} 0 & \lambda(z)/\eta(z) \\ \lambda(z)\eta(z) & 0 \end{bmatrix}$$

and t_z is an operator of ordering in z.

We will apply now the general equations (5.75-78) to analyze electron states at defects of type a, b and c shown in Fig. 5.7. In each of these three cases, at least one local level or, more rigorously, a two-dimensional subband of localized states $E(k_{\perp})$ splits off the bottom of the lowest miniband. Equation (5.77) for the dispersion $E(k_{\perp})$ simplifies considerably if the symmetry or other properties of the defect are taken into account. Indeed, even states localized at an enlarged-well-type defect (Fig. 5.7a) are determined by the equation

$$-\tan(kl/2) + w_{AB}^{(+)} = 0, \tag{5.79}$$

which is obtained from (5.77) after the substitution

$$w^{(+)}(z_{+}) = w_{AB}^{(+)}, \quad w^{(-)}(z_{-}) = w_{BA}^{(-)},$$
$$T_{11} = T_{22} = \cos kl, \quad T_{12} = -T_{21} = \sin kl$$

and after taking into consideration the relation $w_{BA}^{(-)} = -w_{AB}^{(+)}$. The equation for the odd states can be derived from (5.79) by substituting $-\cot(kl/2)$ for $\tan(kl/2)$.

For the heteroboundary-shift-type defect (Fig. 5.7b), the matrix $\hat{T} = \hat{T}_B(d-l)\hat{T}_A(l-a)$ can be conveniently written in the form:

$$\hat{T}_B(b)\hat{T}_B^{-1}(l-a)\hat{T}_A(l-a).$$

This permits presenting (5.77) as

$$w_{AB}^{+} = \frac{T'_{21} + T'_{22}w_{AB}^{-}}{T'_{11} + T'_{12}w_{AB}^{(-)}}, \tag{5.80}$$

where $\hat{T}' = \hat{T}_B^{-1}(l-a)\hat{T}_A(l-a)$. With the equation written this way, one can readily check the validity of the transition $\beta \to 0$, $E(\mathbf{k}_\perp) \to E(0, \mathbf{k}_\perp)$ in the limit $l - a \to +0$, where E(q, $\mathbf{k}_\perp$) is the electron dispersion in a perfect SL.

For a defect of the type *two symmetrically shifted heteroboundaries* we obtain

$$\begin{aligned} w_{AB}^{(+)} &= \frac{1}{2\bar{T}_{12}}\left[\bar{T}_{11} + \bar{T}_{22} \pm \sqrt{\left(\bar{T}_{11} + \bar{T}_{22}\right)^2 - 4\bar{T}_{12}\bar{T}_{21}}\right] \\ &= \left(\bar{T}_{11} \pm 1\right)\bar{T}_{12}, \end{aligned} \tag{5.81}$$

where

$$\hat{\hat{T}} = \hat{T}_B^{-1}(\bar{c})\hat{T}_A(a + 2\bar{c})\hat{T}_B^{-1}(\bar{c}), \quad \bar{c} = (l-a)/2.$$

This expression was derived taking into account the identity (5.76) and the equality of the diagonal components of matrix $\hat{\hat{T}}$. The upper and lower signs in (5.81) correspond to the odd and even states, accordingly.

Figures (5.7a-c) display the calculated binding energies $\varepsilon = E^0 - E$ for the lowest localized state for $\mathbf{k}_\perp = 0$. Shown in the inset of each figure is the type of the defect. The calculations were carried out for the GaAs/$Al_xGa_{1-x}As$ superlattice for $a = b$ and for the composition $x = 0.35$ when $V_0 = 0.25$ eV, $m_B = 1.43m_A$, $m_A \equiv m^*(\text{GaAs}) = 0.067m$. As expected, in all cases $E^0 - E \to 0$ for $l \to a$. The dispersion E($\mathbf{k}_\perp$) is characterized by an effective mass, its value lying between m_A and m_B.

The dashed curves in Fig. 5.7a were calculated by an approximate expression for the binding energy of an electron at a nonstandard-well-type defect

$$E^0 - E = \frac{E^0}{8}\left(\frac{l-a}{a+b}\right)^2 \frac{m_A}{M_\parallel}(\cosh 2\lambda b - 1)\left(\eta + \frac{1}{\eta}\right)^2, \tag{5.82}$$

where $M_\parallel = M_{zz}$ is the electron longitudinal effective mass in the $\nu = 1$ subband, λ and η being defined by (5.65, 73). For $(l-a)/a \ll 1$, the exact and approximate relations $E(l)$ are seen to be close.

5.3 Excitons

We introduce now the states of noninteracting electron-hole pairs

$$|s\mathbf{k}_e, m\mathbf{k}_h\rangle = a^+_{s\mathbf{k}_e} b^+_{m\mathbf{k}_h}|0\rangle, \tag{5.83}$$

where s, m are the indices identifying the conduction and valence band branches, $\mathbf{k}_{e,h}$ is the electron or hole wave vector, $|0\rangle$ the ground state of the crystal in which all the electron states in the valence band are filled, and those in the conduction band are empty, $a^+_{s\mathbf{k}}$ and $b^+_{m\mathbf{k}}$ are the creation operators of the electron in s, $\mathbf{k}$ state, and of the hole, in m, $\mathbf{k}$ state. The many-particle state (5.83) is sometimes presented in the form of a *two-particle wave function*

$$|s\mathbf{k}_e, m\mathbf{k}_h\rangle \to \psi_{s\mathbf{k}_e}(\mathbf{x}_e)\, \psi^{h}_{m\mathbf{k}_h}(\mathbf{x}_h), \tag{5.84}$$

where $\psi_{s\mathbf{k}}(\mathbf{x}_e)$ is the Bloch function for the electron in the conduction band, and $\psi^h_{m\mathbf{k}}(\mathbf{x}_h)$ is that in the hole representation obtained from the corresponding electron Bloch function for the valence band acted upon by the time inversion operator K. With the correspondence between the states in the electron and hole representations properly established, the use of the nomenclature (5.84), rather than presenting any difficulties, offers a number of advantages.

The wave function of the Wannier-Mott exciton can be expanded in states (5.83)

$$\Psi^{\rm exc} = \sum_{s\mathbf{k}_e, m\mathbf{k}_h} C_{s\mathbf{k}_e, m\mathbf{k}_h}|s\mathbf{k}_e, m\mathbf{k}_h\rangle. \tag{5.85}$$

By the envelope wave function of the exciton in the $\mathbf{r}$-representation, one understands the function obtained by inverse Fourier transform

$$\varphi_{sm}(\mathbf{x}_e, \mathbf{x}_h) = \sum_{\mathbf{k}_e\mathbf{k}_h} e^{i(\mathbf{k}_e\cdot\mathbf{x}_e + \mathbf{k}_h\cdot\mathbf{x}_h)} C_{s\mathbf{k}_e, m\mathbf{k}_h}. \tag{5.86}$$

In the effective-mass approximation, this function satisfies the two-particle Schrödinger's equation

$$\sum_{s'm'} \mathcal{H}_{sm,s'm'}\left(\hat{\mathbf{k}}_e, \hat{\mathbf{k}}_h\right) \varphi_{s'm'}(\mathbf{x}_e, \mathbf{x}_h) = E\varphi_{sm}(\mathbf{x}_e, \mathbf{x}_h), \tag{5.87}$$

Here E is the exciton excitation energy, i.e., the energy of the crystal excited state (5.85) measured from the ground-state energy $|0\rangle$, $\hat{\mathbf{k}}_e = -i\partial/\partial\mathbf{x}_e$, $\hat{\mathbf{k}}_h = -i\partial/\partial\mathbf{x}_h$, $H_{sm,sm'}(\mathbf{k}_e, \mathbf{k}_h)$ is the effective Hamiltonian of the electron-hole pair, which for a homogeneous semiconductor has the form

$$\mathcal{H}_{sm,s'm'}(\mathbf{k}_e, \mathbf{k}_h) = \delta_{mm'}\mathcal{H}^e_{ss'}(\mathbf{k}_e) + \delta_{ss'}\mathcal{H}^h_{mm'}(\mathbf{k}_h) + \delta_{ss'}\delta_{mm'}\left[E_g - \frac{e^2}{\kappa\,|\mathbf{x}_e - \mathbf{x}_h|}\right] \tag{5.88}$$

and which can be made more compact

$$\mathcal{H}(\mathbf{k}_e, \mathbf{k}_h) = \mathcal{H}^e(\mathbf{k}_e) + \mathcal{H}^h(\mathbf{k}_h) + E_g - \frac{e^2}{\kappa r}. \tag{5.89}$$

Here $\mathcal{H}^{\mathrm{e}}$ and $\mathcal{H}^{\mathrm{h}}$ are the effective (single-particle) Hamiltonians for the electron and hole, E_{g} is the energy gap, $\mathbf{r} = \mathbf{x}_{\mathrm{e}} - \mathbf{x}_{\mathrm{h}}$, κ is the dielectric constant whose frequency dispersion is neglected. Equation (5.87) with Hamiltonian (5.88) describes the state of the so-called mechanical exciton whose calculation does not include the short-range interaction between the electron and the hole in the exciton. This interaction, as well as the long-range (also called annihilation, or longitudinal-transverse) one along with exciton-photon coupling will be discussed later.

The two-particle Hamiltonian describing the exciton in a heterostructure includes superstructure potentials for the electron and the hole

$$\mathcal{H}(\mathbf{k}_{\mathrm{e}}, \mathbf{k}_{\mathrm{h}}) = \mathcal{H}^{\mathrm{e}}(\mathbf{k}_{\mathrm{e}}) + \mathcal{H}^{\mathrm{h}}(\mathbf{k}_{\mathrm{h}}) + E_{\mathrm{g}} - \frac{e^2}{\kappa r} + V_{\mathrm{e}}(z_{\mathrm{e}}) + V_{\mathrm{h}}(z_{\mathrm{h}})\,. \tag{5.90}$$

At the interface the parameters of the Hamiltonians $\mathcal{H}^e$ and $\mathcal{H}^h$ may change stepwise; however, for the sake of simplicity, this change is neglected in this section. By analogy with the preceding section, we use here for boundary conditions the continuity of the wave function in $\mathbf{x}_{\mathrm{e}}$ and $\mathbf{x}_{\mathrm{h}}$:

$$\begin{aligned} \varphi(\mathbf{x}_{\mathrm{e}}, \mathbf{x}_{\mathrm{h}})\big|_{z_{\mathrm{e}}=z_-} &= \varphi(\mathbf{x}_{\mathrm{e}}, \mathbf{x}_{\mathrm{h}})\big|_{z_{\mathrm{e}}=z_+}\,, \\ \varphi(\mathbf{x}_{\mathrm{e}}, \mathbf{x}_{\mathrm{h}})\big|_{z_{\mathrm{h}}=z_-} &= \varphi(\mathbf{x}_{\mathrm{e}}, \mathbf{x}_{\mathrm{h}})\big|_{z_{\mathrm{h}}=z_+} \end{aligned} \tag{5.91}$$

as well as the conditions taking into account the continuity of particle flow through the boundary between the layers

$$\begin{aligned} \left[v_z^{\mathrm{e}}\left(\hat{\mathbf{k}}_{\mathrm{e}}\right)\varphi\right]_{z_{\mathrm{e}}=z_-} &= \left[v_z^{\mathrm{e}}\left(\hat{\mathbf{k}}_{\mathrm{e}}\right)\varphi\right]_{z_{\mathrm{e}}=z_+}\,, \\ \left[v_z^{\mathrm{h}}\left(\hat{\mathbf{k}}_{\mathrm{h}}\right)\varphi\right]_{z_{\mathrm{h}}=z_-} &= \left[v_z^{\mathrm{h}}\left(\hat{\mathbf{k}}_{\mathrm{h}}\right)\varphi\right]_{z_{\mathrm{h}}=z_+}\,. \end{aligned} \tag{5.92}$$

Here the components $\varphi_{sm}(\mathbf{x}_{\mathrm{e}}, \mathbf{x}_{\mathrm{h}})$ are combined in one many-component vector $\varphi(\mathbf{x}_{\mathrm{e}}, \mathbf{x}_{\mathrm{h}})$, $v_z^{\mathrm{e,h}}$ is the projection on the z-axis of the electron or hole velocity operator

$$\mathbf{v}^{\mathrm{e}}(\mathbf{k}) = \hbar^{-1}\nabla_{\mathbf{k}}\mathcal{H}^{\mathrm{e}}(\mathbf{k}), \quad \mathbf{v}^{\mathrm{h}}(\mathbf{k}) = \hbar^{-1}\nabla_{\mathbf{k}}\mathcal{H}^{\mathrm{h}}(\mathbf{k}), \tag{5.93}$$

and z_- and z_+ are the left- and right-hand interface boundaries.

For nondegenerate bands with effective masses $m^{\mathrm{c}}_{\parallel,\perp}$ and $m^{\mathrm{v}}_{\parallel,\perp}$ the exciton wave function envelope can be presented in the form of a product

$$\varphi_{sm}(\mathbf{x}_{\mathrm{e}}, \mathbf{x}_{\mathrm{h}}) = \Phi_{sm}\phi(\mathbf{x}_{\mathrm{e}}, \mathbf{x}_{\mathrm{h}})\,, \tag{5.94}$$

where Φ depends only on the spin indices s, m. In the equation for $\phi(\mathbf{x}_{\mathrm{e}}, \mathbf{x}_{\mathrm{h}})$, the variables can be partially separated

$$\phi(\boldsymbol{x}_{\mathrm{e}}, \boldsymbol{x}_{\mathrm{h}}) = \mathrm{e}^{\mathrm{i}\boldsymbol{k}_\perp\cdot\mathbf{R}_\perp}\varphi(\boldsymbol{\rho}, z_{\mathrm{e}}, z_{\mathrm{h}})\,, \tag{5.95}$$

where $\mathbf{R}_\perp$ is the in-plane component of the exciton center of mass (x, y plane)

$$\mathbf{R}_\perp = \left(m^{\mathrm{c}}_\perp\mathbf{x}_{\mathrm{e}\perp} + m^{\mathrm{v}}_\perp\mathbf{x}_{\mathrm{h}\perp}\right)/\bar{M}_\perp\,, \tag{5.96}$$

$\boldsymbol{\rho} = \mathbf{x}_{e\perp} - \mathbf{x}_{h\perp}$, and the function $\varphi(\boldsymbol{\rho}, z_e, z_h)$ satisfies the equation

$$\left(\mathcal{H}_0^e + \mathcal{H}_0^h + \mathcal{H}^{eh}\right)\varphi = \left(E - E_g - \frac{\hbar^2 K_\perp^2}{2\bar{M}_\perp}\right)\varphi, \tag{5.97}$$

$$\mathcal{H}_0^e = -\frac{\hbar^2}{2m_\parallel^c}\frac{\partial}{\partial z_e^2} + V_e(z_e), \quad \mathcal{H}_0^h = -\frac{\hbar^2}{2m_\parallel^v}\frac{\partial^2}{\partial z_h^2} + V_h(z_h),$$

$$\mathcal{H}^{eh} = -\frac{\hbar^2}{2\mu_\parallel}\left(\frac{\partial^2}{\partial \rho_x^2} + \frac{\partial^2}{\partial \rho_y^2}\right) - \frac{e^2}{\kappa r}, \tag{5.98}$$

$\bar{M}_\perp = m_\perp^c + m_\perp^v$ is the mass of translational motion of the exciton as a whole in the x, y plane; $\mu_\perp = m_\perp^c m_\perp^v / \bar{M}_\perp$ is the reduced effective mass. The binding energy of the exciton formed of an electron in the lowest subband e1 and of a hole in the upper subband h1 is determined, in analogy with (5.25), as

$$\varepsilon = E_{e1} + E_{h1} - \langle\varphi\left|\mathcal{H}_0^e + \mathcal{H}_0^h + \mathcal{H}^{eh}\right|\varphi\rangle, \tag{5.99}$$

where E_{e1}, E_{h1} are the electron and hole energies at the bottom of the e1, h1 subbands (minibands).

5.3.1 Nondegenerate Bands, Type I Heterostructure

Consider the case of a three-layer *barrier-well-barrier* structure in the limit of very high barriers assuming the electron and hole effective masses m_c, m_v to be isotropic. In a thick well, $a \gg a_B$, one may neglect the distortion of internal motion of the electron-hole pair in the exciton and describe its state by a wave function

$$\varphi(\boldsymbol{\rho}, z_e, z_h) = F(Z)\varphi(\mathbf{r}) \tag{5.100}$$

with the exciton center of mass on the z axis located at

$$Z = (m_c z_e + m_v z_h)/\bar{M}, \quad \bar{M} = m_c + m_v, \tag{5.101}$$

$\varphi(\mathbf{r})$ is the wave function of the electron and hole relative motion in a homogeneous material. For the exciton ground state we have

$$\varphi_{1s}(\mathbf{r}) = \left(\pi a_B^3\right)^{-1/2}\exp\left(-r/a_B\right), \tag{5.102}$$

where the exciton Bohr radius is

$$a_B = \frac{\kappa\hbar^2}{\mu e^2}, \tag{5.103}$$

$\mu = m_c m_v/(m_c + m_v)$. With the boundary conditions F $(\pm\ a/2) = 0$ we have for the excitation energy of the 1s-exciton:

$$E_{1s} = E_g - E_B^{exc} + \frac{\hbar^2 K_\perp^2}{2\bar{M}} + \frac{\hbar}{2\bar{M}}\left(\frac{\pi}{a}\right)^2, \tag{5.104}$$

where the last term describes the quantum confinement energy of the exciton and the exciton Rydberg is

$$E_{\mathrm{B}}^{\mathrm{exc}} = \frac{e^4 \mu}{2\hbar^2 \kappa^2}. \tag{5.105}$$

In terms of this model

$$E_{\mathrm{el}} = \frac{\hbar^2}{2m_{\mathrm{c}}}\left(\frac{\pi}{a}\right)^2, \quad E_{\mathrm{hl}} = \frac{\hbar^2}{2m_{\mathrm{v}}}\left(\frac{\pi}{a}\right)^2 \tag{5.106}$$

and the binding energy in a thick well

$$\varepsilon_{1s} = E_{\mathrm{B}}^{\mathrm{exc}} + \frac{1}{2}\left(\frac{\hbar\pi}{a}\right)^2\left(\frac{1}{\mu} - \frac{1}{\bar{M}}\right). \tag{5.107}$$

As the well width a decreases, the quantum confinement of carriers in the well begins to predominate over the Coulomb interaction, the approximation (5.100) becoming invalid. If (5.97) is solved by the variational technique, it is customary to take as the trial function

$$\varphi(\rho, z_{\mathrm{e}}, z_{\mathrm{h}}) = f_{\mathrm{e}}(z_{\mathrm{e}})\, f_{\mathrm{h}}(z_{\mathrm{h}})\, G(\rho, z), \tag{5.108}$$

where $z = z_{\mathrm{e}} - z_{\mathrm{h}}$, $f_{\mathrm{e,h}}$ are the eigenfunctions of the operators $\mathcal{H}_0^{\mathrm{e}}$ and $\mathcal{H}_0^{\mathrm{h}}$ (3.83, 86). As an illustration, we present a comparatively simple variant of choice of $G(\rho, z)$ for the exciton ground state [5.9]

$$G_{1s}(\rho, z) = C\left(1 + \alpha z^2\right) \mathrm{e}^{-\delta\left(\rho^2 + z^2\right)^{1/2}}, \tag{5.109}$$

where α and δ are the variational parameters and C a normalizing factor.

The left- and right-hand barriers of the quantum well *press* the electron and the hole to one another, with the result that the Coulomb interaction between them increases, as does also the binding energy, which in the case of high barriers varies from $E_{\mathrm{B}}^{\mathrm{exc}}$ in a broad well to the two-dimensional Rydberg $4E_{\mathrm{B}}^{\mathrm{exc}}$ for $a \ll a_{\mathrm{B}}$. In the two-dimensional approximation, we will have for the exciton ground state

$$\varphi(\rho, z_{\mathrm{e}}, z_{\mathrm{h}}) = \begin{cases} \sqrt{\dfrac{2}{\pi}\dfrac{4}{a a_{\mathrm{B}}}} \cos\dfrac{\pi z_{\mathrm{e}}}{a} \cos\dfrac{\pi z_{\mathrm{h}}}{a} \mathrm{e}^{-2\rho/a_{\mathrm{B}}} & \text{for} \quad |z_{\mathrm{e,h}}| < a/2, \\ 0 & \text{for} \quad |z_{\mathrm{e}}| \text{ or } |z_{\mathrm{h}}| > a/2. \end{cases} \tag{5.110}$$

For a finite barrier, the binding energy depends nonmonotonically on a (Fig. 5.8), namely, as the well thickness decreases, the energy ε_{1s} reaches a maximum and then falls off since at small a the electron and the hole bound in the exciton reside predominantly in the region of the barriers. In the limit of very small thicknesses, this state of the 1s-exciton may be considered a three-dimensional exciton in the barrier material *attached* to a thin layer with a potential well for the electron and the hole. The binding energy of such an exciton is close to that of a free exciton in the homogeneous barrier of the material.

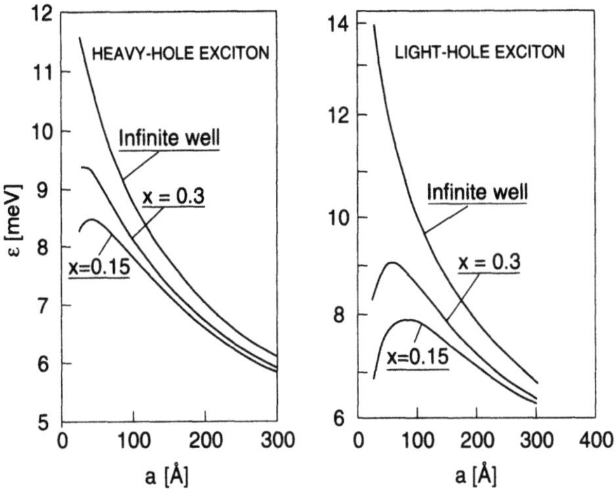

Fig. 5.8. Theoretical dependence of binding energy of e1-hh1 and e1-lh1 exciton on quantum well thickness in GaAs/Al_x Ga_{1-x} As heterostructure with $x = 0.15$ and 0.3, as well as in the infinitely high potential barrier model [5.9]

5.3.2 Nondegenerate Conduction Band and Degenerate Valence Band

Consider now the exciton in a GaAs/AlGaAs-type heterostructure with effective Hamiltonians (5.7) for the electron, and (3.46) for the hole. In this case one can isolate the spinor (5.8) as a multiplier in the functions $\varphi_{sm}(\mathbf{x}_e, \mathbf{x}_h)$ describing the electron spin state:

$$\varphi_{sm}(\mathbf{x}_e, \mathbf{x}_h) = \Phi_s^e \varphi_m(\mathbf{x}_e, \mathbf{x}_h) . \tag{5.111}$$

The functions $\varphi_m(\mathbf{x}_e, \mathbf{x}_h)$ can be conveniently written in the form of a four-component column $\varphi(\mathbf{x}_e, \mathbf{x}_h)$. In the calculation of the exciton states, the hole effective Hamiltonian is written as a sum of diagonal and nondiagonal contributions

$$\mathcal{H}^h(\mathbf{k}) = \mathcal{H}_d(\mathbf{k}) + \mathcal{H}_{nd}(\mathbf{k}) \tag{5.112}$$

and the operator $\mathcal{H}_{nd}$ is taken into account by perturbation theory. In the basis (3.43)

$$\mathcal{H}_d(\mathbf{k}) = F(\mathbf{k}) \begin{bmatrix} 1 & 0 & 0 & 0 \\ 0 & 0 & 0 & 0 \\ 0 & 0 & 0 & 0 \\ 0 & 0 & 0 & 1 \end{bmatrix} + G(\mathbf{k}) \begin{bmatrix} 0 & 0 & 0 & 0 \\ 0 & 1 & 0 & 0 \\ 0 & 0 & 1 & 0 \\ 0 & 0 & 0 & 0 \end{bmatrix}, \tag{5.113}$$

$$\mathcal{H}_{nd}(\mathbf{k}) = \begin{bmatrix} 0 & H & I & 0 \\ H^* & 0 & 0 & I \\ I^* & 0 & 0 & -H \\ 0 & I^* & -H^* & 0 \end{bmatrix}, \tag{5.114}$$

the functions F, G, H and I being introduced in (3.46). We write the first two of these functions in another form:

$$F(\mathbf{k}) = \frac{\hbar^2 k_z^2}{2m_{\parallel}^{\rm hh}} + \frac{\hbar^2 k_{\perp}^2}{2m_{\perp}^{\rm hh}}, \quad G(\mathbf{k}) = \frac{\hbar^2 k_z^2}{2m_{\parallel}^{\rm lh}} + \frac{\hbar^2 k_{\perp}^2}{2m_{\perp}^{\rm lh}}, \tag{5.115}$$

where (for $D = \sqrt{3}B$)

$$\frac{\hbar^2}{2m_{\parallel}^{\rm hh}} = -(A - B), \quad \frac{\hbar^2}{2m_{\parallel}^{\rm lh}} = -(A + B),$$

$$\frac{\hbar^2}{2m_{\perp}^{\rm hh}} = -\left(A + \frac{B}{2}\right), \quad \frac{\hbar^2}{2m_{\perp}^{\rm lh}} = -\left(A - \frac{B}{2}\right). \tag{5.116}$$

In the zero approximation in $\mathcal{H}_{\rm nd}$, we obtain

$$\varphi\left(\mathbf{x}_{\rm e}, \mathbf{x}_{\rm h}\right) = {\rm e}^{{\rm i}\mathbf{K}_{\perp}\cdot\mathbf{R}_{\perp}} \varphi\left(\rho, z_{\rm e}, z_{\rm h}\right) \mathbf{F}^{(m)}, \tag{5.117}$$

where the columns $\mathbf{F}^{(m)}$ are introduced in (5.29), and the function φ satisfies (5.97), where for the effective mass of the hole one should take $m_{\parallel}^{\rm hh}$, $m_{\perp}^{\rm hh}$ for $m = \pm 3/2$, and $m_{\parallel}^{\rm lh}$, $m_{\perp}^{\rm lh}$ for $m = \pm 1/2$. In place of the Coulomb energy $-e^2/\kappa r$, one can conveniently substitute in the equation for $\varphi(\rho, z_{\rm e}, z_{\rm h})$ the Coulomb energy $-e^2/\kappa\rho$ for two-dimensional particles, considering as the perturbation the operator

$$\mathcal{H}' = \mathcal{H}_{\rm nd}\left(\hat{\mathbf{k}}_{\rm h}\right) + \frac{e^2}{\kappa}\left(\frac{1}{\rho} - \frac{1}{r}\right). \tag{5.118}$$

In the zero approximation, the Hamiltonian allows the separation of the variables $z_{\rm e}$, $z_{\rm h}$, ρ, the set of the eigenfunctions having the form

$$\varphi\left(\rho, z_{\rm e}, z_{\rm h}\right) = f_{\rm e}\left(z_{\rm e}\right) f_{\rm h}\left(z_{\rm h}\right) G(\rho), \tag{5.119}$$

where $G(\rho)$, are the eigenfunctions of the two-dimensional exciton describing the discrete and continuum states. For discrete states characterized by a pair of quantum numbers $n = 1, 2, \ldots$ and $l = 0, \pm 1, \ldots$ we obtain

$$\varepsilon_{n,l} = E_{\rm B}^{\rm exc} \Big/ \left(n - \frac{1}{2}\right)^2,$$

$$G_{n,l}(\rho) = \frac{1}{\sqrt{2\pi}} {\rm e}^{{\rm i}l\varphi} R_{nl}^{\rm 2D}(\rho),$$

$$R_{nl}^{\rm 2D}(\rho) = N_{n,|l|} \exp\left(-\xi_n\right) \left(2\xi_n\right)^{|l|} L_{n+|l|-1}^{2|l|}\left(2\xi_n\right), \tag{5.120}$$

where

$$\xi_n = \frac{1}{n - \dfrac{1}{2}} \frac{\rho}{a_{\rm B}},$$

$L_n^{\alpha}(\xi)$ is the associated Laguerre polynomial and $N_{n,|l|}$ a normalization factor. Using the set of zero-approximation functions (5.119), one can determine the exciton eigenfunctions taking into account the mixing (hybridization) of heavy and light hole states, i.e., of states with $m = \pm 3/2$ and $m = \pm 1/2$.

To calculate the matrix elements of optical transitions into exciton states, one can expand the two-particle exciton wave function into the states of non-interacting particles

$$\Psi^{\text{exc}} = \frac{1}{\sqrt{S}} \sum_{\nu' j \mathbf{k}_\perp} G_{\nu\nu' j}(\mathbf{k}_\perp)\, \psi_{\nu s \mathbf{k}_\perp}(\mathbf{x}_\text{e})\, \psi^\text{h}_{\nu' j, -\mathbf{k}_\perp + \mathbf{K}_\perp}(\mathbf{x}_\text{h}) \,. \tag{5.121}$$

Here S is the area of the structure which we consider, as a rule, to be a unit, $\psi_{\nu s \mathbf{k}_\perp}$ and $\psi^\text{h}_{\nu' j \mathbf{k}_\perp}$ are the electron and hole wave functions in a quantum-well structure, accordingly, in the ν and ν' subbands

$$\Psi_{\nu s}\mathbf{k}_\perp(\mathbf{x}) = \text{e}^{\text{i}\mathbf{k}_\perp \cdot \mathbf{x}_\perp} f^\text{e}_{\nu \mathbf{k}_\perp}(z) u^{(0)}_{cs}(\mathbf{x}), \tag{5.122a}$$

$$\Psi^\text{h}_{\nu' j \mathbf{k}_\perp}(\mathbf{x}) = \text{e}^{\text{i}\mathbf{k}_\perp \cdot \mathbf{x}_\perp} \sum_m f^\text{h}_{\nu' j \mathbf{k}_\perp, m}(z) u^{(0)}_{\text{v}m}(\mathbf{x}), \tag{5.122b}$$

and $u^{(0)}_{cs}(\mathbf{x})$, $u^{(0)}_{\text{v}m}(\mathbf{x})$ are three-dimensional Bloch functions at the Γ point. Since the mixing of electron states in the various conduction subbands is neglected in (5.121), there is no summation in ν. As shown in Chap. 3, the inclusion of the nondiagonal contribution $\mathcal{H}_\text{nd}$ in the effective hole Hamiltonian results in (5.122b) in a hybridization of the light and heavy holes. For instance, in the first order in $k_z k_\perp$-mixing between the lh1 and hh2 states, we have

$$f^\text{h}_{1,\pm 1/2, \mathbf{k}_\perp; \pm 3/2}(z) = \pm \frac{D k_z^{(2,1)} k_\pm}{E^0_{\text{lh}1} - E^0_{\text{hh}2}} f^\text{h}_{2,\pm 3/2, 0; \pm 3/2}(z) \tag{5.123}$$

where $E^0_{\text{lh}1}$ or $\text{E}^0_{\text{hh}2}$ is the hole energy for $\mathbf{k}_\perp = 0$,

$$k_z^{(2,1)} = \int f^\text{h}_{\text{hh}2}(z) \left(-\text{i}\frac{\partial}{\partial z}\right) f^\text{h}_{\text{lh}1}(z)\, \text{d}z.$$

In a symmetric quantum well the functions f^e, f^h possess certain symmetry properties

$$\begin{aligned} f^\text{e}_{\nu, \mathbf{k}_\perp}(-z) &= (-1)^{\nu+1} f^\text{e}_{\nu, \mathbf{k}_\perp}(z), \\ f^\text{e}_{\nu, -\mathbf{k}_\perp}(z) &= f^\text{e}_{\nu, \mathbf{k}_\perp}(z); \end{aligned} \tag{5.124}$$

$$f_{\nu j \mathbf{k}_\perp, m}(-z) = (-1)^{\nu + j - m + 1} f^\text{h}_{\nu j \mathbf{k}_\perp, m}(z),$$

$$f^\text{h}_{\nu j, -\mathbf{k}_\perp, m}(z) = (-1)^{j-m} f^\text{h}_{\nu j \mathbf{k}_\perp, m}(z). \tag{5.125}$$

In the particular case of an exciton at rest ($K_\perp = 0$) the functions $G_{\nu\nu' j}(\mathbf{k}_\perp)$ in (5.121) satisfy the system of equations

$$\left(E_{\text{c}\nu\mathbf{k}} - E_{\upsilon\nu' j, -\mathbf{k}} - E\right) G_{\nu\nu' j}(\mathbf{k}) + \sum_{\nu'_1 j_1 \mathbf{k}_1} G_{\nu\nu'_1 j_1}(\mathbf{k}_1)\, V_{\nu' j, \nu'_1 j_1}(q_\perp) = 0, \tag{5.126}$$

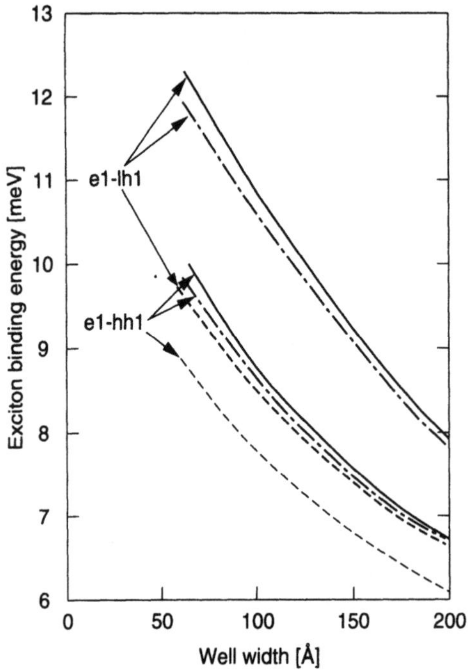

Fig. 5.9. Dependence of the binding energy of e1-hh1 and e1-lh1 exciton on quantum well thickness in GaAs/$Al_{0.4}Ga_{0.6}$As heterostructure. Calculation made neglecting (dashed) and with inclusion of (dot-and-dash lines) light and heavy hole hybridization, as well as with inclusion of conduction band nonparabolicity (solid lines) [5.10]

where $q_\perp = |\mathbf{k} - \mathbf{k}_1|$,

$$V_{\nu' j, \nu_1' j_1}(q_\perp) = -\frac{2\pi e^2}{\kappa q_\perp} \int\int \mathrm{d}z_\mathrm{e}\, \mathrm{d}z_\mathrm{h} \mathrm{e}^{-q_\perp |z_\mathrm{e} - z_\mathrm{h}|}$$
$$\left[f_\nu^\mathrm{e}(z_\mathrm{e})\right]^2 \sum_m f^{\mathrm{h}*}_{\nu' j, -\mathbf{k}, m}(z_\mathrm{h}) f^{\mathrm{h}}_{\nu_1' j_1, -\mathbf{k}, m}(z_\mathrm{h}). \tag{5.127}$$

Here for the sake of brevity the index $\perp$ of the vector $\mathbf{k}_\perp$ is dropped, and the dependence of $f^\mathrm{e}_{\nu \mathbf{k}_\perp}(z)$ on $\mathbf{k}_\perp$ is neglected. In the derivation of (5.126), (5.127) it was taken into account that the two-dimensional integral

$$\int \frac{\mathrm{d}\rho}{\sqrt{\rho^2 + z^2}} \mathrm{e}^{\mathrm{i}\mathbf{q} \cdot \rho} = \frac{2\pi}{q} \mathrm{e}^{-q|z|} \tag{5.128}$$

and the three-dimensional Bloch functions $u^{(0)}_{\nu m}(\mathbf{x})$ are orthogonal for $m \neq m'$. In the two-subband approximation, the exciton wave function is made up of electron states in one conduction subband ν and hole states in one subband $\nu' j$, the sum (5.126) retaining only one term with $\nu_1' = \nu'$, $j_1 = j$.

The binding-energy calculations for the hh1 and lh1 excitons in a quantum-well structure are illustrated by Fig. 5.9.

5.3.3 Type II Heterostructure

In a heterostructure with a single quantum well for the electron and a single potential barrier for the hole in the same layer $|z| < a/2$, the exciton wave function $\varphi(\mathbf{x}_e, \mathbf{x}_h)$ has a maximum in the region of z_e and z_h values adjoining the $z = \pm a/2$ interfaces from the side of the barrier and the well, respectively. In the limit of infinite barriers, V_0^e, V_0^h, the simplest trial function in a variational calculation is

$$\varphi(\rho, z_e, z_h) = C f_e(z_e)\, g(z_e)\, h(z_h)\, G(\boldsymbol{\rho}, \mathbf{z}) \tag{5.129}$$

with the function $f_e(z_e)$ defined in (5.12),

$$g(z) = \exp\left[-\beta_e\left(\frac{a}{2} - z\right)\right], \tag{5.130}$$

$$h(z) = \begin{cases} \left(z - \frac{a}{2}\right)\exp\left[-\beta_h\left(z - \frac{a}{2}\right)\right] & \text{for } z > a/2, \\ 0 & \text{for } z < a/2, \end{cases} \tag{5.131}$$

and the function $G(\rho, z)$ for the exciton ground state chosen in the form (5.49). With such a choice, the exciton energy depends on four variational parameters $\beta_e, \beta_h, a_\parallel$ and $a_\perp$, the function $\varphi(\rho, z_e, z_h)$ reaching a maximum at $z_e \simeq (a/2) - \beta_e^{-1}$ and $z_h = (a/2) + \beta_h^{-1}$.

5.3.4 Periodic Heterostructure

The free-exciton state in a periodic heterostructure is characterized by one more quantum number, namely, the longitudinal component of the wave vector K_z:

$$\varphi_{sm}(\rho, z_e + d, z_h + d) = e^{iK_z d}\varphi_{sm}(\rho, z_e, z_h). \tag{5.132}$$

We will limit ourselves to a consideration of the $n = 1$ exciton ground state for $K_\perp = K_z = 0$ in a semiconductor with nondegenerate bands. In short-period superlattices corresponding to broad energy minibands, the function $\varphi(\rho, z_e, z_h)$ in (5.132) can be conveniently expanded in two-particle miniband states

$$\varphi(\rho, z_e, z_h) = \sum_{\mathbf{k}} G_{\mathbf{k}} \chi_{\nu\mathbf{k}}^e(\mathbf{x}_e)\, \chi_{\nu', -\mathbf{k}}^h(\mathbf{x}_h), \tag{5.133}$$

where $\chi^{e,h}$ is the super-Bloch envelope of the electron or hole wave function defined by (5.46).

Chomette et al. [5.11] used the following trial function in a variational calculation of the exciton ground state in a short-period SL

$$\varphi(\rho, z_e, z_h) = C e^{-\rho/\bar{\lambda}} \sum_{k_z} e^{-k_z^2/2\beta^2} \chi_{k_z}^e(z_e)\, \chi_{-k_z}^h(z_h), \tag{5.134}$$

where

$$\chi_{k_z}^{e,h}(z) = e^{ik_z z} U_{1k_z}^{e,h}(z; k_\perp = 0).$$

The function (5.134) is obtained from (5.133) when the dependence of the super-Bloch amplitude (5.46) on $\mathbf{k}_\perp$ is neglected, and $G_{\mathbf{k}}$ is chosen such that

$$G(\rho, k_z) = \sum_{\mathbf{k}_\perp} G_{\mathbf{k}} e^{i\mathbf{k}_\perp \cdot \rho} = C e^{-\rho/\bar{\lambda}} \bar{e}^{k_z^2/2\beta^2},$$

i.e.,

$$G(\rho, z) \equiv \sum_{\mathbf{k}} G_{\mathbf{k}} e^{i\mathbf{k}\cdot\mathbf{x}} = C' e^{-\rho/\bar{\lambda}} e^{-\beta^2 z^2/2}.$$

The function $G(\rho, z)$ obtained from $G_{\mathbf{k}}$ by inverse Fourier transformation and describing the relative motion of the electron and the hole can be calculated in the effective-mass approximation where this function satisfies Schrödinger's equation

$$\left[-\frac{\hbar^2}{2\mu_\perp}\left(\frac{\partial^2}{\partial\rho_x^2}+\frac{\partial^2}{\partial\rho_y^2}\right)-\frac{\hbar^2}{2\mu_\parallel}\frac{\partial^2}{\partial z^2}-\frac{e^2}{\kappa r}\right]G(\rho, z) = -\varepsilon G(\rho, z). \quad (5.135)$$

Here $\mu_{\parallel,\perp}$ are the longitudinal and transverse reduced effective masses of the electron-hole pair. For the e1-hh1 exciton in the GaAs/AlGaAs SL, hole tunneling into the barrier may be neglected, the longitudinal hole mass $M_\parallel^{\mathrm{h}}$ is large (SLs with ultrashort periods excluded), and $\mu_\parallel$ practically coincides with the longitudinal electron mass $M_\parallel^{\mathrm{e}}$, which can be calculated by (3.93). For the transverse mass $\mu_\perp$ in this case we have

$$\mu_\perp = \frac{M_\perp^{\mathrm{e}} m_\perp^{\mathrm{hh}}}{M_\perp^{\mathrm{e}} + m_\perp^{\mathrm{hh}}}, \quad (5.136)$$

where $M_\perp^{\mathrm{e}}$ is also calculated by the Kronig-Penney method. According to (5.116), and neglecting the mixing of heavy and light hole states in the e1-hh1 exciton, we have

$$\frac{1}{m_\perp^{\mathrm{hh}}} = -\frac{2}{\hbar^2}\left(A + \frac{B}{2}\right) = \frac{1}{4}\left(\frac{3}{m_{lh}} + \frac{1}{m_{\mathrm{hh}}}\right), \quad (5.137)$$

where $m_{\mathrm{hh}}, m_{\mathrm{lh}}$ are the effective masses of the heavy and light holes in bulk GaAs. By choosing (5.49) as the trial function in solving (5.135) by variational technique, one can calculate the exciton binding energy ε in a short-period SL with an accuracy higher than that obtained using the trial function (5.134).

For a multiple quantum-well structure, i.e., for a thick-barrier SL, the tight-binding approximation is valid, where the exciton wave function φ_{MQW} is expressed in terms of the exciton wave function φ_{SQW} in a single quantum well (5.108) in the form

$$\varphi_{\mathrm{MQW}}(\rho, z_{\mathrm{e}}, z_{\mathrm{h}}) = \frac{1}{\sqrt{N}} \sum_l \varphi_{\mathrm{SQW}}(\rho, z_{\mathrm{e}} - \bar{z}_l, z_{\mathrm{h}} - \bar{z}_l) \quad (5.138)$$

where $\bar{z}_l$ is the position of the center of the lth well, N is the number of periods in SL. For $K_z \neq 0$, the factor $\exp(\mathrm{i}K_z\bar{z}_l)$ should be put under the sum sign.

The binding energy of the e1-hh1 is exciton in a periodic heterostructure as a function of the period d is presented in Figure 5.10

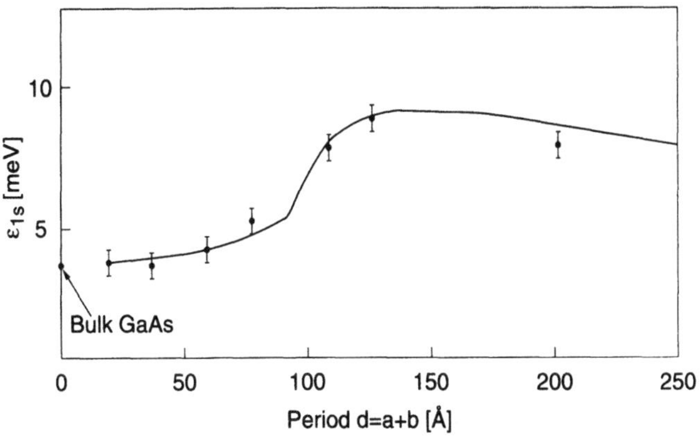

Fig. 5.10. Dependence of the binding energy of e1-hh1 (1s) exciton in the periodic heterostructure GaAs/$Al_{0.3}Ga_{0.7}$ As with $a = b$ on lattice period. Solid line: theory; dots with error bars: experiment [5.11]

5.3.5 Biexciton

In his analysis of the biexciton in a single quantum-well structure, *Kleinman* [5.12] chose a trial function containing six variational parameters k, β, ν, ρ, λ, τ:

$$\begin{aligned}
\varphi_{\text{biexc}} &= \Phi(kr) f_{\text{e}}(z_{\text{e1}}) f_{\text{e}}(z_{\text{e2}}) f_{\text{h}}(z_{h1}) f_{\text{h}}(z_{h2}), \\
\Phi(r) &= \psi(\beta; r)\chi\left(\nu, \rho, \lambda, \tau; r_{\text{h1,h2}}\right), \\
\chi(\nu, \rho, \lambda, \tau; u) &= u^{\nu}\text{e}^{-\rho u} + \lambda\text{e}^{-\tau u}, \\
\psi(\beta; r) &= \exp\left[-(s_1 + s_2)/2\right] \cosh\left[\beta(t_1 - t_2)/2\right], \\
s_1 &= r_{\text{e1,h1}} + r_{\text{e1,h2}}, s_2 = r_{\text{e2,h1}} + r_{\text{e2,h2}},
\end{aligned} \tag{5.139}$$

where $r_{li,l'i'} = |\mathbf{x}_{li} - \mathbf{x}_{l'i'}|$, $l = \text{e}, \text{h}$, $i = 1, 2$. The binding energy $\varepsilon_{\text{biexc}}$ of excitons in the biexciton is defined as the difference $2E_{\text{exc}} - E_{\text{biexc}}$, where E_{exc} and E_{biexc} are the exciton and biexciton excitation energies, accordingly. Calculations show that in the two-dimensional limit $\varepsilon_{\text{biexc}}$ exceeds the biexciton binding energy in bulk GaAs by a factor of 3 to 4.

5.4 Exchange Splitting of Exciton Levels

Consider an exciton made up of an electron in a conduction band of symmetry D_{c} and a hole in a valence band of symmetry D_{v}. The wave functions in the $n = 1$ ground-state exciton transform according to the representation $D_{\text{c}} \times D_{\text{v}}$. In general, this representation is reducible and can be expanded in irreducible representations. The exchange interaction between the electron and the hole appearing in the theory of Wannier-Mott excitons, when including corrections to the effective-mass approximation, results in a partial lifting of degeneracy of the exciton ground state and splits it into the corresponding irreducible representations. Figure 5.11 shows the exchange splitting pattern of the $\Gamma_6 \times \Gamma_8$, $\Gamma_6 \times \Gamma_7$ excitons in A_3B_5 or A_2B_6 crystals (class T_d), and $\Gamma_7 \times \Gamma_9$, $\Gamma_7 \times \Gamma_7$ excitons in A_2B_6 hexagonal crystals (class C_{6v}). The Hamiltonian $\mathcal{H}_{ex}$ describing this splitting can be constructed using the method of invariants. To do this, one has to find all linearly independent invariants I_i^{cv} composed of products of basis matrices $X_j^c X_{j'}^v$, and to construct their linear combination

$$\mathcal{H}_{\mathrm{ex}} = \sum_i \Delta_i I_i^{\mathrm{cv}}. \tag{5.140}$$

Consider particular examples. The ground state of the $\Gamma_6 \times \Gamma_8$ exciton in a cubic crystal is eightfold degenerate. The spin indices of the smooth function (5.86) run through the values $s = \pm 1/2$ and $m = \pm 3/2, \pm 1/2$. Exchange interaction splits the $\Gamma_6 \times \Gamma_8$ state into three levels

$$\Gamma_6 \times \Gamma_8 = \Gamma_{12} + \Gamma_{15} + \Gamma_{25}. \tag{5.141}$$

The Hamiltonian describing this splitting is determined in the general case by two constants

$$\mathcal{H}_{\mathrm{ex}} = \frac{\Delta_0}{4} \left(\sigma^e \cdot \mathbf{J}\right) + \Delta_1 \sum_\lambda \sigma_\lambda^e J_\lambda^3, \tag{5.142}$$

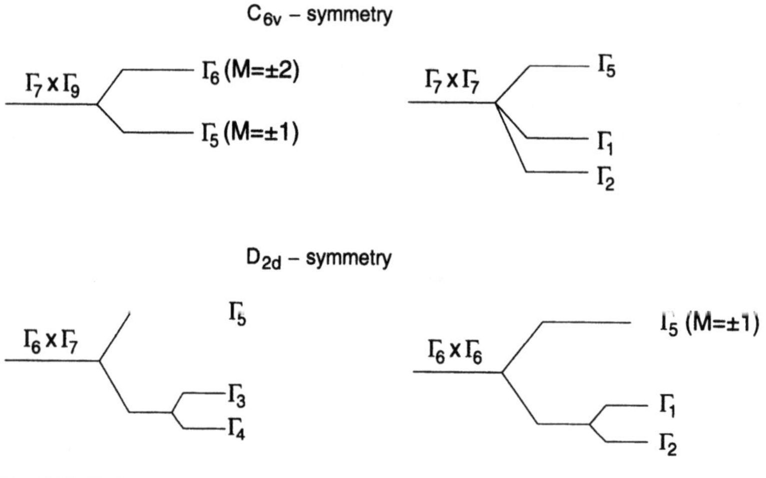

Fig. 5.11. Exchange splitting of $\Gamma_6 \times \Gamma_8$ and $\Gamma_6 \times \Gamma_7$ exciton states in direct gap zincblende crystals (T_d symmetry), of $\Gamma_7 \times \Gamma_9$ and $\Gamma_7 \times \Gamma_7$ excitons in hexagonal wurtzite crystals, and e1-hh1 (1s) and e1-lh1(1s) excitons in GaAs/Al_xGa_{1-x} As (001) heterostructures

where $\sigma_\lambda^{\mathrm{e}}$ are the Pauli matrices acting upon the electron-spin index s, the matrices J_λ introduced in (3.40) acting on the hole spin index m. Applying the momentum summation rule to $s = 1/2$ and $j = 3/2$, we find that the triplet level Γ_{15} corresponds to the total angular momentum $J = 1$ with projections $M = 1, 0, -1$. The constant Δ_1 differs from zero to the extent of spin-orbit admixture of states from other bands to the (v, Γ_8) state. In the spherical approximation, $\Delta_1 = 0$, and the exchange interaction does not remove degeneracy between the Γ_{12} and Γ_{25} states. These states correspond to the angular momentum $J = 2$ and are shifted by Δ_0 relative to Γ_{15}. The selection rules for optical transitions involving excitation of the Γ_{15} exciton are similar to those for the $1s \to 2p$ transitions in the hydrogen atom. The $J = 2$ states are optically inactive in the dipole approximation.

The $\Gamma_6 \times \Gamma_7$ exciton ground state is fourfold degenerate, the exchange interaction splitting it into a dipole-allowed triplet level Γ_{15} with an angular momentum $J = 1$ and an optically inactive singlet Γ_2 with $J = 0$. The exchange-interaction operator

$$\mathcal{H}_{\mathrm{ex}} = \frac{\Delta'}{2} \sum_\lambda \sigma_\lambda^{\mathrm{e}} I_\lambda^{\mathrm{h}},$$

where I_λ^{h} is the matrix of the hole angular-momentum projection operator in the Γ_7 basis. It is appropriate to choose this basis in the form of the functions $\psi_{1/2}^{1/2}, \psi_{-1/2}^{1/2}$ defined by (3.43). Then we will have $I_\lambda^{\mathrm{h}} = \sigma_\lambda^{\mathrm{h}}/2$, and $\mathcal{H}_{\mathrm{ex}} = (\Delta'/4)(\sigma^{\mathrm{e}} \cdot \sigma^{\mathrm{h}})$. In this basis, the spin wave functions Φ_{sm} in (5.94) have the form

$$\begin{cases} \Phi_{sm}^{(1,1)} = \alpha_s^{\mathrm{e}} \alpha_m^{\mathrm{h}}, & \Phi_{sm}^{(1,-1)} = \beta_s^{\mathrm{e}} \beta_m^{\mathrm{h}}, \\ \Phi_{sm}^{(1,0)} = \left(\alpha_s^{\mathrm{e}} \beta_m^{\mathrm{h}} + \beta_s^{\mathrm{e}} \alpha_m^{\mathrm{h}}\right)/\sqrt{2}, & \end{cases} \quad (\Gamma_{15})$$

$$\Phi_{sm}^{(0,0)} = \left(\alpha_s^{\mathrm{e}} \beta_m^{\mathrm{h}} - \beta_s^{\mathrm{e}} \alpha_m^{\mathrm{h}}\right)/\sqrt{2}, \quad (\Gamma_2), \tag{5.143}$$

where

$$\alpha_s^{\mathrm{e}} = \delta_{s,1/2}, \beta_s^{\mathrm{e}} = \delta_{s,-1/2}, \alpha_m^{\mathrm{h}} = \delta_{m,1/2}, \beta_m^{\mathrm{h}} = \delta_{m,-1/2}, s, m = \pm 1/2. \tag{5.144}$$

In hexagonal crystals of CdS, CdSe-type there are three exciton series: $A(\Gamma_7 \times \Gamma_9)$, $B(\Gamma_7 \times \Gamma_7)$, and $C(\Gamma_7 \times \Gamma_7)$. The ground state of each of these excitons is fourfold degenerate. Exchange interaction splits the $A(n = 1)$ level into two levels Γ_5 and Γ_6, and the level $B(n = 1)$ or $C(n = 1)$, into three levels Γ_1, Γ_2 and Γ_5. The states Γ_1 and Γ_2 correspond to the angular momentum projection $M = 0$, the levels Γ_5 and Γ_6, to the projections $M = \pm 1$ and $M = \pm 2$. The allowed optical transitions are to the Γ_1 state in $\mathcal{I} \parallel z$ polarization, and to Γ_5 in $\mathcal{I} \perp z$ polarization. We use for the A exciton the exchange Hamiltonian

$$\mathcal{H}_{\mathrm{ex}} = \frac{\Delta_0}{2} \sigma_z^{\mathrm{e}} \sigma_z^{\mathrm{h}}$$

and the corresponding spin functions

$$\Phi^{(5+)}_{sm} = \beta^{\rm e}_s \alpha^{\rm h}_m, \quad \Phi^{(5-)}_{sm} = \alpha^{\rm e}_s \beta^{\rm h}_m,$$
$$\Phi^{(6+)}_{sm} = \alpha^{\rm e}_s \alpha^{\rm h}_m, \quad \Phi^{(6-)}_{sm} = \beta^{\rm e}_s \beta^{\rm h}_m, \tag{5.145}$$

where $\alpha^{\rm e}_s$, $\beta^{\rm e}_s$ were introduced in (5.144), and $\alpha^{\rm h}_m = \delta_{m,3/2}$, $\beta^{\rm h}_m = \delta_{m,-3/2}$ since to the hole in subband A correspond states with the angular-momentum component $m = \pm 3/2$. The states $\Gamma_{5\pm}$ are excited by circularly polarized light $\sigma_\pm$ with the projection of the photon angular momentum on the z axis ± 1. It may sometimes become appropriate to use linear combinations

$$|\Gamma_{5x}\rangle = (|\Gamma_{5+}\rangle + |\Gamma_{5-}\rangle)/\sqrt{2},$$
$$|\Gamma_{5y}\rangle = {\rm i}\,(|\Gamma_{5+}\rangle - |\Gamma_{5-}\rangle)/\sqrt{2}, \tag{5.146}$$

which are dipole-active in the $\mathcal{I} \parallel x$ or $\mathcal{I} \parallel y$ polarization, respectively.

There is a certain analogy, albeit incomplete, between excitons of series A and B in hexagonal crystals and e1-hh1, e1-lh1 excitons in quantum-well structures or superlattices made up of A_3B_5 semiconductor layers and having D_{2d} symmetry. In accordance with the multiplication laws for group representations

$$\Gamma_6 \times \Gamma_6 = \Gamma_1 + \Gamma_2 + \Gamma_5, \quad \Gamma_6 \times \Gamma_7 = \Gamma_3 + \Gamma_4 + \Gamma_5 \tag{5.147}$$

exchange interaction splits the ground state of the e1-hh1 exciton into a radiative doublet and two dipole-forbidden singlet levels Γ_1 and Γ_2, and that of the e1-lh1 exciton into a Γ_5 doublet which is dipole-allowed in the $\mathcal{I} \perp z$ polarization, a singlet Γ_3 excited by light of $\mathcal{I} \parallel z$ polarization, and an optically inactive singlet Γ_4. Consider in more detail the exchange interaction in the case of the $\Gamma_6 \times \Gamma_6$ exciton, choosing as electron and hole basis states functions transforming in the group D_{2d} according to equivalent representations as spin columns α, β or spinor functions $\beta(X - {\rm i}Y)$, $-\alpha(X + {\rm i}Y)$, where X, Y, Z are the basis functions of the representation Γ_{15} of the orthogonal group K_h. In this case, the basis matrices in the conduction and valence bands have the same form:

I^l (representation Γ_1), $\sigma^l_z(\Gamma_2), \sigma^l_x, -\sigma^l_y(\Gamma_5)$, where l = e, h, and to the elements of the first row (column) in the hole matrices corresponds $m = -3/2$, with $m = 3/2$ corresponding to the elements of the second row (column). The first of these four matrices is even; the other three, odd under time inversion. The method of invariants yields

$$\mathcal{H}_{\rm ex} = \frac{\Delta_0}{2}\sigma^{\rm e}_z \sigma^{\rm h}_z + \frac{\Delta_1}{4}\left(\sigma^{\rm e}_x \sigma^{\rm h}_x + \sigma^{\rm e}_y \sigma^{\rm h}_y\right), \tag{5.148}$$

where the factors 1/2 and 1/4 are introduced for convenience. To the level Γ_5 of energy $\Delta_0/2$ correspond the exciton states

$$|\Gamma_{5+}\rangle \equiv |-1/2, 3/2\rangle, \quad |\Gamma_{5-}\rangle \equiv |1/2, -3/2\rangle, \tag{5.149}$$

which are dipole-allowed in the σ_+ or σ_- polarization, or superpositions

$$|\Gamma_{5x}\rangle = (|-1/2, 3/2\rangle + |1/2, -3/2\rangle)/\sqrt{2},$$
$$|\Gamma_{5y}\rangle = -{\rm i}\,(|-1/2, 3/2\rangle - |1/2, -3/2\rangle)/\sqrt{2} \tag{5.150}$$

which are allowed in the $\mathcal{I}\|x$ and $\mathcal{I}\|y$ polarization, respectively. To the levels Γ_1 and Γ_2 with energy $-(\Delta_0/2) \pm (\Delta_1/4)$ correspond the states

$$|\Gamma_1\rangle = (|1/2, 3/2\rangle + |-1/2, -3/2\rangle)/\sqrt{2},$$
$$|\Gamma_2\rangle = -\mathrm{i}\,(|1/2, 3/2\rangle - |-1/2, -3/2\rangle)/\sqrt{2}. \tag{5.151}$$

It is established [5.13] that the exciton symmetry in type II short-period GaAs/AlAs (001) superlattices is lower than D_{2d}. The radiative doublet turns out to be split into states polarized along $x'\|[1\bar{1}0]$ and $y'\|[110]$. The same sample was found to have two classes of excitons whose anisotropic exchange splitting

$$\bar{\Delta} = E(\Gamma_{5x'}) - E(\Gamma_{5y'}) \tag{5.152}$$

coincides in modulus while differing in sign. To obtain such splitting, one has to add to (5.148) the term

$$\Delta\mathcal{H}_{\mathrm{ex}} = \frac{\bar{\Delta}}{4}\left(\sigma_x^{\mathrm{e}}\sigma_y^{\mathrm{h}} + \sigma_y^{\mathrm{e}}\sigma_x^{\mathrm{h}}\right) = \frac{\bar{\Delta}}{4}\left(-\sigma_{x'}^{\mathrm{e}}\sigma_{x'}^{\mathrm{h}} + \sigma_{y'}^{\mathrm{e}}\sigma_{y'}^{\mathrm{h}}\right), \tag{5.153}$$

where $\sigma_{x'}, \sigma_{y'}$ are Pauli matrices in the coordinate system x', y'. The term (5.153) arises when the symmetry is lowered, $D_{2d} \to C_{2v}$, since in the C_{2v} group the matrices $I^l, \sigma_z^l, \sigma_{x'}^l, \sigma_{y'}^l$ transform according to inequivalent one-dimensional representations, each of the products $\sigma_z^{\mathrm{e}}\sigma_z^{\mathrm{h}}, \sigma_{x'}^{\mathrm{e}}\sigma_{x'}^{\mathrm{h}}, \sigma_{y'}^{\mathrm{e}}\sigma_{y'}^{\mathrm{h}}$ being itself an invariant. It should be pointed out that the operator (5.153) does not mix the states Γ_1 and Γ_2 with one another or with the state Γ_5, and, in particular, does not affect the splitting of the levels Γ_1, Γ_2.

6 Interband Optical Transitions

The property of translational invariance of a periodic medium provides a basis for searching for normal light waves propagating in such a medium in the form of Bloch solutions. We discriminate between optical superlattices and short-period regular heterostructures. In an optical superlattice, adjacent layers have different dielectrical constants. Apart from this, the layers are assumed to be thick enough to be able to neglect electron and hole quantum-confinement effects. Under these conditions the quantum confinement of the exciton as a whole within one layer may be taken into account in terms of the macroscopic theory of additional light waves, in which Maxwell's boundary conditions are complemented by the boundary conditions for an excitonic contribution to the dielectric polarization of the medium (the so-called *additional boundary conditions*).

Light propagation in a short-period quantum-well structure or superlattice may be described in terms of the effective homogeneous-medium approximation, similar to the way it is done in the description of the dielectric response of a three-dimensional crystal. In particular, just as in crystal optics, each optically active exciton makes a pole contribution to the effective dielectric-permittivity tensor, which is determined by the resonance frequency, oscillator strength (or longitudinal-transverse splitting), and exciton decay in the heterostructure. These three parameters can be derived from experimental data, for example, reflection or transmission spectra. This chapter focusses on the calculation of the exciton oscillator strength, a parameter which is very sensitive to relative electron and hole density redistribution in the exciton. Theory is compared with experimental data on resonant interference reflection from single-quantum-well structures, a regular system of isolated quantum wells, and a superlattice. The effect of the electric or magnetic field on interband optical transitions in heterostructures is discussed.

6.1 Optical Superlattices

To describe optical phenomena in terms of macroscopic electrodynamics, one has to write the equations relating the displacements $\mathbf{D}, \mathbf{B}$ and the fields $\mathcal{I}, \mathbf{H}$. There is a certain arbitrariness in formulating these equations. We will use the form of material relation which sets

$$\mathbf{B} = \mathbf{H},$$

and where the properties of the medium are given by the dielectric constant tensor κ:

$$D_\alpha(\omega, \mathbf{q}) = \kappa_{\alpha\beta}(\omega, \mathbf{q})\mathcal{I}_\beta(\omega, \mathbf{q}), \tag{6.1}$$

ω being the frequency and $\mathbf{q}$ the wave vector of light. Unless otherwise specified, we will neglect in what follows spatial dispersion (or nonlocality) of the dielectric constant, i.e., the dependence of the tensor κ on $\mathbf{q}$.

We will start the analysis of the linear dielectric response in multilayered structures with the optical superlattice which is made up of a sequence of double layers having isotropic dielectric constants: $\kappa^a_{\alpha\beta} = \kappa^a\delta_{\alpha\beta}$ (layer of thickness a) and $\kappa^b_{\alpha\beta} = \kappa^b\delta_{\alpha\beta}$ (layer of thickness b). Taking into account the periodic nature of the superlattice, we will look for solutions of the Maxwell equations, called also normal light waves, in the form of Bloch functions

$$\begin{aligned} &\mathcal{I}_\mathbf{Q}(\mathbf{x}) = \exp(\mathrm{i}\mathbf{Q}\cdot\mathbf{x})\mathcal{I}_\mathbf{Q}(z), \\ &\mathcal{I}_\mathbf{Q}(z+d) = \mathcal{I}_\mathbf{Q}(z), \quad d = a+b. \end{aligned} \tag{6.2}$$

In a structure with isotropic layers, there are two types of Bloch solutions, namely, TE waves, that is, s-polarized waves whose vector $\mathcal{I}_\mathbf{Q} \perp \mathbf{Q}, z$, and the vector $\mathbf{H}_\mathbf{Q}$ lies in the $(\mathbf{Q}, z)$ plane, and TM, or p-polarized waves with $\mathbf{H}_\mathbf{Q} \perp \mathbf{Q}, z$. The dispersion relation for these waves is similar to the relation (3.91) for the Bloch electrons in a SL or to (4.74) for the interface modes:

$$\begin{aligned} \cos\left[Q_z(a+b)\right] &= \cos(k_a a)\cos(k_b b) \\ &- \frac{1}{2}\left(\eta + \frac{1}{\eta}\right)\sin(k_a a)\sin(k_b b)\,, \end{aligned} \tag{6.3}$$

where for the TE and TM waves, accordingly,

$$\eta_s = \frac{k_b}{k_a}, \quad \eta_p = \frac{\kappa_a}{\kappa_b}\frac{k_b}{k_a} \tag{6.4}$$

with the nomenclature

$$\begin{aligned} &k_a = \left(\kappa^a k_0^2 - Q_\perp^2\right)^{1/2}, \quad k_b = \left(\kappa^b k_0^2 - Q_\perp^2\right)^{1/2}, \\ &k_0 = \omega/c, \quad Q_\perp^2 = Q_x^2 + Q_y^2. \end{aligned} \tag{6.5}$$

A few words concerning the similarity of the equations for the TE and TM waves. As follows from Maxwell's equations for a homogeneous medium with a dielectric constant κ, the fields $\mathcal{I}$ and $\mathbf{H}$ for a plane monochromatic wave satisfy the relations

$$\mathbf{H} = [\mathbf{n}\mathcal{I}], \quad \mathcal{I} = -\left[\frac{\mathbf{n}}{n^2}\mathbf{H}\right], \tag{6.6}$$

where the refraction vector $\mathbf{n} = c\mathbf{q}/\omega$, $\mathbf{n}^2 = \kappa$. Therefore the boundary conditions for TE waves expressed in terms of electric field amplitudes and those for TM waves expressed in terms of magnetic field amplitudes go into one another under a formal replacement $n_z \to -n_z/n^2$. As a result, the dispersion relations for these two types of waves differ simply in the replacement of k_b/k_a by $(k_b\kappa_b^{-1})/(k_a\kappa_a^{-1})$.

To analyze the spectrum in the vicinity of the extremum, (6.3) can be conveniently rewritten in the form

$$1 - \cos Q_z d = \frac{1}{2} \sin k_a a \sin k_b b \, (\tan\phi_a + \eta \tan\phi_b) \times \left(\frac{1}{\eta} \cot\phi_a + \cot\phi_b \right) \qquad (6.7)$$

or

$$1 + \cos Q_z d = \frac{1}{2} \sin k_a a \sin k_b b \, (\tan\phi_a - \eta \cot\phi_b) \times \left(-\frac{1}{\eta} \cot\phi_a + \tan\phi_b \right), \qquad (6.8)$$

where $\phi_a = k_a a/2$, $\phi_b = k_b b/2$. The frequencies of normal light waves for $Q_z = 0$ or $Q_z = \pm\pi/d$ are found from the condition of vanishing of one of the two parentheses on the right-hand side of (6.7, 8), respectively.

In the long-wavelength approximation, when $|\phi_a|$, $|\phi_b| \ll 1$, the electromagnetic fields $\mathcal{I}_x$, $\mathcal{I}_y$ and the displacement D_z change very little on the length scale of the order of the period d. In this case, the optical superlattice may be considered a homogeneous medium with an effective dielectric constant (4.75a):

$$\begin{aligned} &\kappa_{\alpha\beta} = \kappa_{\alpha\alpha}\delta_{\alpha\beta}, \\ &\kappa_{xx} = \kappa_{yy} \equiv \kappa_\perp = p\kappa^a + (1-p)\kappa^b, \\ &\frac{1}{\kappa_{zz}} \equiv \frac{1}{\kappa_\parallel} = p\frac{1}{\kappa^a} + (1-p)\frac{1}{\kappa^b}, \end{aligned} \qquad (6.9)$$

where $p = a/(a+b)$. Whence, in particular, it follows that in this medium $\kappa_\perp \geqslant \kappa_\parallel$ for real positive κ^a and κ^b.

For a structure formed of layers of anisotropic materials, the components of the effective dielectric tensor are defined in the long-wavelength approximation by the relations

$$\kappa_{\alpha\beta} - \frac{\kappa_{\alpha 3}\kappa_{\beta 3}}{\kappa_{33}} = p\left(\kappa^a_{\alpha\beta} - \frac{\kappa^a_{\alpha 3}\kappa^a_{\beta 3}}{\kappa^a_{33}}\right) + (1-p)\left(\kappa^b_{\alpha\beta} - \frac{\kappa^b_{\alpha 3}\kappa^b_{\beta 3}}{\kappa^b_{33}}\right),$$

$$\frac{\kappa_{\alpha 3}}{\kappa_{33}} = \frac{\kappa_{3\alpha}}{\kappa_{33}} = p\frac{\kappa^a_{\alpha 3}}{\kappa^a_{33}} + (1-p)\frac{\kappa^b_{\alpha 3}}{\kappa^b_{33}}, \quad \frac{1}{\kappa_{33}} = p\frac{1}{\kappa^a_{33}} + (1-p)\frac{1}{\kappa^b_{33}} \qquad (6.10)$$

where axis $3 \| z$.

6.2 Interband Transitions and Dielectric Susceptibility of a Periodic Heterostructure

6.2.1 Multiple Quantum-Well Structure

For convenience of comparison, we are presenting side-by-side expressions for the dielectric tensor in a homogeneous intrinsic semiconductor and a periodic Multiple Quantum-Well (MQW) structure considered an effective homogeneous medium:

$$\kappa_{\alpha\beta}(\omega) = \bar{\kappa}_{\alpha\beta} + \frac{4\pi e^2}{m^2\omega^2 V} \sum_{sj\mathbf{k}} \frac{p^{\alpha}_{\mathrm{v}j,\mathrm{c}s}(\mathbf{k})\, p^{\beta}_{\mathrm{c}s,\mathrm{v}j}(\mathbf{k})}{E_{\mathrm{c}\mathbf{k}} - E_{\mathrm{v}j\mathbf{k}} - \hbar\omega - i\hbar\Gamma_{\mathrm{cv}}}, \tag{6.11}$$

$$\kappa_{\alpha\beta}(\omega) = \bar{\kappa}_{\alpha\beta} + \frac{4\pi e^2}{m^2\omega^2 (a+b) S} \sum_{\substack{sj\mathbf{k}_\perp \\ \nu,\nu'}} \frac{p^{\alpha}_{\mathrm{v}\nu' j,\mathrm{c}\nu s}(\mathbf{k}_\perp)\, p^{\beta}_{\mathrm{c}\nu s,\mathrm{v}\nu' j}(\mathbf{k}_\perp)}{E_{\mathrm{c}\nu\mathbf{k}_\perp} - E_{\mathrm{v}\nu' j\mathbf{k}_\perp} - \hbar\omega - i\hbar\Gamma_{\mathrm{cv}}}. \tag{6.12}$$

Here, $\bar{\kappa}$ is the nonresonant contribution weakly dependent on frequency, V is the crystal volume, S is the heterostructure area in the (x, y) plane, $E_{\mathrm{c}\mathbf{k}}$ and $E_{\mathrm{v}j\mathbf{k}}$ are the electron energies in the conduction (c) and valence (vj) bands, respectively, $\mathbf{k}$ and $\mathbf{k}_\perp$ are the three- and two-component electron wave vectors, $p^{\alpha}_{\mathrm{c}s,\mathrm{v}j}(\mathbf{k})$ and $p^{\alpha}_{\mathrm{c}\nu s,\mathrm{v}\nu' j}(\mathbf{k}_\perp)$ the momentum operator matrix elements calculated, accordingly, between the Bloch states $|\mathrm{c}s\mathbf{k}\rangle$ and $|\mathrm{v}j\mathbf{k}\rangle$ and between the quasi-2D states $|\mathrm{c}\nu s\mathbf{k}_\perp\rangle$ and $|\mathrm{v}\nu' j\mathbf{k}_\perp\rangle$ in a single quantum well, ν and ν' are the indices of the conduction and valence subbands originating from quantum confinement and Γ_{cv} is the damping. Both in (6.11 and 12), the Coulomb interaction between the electron and the hole created by an absorbed photon is neglected for the sake of simplicity. Formally, (6.11, 12) differ in that the volume V is replaced by a product of the period $d = a + b$ by the area S and in the transition from summation in the three-dimensional vector $\mathbf{k}$ to that in the two-dimensional vector $\mathbf{k}_\perp$ and the quantum-well subband indices.

For simplicity, our consideration in the main part of this section will be carried out in terms of the simplest band-structure model, by which the compositional semiconducting materials are of cubic symmetry, the band extrema are at the Γ point, the carrier dispersion is characterized by the effective electron mass $m_{A,B}$ in the conduction band, and effective hole mass $m^{\mathrm{h}}_{A,B}$, where A and B are the indices of the well and the barrier layers, respectively, and interband optical transitions at the extremum point are allowed.

This model is fully applicable, for instance, to the pair of bands c, Γ_6 and v, Γ_7 in semiconductors of the class T_d. We will point out when necessary the peculiarities arising with a more complex band structure.

In a crystal of cubic symmetry with nondegenerate bands, calculation by (6.11) yields the following resonant contribution to the dielectric tensor

$$\kappa^{(\mathrm{res})}(\omega) = \left(\frac{eP_{\mathrm{cv}}}{m\omega}\right)^2 \left(\frac{2\mu_{\mathrm{cv}}}{\hbar^2}\right)^{3/2} \begin{cases} \sqrt{E_0 - \hbar\omega} & \text{for } \hbar\omega < E_0, \\ \mathrm{i}\sqrt{\hbar\omega - E_0} & \text{for } \hbar\omega > E_0. \end{cases} \tag{6.13}$$

Here, E_0 is the energy gap at the extremum point $\mathbf{k} = 0$ between the bands in question, $\mu_{\mathrm{cv}} = m_{\mathrm{c}} m_{\mathrm{v}}/(m_{\mathrm{c}} + m_{\mathrm{v}})$ is the reduced effective mass of the electron and the hole,

$$P_{\mathrm{cv}}^2 = \sum_{sj} |p^{\alpha}_{\mathrm{c}s,\mathrm{v}j}(0)|^2 \tag{6.14}$$

where α is one of the cordinates x, y, or z.

For interband transitions from the valence band Γ_7 split off by spin-orbit interaction from the band Γ_8, we have

$$E_0 = E_g + \Delta, \quad P_{\mathrm{cv}}^2 = \frac{2}{3}|\langle S|P_x|X\rangle|^2. \tag{6.15}$$

Expression (6.13) is valid also for transitions from heavy and light hole subbands (in the isotropic approximation) if by m_{v} one understands the corresponding effective mass (m_{hh} or m_{lh}). As for P_{cv}^2, it is defined here by the same expression (6.15).

We write now an expression for the resonant contribution to the dielectric tensor for a MQW structure:

$$\kappa^{(\mathrm{res})}(\omega) = \left(\frac{eP_{\mathrm{cv}}}{\hbar\omega m}\right)^2 \frac{2\mu_{\mathrm{cv}}}{a+b} \sum_{\nu} \left[\ln\frac{\bar{E}}{|E_0^{\nu\nu} - \hbar w|} + \mathrm{i}\pi\theta\left(\hbar\omega - E_0^{\nu\nu}\right)\right]. \tag{6.16}$$

Here $E_0^{\nu,\nu'}$ is the energy gap at the point $k_\perp = 0$ between the νth subband of the conduction band, and the ν'th subband of the valence band. The calculation was performed in the infinite barrier limit where

$$\begin{aligned} \mathbf{p}_{\mathrm{c}s\nu,\mathrm{v}m\nu'} &= \mathbf{p}_{\mathrm{c}s,\mathrm{v}m} \int f^*_{\mathrm{e}\nu}(z) f_{\mathrm{h}\nu'}(z)\,\mathrm{d}z \\ &= \mathbf{p}_{\mathrm{c}s,\mathrm{v}m}\delta_{\nu\nu'}. \end{aligned} \tag{6.17}$$

The energy $\bar{E}$ in (6.16) was introduced to make the argument of the logarithmic function dimensionless.

In the case of finite potential barriers the selection rules (6.17) for transitions between states with ν and ν' of the same parity break down. However, for semiconductors with nondegenerate bands and for symmetric quantum wells, the probability of additionally allowed transitions is insignificant, making (6.16) applicable. As an illustration, consider the quantity

$$\Lambda_{e\nu,h\nu'} = \left[\int f_{e\nu}(z) f_{h\nu'}(z)\, dz \right]^2 , \tag{6.18}$$

which characterizes the intensity of the optical transitions $h\nu' \to e\nu$ in the case where the potential barriers for holes may be assumed infinite, as before, and the functions $f_{h\nu'}(z)$ are defined by (3.83). For transitions to the lower subband of the conduction band e1, we have

$$\Lambda_{e1,h\nu} = \frac{[1-(-1)^\nu]\, 4(\nu\pi)^2(1+\cos ka)}{\left(1+\dfrac{\sin ka}{ka}+\dfrac{1+\cos ka}{\lambda a}\right)\left[(\nu\pi)^2-(ka)^2\right]^2} . \tag{6.19}$$

Taking the conduction band parameters of the GaAs/Al_x Ga_{1-x} As quantum well for $x = 0.35$, $a = 100$Å, we obtain for $\Lambda_{e1,h\nu}$ the values 0.92, 2.5×10^{-2}, 8×10^{-3}, respectively, for $\nu = 1, 3, 5$. In our approximation

$$D_h(z, z') \equiv \sum_\nu f_{h\nu}(z) f_{h\nu}(z') = \theta\left(\frac{a}{2} - |z|\right) \delta(z - z') . \tag{6.20}$$

Therefore the sum

$$\sum_{\nu'} \Lambda_{e\nu,h\nu'} = \iint dz\, dz' f_{e\nu}(z) D_h(z, z') f_{e\nu}(z')$$

$$= \int_{-a/2}^{a/2} f_{e\nu}^2(z)\, dz = \left(1 + \frac{1}{\lambda a} \frac{1+\cos ka}{1+\dfrac{\sin ka}{ka}}\right)^{-1}$$

is less than unity. In particular, for the above parameters this sum is 0.967. For a finite potential barrier V_0^h, the functions $f_{h\nu'}(z)$ are nonzero in the region $|z| > a/2$ as well, so that we have $D_h(z, z') = \delta(z - z')$, and

$$\sum_{\nu'} \Lambda_{e\nu,h\nu'} = 1 .$$

For transitions originating from the valence band Γ_7 split off by spin-orbit coupling, the quantity P_{cv}^2 is still defined by (6.15). As for the $hh\nu \to e\nu$ and $lh\nu \to e\nu$ transitions from the subbands resulting from quantization of states in the Γ_8 band of class T_d crystals, their contribution makes the dielectric tensor anisotropic ($\kappa_{xx} = \kappa_{yy} \neq \kappa_{zz}$):

(i) transitions $hh\nu \to e\nu$

$$P_{cv}^2(\mathbf{e} \perp z) = |\langle S|p_x|X\rangle|^2, \quad P_{cv}^2(\mathbf{e} \| z) = 0, \tag{6.21}$$

(ii) transitions $lh\nu \to e\nu$

$$P_{cv}^2(\mathbf{e} \perp z) = \frac{1}{3}|\langle S|p_x|X\rangle|^2, \quad P_{cv}^2(\mathbf{e} \| z) = \frac{4}{3}|\langle S|p_x|X\rangle|^2, \tag{6.22}$$

where $\mathbf{e}$ is the unit vector of light polarization.

Note that the expressions (6.13, 16) were derived in the limit as $\Gamma_{cv} \to 0$. In this case, the root singularity of the real (imaginary) part of the dielectric constant of a three-dimensional crystal exists only for $\hbar\omega < E_0$ ($\hbar\omega > E_0$). In a multiple quantum-well structure, the logarithmic singularity of the real part of $\kappa(\omega)$ exists both for $\hbar\omega < E_0$ and for $\hbar\omega > E_0$. Finite damping of Γ_{cv} washes out these singularities.

Using the expression for the absorption coefficient

$$K(\omega) = \frac{\omega}{c}\frac{\kappa''}{\sqrt{\kappa'}}, \tag{6.23}$$

we can obtain from (6.13) for a bulk crystal

$$K^{(3D)}(\omega) = 2\frac{e^2}{\hbar c n_\omega}\frac{\mu_{cv}}{m}\frac{P_{cv}^2}{m\hbar\omega}k_\omega,$$
$$k_\omega = \left[2\mu_{cv}\left(\hbar\omega - E_0\right)/\hbar^2\right]^{1/2} \tag{6.24}$$

and from (6.16) for a multiple quantum-well structure:

$$K^{MQW}(\omega) = 2\frac{e^2}{\hbar c n_\omega}\frac{\mu_{cv}}{m}\frac{P_{cv}^2}{m\hbar\omega}\frac{\pi}{a+b}\sum_\nu \theta\left(\hbar\omega - E_0^{\nu\nu}\right), \tag{6.25}$$

where $n_\omega = \sqrt{\kappa'}$. In some cases, in the determination of the light absorption coefficient for a periodic MQW structure, one takes in place of the true length only the total length of the quantum wells. The absorption coefficient K^{SQW} introduced in this way is related to (6.25) through

$$K^{SQW} = \frac{a+b}{a}K^{MQW}. \tag{6.26}$$

The difference in the frequency behavior of (6.24, 25) comes from the difference in the dependence of the density of states on energy in the three- and two-dimensional cases:

$$g^{(3D)}(\varepsilon) = \frac{2}{V}\sum_{\mathbf{k}}\delta\left(\frac{\hbar^2k^2}{2m^*} - \varepsilon\right) = \frac{1}{2\pi^2}\left(\frac{2m^*}{\hbar^2}\right)^{3/2}\theta(\varepsilon)\varepsilon^{1/2}, \tag{6.27}$$

$$g^{(2D)}(\varepsilon) = \frac{2}{S}\sum_{\mathbf{k}_\perp}\delta\left(\frac{\hbar^2k_\perp^2}{2m^*} - \varepsilon\right) = \frac{m^*}{\pi\hbar^2}\theta(\varepsilon). \tag{6.28}$$

6.2.2 Short-Period SL

For SLs with three-dimensional electron and hole subbands, the tensor κ is calculated by (6.11) where the indices c and v include the miniband number. It is of interest to consider a mixed case realized under the condition $m_c \ll m_v$, where the tunneling of holes between wells may be neglected and the electron motion in the conduction band is three-dimensional. In this case, for the contribution of the h1 → e1 transitions, for instance, we obtain

$$\kappa^{(\text{res})}_{\text{e1-h1}}(\omega) = \left(\frac{eP_{\text{cv}}}{m\omega}\right)^2 \frac{2\mu_\perp}{\hbar^2}\left(\frac{2M^{\text{e}}_{\parallel}}{\hbar^2}\right)^{1/2} \Lambda_{\text{e1,h1}}\,(E_0-\hbar\omega)^{1/2}\,, \tag{6.29}$$

where $E_0 = E_{\text{g}}+E_{\text{e1}}+E_{\text{h1}}$, $\mu_\perp = M^{\text{e}}_\perp m_{\text{v}}/(M^{\text{e}}_\perp + m_{\text{v}})$, $M^{\text{e}}_{\parallel,\perp}$ are the longitudinal and transverse electron masses at the subband bottom,

$$\Lambda_{\text{e1,h1}} = \left[\int \mathrm{d}z \frac{U^{\text{e}}_0(z)}{\sqrt{a+b}} f_{\text{h1}}(z)\right]^2 = \frac{4a}{a+b} C_a^2 \frac{\pi^2(1+\cos ka)}{\left[\pi^2-(ka)^2\right]^2}\,. \tag{6.30}$$

Here C_a and k are defined in accordance with (5.64, 67).

6.3 Coulomb Interaction Between the Electron and the Hole

To include the electron-hole Coulomb interaction in the dielectric response we write the expression for the tensor $\kappa^{(\text{res})}$ in an intrinsic semiconductor in a more general form, compared to (6.11),

$$\kappa^{(\text{res})}_{\alpha\beta}(\omega,\mathbf{q}) = \frac{4\pi}{\hbar\omega^2 V}\sum_{f\neq 0}\frac{j^{\alpha}_{0,f}(-\mathbf{q})\,j^{\beta}_{f,0}(\mathbf{q})}{\omega_f-\omega-\mathrm{i}\Gamma_f}\,, \tag{6.31}$$

where $|0\rangle$ is the ground state of the crystal, $|f\rangle$ are the excited states including both the excitons of the discrete spectrum (of energy $\hbar\omega_f < E_{\text{g}}$) and the electron-hole pairs whose relative motion is influenced by Coulomb interaction ($\hbar\omega_f > E_{\text{g}}$), $\mathbf{j}_{f,0}(\mathbf{q})$ is the matrix element of the many-particle current-density operator:

$$\hat{\mathbf{j}}(\mathbf{q}) = \frac{-e}{2m}\sum_l\left(\hat{\mathbf{p}}_l \mathrm{e}^{\mathrm{i}\mathbf{q}\cdot\mathbf{x}_l} + \mathrm{e}^{\mathrm{i}\mathbf{q}\cdot\mathbf{x}_l}\hat{\mathbf{p}}_l\right), \tag{6.32}$$

the index 1 running over the electrons. In general, (6.31) includes spatial dispersion which we continue to neglect in what follows.

Disregarding Coulomb interaction for the optical transition $|0\rangle \to |f\rangle = |\text{c}s, \text{v}\bar{m}\rangle$ involving excitation of an electron cs, $\mathbf{k}=0$ and of a hole v$\bar{m}$, $\mathbf{k}=0$ (the state $\bar{m}$ is related to m by the time inversion operation)

$$j^{\alpha}_{f,0} = -\frac{e}{m}p^{\alpha}_{\text{c}s,\text{v}m}\,. \tag{6.33}$$

For the optical transitions $|0\rangle \to |f\rangle = |\text{c}s, \text{v}\bar{m}, \mu\rangle$ involving excitation of an exciton μ (in the discrete spectrum or continuum)

$$j^{\alpha}_{f,o}(O) = -\frac{e}{m}p^{\alpha}_{\text{c}s,\text{v}m}\sqrt{V}\varphi^*_\mu(\mathbf{x}_{\text{e}}-\mathbf{x}_{\text{h}}=0)\,, \tag{6.34}$$

where $\varphi_\mu(\mathbf{r})$ is the envelope function describing the relative motion of the electron and hole in the two-particle exciton wave function

$$\Psi_{s\bar{m}\mu} = \frac{\mathrm{e}^{\mathrm{i}\mathbf{k}\cdot\mathbf{R}}}{\sqrt{V}}\varphi_\mu(\mathbf{r})u^{(0)}_{cs}(\mathbf{x}_{\text{e}})u^{(0)}_{\text{v}\bar{m}}(\mathbf{x}_{\text{h}})\,. \tag{6.35}$$

Here $u^0(\mathbf{x})$ is the three-dimensional Bloch periodic amplitude at the Γ point.

Substituting (6.34) in (6.31) we obtain after some algebra

$$\kappa(\omega) = \bar{\kappa} - \sum_{n=1}^{\infty} \frac{\Omega_n^{(3D)}}{\omega_g - \left(E_B/\hbar n^2\right) - \omega - i\Gamma_n} + \frac{2\pi e^2}{m^2\omega^2} P_{cv}^2 \int_0^{\bar{E}} \frac{g_{cv}^{(3D)}(\varepsilon)\eta^{(3D)}(\varepsilon)}{E_g - \hbar\omega + \varepsilon - i\hbar\Gamma} d\varepsilon. \tag{6.36}$$

Here $\bar{\kappa}$ is the nonresonant contribution, $\omega_g = E_g/\hbar$, and the oscillator strength of the ns-exciton $\Omega_n^{(3D)} = \Omega_1^{(3D)}/n^3$,

$$\Omega_1^{(3D)} = \frac{4e^2}{a_B^3 m\omega} \frac{P_{cv}^2}{m\hbar\omega}, \tag{6.37}$$

$g_{cv}^{(3D)}(\varepsilon)$ is the reduced density of states derived from (6.27) by substituting μ_{cv} for m^*, E_B the exciton Rydberg, a_B the Bohr radius of the exciton (5.103), and the Sommerfeld factor

$$\eta^{(3D)}(\varepsilon) = \frac{X}{1 - \exp(-X)}, \quad X = 2\pi \left(\frac{E_B}{\varepsilon}\right)^{1/2}. \tag{6.38}$$

Note that in (6.36) and in what follows we define the oscillator strength of the n-exciton as

$$\Omega_n = \operatorname{Im}\left\{\frac{1}{\pi}\int \kappa_n^{(res)}(\omega)\, d\omega\right\}, \tag{6.39}$$

where $\kappa_n^{(res)}(\omega)$ is the resonant contribution of this exciton to the dielectric constant. The oscillator strength introduced this way is related to the longitudinal-transverse exciton splitting through the expression

$$\omega_{LT}^{(n)} = \Omega_n/\kappa_b, \tag{6.40}$$

where κ_b is the background dielectric constant including the non-resonant contribution $\bar{\kappa}$ and the contributions to κ coming from $n' \neq n$ excitons and electron-hole pairs. In deriving (6.40) we used the inequality $\omega_{LT}^{(n)} \ll \omega_g$, which is valid for excitons in semiconductors.

As follows from (6.36), for $\hbar\omega > E_g$ the absorption coefficient including Coulomb interaction is obtained from (6.24) by multiplying it by the Sommerfeld factor (6.38):

$$K^{(3D)}(\omega) = 4\pi \frac{e^2}{\hbar c n_\omega} \frac{\mu_{cv}}{m} \frac{p_{cv}^2}{m\hbar\omega} \frac{1}{a_B} \frac{1}{1 - \exp(-X)}. \tag{6.41}$$

For a periodic MQW structure, the dielectric constant including Coulomb interaction assumes the form

$$\kappa(\omega) = \bar{\kappa} + \sum_{\beta} \frac{\Omega_\beta^{MQW}}{\omega_\beta - \omega - i\Gamma},$$

$$\Omega_\beta^{\mathrm{MQW}} = \Omega_1^{(3\mathrm{D})} \frac{\pi a_{\mathrm{B}}^3}{a+b} \left| \int \varphi_\beta(0, z, z)\, \mathrm{d}z \right|^2 . \tag{6.42}$$

Here $\varphi_\beta(\boldsymbol{\rho}, z_{\mathrm{e}}, z_{\mathrm{h}})$ is the envelope of the exciton wave function (5.95), which determines the exciton wave function in a single quantum-well structure

$$\Psi_{sm\beta} = \frac{\mathrm{e}^{\mathrm{i}K_\perp \cdot R_\perp}}{\sqrt{S}} \varphi_\beta(\boldsymbol{\rho}, z_{\mathrm{e}}, z_{\mathrm{h}}) u_{\mathrm{cs}}^{(0)}(\mathbf{x}_e)\, u_{\mathrm{v}m}^{(0)}(\mathbf{x}_h),$$

$$\int \left|\varphi_\beta(\boldsymbol{\rho}, z_{\mathrm{e}}, z_{\mathrm{h}})\right|^2 \mathrm{d}\boldsymbol{\rho}\, \mathrm{d}z_{\mathrm{e}}\, \mathrm{d}z_{\mathrm{h}} = 1. \tag{6.43}$$

Choosing the trial function of the type (5.108), we obtain

$$\int \varphi_{1s}(0, z, z)\, \mathrm{d}z = G(0,0) \int f_{\mathrm{e}1}(z) f_{\mathrm{h}1}(z)\, \mathrm{d}z. \tag{6.44}$$

Taking into account Coulomb interaction in expression (6.25) for the light absorption coefficient in a periodic MQW structure for $\hbar\omega > E_0^{\nu\nu}$, the function $\theta(\hbar\omega - E_0^{\nu\nu})$ has to be multiplied by a factor $\eta_\nu^{\mathrm{QW}}(\hbar\omega - E_0^{\nu\nu})$. No analytic form of this factor has been found for a quantum well of arbitrary thickness. In making an evaluation it may be useful to keep in mind that, in an ideal two-dimensional case, the Sommerfeld factor can be written as

$$\eta^{(2\mathrm{D})}(\varepsilon) = \frac{2}{1+\exp(-X)}, \tag{6.45}$$

where X is defined according to (6.38). The values of $\eta^{(2\mathrm{D})}$ range from 1 (for $\varepsilon \to 0$) to 2 (for $\varepsilon \to \infty$), whereas $\eta^{(3\mathrm{D})}(\varepsilon) \to \infty$ for $\varepsilon \to 0$. Hence, quantum confinement effects reduce the effect of Coulomb interaction on the light absorption spectrum in the continuum ($\hbar\omega > E_0^{\nu\nu}$). At the same time, the oscillator strength for exciton excitation in the discrete spectrum due to quantum confinement increases. To demonstrate this, let us write $\varphi(\boldsymbol{\rho}, z_{\mathrm{c}}, z_{\mathrm{h}})$ in the approximation of two-dimensional Coulomb interaction between the electron and the hole (5.110, 120). For the ground state (5.110) of the two-dimensional exciton we obtain

$$\left[\int \varphi_{n=1}(0, z, z,)\, \mathrm{d}z\right]^2 = \frac{8}{\pi a_{\mathrm{B}}^2}.$$

Therefore it follows from (6.42) that the contribution of this exciton to the dielectric constant of an effective medium can be presented as

$$\kappa_{n=1}(\omega) = \frac{\Omega_1^{(2\mathrm{D})}}{\omega_1 - \omega - \mathrm{i}\Gamma_1}, \quad \Omega_1^{(2\mathrm{D})} = \frac{8a_{\mathrm{B}}}{a+b}\Omega_1^{(3\mathrm{D})}. \tag{6.46}$$

The two-dimensional approximation used in the derivation of (6.46) assumes the inequality $a < a_{\mathrm{B}}$ to be met, so that for comparable a and b we have indeed $\Omega_1^{(2\mathrm{D})} > \Omega_1^{(3\mathrm{D})}$. Physically, this inequality is accounted for by the fact that the barriers *press* an electron and a hole in the exciton excited in a quantum well, thus increasing the probability to find them at the same point, which, in turn, increases the exciton-photon coupling constant. Adding to (6.46) the resonant contributions of the other electron-hole excitations $e\nu - h\nu$, we obtain for the resonant contribution to κ in the two-dimensional approximation

$$\kappa_{\rm res}^{(2D)}(\omega) = \sum_{n=1}^{\infty} \frac{\Omega_n^{(2D)}}{\omega_n - \omega - {\rm i}\Gamma_n} + \frac{2\pi}{a+b}\left(\frac{eP_{\rm cv}}{m\omega}\right)^2 \int_0^{\bar{E}} \frac{g^{(2D)}(\varepsilon)\eta^{(2D)}(\varepsilon)}{E_0^{\nu\nu} - \hbar\omega + \varepsilon - {\rm i}\hbar\Gamma}{\rm d}\varepsilon$$

$$= \Omega_1^{(2D)} \left(\sum_{n=1}^{\infty} \frac{(2n-1)^{-3}}{\omega_n - \omega - {\rm i}\Gamma_n} + \frac{\hbar}{32E_{\rm B}} \times \int_0^{\bar{E}} \frac{{\rm d}\varepsilon}{E_0^{\nu\nu} - \hbar\omega + \varepsilon - {\rm i}\hbar\Gamma} \frac{1}{1+\exp(-X)} \right), \tag{6.47}$$

where

$$\hbar\omega_n = E_0^{\nu\nu} - \frac{4E_{\rm B}}{(2n-1)^2}, \quad \Omega_n^{(2D)} = \frac{\Omega_1^{(2D)}}{(2n-1)^3}.$$

As the period decreases, the barriers become thinner and the motion of the electron or hole (primarily of the carrier with a smaller effective mass) three-dimensional. This means that, in a periodic heterostructure with $a \sim b$, the exciton oscillator strength reaches a maximum with decreasing period, otherwise it begins to fall off. For illustration, we present here an expression for the oscillator strength of the exciton ground state in a SL, where the electron motion is three-dimensional and the hole states in neighboring wells practically do not overlap:

$$\Omega_1^{\rm SL} = \Omega_1^{(3D)} \frac{a_{\rm B}^3}{a_\parallel a_\perp^2} \Lambda_{\rm e1,h1}. \tag{6.48}$$

Here $a_\parallel$ and $a_\perp$ are the variational parameters of the trial function (5.51) related through expressions (5.49) with the parameters

$$\gamma = \mu_\perp / M_\parallel^{\rm e} \quad \text{and} \quad a_{\rm B}^\perp = \hbar^2\kappa/\mu_\perp e^2,$$

the masses $\mu_\perp$ and $M_\parallel^{\rm e}$ are introduced in (6.29) and the coefficient $\Lambda_{\rm e1,h1}$ is determined by (6.30).

Using the general presentation of the exciton wave function in a single quantum well structure of the type GaAs/AlGaAs as an expansion (5.121) in states of the noninteracting electron and hole, we obtain, in place of (6.34), the following expression for the matrix element of the current operator

$$j_{f,O}^{\alpha}(\mathbf{q}=0) = \sqrt{S}\frac{-e}{m}\sum_{m'} p_{{\rm cs},\nu m'}^{\alpha}$$

$$\times \left[\sum_{\nu' j \mathbf{k}_\perp} G_{\nu\nu' j}(\mathbf{k}_\perp) \int f^{\rm e}_{\nu \mathbf{k}_\perp}(z) f^{\rm h}_{\nu' j, -\mathbf{k}_\perp, \bar{m}'}(z) {\rm d}z \right]^* . \quad (6.49)$$

For a semiconductor with nondegenerate bands, when $f^{\rm h}_{\nu j \mathbf{k}_\perp, m} \propto \delta_{mj}$ and the dependence of the functions $f^{\rm e,h}_{\nu \mathbf{k}_\perp}(z)$ on $\mathbf{k}_\perp$ and the mixing of hole states from different subbands may be disregarded, (6.49) simplifies to

$$j^\alpha_{f,O}(\mathbf{q}=0) = \sqrt{S}\frac{-e}{m} p^\alpha_{{\rm cs},{\rm v}\bar{j}} \left[G_{\nu\nu'}(\rho = 0) \int f_{{\rm e}\nu}(z) f_{{\rm h}\nu}(z)\, {\rm d}z \right] , \quad (6.50)$$

where

$$G_{\nu\nu'}(\rho) = \sum_{\mathbf{k}_\perp} e^{{\rm i}\mathbf{k}_\perp \cdot \rho} G_{\nu\nu'}(\mathbf{k}_\perp).$$

A similar expression has already been used in the calculation of the oscillator strength $\Omega^{\rm MQW}_\beta$ (6.44, 46).

For a semiconductor with degenerate valence band Γ_8, the hole in state (5.122b) and the exciton in state (5.121) can be assigned in the isotropic approximation ($D = \sqrt{3}B$) clearly defined projections j and M of the angular momentum on the z axis, where

$$M = s + j + l, \quad (6.51)$$

l is the projection of the angular momentum of relative motion of the electron and hole, which determines the angular dependence of the function

$$G_{\nu\nu' j}(\mathbf{k}_\perp, l) = \left(k_x + {\rm i}k_y \,{\rm sign}\, l\right)^{|l|} R_{\nu\nu' j}\left(k^2_\perp, |l|\right) \quad (6.52)$$

found in the two-subband approximation, where in (5.121) one assumes $\nu'_1 = \nu'$ and $j_1 = j$. For the s- and $p_{\pm 1}$-excitons eν − h$\nu' j$ created with participation of a hole from the v$\nu' j$ subband, 1 = 0, ±1, accordingly. Note that the functions

$$f^{\rm h}_{\nu j \mathbf{k}_\perp, m} \propto \left[k_x + {\rm i}k_y \,{\rm sign}(j - m)\right]^{|j-m|} \quad (6.53)$$

have a structure similar to (6.52). One can readily check that (5.123), which illustrates the hybridization of heavy and light holes in the exciton, agrees with the angular dependence (6.53). As follows from (6.52, 53), integration in the direction of the vector $\mathbf{k}_\perp$ in (6.49) leaves only terms with $1 = m - j$ nonzero. Therefore optical excitation of the exciton leaves the total projection of the angular momentum unchanged, so that

$$M = P,$$

where $P = \pm 1$ for light polarized in the (x, y) plane counterclock-wise and clockwise, respectively, and $P = 0$ for light polarized along the z axis.

Since the potential energy of the Coulomb interaction between the electron and the hole is invariant under symmetry transformations of the heterostructure, relations (5.125) forbid mixing of s-excitons belonging to valence subbands ν' of different parity. This can be checked using expression (5.127) for the matrix

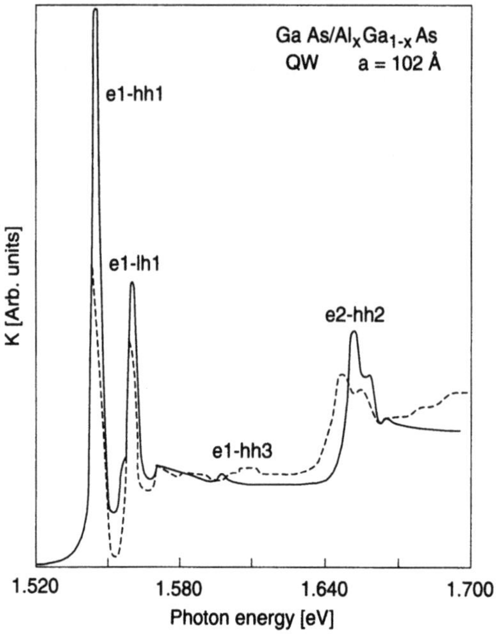

Fig. 6.1. Theoretical absorption spectrum (solid line) and experimental photoluminescence excitation spectrum measured [6.1] on GaAs/$Al_{0.27}Ga_{0.73}As$ structure with 102 Å-thick quantum well (from [6.2])

element $V_{\nu' j, \nu'_1 j_1}(\mathbf{q}_\perp)$. Therefore in the case of arbitrary mixing of heavy and light hole states in the function $\psi^{\mathrm{h}}_{\nu j \mathbf{k}_\perp}(\mathbf{x})$ one can propose the following selection rules: (i) for ν and ν' of opposite parity, transitions involving excitation of e$\nu-$hhν' or e$\nu-$lhν' s-excitons are forbidden in the dipole approximation while those involving excitation of p-excitons are allowed; (ii) in the $\mathbf{e}\|z$ polarization, s-excitons eν $-$ hhν' are not excited at any ν, ν'. For a structure with $z\|[001]$, these rules are applicable in the anisotropic case $D \neq \sqrt{3}B$ as well. In an asymmetric well or in an external electric field $\mathcal{I}\|z$, the properties (5.125) break down, and the constraints (i), (ii) are removed.

Figure 6.1 presents a spectrum of light absorption in interband transitions in a GaAs/$Al_{0.27}Ga_{0.73}As$ structure with a 102 Å quantum well. The spectrum reveals clearly pronounced absorption peaks associated with the e1-hh1 and e1-lh1 1s excitons. Note the doublet structure of the feature near the resonance frequency of the e2-hh2 exciton; it originates from the excitation of e2-hh2 1s- and e1-lh1 2p-excitons. The calculations presented by the solid line include the heavy and light-hole state mixing, with the coupling between the e1-lh1 discrete exciton states and the continuum of e2-hh2 states. The theoretical spectrum reveals a weak feature due to the e1-hh3 exciton. This validates the selection rules (6.17), by which the principal contribution to absorption comes from transitions preserving the subband index ν. Expressions (6.25, 26) predict a plateau-like domain in the $K^{\mathrm{MQW}}(\omega)$ or $K^{\mathrm{SQW}}(\omega)$ spectrum immediately after the exciton peak. In full agreement with this prediction, the spectrum in Fig. 6.1 is practically horizontal in the region between the e1-lh1 and e2-hh2 peaks.

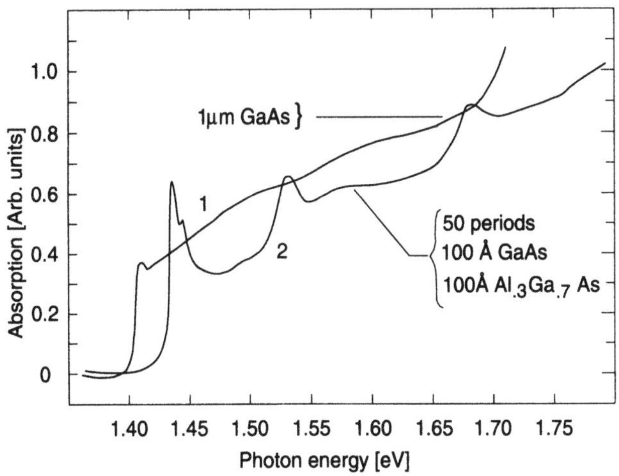

Fig. 6.2. Comparison of absorption spectra of (1) high quality bulk GaAs, and (2) GaAs/Al_xGa_{1-x} As structure with 100 Å-thick quantum wells, measured at room temperature [6.3]

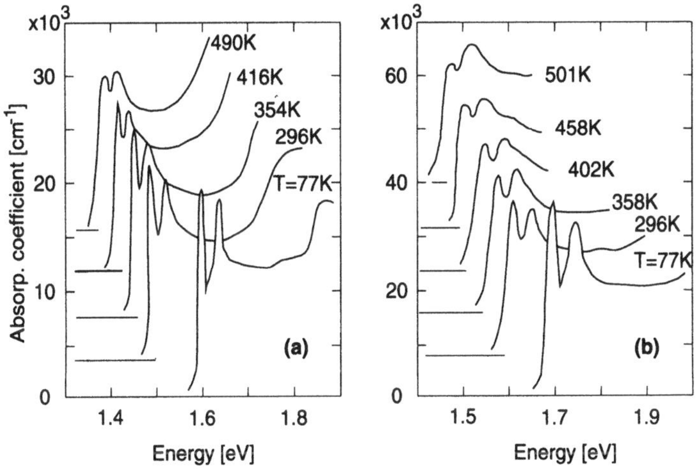

Fig. 6.3a,b. Absorption spectra for periodic heterostructure GaAs/AlAs with **a** $a = 73$ Å, $b = 29$ Å, and **b** $a = 46$ Å, $b = 53$ Å obtained at different temperatures. The absorption coefficient determined by (6.26) [6.4]

For a convenient comparison, Fig. 6.2 presents absorption spectra of bulk GaAs and of a MQW structure GaAs/$Al_{0.3}Ga_{0.7}$As. We see $e\nu - hh\nu$, $e\nu - lh\nu$ exciton absorption peaks for $\nu = 1, 2, 3$ (for transitions with $\nu = 2, 3$ the peaks due to heavy and light excitons merge). In contrast to bulk GaAs for which the absorption coefficient increases monotonically for $\hbar\omega > E_g$, MQW structures exhibit a plateau region after the exciton peak. The fact that the excitons are much more clearly pronounced on curve 2 is due to the greater oscillator strength and the stronger binding energy of the quasitwo-dimensional excitons. For the same reason, the exciton absorption peaks for the GaAs/Al_xGa_{1-x}As or GaAs/AlAs heterostructures are observed not only at room temperature, but even at $T \geqslant 500$ K (Fig. 6.3).

The calculated absorption spectra for three periodic heterostructures with the same well thickness and different barrier widths are presented in Fig. 6.4. Just as in Figs. 6.1, 6.3, the main absorption peaks originate from the e1-hh1 and e1-lh1 1s-exciton transitions. The vertical H and L lines identify the energy gaps between the e1, hh1 and e1, lh1 minibands for $k = 0$ (a type M_0 singularity in the reduced density of states). The lines H' and L' specify the position of the M_1 singularity at the saddle point $\mathbf{k} = (0, 0, \pi/d)$. The energy difference $H - H'$ or $L - L'$ determines the total width of the miniband spectrum of electrons and holes in SL in the direction $\mathbf{k} \| z$. In a wide barrier ($b = 140$ Å) structure, there is practically no subband dispersion in k_z, and in calculations of the absorption coefficient one may neglect overlap of carrier wave functions in neighboring wells. In a structure with $b = 65$ Å or 56 Å,

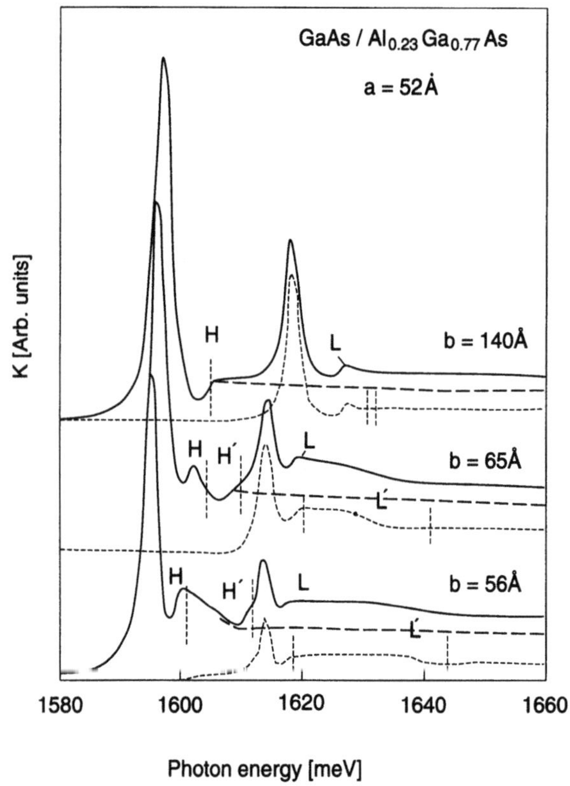

Fig. 6.4. Calculated absorption spectra of GaAs/$Al_{0.23}Ga_{0.77}$As periodic heterostructures with 52 Å-thick wells and barrier width of 140, 65 and 56 Å. Dashed line: contribution of e1-hh1 transitions; dotted line: contribution of e1-lh1 transitions; solid line: total absorption coefficient. In the region of the e1-hh1 exciton peak the solid and dashed lines merge (from [6.2])

a noticeable contribution to absorption comes already from the 2s-exciton, the absorption curve in the regions $H - H'$ and $L - L'$ being no longer horizontal if miniband dispersion is included.

6.4 Exciton Polaritons in an Optical Superlattice

Consider the propagation of exciton polaritons in an optical SL (Fig. 6.5), where a layer of thickness b has a dielectric constant κ^b independent of the wave vector, and a layer of thickness a is characterized by an isotropic dielectric function with a resonant exciton contribution

$$\kappa^a(\omega, \mathbf{q}) = \kappa_0^a \left(1 + \frac{\omega_{\mathrm{LT}}}{\omega_0(\mathbf{q}) - \omega}\right),$$
$$\omega_0(\mathbf{q}) = \omega_0 + \hbar q^2/2M. \tag{6.54}$$

This model is valid for a semiconductor SL with thick wells $a \gg a_{\mathrm{B}}$ and impenetrable barriers, where the envelope of the wave function of the mechanical exciton φ_{MQW} can be written in the form (5.138), with φ_{SQW} defined by (5.100).

Within the layer of type B, the Bloch amplitude of a normal light wave (6.3) represents a superposition of the direct and back-scattered transverse plane waves

$$\mathcal{I}_{\mathrm{Q}}(z) = e^{\mathrm{i}Q_z\bar{z}_a}\left(\mathcal{I}^+ \mathrm{e}^{\mathrm{i}k_b(z-\bar{z}_b)} + \mathcal{I}^- e^{-\mathrm{i}k_b(z-\bar{z}_b)}\right), \tag{6.55}$$

where k_b was introduced in (6.5), and $\bar{z}_a$, $\bar{z}_b$ are the coordinates of the centers of the adjacent layers A and B. The dielectric constant of material A has a resonant spatial dispersion. Therefore, within the layer of type A, one has to take into account the additional light waves, so that

$$\mathcal{I}_{\mathrm{Q}}(z) = \mathrm{e}^{\mathrm{i}Q_z\bar{z}_a} \sum_{l=1,2,3} \left(\mathcal{I}_l^+ \mathrm{e}^{\mathrm{i}k_l(z-\bar{z}_a)} + \mathcal{I}_l^- \mathrm{e}^{-\mathrm{i}k_l(z-\bar{z}_a)}\right), \tag{6.56}$$

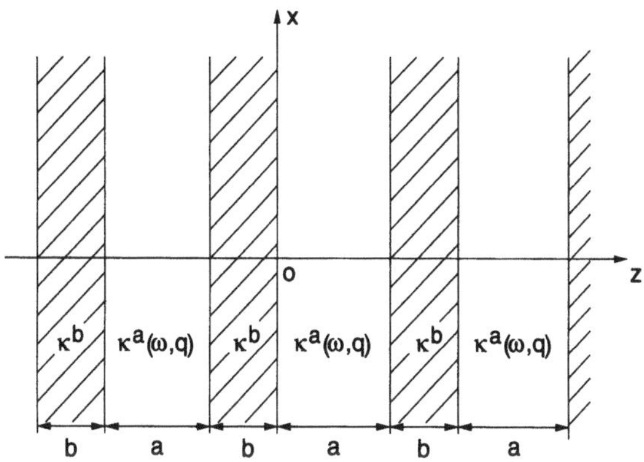

Fig. 6.5. Optical superlattice with dielectric constant κ^b in layers of type B (with thickness b), and nonlocal dielectric constant $\kappa^a(\omega, \mathbf{q})$ in layers of type A (with thickness a)

where the index l identifies the branches of the transverse ($l = 1, 2$) and longitudinal ($l = 3$) normal waves in a homogeneous medium with the dielectric constant (6.54). The wave vector $\mathbf{q} = (Q_x, Q_y, k_l)$ of these waves satisfies the dispersion relation

$$q^2 = k_0^2 \kappa^a(\omega, q) \quad \text{for} \quad l = 1, 2,$$
$$\kappa^a(\omega, q) = 0 \quad \text{for} \quad l = 3, \tag{6.57}$$

where $q^2 = k_l^2 + Q_\perp^2$, $k_0 = \omega/c$. The contribution of the exciton resonance ω_0 in question to the polarization of the medium in the layer of type A is related to the amplitudes $\mathcal{I}_l^\pm$ through the expression

$$4\pi \mathbf{P}_\mathbf{Q}(z) = \mathrm{e}^{\mathrm{i}Q_z \bar{z}_a} \left[\sum_{l=1,2} \left(n_l^2 - \kappa_0^a\right) \left(\mathcal{I}_l^+ \mathrm{e}^{\mathrm{i}k_l(z-\bar{z}_a)} \right.\right.$$
$$\left.\left. + \mathcal{I}_l^- \mathrm{e}^{-\mathrm{i}k_l(z-\bar{z}_a)}\right) - \kappa_0^a \left(\mathcal{I}_3^+ \mathrm{e}^{\mathrm{i}k_3(z-\bar{z}_a)} + \mathcal{I}_3^- \mathrm{e}^{-\mathrm{i}k_3(z-\bar{z}_a)}\right)\right], \tag{6.58}$$

where the refractive index $n_l = q_l/k_0 = (k_l^2 + Q_\perp^2)^{1/2}/k_0$. Substitution of the fields (6.55, 56) into Maxwell's boundary conditions for the tangential components of $\mathcal{I}$ and $\mathbf{H}$, and into the additional boundary conditions (ABC) for $\mathbf{P}_\mathbf{Q}(z)$ for $z = \bar{z}_a \pm a/2$ with the inclusion of the Bloch character of the solutions results in a set of algebraic equations for the determination of the amplitudes $\mathcal{I}^\pm$ and $\mathcal{I}_l^\pm$. The condition of its solvability yields a dispersion relation. With the Pekar ABC

$$\mathbf{P}(\bar{z}_a \pm a/2) = 0,$$

this equation for the TE waves assumes the form [6.5]

$$\cos Q_z(a+b) = \left[s_2 \tilde{n}_1(\eta_2 - 1) + s_1 \tilde{n}_2(\eta_1 - 1)\right]^{-1}$$
$$\times \left\{ c \left[c_1 s_2 \tilde{n}_1(\eta_2 - 1) + c_2 s_1 \tilde{n}_2(\eta_1 - 1) + s \tilde{n}_1 \tilde{n}_2 (1 - c_1 c_2)\right.\right.$$
$$\left.\left. + s s_1 s_2 \left[1 - \frac{1}{2}\eta_2(\tilde{n}_1^2 + 1) - \frac{1}{2}\eta_1\left(\tilde{n}_2^2 + 1\right)\right]\right]\right\}, \tag{6.59}$$

where

$$\tilde{n}_l = k_l/k_b, \quad s = \sin k_b b, \quad c = \cos k_b b,$$
$$s_l = \sin k_l a, \quad c_l = \cos k_l a, \quad \eta_1 = \eta_2^{-1} = (n_1^2 - \kappa_0^a)/(n_2^2 - \kappa_0^a).$$

In the nonresonance frequency region satisfying the inequality $|\omega - \omega_0| > \omega_{\mathrm{LT}}$ so that $|n_1/n_2|^2 \ll 1$ or $|n_2/n_1|^2 \ll 1$, (6.59) transforms to (6.3), where one has to set $k_a = k_1$ for $\omega < \omega_0$ and $k_a = k_2$ for $\omega > \omega_\mathrm{L} = \omega_0 + \omega_{\mathrm{LT}}$.

In the long-wavelength approximation, where

$$Q_z(a+b), k_b b, k_2 a \ll 1,$$

the dispersion relation for TE waves reduces to

$$(Q_z/k_0)^2 = \kappa_\perp(\omega, \mathbf{Q}_\perp),$$

where the effective dielectric constant

$$\kappa(\omega, \mathbf{Q}_\perp) = p\kappa_0^a + (1-p)\kappa^b + \kappa_0^a \sum_{N=0}^{\infty} \frac{\omega_{\text{LT}}^{(2N+1)}}{\omega_{2N+1}(\mathbf{Q}_\perp) - \omega}, \tag{6.60}$$

$$p = a/(a+b),$$

$$\omega_j(\mathbf{Q}_\perp) = \omega_0 + \frac{\hbar}{2M}\left(j\frac{\pi}{a}\right)^2 - \frac{\hbar Q_\perp^2}{2M},$$

$$\omega_{\text{LT}}^{(2N+1)} = \frac{8p}{\pi^2(2N+1)^2}\omega_{\text{LT}}. \tag{6.61}$$

Expression (6.61) can be derived also from (6.42) in the particular case $a \gg a_{\text{B}}$, where the difference $\hbar\omega_j(0) - \hbar\omega_0$ is equal to the quantum confinement energy of the 1s exciton as a whole in a well of width a,

$$\int \varphi_j(0, z, z)\mathrm{d}z = \varphi(0) \int F_j(Z)\mathrm{d}Z = \left(\frac{2a}{\pi a_{\text{B}}^3}\right)^{1/2} \frac{1-(-1)^j}{\pi j}$$

and the oscillator strength

$$\Omega_{2N+1} = \frac{8}{\pi^2} p \Omega_1^{(3\text{D})}.$$

Equation (6.60) includes spatial dispersion, i.e., dependence of $\kappa_\perp$ on $\mathbf{Q}_\perp$. Note that in the model of the optical SL considered here a layer of thickness b, there is an infinite barrier for the exciton, so that *mechanical* motion of the exciton is possible only within one layer. Therefore the dielectric constant (6.60) does not depend on Q_z.

Figure 6.6 presents dispersion curves of exciton polaritons in an optical SL with $\kappa_0^a = \kappa^b \equiv \kappa_0$ for two values of thickness a for which $ak_0\sqrt{\kappa_0} = 1$ and 2, respectively. The calculation was carried out for normally propagating waves ($Q_x = Q_y = 0$, $Q_z = Q$). For simplicity, a SL with thin barriers $b \ll a$, which were assumed, as before, impenetrable for free carriers, was considered. In the limit as $b \to 0$, the dispersion relation (6.59) transforms to

$$\cos Qa = \frac{c_1 s_2 n_1 (n_2^2 - \kappa_0) - c_2 s_1 n_2 (n_1^2 - \kappa_0)}{s_2 n_1 (n_2^2 - \kappa_0) - s_1 n_2 (n_1^2 - \kappa_0)}. \tag{6.62}$$

A comparison of the dispersion curves calculated by the exact and approximate formulas shows that for $ak_0\sqrt{\kappa_0} \lesssim 1$ one can use the long-wavelength approximation. For an approximate description of the spectrum in the region of dipole-forbidden states with $j = 2N$, one has to include in (6.60) additional resonant terms

$$\kappa_0^a \tilde{\omega}^{(2N)}(Q)(\omega_{2N} - \omega)^{-1},$$

where

$$\tilde{\omega}^{(2N)}(Q) = (aQ)^2 \omega_{\text{LT}}^{(1)}/(4N)^2.$$

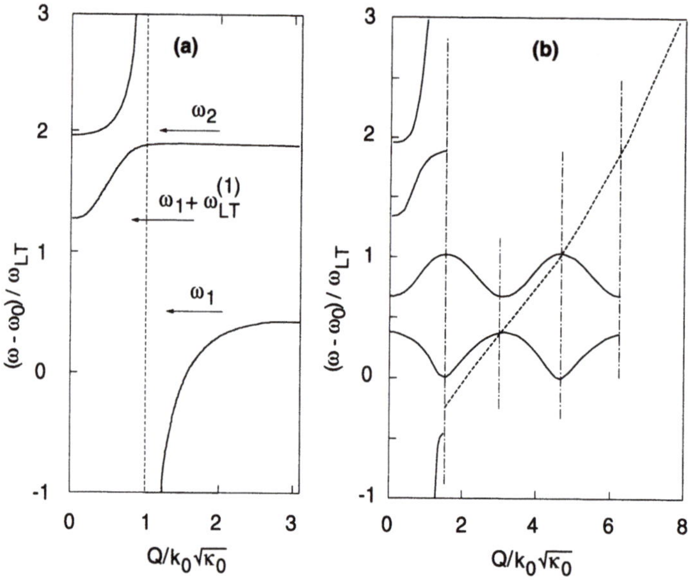

Fig. 6.6a,b. Dispersion curves for transverse polaritons propagating perpendicular to the layers of optical superlattice with $b \ll a$, for **a** $ak_0\sqrt{\kappa_0} = 1$ and **b** $ak_0\sqrt{\kappa_0} = 2$. For convenience, two branches in **b** are shown in the extended zone scheme. Dot-and-dash vertical lines specify the values of wave vector $Q = m\pi/a$ ($m = 1$–4), multiples of one half of reciprocal lattice vector $G_1 = 2\pi/a$ [6.5]

For $ak_0\sqrt{\kappa_0} > 1$, the conditions of applicability of the long-wavelength approximation break down, and the energy spectrum of normal waves near the longitudinal frequency ω_L should be calculated by (6.59, 62).

In an optical SL with a two-dimensional defect of the "non-standard-layer" type of a different thickness or with a different dielectric constant there are states of exciton polaritons localized in one direction (the principal SL axis). The scheme of calculation of such states is similar to that of the solution of the electron-localization problem discussed in Sect. 5.2.

6.5 Light Reflection

We will consider here successively the reflection of light from a single quantum-well structure, a multiple quantum-well structure (thick barrier SL) and a short-period SL.

6.5.1 Single Quantum-Well (SQW) Structure (Figure 6.7)

The amplitude coefficient $r = \mathcal{I}_r/\mathcal{I}_0$ for reflection from the structure shown in Fig. 6.7 is connected with the coefficient r_{123} of reflection from layer 2

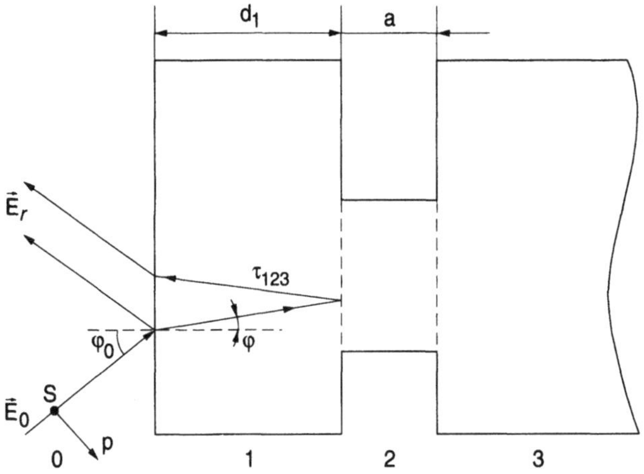

Fig. 6.7. Schematic representation of interference reflection from a single quantum well structure

(quantum well) by the relation

$$r = r_{01} + \frac{t_{01}t_{10}\mathrm{e}^{2\mathrm{i}\phi_1}}{1 - r_{10}r_{123}\mathrm{e}^{2\mathrm{i}\phi_1}} r_{123} = \frac{r_{01} + r_{123}\mathrm{e}^{2\mathrm{i}\phi_1}}{1 - r_{10}r_{123}\mathrm{e}^{2\mathrm{i}\phi_1}}. \tag{6.63}$$

Here r_{ij} and t_{ij} are the amplitude coefficients of reflection and transmission for light incident from a half-infinite medium i ($i = 0$ in vacuum and $i = 1$ in layer 1) on a half-infinite medium j, $\phi_1 = k_{1z}d_1$ is the phase shift of the light wave after its passage through layer 1,

$$k_{1z} = k_0 \left(\kappa_0 - \sin^2 \varphi_0\right)^{1/2}, \tag{6.64}$$

φ_0 is the incidence angle and κ_0 the dielectric constant of the barrier material whose frequency dependence is neglected since one considers a narrow spectral region near the resonance frequency of the e1 − h1 1s-exciton in the quantum well. It is assumed for simplicity that the background (local) dielectric constant of the quantum well is also equal to κ_0.

By (6.63), the reflectance

$$R = \frac{r_{01}^2 + 2\mathrm{Re}\left\{r_{01}r_{123}\mathrm{e}^{2\mathrm{i}\phi_1}\right\} + |r_{123}|^2}{1 + 2\mathrm{Re}\left\{r_{01}r_{123}\mathrm{e}^{2\mathrm{i}\phi_1}\right\} + r_{01}^2\,|r_{123}|^2}, \tag{6.65}$$

where it is included that $r_{10} = -r_{01}$ and the imaginary part of r_{01} is disregarded.

To derive an expression for r_{123}, we use the general expression for the electric field of the reflected wave propagating in layer 1 towards the vacuum

$$\mathcal{I}(\mathbf{x}) = 2\pi\mathrm{i}\frac{k_0^2}{k_{1z}} \int \mathrm{d}z' \tilde{\mathbf{P}}_{\mathrm{tr}}\left(z', \mathbf{Q}_\perp\right) \exp\left[\mathrm{i}k_{1z}(z' - z) + \mathrm{i}\mathbf{Q}_\perp \mathbf{x}_\perp\right]. \tag{6.66}$$

Here $\mathbf{Q}_\perp$ is the light-wave vector component in the (x, y) plane, which is preserved under reflection from the boundary ($Q_\perp = k_0 \cos\varphi_0$). $\tilde{\mathbf{P}}$ is the polarization excited in the quantum well by the initial wave and $\tilde{\mathbf{P}}_{\mathrm{tr}}$ is its component perpendicular to the wave vector $\mathbf{k}_{1r} = (Q_x, Q_y, -k_{1z})$. To derive (6.66), it is sufficient to take into account that the equation

$$\left(-\frac{\partial^2}{\partial z^2} - k^2\right) G(z) = \delta(z - z')$$

is satisfied by the function

$$G(z) = \frac{\mathrm{i}}{2k} \mathrm{e}^{\mathrm{i}k|z-z'|} .$$

The dielectric response of the quantum well is nonlocal. In particular, for the resonant contribution of 1s-exciton to $\tilde{\mathbf{P}}$, we have

$$\tilde{\mathbf{P}}^{(\mathrm{res})}(z) = \int \mathrm{d}z' \chi_{\mathrm{SQW}}^{(\mathrm{res})}(z, z') \mathcal{I}(z'), \tag{6.67}$$

$$4\pi \chi_{\mathrm{SQW}}^{\mathrm{res}}(z, z') = \frac{\Omega^{(3D)} \pi a_{\mathrm{B}}^3}{\omega_0 (Q_\perp) - \omega - \mathrm{i}\Gamma} \varphi_{1s}(0, z, z) \varphi_{1s}^*(0, z', z'). \tag{6.68}$$

Here

$$\omega_0(\mathbf{Q}_\perp) = \omega_0 + \left(\hbar Q_\perp^2 / 2M_\perp\right),$$

ω_0 is the resonance frequency, Γ the 1s-exciton damping in a single quantum well and the function $\varphi(\rho, z_{\mathrm{e}}, z_{\mathrm{h}})$ was introduced in (5.95), $\varphi_{1s}^* = \varphi_{1s}$. The right-hand side of (6.68) contains the parameters of the three-dimensional exciton in a bulk semiconductor: the Bohr radius (5.103) and the oscillator strength (6.37) (index 1 in $\Omega^{(3D)}$ is dropped). Note that the product $\Omega^{(3D)} a_{\mathrm{B}}^3$ does not depend on a_{B} and, hence, on the choice of the electron and hole effective masses. Substituting (6.67, 68) in (6.66), we find the coefficient of reflection from a quantum well

$$r_{123} = \frac{\mathrm{i}\tilde{\omega} \mathrm{e}^{\mathrm{i}\phi_2}}{\omega_0 - \omega - \mathrm{i}\Gamma}, \tag{6.69}$$

where

$$\tilde{\omega} = C \frac{\pi k_0^2}{2k_{1z}} \Omega^{(3D)} a_{\mathrm{B}}^3 \left[\int \mathrm{d}z \varphi_{1s}(0, z, z) \mathrm{e}^{\mathrm{i}k_{1z}\tilde{z}}\right]^2 . \tag{6.70}$$

Here $\tilde{z} = z - \bar{z}_a$, $\bar{z}_a$ is the center of the quantum well, $\phi_2 = k_{1z} a$, and integration in z is performed inside the quantum well, as well as over the region of the barrier where the function $\varphi_{1s}(0, z, z)$ is noticeably nonzero. In a symmetrical quantum well, the quantity $\tilde{\omega}$ is real. For light of s-polarization the factor $C = 1$, for p-polarized light in the nondegenerate band model $C = \cos 2\varphi = 1 - (2 \sin^2 \varphi_0 / \kappa_0)$, and for e1 - hh1 and e1 - lh1 1s-excitons in a GaAs/AlGaAs-type heterostructure, $C = \cos^2 \varphi$ and $C = 1 - 5 \sin^2 \varphi$, respectively. Note that in expression (6.37) for P_{cv}^2 one should substitute the quantity $P_{\mathrm{cv}}^2(\mathbf{e} \perp z)$ introduced in (6.21, 22) for $\Omega^{(3D)}$. Spatial dispersion of $\omega_0(\mathbf{Q}_\perp)$ in (6.69) is neglected, which is possible if

$$\frac{\hbar Q_\perp^2}{2M_\perp} = \frac{\hbar k_0^2}{2M_\perp}\sin^2\varphi_0 \ll \Gamma.$$

It is instructive to compare (6.69, 70) with the reflection coefficient r_{123}^{eff} from a layer of thickness a having a local dielectric constant

$$\kappa(\omega) = \kappa_0 + \frac{\kappa_0\omega_{\text{LT}}^{\text{eff}}}{\omega_0 - \omega - \text{i}\Gamma}, \tag{6.71}$$

and placed in a medium with dielectric constant κ_0. For $\kappa_0 \gg 1$ and $\omega_{\text{LT}}^{\text{eff}} \ll \Gamma$, we obtain

$$r_{123}^{\text{eff}} = \frac{\text{i}}{2}\frac{\omega_{\text{LT}}^{\text{eff}}}{\omega_0 - \omega - \text{i}\Gamma}\sin\phi_2 \text{e}^{\text{i}\phi_2}, \tag{6.72}$$

where $\phi_2 = k_0 a\left[\kappa(\omega) - \sin^2\varphi_0\right]^{1/2}$. In the limit $|\phi_2| \ll 1$, the coefficients r_{123} and r_{123}^{eff} coincide if we set

$$\omega_{\text{LT}}^{\text{eff}} = \omega_{\text{LT}}\frac{\pi a_{\text{B}}^3}{a}\left[\int \varphi_{1s}(0, z, z)\,\text{d}z\right]^2, \tag{6.73}$$

where $\omega_{\text{LT}} = \Omega^{(3\text{D})}/\kappa_0$ is the longitudinal–transverse splitting of the three-dimensional exciton. Comparing (6.72) with the earlier expression (6.69) for the 1s-exciton oscillator strength Ω^{MQW} in a periodic MQW structure yields

$$\omega_{\text{LT}}^{\text{eff}} \equiv \omega_{\text{LT}}^{\text{SQW}} = \frac{a+b}{a}\frac{\Omega^{\text{MQW}}}{\kappa_0} \equiv \frac{a+b}{a}\omega_{\text{LT}}^{\text{MQW}}. \tag{6.74}$$

It should be pointed out that (6.69, 70) are valid provided

$$|\tilde{\omega}| \ll \Gamma. \tag{6.75}$$

For completeness, we are presenting an expression for r_{123} for an arbitrary relation between the values of $\tilde{\omega}$ and Γ (in the case of normal incidence)

$$r_{123} = \frac{\text{i}\Gamma_0}{\omega_0 + \Delta\omega_0 - \omega - \text{i}(\Gamma + \Gamma_0)}, \tag{6.76}$$

where the radiative damping

$$\Gamma_0 = k_1 a\omega_{\text{LT}}^{\text{SQW}}/2,$$

$$\omega_{\text{LT}}^{\text{SQW}} = \omega_{\text{LT}}\left(\pi a_{\text{B}}^3/a\right)\left[\int \varphi_{1s}(0, z, z)\cos k_1 z\,\text{d}z\right]^2.$$

This expression takes into account the renormalization of the resonance frequency

$$\Delta\omega_0 = \frac{1}{2}\omega_{\text{LT}}\pi a_{\text{B}}^3 k_1 \iint \varphi_{1s}(0, z, z)\varphi_{1s}(0, z'z')\sin k_1|z - z'|\,\text{d}z\text{d}z'. \tag{6.77}$$

Expanding (6.65) in powers of r_{123}, we obtain in the linear approximation

$$R = \bar{R}\left[1 + 2\frac{t_{01}t_{10}}{r_{01}}\mathrm{Re}\left\{r_{123}e^{2i\phi_1}\right\}\right], \tag{6.78}$$

where $\bar{R} = r_{01}^2$. Under normal incidence

$$\bar{R} = \left(\frac{n-1}{n+1}\right)^2, \quad \frac{t_{01}t_{10}}{r_{01}} = -\frac{4n}{n^2-1}, \quad n = \sqrt{\kappa_0}. \tag{6.79}$$

Under oblique incidence

$$R = \bar{R}\left(\varphi_0\right)\left[1 + S(\varphi_0)f(x,\Phi)\right], \tag{6.80}$$

$$S\left(\varphi_0\right) = \frac{8\sqrt{\kappa_0 - \sin^2\varphi_0}}{\kappa_0 - 1} \times \begin{cases} \cos\varphi_0 & \text{for } s\text{-polarization,} \\ \dfrac{\kappa_0\cos\varphi_0}{\kappa_0\cos^2\varphi_0 - \sin^2\varphi_0} & \text{for } p\text{-polarization,} \end{cases} \tag{6.81}$$

$$r_{01}^s\left(\varphi_0\right) = \frac{\cos\varphi_0 - \sqrt{\kappa_0 - \sin^2\varphi_0}}{\cos\varphi_0 + \sqrt{\kappa_0 - \sin^2\varphi_0}}, \tag{6.82}$$

$$r_{01}^p\left(\varphi_0\right) = \frac{\sqrt{\kappa_0 - \sin^2\varphi_0} - \kappa_0\cos\varphi_0}{\sqrt{\kappa_0 - \sin^2\varphi_0} + \kappa_0\cos\varphi_0} = -\frac{(\kappa_0 - 1)\left(\kappa_0\cos^2\varphi_0 - \sin^2\varphi_0\right)}{\left(\sqrt{\kappa_0 - \sin^2\varphi_0} + \kappa_0\cos\varphi_0\right)^2}, \tag{6.83}$$

$$f(x,\Phi) = \frac{\tilde{\omega}}{\Gamma}\frac{\sin\Phi + x\cos\Phi}{1+x^2}, \tag{6.84}$$

where $x = (\omega - \omega_0)/\Gamma$, $\Phi = 2\phi_1 + \phi_2 + (\pi/2)$.

Note that absorbance, i.e., the fraction of incident energy absorbed in a quantum well, can be written in the linear-in-$\tilde{\omega}/\Gamma$ approximation in the form

$$A \simeq 1 - |t_{123}|^2 = \frac{2\tilde{\omega}\Gamma}{(\omega_0 - \omega)^2 + \Gamma^2}. \tag{6.85}$$

The reflection coefficient (6.63) can be conveniently presented in the form

$$r = r_{01} + r'. \tag{6.86}$$

For p-polarized light incident on the structure at close to the Brewster angle

$$\varphi_{Br} = \arctan\sqrt{\kappa_0} \tag{6.87}$$

the reflection coefficient at the external boundary is small and the contribution from the quantum well r' can be comparable to r_{01} even for $|r_{123}| \ll 1$. We are presenting below an expression for R applicable for $r_{01}^2 \ll 1$ and an arbitrary relative magnitude of r_{01} and r', which includes the contributions of two close exciton resonances (say, e1 – hh1 and e1 – lh1):

$$R = \bar{R}(\varphi_0)\left[1 + 2\mathrm{Re}\left\{r'/r_{01}\right\}\right] + |r'|^2, \tag{6.88}$$

$$\mathrm{Re}\left\{\frac{r'}{r_{01}}\right\} = \frac{2\bar{n}\delta}{\bar{n}^2 - \sin^2\varphi_0}\sum_{\sigma=\mathrm{h,l}}\frac{\omega_{\mathrm{LT}}^{\sigma}}{\Gamma_\sigma}\frac{\sin\Phi + x_\sigma\cos\Phi}{1 + x_\sigma^2}, \tag{6.89}$$

$$|r'|^2 = \frac{4\bar{n}^2\delta^2}{(\bar{n}+1)^4}\left[\sum_{\sigma=\mathrm{h,l}}\left(\frac{\omega_{\mathrm{LT}}^{\sigma}}{\Gamma_\sigma}\right)^2\frac{1}{1+x_\sigma^2} + 2\frac{\omega_{\mathrm{LT}}^{\mathrm{h}}}{\Gamma_{\mathrm{h}}}\frac{\omega_{\mathrm{LT}}^{\mathrm{l}}}{\Gamma_{\mathrm{l}}}\frac{1 + x_{\mathrm{h}}x_{\mathrm{l}}}{\left(1+x_{\mathrm{h}}^2\right)\left(1+x_{\mathrm{l}}^2\right)}\right]. \tag{6.90}$$

Here $\bar{n} = \sqrt{\kappa_0}\cos\varphi_0$, $\delta = \sin\phi_2$, the index σ identifies the e1 – hh1 and e1 – lh1 1s-excitons, $x_\sigma = (\omega - \omega_0^\sigma)/\Gamma_\sigma, \Gamma_\sigma$ is the damping, and $\omega_{\mathrm{LT}}^{\sigma}$ is the longitudinal-transverse splitting determining the exciton contribution to the effective dielectric constant of a quantum-well layer (6.71, 73):

$$\kappa(\omega) = \kappa_0 + \sum_{\sigma=\mathrm{h,l}}\frac{\kappa_0\omega_{\mathrm{LT}}^{\sigma}}{\omega_0^\sigma - \omega - \mathrm{i}\Gamma_\sigma}. \tag{6.91}$$

When the inequality

$$|\varphi_0 - \varphi_{\mathrm{Br}}| \gg \frac{\omega_{\mathrm{LT}}^{\sigma}\delta}{2\Gamma_\sigma}\cos\varphi_{\mathrm{Br}} \tag{6.92}$$

is met, the term $|r'|^2$ in (6.88) is negligible and (6.88) reduces to (6.78).

6.5.2 Multiple Quantum-Well Structure

Using (6.68) for the dielectric constant of a single quantum well, one can determine the boundaries of applicability of the effective homogeneous medium approximation to the description of light propagation in a periodic multiple quantum well structure (6.42). The nonlocal dielectric response of such a structure can be written as

$$\chi_{\mathrm{MQW}}(z, z') = \sum_l \chi_{\mathrm{SQW}}(z, z' - \bar{z}_l). \tag{6.93}$$

Expand in a Fourier series the periodic amplitude $\mathcal{I}_{\mathbf{Q}}$ of the Bloch solution (6.2) for the electric field then

$$\mathcal{I}_{\mathbf{Q}}(z) = \sum_m \mathcal{I}_{\mathbf{Q}}^{(m)}\mathrm{e}^{\mathrm{i}G_m z}, \quad G_m = \frac{2\pi m}{a+b}. \tag{6.94}$$

In general, the spatial harmonic of displacement $\mathbf{D}_{\mathbf{Q}}^{(m)}$ is linearly related to all spatial harmonics of the electric field

$$\mathbf{D}_{\mathbf{Q}}^{(m)} = \sum_{m'}\kappa^{(m,m')}(\omega, \mathbf{Q})\mathcal{I}_{\mathbf{Q}}^{(m')}. \tag{6.95}$$

Equations (6.68, 93, 94) can be used to obtain the following expression for the spectral region of the 1s-exciton resonance

$$\kappa^{(m,m')}(\omega, \mathbf{Q}) = \kappa_0 \delta_{mm'} + \frac{\Omega^{(3D)} \pi a_B^3}{a+b} \frac{I^{(m)*}(Q_z) I^{(m')}(Q_z)}{\omega_0(\mathbf{Q}_\perp) - \omega - \mathrm{i}\Gamma}, \tag{6.96}$$

where

$$I^{(m)}(K) = \int \mathrm{d}z \varphi_{1s}(0, z, z) \mathrm{e}^{\mathrm{i}(G_m + K)z}. \tag{6.97}$$

If the dielectric constant of the barrier κ^b does not coincide with the background constant of the quantum well κ_0^a, one more term will appear in (6.96):

$$\left(\kappa_0^a - \kappa^b\right) \frac{a}{a+b} F(G_{m'-m} a/2),$$

where $F(t) = (\sin t)/t$. Note also that the dependence of $I^{(m)}$ on Q_z for $Q_z a \ll 1$ may be neglected.

In the long-wavelength approximation, we have

$$\mathcal{I}_{\mathbf{Q}}^{(m)} \simeq (k_0/G_m)^2 \kappa^{(m,0)} \mathcal{I}_{\mathbf{Q}}^{(0)}$$

and the coefficient $\kappa_t(\omega, Q)$ relating the mean fields $\mathbf{D}_{\mathbf{Q}}^{(0)}$ and $\mathcal{I}_{\mathbf{Q}}^{(0)}$ can be presented in the following form taking into account the higher spatial harmonics:

$$\kappa_t = \kappa(\omega) + \Delta\kappa, \tag{6.98}$$

where

$$\kappa(\omega) \equiv \kappa^{(0,0)}(\omega) = \kappa_0 + \frac{\kappa_0 \omega_{LT}^{MQW}}{\omega_0 - \omega - \mathrm{i}\Gamma}, \tag{6.99}$$

$$\Delta\kappa = \sum_{m \neq 0} (k_0/G_m)^2 \left|\kappa^{(m,0)}\right|^2 \tag{6.100}$$

with ω_{LT}^{MQW} defined in (6.74). Since the integrals $I^{(m)}$ do not exceed $I^{(0)}$, the contribution of the $m \neq 0$ spatial harmonics to κ_t can be estimated as

$$\frac{\Delta\kappa}{\kappa_t} \sim \left(\frac{a+b}{\lambda}\right)^2 \frac{[\kappa(\omega) - \kappa_0]^2}{\kappa(\omega)},$$

where λ is the wavelength of light in vacuum, and the correction $\Delta\kappa$ may be neglected compared to the resonant contribution $\kappa(\omega) - \kappa_0$ provided

$$\left(\frac{a+b}{\lambda}\right)^2 \kappa_0 \frac{\omega_{LT}^{MQW}}{\Gamma} \ll 1. \tag{6.101}$$

For the sake of completeness, we are presenting here the dispersion relation for light waves at normal incidence with $Q_x = Q_y = 0$ for an arbitrary value of the parameter (6.101):

$$\cos Q_z d = \cos k_1 d - \frac{1}{2} k_1 d \sin k_1 d \frac{\omega_{LT}^{MQW}}{\omega_0 + \Delta\omega_0 - \omega - \mathrm{i}\Gamma} \tag{6.102}$$

with $\Delta\omega_0$ defined by (6.77).

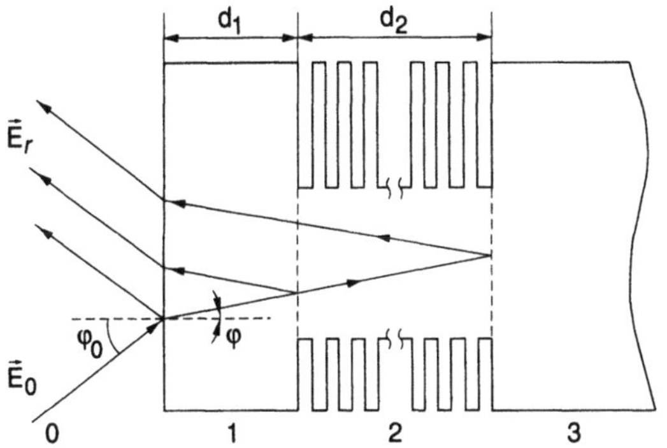

Fig. 6.8. Schematic representation of interference reflection from a multiple quantum well structure or superlattice

The amplitude reflection coefficient and reflectance for the structure shown in Fig. 6.8 are defined by expressions (6.63, 65) where one has to set

$$r_{123} = \frac{1 - e^{2i\phi_2}}{1 - r_{12}^2 e^{2i\phi_2}} r_{12}, \tag{6.103}$$

with $\phi_2 = k_{2z} d_2$, $k_{2z} = k_0[\kappa(\omega) - \sin^2 \varphi_0]^{1/2}$. For $\kappa_0 \gg 1$ one can take for r_{12} the reflection coefficient at normal incidence:

$$r_{12} = \frac{n_1 - n_2}{n_1 + n_2}. \tag{6.104}$$

Here $n_1 = \sqrt{\kappa_0}$, $n_2 = \sqrt{\kappa(\omega)}$. Further simplifications are possible if $\omega_{LT}^{MQW} \ll \Gamma$, then we will have, respectively:

$$\begin{aligned} n_2 &= n_1 \left(1 + \frac{1}{2} \frac{\omega_{LT}}{\omega_0 - \omega - i\Gamma}\right), \\ r_{12} &= -\frac{1}{4} \frac{\omega_{LT}}{\omega_0 - \omega - i\Gamma}, \\ r_{123} &= \frac{1}{4} \left(e^{2i\phi_2} - 1\right) \frac{\omega_{LT}}{\omega_0 - \omega - i\Gamma} \end{aligned} \tag{6.105}$$

and

$$R = \bar{R}(\varphi_0) \left\{ 1 + \frac{t_{01} t_{10}}{2 r_{01}} \left[f_{SL}(x, 2\phi_1) - e^{-K_2 d_2} f_{SL}(x, \Phi_2) \right] \right\} \tag{6.106}$$

where $x = (\omega - \omega_0)/\Gamma$, $\Phi_2 = 2(\phi_1 + \mathrm{Re}\{\phi_2\})$, $K_2 = 2\ \mathrm{Im}\ \{k_{2z}\}$,

$$f_{SL}(x, \Phi) = \frac{\omega_{LT}}{\Gamma} \frac{\sin \Phi + x \cos \Phi}{1 + x^2}.$$

For compactness, the top index "MQW" of ω_{LT} is dropped. Note that by (6.46), in the two-dimensional approximation, one can write

$$\omega_{LT}^{MQW} = \frac{8 a_B}{a + b} \omega_{LT}, \tag{6.107}$$

where ω_{LT} is the longitudinal-transverse splitting of the 1s-exciton in a bulk crystal ($\omega_{LT}^{(1)}$ in the notation of (6.40)).

6.5.3 Short-Period Superlattice

Expressions (6.103–106) are valid also for the description of light reflection from a structure with a short period SL. In this case, however, the electron motion is three-dimensional, and the longitudinal-transverse splitting ω_{LT}^{SL} cannot be expressed by the simple relation (6.74) in terms of ω_{LT}^{SQW}. Neglecting hole tunneling through the barrier, (6.40, 48) yield

$$\omega_{LT}^{SL} = \omega_{LT} \frac{a_B^3}{a_{\|} a_{\perp}^2} \Lambda_{e1,h1} \,. \tag{6.108}$$

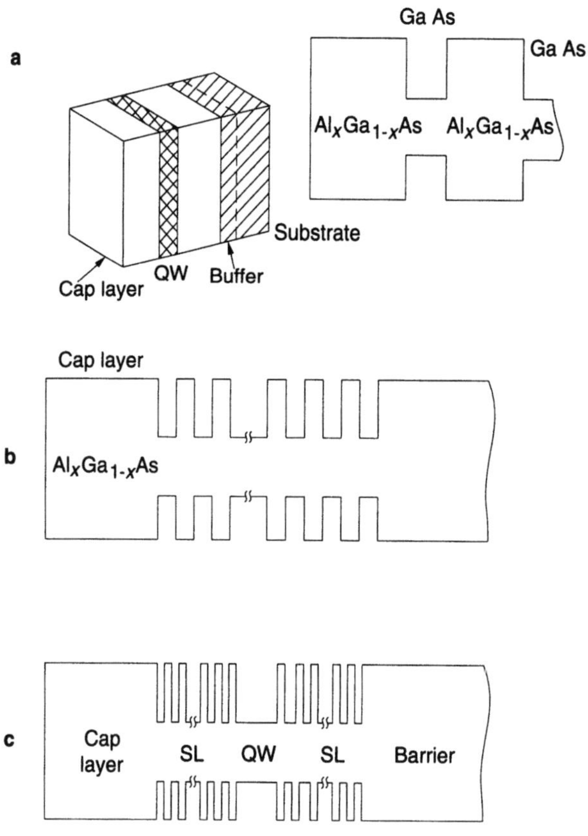

Fig. 6.9a-c. Schematic representation of GaAs/AlGaAs heterostructures with **a** single quantum well, **b** multiple quantum wells or superlattice, **c** single quantum well at the center of a short-period superlattice

6.5.4 Experiment

Figure 6.9 shows schematically some types of GaAs-based heterostructures whose optical properties have been studied experimentally. Figure 6.10 presents experimental reflectance spectra for a heterostructure which includes a single quantum well, sandwiched between two identical superlattices, and two thick barrier layers on the left and right (Fig. 6.9c). The arrows identify the resonance frequencies of the e1-hh1 and e1-lh1 excitons in the quantum well and the SL. In the vicinity of these frequencies, one observes a characteristic resonant modulation of the spectrum (resonance interference reflection). By (6.78, 106), at normal incidence the relative amplitude of modulation, $(R_{\text{max}} - R_{\text{min}})/\bar{R}$ is of order $(8/\sqrt{\kappa_0})(\tilde{\omega}/\Gamma)$ for excitons in a quantum well, and $(2/\sqrt{\kappa_0})(\omega_{\text{LT}}/\Gamma)$ for excitons in a SL. To increase the relative modulation of the reflectance spectrum, one can conveniently use an oblique incidence geometry with an angle close to the Brewster angle (6.87). As seen from Fig. 6.10, the reflectance spectrum obtained in p-polarization at $\varphi_0 = 73.5°$ (for $\kappa_0 = 13, \varphi_{\text{Br}} = 74.5°$), indeed reveals giant resonance modulation. This permits reliable determination of the exciton parameters ω_0, Γ and ω_{LT} from measured interference reflectance spectra. In particular, the damping Γ is evaluated from the width of the

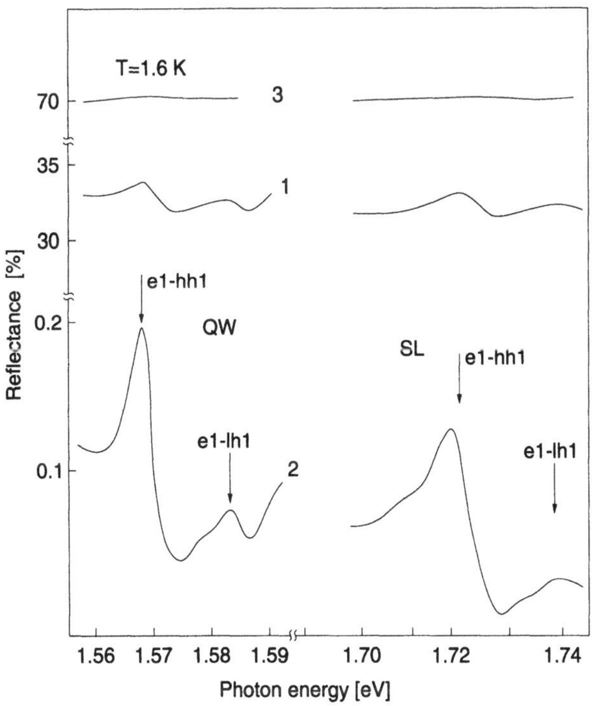

Fig. 6.10. Reflection spectra measured on the sample shown in Fig. 6.9c at (1) normal, and (2,3) oblique incidence ($\varphi_0 = 73.5°, \varphi_{\text{Br}} = 74.5°$) in (2) p- and (3) s-polarization. Arrows indicate positions of exciton resonances. Quantum well thickness $\simeq$ 70 Å, superlattice period 28 Å[6.6]

resonance modulation spectrum: indeed, the function $f(x, \Phi)$ in (6.80) has for $\Phi \neq (2N+1)\pi/2$ its maximum and minimum at the points $\omega_{\max}$ and $\omega_{\min}$ related to Γ through

$$\Gamma = \frac{1}{2}\left|\cos\Phi(\omega_{\max} - \omega_{\min})\right| .$$

The phase Φ, which depends linearly on the distance d_1 between the outer plane of the sample and the quantum well, determines the shape of the resonance profile. Figure 6.11 presents three reflectance spectra obtained on samples with three different values of d_1, which permits us to follow the evolution of the normal incidence reflectance spectrum from a *single minimum*-type profile ($\Phi \approx 3.5\pi$) to a *weak minimum and a clearly pronounced maximum* profile ($\Phi \approx 4.5\pi$).

The linear-approximation expressions (6.78, 106) can be used to analyze the spectra obtained on SQW structures if the condition (6.92) is met, or on superlattices if

$$|\varphi_0 - \varphi_{\rm Br}| \gg \frac{\omega_{\rm LT}^{\rm SL}}{\sqrt{\kappa_0}\Gamma} .$$

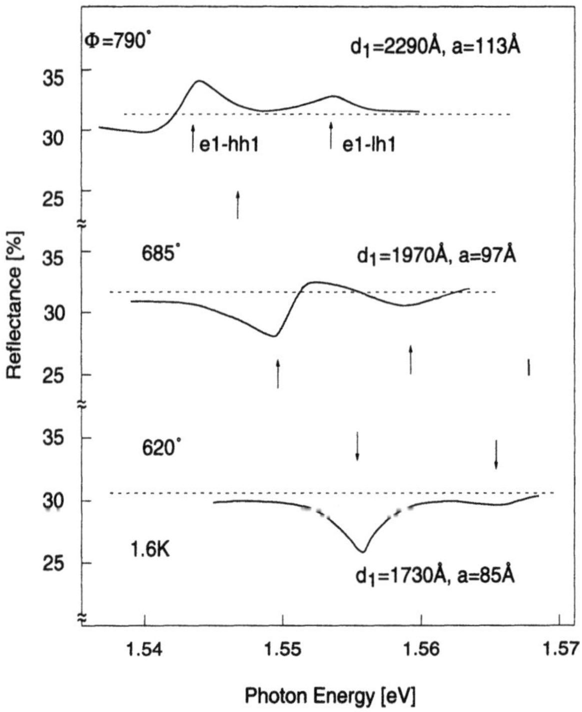

Fig. 6.11. Effect of cap layer thickness d_1 on phase Φ of exciton reflection at normal incidence [6.6]

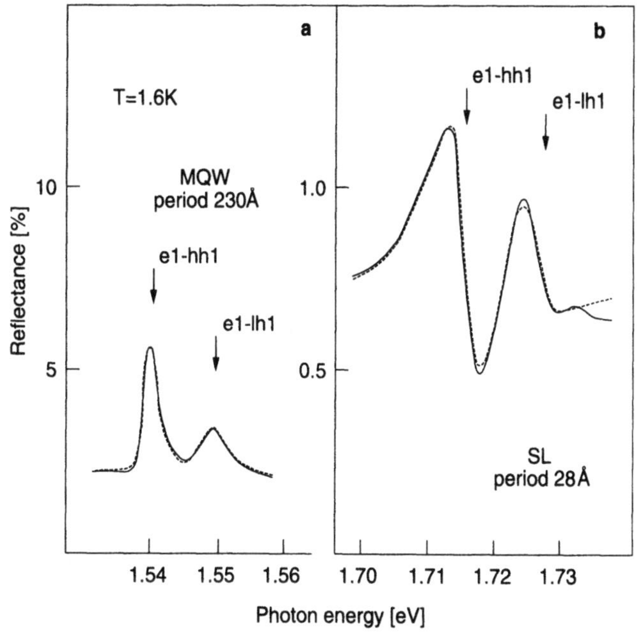

Fig. 6.12a,b. Calculated (dots) and experimental (solid lines) reflectance spectra for **a** SQW structure and **b** superlattice with equal thicknesses of wells and barriers. Measurements performed in Brewster geometry at $\varphi_0 = 69°$ and 72.5°, accordingly [6.6]

If this is not the case, one should use exact formulas of the type (6.88–90). Figure 6.12 displays experimental and calculated interference reflectance spectra for (a) MQW structure, and (b) SL.

The dependence of the longitudinal-transverse splitting of the e1 – hh1 1s-exciton on period d of the GaAs/AlGaAs hetero-structure with layers A and B of equal thickness is shown in Fig. 6.13. The branches of the theoretical curve were calculated by (6.107, 108). In agreement with theoretical predictions (sect. 6.3), the $\omega_{\mathrm{LT}}^{\mathrm{h}}(d)$ dependence is nonmonotonic. As the well width a decreases, the probability to find the electron and the hole of a quasitwo-dimensional exciton at the same point in a thick-barrier SL increases, and $\omega_{\mathrm{LT}}^{\mathrm{h}}$ increases. As the period decreases still more, the overlap of electron states in neighboring wells increases, minibands appear, and the electron motion in the structure becomes three-dimensional. Therefore $\omega_{\mathrm{LT}}^{\mathrm{h}}$ reaches a maximum at a certain value d = 100–110 Å and then starts to fall off. Thus nonmonotonic behavior of $\omega_{\mathrm{LT}}^{\mathrm{h}}(d)$ is actually a manifestation of the 2D-3D transition in a periodic heterostructure.

Since at $\varphi_0 = \varphi_{\mathrm{Br}}$ the coefficient r_{01}^{p} reverses its sign, the passage through the Brewster angle as the incidence angle is varied should be accompanied by a reversal of the sign of the quantities $2\mathrm{Re}\{r_{01}r_{123}\mathrm{e}^{2\mathrm{i}\phi_1}\}$ in (6.65) or $(t_{01}t_{01}/r_{01})$ in (6.78, 106) and, hence, by inversion of the resonance profile of interference reflectance. The line shape inversion effect is illustrated by Fig. 6.14 showing

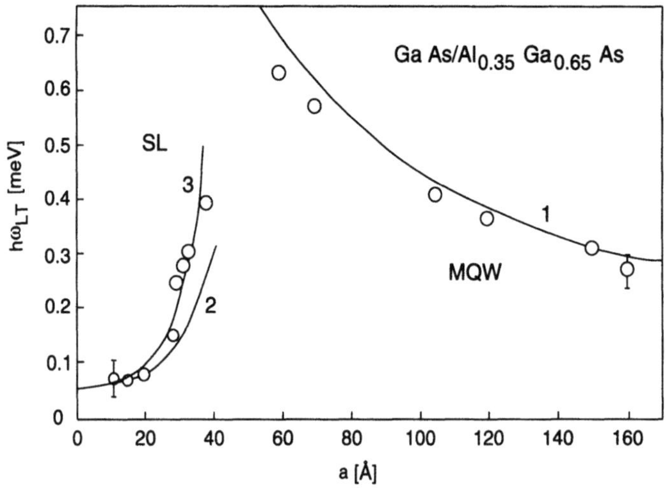

Fig. 6.13. Longitudinal-transverse splitting of heavy hole excitons vs. period of multilayered $GaAs/Al_{0.35}Ga_{0.65}As$ structure with equal thickness of wells and barriers. Curves 1 and 2 calculated by (6.107) and (6.108), respectively [6.6]. Curve 3 calculated with inclusion of nonparabolicity of the electron miniband spectrum [6.7]

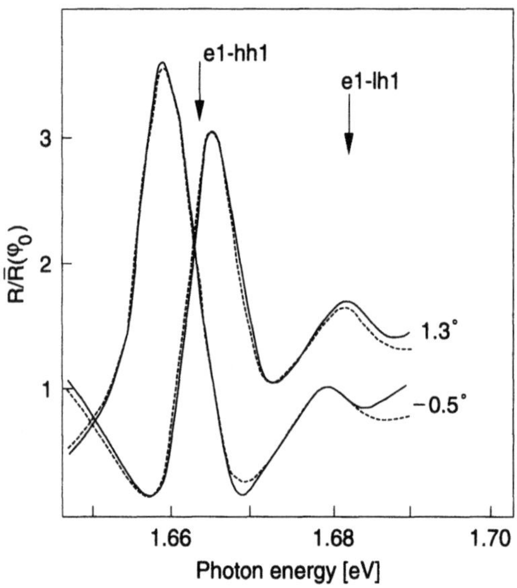

Fig. 6.14. Evolution of exciton reflectance as the incidence angle φ_0 passes through Brewster angle φ_{Br}. The spectra measured on a structure with a superlattice with a period of 70 Å. The angle difference $\varphi_0 - \varphi_{Br}$ and exciton resonance positions are specified. The quantity $\bar{R}(\varphi_0)$ is defined by (6.106). Solid lines: experiment, dotted lines: theory [6.8]

reflectance spectra obtained at $\varphi_0 - \varphi_{Br} = -0.5°$ and $+1.3°$. We see that, in the region of the exciton resonance ω_0^h, the frequency corresponding to the maximum of reflectance at one incidence angle is close to that of minimum reflectance at the other incidence angle.

6.6 Electro-Optical Effects in Interband Transitions

The effects of the longitudinal ($\mathcal{I} \parallel z$) and transverse ($\mathcal{I} \perp z$) electric fields on current carriers in heterostructures are radically different because in a SQW structure electron transport is possible only in the layer plane, while in a periodic MQW structure with not very thin barriers transport along a normal to the layers is of the hopping nature and, thus, difficult. We are going to analyze here how the electric field affects the exciton state in a quantum well. Figure 6.15a shows the potential energy

$$U(\mathbf{r}) = -\frac{e^2}{\kappa r} + e\mathcal{I} \cdot \mathbf{r} \tag{6.109}$$

of the electron and hole in a field $\mathcal{I} \parallel x$ as a function of $\mathbf{r} = \mathbf{x}_e - \mathbf{x}_h$. With the field applied in this direction the exciton state becomes quasistationary, just as it does in a bulk semiconductor because of the possibility for either of the particles (primarily for the lower mass one) to tunnel through the barrier. For the 1s-exciton in weak fields, the height of the barrier $\sim E_B$ and its width $\Delta x \sim (E_B/e\mathcal{I})$. One can therefore write the following expressions for the transparency T and exciton level halfwidth Γ:

$$\ln T^{-1}, -\ln \Gamma \propto \frac{\Delta x}{a_B} \sim \frac{E_B}{e\mathcal{I}a_B}. \tag{6.110}$$

The broadening of the exciton absorption peak (determined by Γ) occurs in moderate fields $\mathcal{I} \sim E_B/ea_B \sim (10^3 - 10^4)$ V/cm. No noticeable shift of the absorption peak is observed to occur.

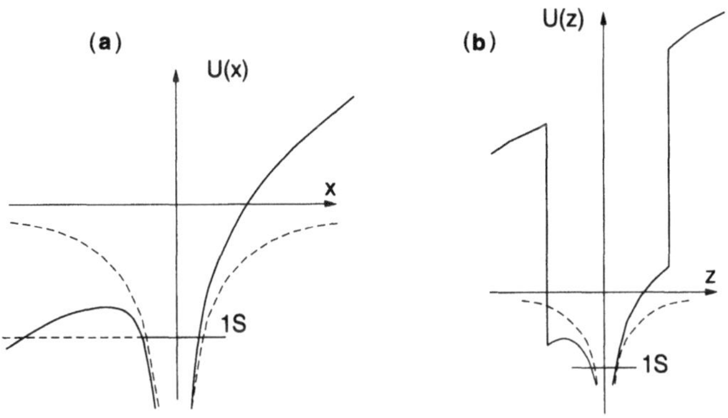

Fig. 6.15a,b. Potential energy of the electron in SQW structure in an external uniform electric field **a** $\mathcal{I} \perp z$ and **b** $\mathcal{I} \parallel z$ and in the Coulomb field of a hole at the well's center. The Coulomb field is shown by the dotted line. The horizontal line identifies the exciton ground state level in the quantum well

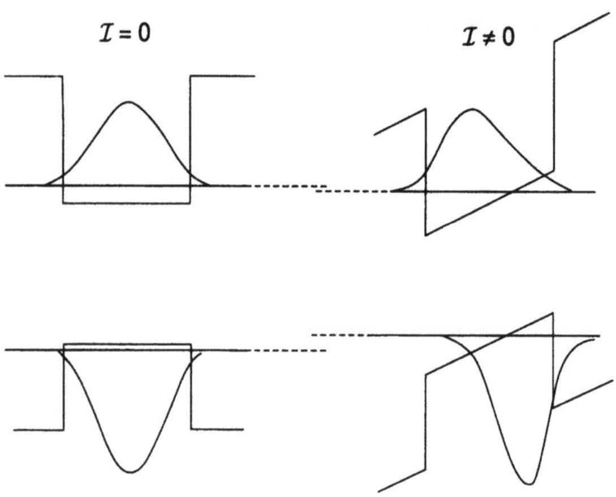

Fig. 6.16. Schematic representation of electron and hole wave functions for the lowest confined states in the absence of and with electric field $\mathcal{I} \parallel z$

The exciton behaves differently in a field applied across the heterolayers (Fig. 6.15b). The height of the barrier for the tunneling decay of excitons is equal in this case to that of the heterostructure barrier V_0 and the exciton level is well defined even in fields $\sim 10^5$ V/cm in which the levels of confined free carriers shift by an amount in excess of the exciton Rydberg. Indeed, in a GaAs/$Al_{0.35}Ga_{0.65}As$ quantum well the quantum-confined Stark effect shifts the lowest electron and hole states in the field $\mathcal{I} = 10^5$ V/cm by 6 and 15 meV, respectively, for a well 100 Å-thick, and by 55 and 81 meV for $a = 250$ Å [6.9]. At the same time, the envelope wave functions $f_e(z)$ and $f_h(z)$ undergo deformation as is shown schematically in Fig. 6.16. In low fields, $e\mathcal{I}a \ll (\hbar\pi/a)^2/2m^*$ the level shift ΔE is quadratic in field, and the shift of the center of mass $\bar{z}_{e,h} \propto \mathcal{I}$. The existence of the barriers prevents the breakup of the exciton. Therefore, as the field increases, the electron and the hole are pressed close to the interfaces and the $\Delta E(\mathcal{I})$ and $\bar{z}(\mathcal{I})$ saturate. While the electron and hole binding energy in the exciton is lowered, its change is small compared to ΔE, the shift of the exciton absorption peak in a field $\mathcal{I} \parallel z$ being determined primarily by the total shift $\Delta E_{e1} + \Delta E_{h1}$. In contrast to the binding energy of the exciton, its oscillator strength does not saturate with increasing electric field, since the field tends to draw the electron and the hole toward opposite boundaries of the well, and the envelope overlap integral $\int f_{ev}(z) f_{hv'}(z) dz$ decreases to zero for arbitrary v and v' in the limit of very strong fields. The oscillator strength of the e1 − h1 exciton decreases monotonically. At the same time, the $vv' \rightarrow ev$ interband transitions, forbidden for $= 0$ by parity conservation, become allowed. In weak fields, the oscillator strength $\propto$ ^{2} for the transitions v2→c1 or v1→c2.

6.6.1 Interference Reflection in an Electric Field

The main contribution to the field-induced change in the resonant dielectric response (6.68) of a quantum well comes from the exciton resonance frequency shift $\Delta\omega_0$, which is due to the quantum-confined Stark effect. Therefore, in the approximation linear in $\Delta\omega_0$, the electroreflectance of a quantum-well structure is described by the expression (6.80)

$$\Delta R = -\bar{R}(\varphi_0)\, S(\varphi_0)\, \frac{\partial f(x,\Phi)}{\partial x} \frac{\Delta\omega_0}{\Gamma}. \tag{6.111}$$

According to (6.84), the first derivative of the function $f(x,\Phi)$

$$\frac{\partial f(x,\Phi)}{\partial x} = \frac{\tilde{\omega}}{\Gamma} \frac{\left(1-x^2\right)\cos\Phi - 2x\sin\Phi}{\left(1+x^2\right)^2}. \tag{6.112}$$

Since $\Delta\omega_0 < 0$, the electroreflectance signal represents for $\Phi = 2\pi N$ (or $\Phi = (2N+1)\pi$) a maximum (minimum) at the resonance frequency and to minima (maxima) at $\omega - \omega_0 = \pm\Gamma$. At $\Phi = (2N+1)\pi/2$ there is no electroreflectance at the resonance frequency, the extrema (minimum and maximum) in the electroreflectance spectra being shifted relative to ω_0 by $\pm\Gamma/\sqrt{3}$. The minimum width of the cap layer at which one can change the phase Φ by π by properly varying the incidence angle and, thus, invert the electroreflectance profile of s-polarized light is defined by the relation

$$2\frac{\omega}{c} d_1 \left(\sqrt{\kappa_0} - \sqrt{\kappa_0 - 1}\right) = \pi$$

or (for $\kappa_0 \gg 1$)

$$d_1 = \sqrt{\kappa_0}\frac{\lambda}{2}.$$

For $\kappa_0 = 12.5$ and $\lambda = 8050$ Å we obtain $d_1 = 1.4\ \mu$m.

In the case of a GaAs/AlGaAs-type structure, one should take into account in (6.111) two contributions corresponding to the e1-hh1 and e1-lh1 1s-excitons. Figure 6.17 presents an experimental and a calculated spectrum of electroreflectance for oblique incidence of s-polarized light on a SQW structure.

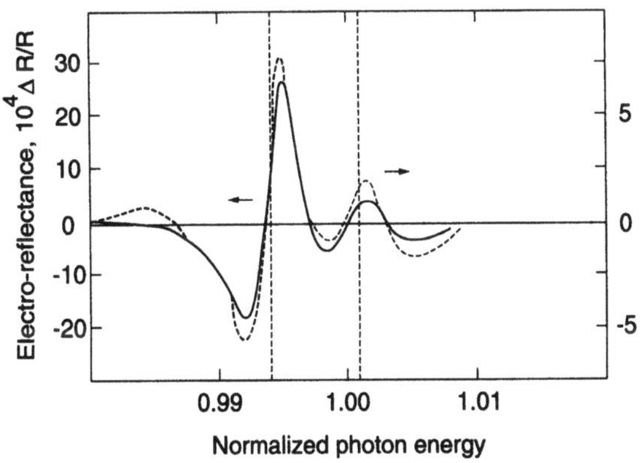

Fig. 6.17. Electroreflectance spectrum of s-polarized light incident at 45° on GaAs/Al$_x$Ga$_{1-x}$As SQW structure. Solid line: theory; dotted line: experiment. Dashed vertical lines identify the positions of the e1-hh1 and e1-lh1 excitons [6.10]

6.6.2 Stark Ladder in a Superlattice

As already pointed out in Chap. 3, in an electric field $\mathcal{I} \parallel z$, each electron miniband in a SL splits into a series of subbands separated in energy by $e\mathcal{I}d$. The overlap of the SQW electron wave functions is governed by the parameter:

$$t = \frac{\Delta}{e\mathcal{I}d}. \tag{6.113}$$

For $\mathcal{I} \neq 0$, the electron wave function (with a real wave vector) preserves the Bloch character only for in-plane motion, the electron states becoming localized in the z-direction. These states are characterized not by k_z, but by another quantum number, namely, the number l of the well where the probability to find an electron is the highest. As one moves away from this well, the probability decreases (Fig. 6.18) but remains non-negligible within t wells, if $t \gg 1$, or, only within the lth well, if $t < 1$.

In principle, the Stark ladder can be produced in a bulk crystal as well. It was reliably observed, however, only in a SL with a period exceeding by far the host lattice constant. Indeed, in a GaAs/Al_xGa_{1-x}As SL, for $a = 40$ Å, $b = 20$ Å, $x = 0.35$ the width of the lowest electron miniband $\Delta \sim 0.07$ eV, the parameter t is of the order unity for $\mathcal{I} \sim 10^5$ V/cm. Figure 6.19 presents spectra of the photocurrent induced in a GaAs/AlGaAs SL at various electric fields. Taking into account the wave-function overlap, optical excitation can produce interband transitions $\mathrm{v}\nu' l' \to \mathrm{c}\nu l$ both with and without a change in the quantum number l. Therefore the absorption spectra obtained with $\mathcal{I}_z \neq 0$

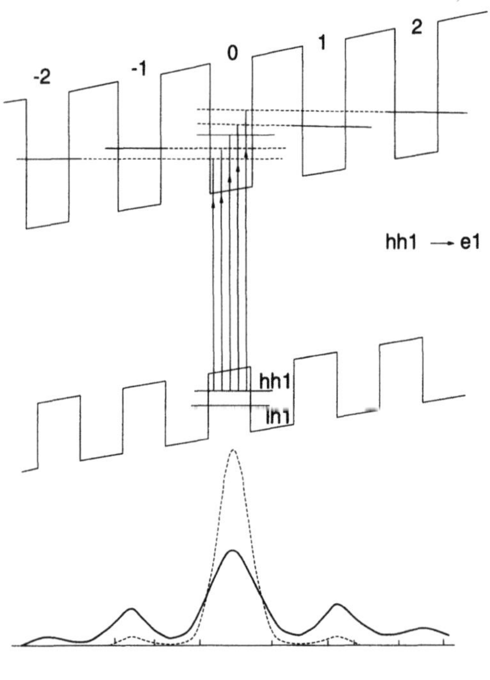

Fig. 6.18. Schematic representation of a series of hh10 → c1l transitions ($l = 0, \pm1, \pm2$) in GaAs/Al_xGa_{1-x}As superlattice in an electric field $\mathcal{I} \parallel z$. Solid line: electron wave function for c10 state. Dotted line: electron wave function in a strong electric field when the electron is localized practically in one well

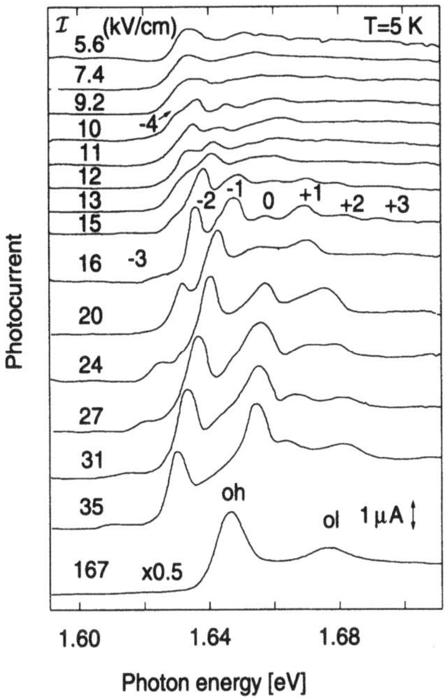

Fig. 6.19. Low-temperature photocurrent spectra in GaAs/$Al_{0.35}Ga_{0.65}$As superlattice for different electric fields $\mathcal{I} \parallel z$. For convenience, the spectra are displaced vertically with respect to one another. The numbers refer to the peaks corresponding to hh$1l' \to$ c$1l$ transitions ($l - l' = 0, \pm 1, \pm 2, \ldots$) in an intermediate field. Symbols 0h and 0l identify the peaks corresponding to transitions in a strong field, when both the electron and hole states are localized within one well [6.9]

reveal not one (intra-well) energy gap $E^{(\nu\nu')}$ but rather a series of threshold energies

$$E_{ll'}^{(\nu\nu')} = E_{oo}^{(\nu\nu')} + (l - l')e\mathcal{I}d. \tag{6.114}$$

The numbers $0, \pm 1, \ldots$ in Fig. 6.19 specify the spectral peaks corresponding to the hh$1l' \to$ c$1l$ transitions with the corresponding difference $l - l'$. Since the heavy hole is practically localized in one well, these peaks can be identified with a Stark ladder of electron states in the conduction subband c1. In a strong electric field, the electron states in neighboring wells are so widely separated in energy ($t < 1$) that the electrons turn out to be localized practically completely within one well and the overlap integrals for the $l \neq l'$ interband transitions become very small, with the result that only the intrawell transitions can be observed in the frequency spectrum of the photocurrent.

6.7 Magneto-Optical Spectra

Figure 6.20 presents light absorption spectra obtained on a MQW structure in a magnetic field $\mathbf{B} \parallel z$. The spectrum measured in the absence of magnetic field, just as in Figs. 6.1, 6.3, is dominated by excitation peaks of the e1-hh1 and e1-lh1 1s-excitons. In the presence of magnetic field, they shift toward higher energies and, in fields of a few Tesla and higher, new absorption peaks due to optical transitions between the Landau levels with quantum

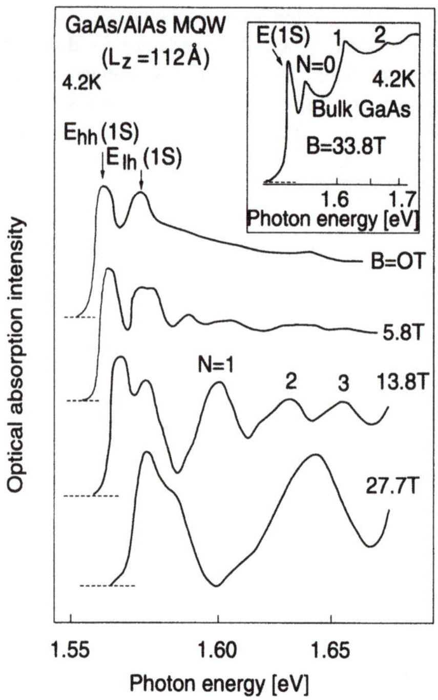

Fig. 6.20. Spectra of optical absorption in a MQW structure with 112 Å-thick wells in a magnetic field $\mathbf{B} \parallel z$. The inset shows absorption spectrum for bulk GaAs in a field $B = 33.8$ T. The arrows identify the ground state excitation energies of heavy and light holes in a quantum well and the exciton 1s-state in bulk GaAs. The dashed lines indicate the zero level. The absorption peaks corresponding to particular Landau levels are specified by the level number N [6.11]

numbers $N = 1, 2, \ldots$ appear. For comparison, the inset of Fig. 6.20 shows a magnetoabsorption spectrum for bulk GaAs. We see immediately that the $N \geqslant 1$ peaks are symmetric for the structure and asymmetric for the homogeneous semiconductor. It is only natural to attribute this difference to the difference in the dimension of the electron or hole states in a quantizing magnetic field in the materials in question.

Current carriers in a quantum well bounded by wide barriers can move freely only in two directions. In a strong magnetic field $\mathbf{B} \parallel z$ the cyclotron motion of the carriers is likewise quantized, their states becoming localized in all three directions. The density of states in this case is a sum of δ-functions, each of them corresponding to a separate quantum level. For the same reason, the reduced density of states also consists of separate peaks whose width is determined by the homogeneous and inhomogeneous broadening of a pair of electron and hole levels. Since this broadening occurs in equal measure toward higher and lower energies, the interband magneto-optical absorption peaks in a MQW are of symmetric shape.

In a homogeneous material, a quantizing magnetic field does not suppress free-carrier motion in one direction, so that the reduced density of states with coinciding Landau-level numbers is the sum of asymmetric peaks

$$g_{\mathrm{cv}}^{(3\mathrm{D})}(\hbar\omega) = \frac{1}{\pi^2}\frac{eB}{\hbar^2 c}\left(\frac{\mu_{\mathrm{cv}}}{2}\right)^{1/2}\sum_{N=0}^{\infty}\frac{\theta(\varepsilon_N)}{\sqrt{\varepsilon_N}},$$

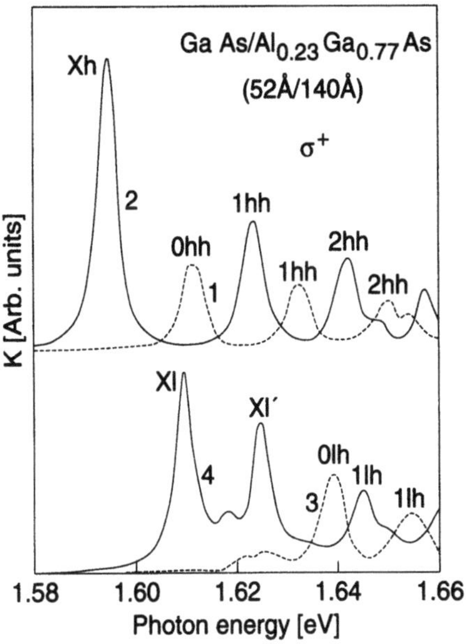

Fig. 6.21. Absorption spectra of periodic GaAs/$Al_xGa_{1-x}As$ heterostructure with $a = 52$ Å, $b = 140$ Å calculated for a magnetic field $B = 10$ T for σ_+ polarization with (solid curve) and without inclusion (dotted curve) of the exciton effects. The upper and lower curves describe the contributions to absorption coming from the hh1 $\rightarrow$ c1 and lh1 $\rightarrow$ c1 transitions, respectively [6.12]

$$\varepsilon_N = \hbar\omega - E_g - \left(N + \frac{1}{2}\right)\hbar\tilde{\Omega}_c, \tag{6.115}$$

where $\tilde{\Omega}_c$ is the cyclotron frequency for the reduced mass μ_{cv}. The broadening of the function $\varepsilon_N^{-1/2}\theta(\varepsilon_N)$ does not remove its asymmetry. For this reason, the shape of the $N = 0$ and 1 absorption peaks in the inset of Fig. 6.20 is characterized by a steep rise and a smooth decay with increasing energy.

The effect of Coulomb interaction on magneto-optical absorption spectra is illustrated by the display in Fig. 6.21 of the calculated contributions to absorption of interband transitions involving heavy and light-hole states in a periodic heterostructure $GaAs/Al_{0.23}Ga_{0.77}As$. Note that the appearance of an additional structure in spectrum 3 at energies below the 0lh peak is due to the hybridization of heavy and light-hole states in the heterostructure. The inclusion of Coulomb interaction shifts the absorption peaks to lower energies and results in a redistribution of oscillator strength between the peaks. The lower peaks are labeled by Xh and Xl rather than by 0hh and 0lh, since the corresponding states contain a noticeable admixture of other Landau levels. The anomalous peak Xl′ arises due to a strong admixture of the 0lh Landau level to heavy-hole states with a nonzero projection of the orbital moment.

6.7.1 Magnetoreflection

Measurement of reflectance spectra in a magnetic field can be used, in particular, in the determination of the exciton g-factor. In the linear-in-field approximation, one can neglect the diamagnetic shift of the exciton excitation energy and take into account only the Zeeman splitting of the exciton levels. In a semiconductor with nondegenerate bands, the triplet level $J = 1$ of the 1s-exciton splits in a $\mathbf{B} \parallel z$ field into sublevels with $J_z = -1, 0, 1$, and the radiative doublet Γ_5 of the e1-hh1 1s-exciton in a type GaAs/AlGaAs (001) heterostructure splits into the sub-levels $\Gamma_{5\pm}$ with $J_z = \pm 1$, which are optically active in the σ_+ and σ_- polarizations, respectively. In the case where this splitting, $\Delta\omega = \omega_+ - \omega_- \equiv g\mu_B B_z$, exceeds the exciton damping Γ, it can be readily measured spectroscopically. In the other limiting case, $|\Delta\omega| \ll \Gamma$, the splitting $\Delta\omega$ can be found by studying normal-incidence reflection of nonpolarized or linearly polarized light and measuring the spectral behavior of the degree of circular polarization of reflected light

$$P_{\mathrm{circ}}(\omega) = \frac{R(\sigma_+) - R(\sigma_-)}{R(\sigma_+) + R(\sigma_-)} = -\frac{\Delta\omega}{2}\frac{1}{R(\omega)}\frac{\partial R(\omega)}{\partial\omega}, \tag{6.116}$$

where $R(\omega)$ is the reflection coefficient for zero magnetic field.

7 Intraband Transitions

In quantum-mechanical terms, the cyclotron and electron-spin resonances originate from optical transitions of carriers between the Landau levels and magnetic-field-splitted spin sublevels. Measurement of the dependence of the cyclotron-resonance frequency on the magnitude and direction of magnetic field provides a direct and reliable way for determining the electron (or hole) effective mass, as well as for studying the nonparabolicity and nonsphericity of an electronic band in a semiconductor. In connection with this, we derive in Sec. 7.1 expressions for the longitudinal and transverse electron mass in a superlattice at the miniband bottom, analyze how the choice of the boundary conditions for the envelopes at the interfaces affects these masses, and discuss the nonparabolicity of the miniband spectrum.

The electron-spin resonance frequency is determined by the electron g-factor. Therefore a separate section deals with the calculation of this essential band parameter, which can show up in other optical phenomena as well, for example, in light scattering or photoluminescence in a magnetic field.

Quantum-confinement effects transform the free-carrier energy spectrum into a series of subbands (or minibands). This permits observation of the optical absorption due to optical transitions between subbands in doped heterostructures. We are going to analyze the corresponding selection rules and calculate the absorption coefficient associated with intersubband transitions. Besides the intraband carrier transitions, we shall consider also IR reflection from an undoped structure with inclusion of the light interaction with optical vibrations in the superlattice.

7.1 Cyclotron Resonance and Effective Electron Mass

In a bulk semiconductor with a parabolic conduction band

$$E_{\mathbf{k}} = \frac{\hbar^2}{2}\left(\frac{k_1^2}{m_1} + \frac{k_2^2}{m_2} + \frac{k_3^2}{m_3}\right), \tag{7.1}$$

Schrödinger's equation for the electron envelope wave function in a homogeneous magnetic field **B** can be reduced by substitution of the variables (3.139):

$$x_i \rightarrow \tilde{x}_i = (m_i/\bar{m})^{1/2}\, x_i$$

to an equivalent equation for the electron with an isotropic mass $\bar{m} = (m_1 m_2 m_3)^{1/3}$ in an effective field **B** with components

$$\tilde{B}_i = \sqrt{\frac{m_i}{\bar{m}}} B_i. \tag{7.2}$$

Therefore, the energy separation between the closest Landau levels which determines the cyclotron resonance frequency (3.132)

$$\hbar\omega_c = \frac{e\tilde{B}}{\bar{m}c} \equiv \frac{eB}{m_c c}, \tag{7.3}$$

where the cyclotron mass

$$m_c = \bar{m} \left(\sum_i \frac{m_i}{\bar{m}} \cos^2\theta_i \right)^{-1/2}, \tag{7.4}$$

θ_i is the angle between **B** and the axis $i = 1, 2, 3$. If two of the three m_i masses coincide, for instance, if $m_1 = m_2 \equiv m_\perp$, the expression for the cyclotron mass simplifies to

$$m_c = \left(\frac{\sin^2\theta_3}{m_\perp m_\parallel} + \frac{\cos^2\theta_3}{m_\perp^2} \right)^{-1/2}, \tag{7.5}$$

where $m_\parallel \equiv m_3$. By measuring the cyclotron resonance in the $\mathbf{B} \parallel 3$ and $\mathbf{B} \perp 3$ geometries, one can determine the cyclotron masses $m_{c,\parallel} \equiv m_c(\theta_3 = 0) = m_\perp$, and $m_{c,\perp} \equiv m_c(\theta_3 = \pi/2) = (m_\parallel m_\perp)^{1/2}$ and, hence, the longitudinal mass

$$m_\parallel = \frac{m_{c,\perp}^2}{m_{c,\parallel}}. \tag{7.6}$$

In a semiconductor with an anisotropic nonparabolic spectrum

$$E_\mathbf{k} = \frac{\hbar^2 k_\perp^2}{2m_\perp} + \frac{\hbar^2 k_z^2}{2m_\parallel} + \Lambda k_z^4, \tag{7.7}$$

the electron energy in the presence of a magnetic field is determined in a linear-in-Λ approximation by the following expressions (where the spin splitting is neglected):

(i) $\mathbf{B} \parallel z$

$$E_{nk_z} = \hbar\omega_{c,\parallel} \left(n + \frac{1}{2} \right) + \frac{\hbar^2 k_z^2}{2m_\parallel} + \Lambda k_z^4, \tag{7.8}$$

(ii) $\mathbf{B} \parallel y$

$$E_{nk_y} = \hbar\omega_{c,\perp} \left[n + \frac{1}{2} + \frac{\hbar^2 k_y^2}{2m_\perp} \frac{1}{\hbar\omega_{c,\perp}} + \Lambda \left(\frac{m_\parallel}{\hbar^2} \right)^2 \hbar\omega_{c,\perp} \frac{3}{4} (2n^2 + 2n + 1) \right], \tag{7.9}$$

where n is the number of the Landau level, the frequencies $\omega_{c,\parallel}$ and $\omega_{c,\perp}$ being related to the cyclotron masses $m_{c,\parallel}$ and $m_{c,\perp}$ through the relation (7.3). In a longitudinal field, the levels are equally spaced, whereas in a transverse field the distance between the closest levels $n+1$ and n depends on n

$$\omega_{n+1,n}(\mathbf{B} \parallel y) = \hbar\omega_{c,\perp}\left[1 + \Lambda\left(\frac{m_\parallel}{\hbar^2}\right)^2 \hbar\omega_{c,\perp}3(n+1)\right]. \tag{7.10}$$

In the case where the electron dispersion $E_\mathbf{k}$ has a more complex form than (7.1, 7), one can use for the determination of the cyclotron mass m_c the quasiclassical approximation (valid for $n \gg 1$)

$$m_c(E, k_{z'}) = \frac{\hbar^2}{2\pi}\iint dk_{x'}\, dk_{y'}\delta(E - E_\mathbf{k}), \tag{7.11}$$

where the z' axis of the Cartesian coordinate frame x', y', z' is oriented along the magnetic field. Of the states with a fixed energy E, the main contribution to the cyclotron resonance comes from those characterized by extremal orbits to which the maximum or minimum in the dependence of m_c on k'_z corresponds.

7.1.1 Boundary Conditions and Effective Electron Mass in a Superlattice

We expand the electron energy in the νth miniband of the semiconductor in powers of the wave vector

$$E_{\nu\mathbf{Q}} = E_\nu^0 + \frac{\hbar^2 Q_z^2}{2M_\parallel} + \frac{\hbar^2 Q_\perp^2}{2M_\perp} + \Lambda_1 Q_z^4 + \Lambda_2 Q_\perp^4 + \Lambda_3 Q_z^2 Q_\perp^2 + \ldots . \tag{7.12}$$

To calculate the expansion coefficients and determine the cyclotron resonance frequencies, one can use the dispersion relation (3.91). It can be conveniently represented in the form similar to (6.7) for normal light waves in an optical SL:

$$1 - \cos Q_z d = \frac{1}{2}\sin(ka)\sinh(\lambda b) f_1 f_2. \tag{7.13}$$

If in homogeneous materials A and B the effective electron masses m_A and m_B are isotropic, then

$$k = \left(\frac{2m_A}{\hbar^2}E - Q_\perp^2\right)^{1/2}, \quad \lambda = \left[\frac{2m_B}{\hbar^2}(V_0 - E) + Q_\perp^2\right]^{1/2}, \tag{7.14}$$

where the energy E is reckoned from the bottom of the conduction band of semiconductor A. For the boundary conditions (5.5, 9)

$$\begin{aligned} f_1 &= \tan\phi_a - \eta\tanh\phi_b, \\ f_2 &= \eta^{-1}\cot\phi_a + \coth\phi_b, \end{aligned} \tag{7.15}$$

where $\phi_a = ka/2$, $\phi_b = \lambda b/2$, $\eta = (m_A/m_B)(\lambda/k)$.

Consider boundary conditions of a general form (3.23)

$$\begin{pmatrix} \varphi_A \\ \tilde{\varphi}_A \end{pmatrix} = \hat{T}_{AB} \begin{pmatrix} \varphi_B \\ \tilde{\varphi}_B \end{pmatrix} = \begin{pmatrix} t_{11} & t_{12} \\ t_{21} & t_{22} \end{pmatrix} \begin{pmatrix} \varphi_B \\ \tilde{\varphi}_B \end{pmatrix}, \tag{7.16}$$

where

$$\tilde{\phi}_A = l \left(\frac{\partial \phi}{\partial z} \right)_A, \quad \tilde{\phi}_B = l \frac{m_A}{m_B} \left(\frac{\partial \phi}{\partial z} \right)_B,$$

l is an arbitrarily chosen and fixed length, the matrix $\hat{T}_{AB}$ satisfying the condition of particle-flux conservation at the heteroboundary (3.24a)

$$\det \hat{T}_{AB} = t_{11}t_{22} - t_{12}t_{21} = 1.$$

The matrix $\hat{T}_{BA}$ for a boundary of type BA differs from $\hat{T}_{AB}$ in the sign of the nondiagonal elements. For the Bloch solutions (5.46), the matrix representing transfer across the period

$$\hat{T} = \hat{T}_{AB}\hat{T}_B\hat{T}_{BA}\hat{T}_A$$

satisfies the relation

$$\det \parallel T_{ij} - \mathrm{e}^{\mathrm{i}Q_z d}\delta_{ij} \parallel = 0$$

or

$$2\cos Q_z d = T_{11} + T_{22}, \tag{7.17}$$

where the identity $\det \hat{T} = 1$ is taken into account. Equation (7.17) reduces to the form (7.13) with

$$f_1 = (t_{11} - t_{12}\rho \tanh \phi_b) \tan \phi_a - \eta \left(t_{22} \tanh \phi_b - t_{21}\rho^{-1} \right),$$
$$f_2 = \eta^{-1} (t_{11} - t_{12}\rho \coth \phi_b) \cot \phi_a + t_{22} \coth \phi_b - t_{21}\rho^{-1}). \tag{7.18}$$

where

$$\rho = \frac{m_A}{m_B} \lambda l.$$

The representation (7.13) can be conveniently used to analyze the electron spectrum for small Q. In particular, from (7.13) it follows that the electron energy E_ν^0 for $Q = 0$ satisfies one of the equations

$$f_1 = 0 \quad \text{or} \quad f_2 = 0 \tag{7.19}$$

for the solutions $\chi_{\nu\mathbf{Q}}(\mathbf{x})$ which are, respectively, even or odd under reflection in the (x, y) plane passing through the center of the well or barrier.

To find the longitudinal effective mass $M_\parallel$, expand the left-hand side of (7.13) in powers of $Q_z d$, and the right-hand side which we denote by $\tilde{F}$, in powers of $E - E^0$. We obtain

$$M_\parallel = \left(\frac{\hbar}{\mathrm{d}} \right)^2 \left(\frac{\partial \tilde{F}}{\partial E} \right)_0, \tag{7.20}$$

where the index "0" identifies the value of the derivative at $E = E^0$. Note that the function $\tilde{F}$ is connected with the function F in (73.91) through the relation $\tilde{F} = 1 - F$. As follows from (7.13), in order to calculate $(\partial \tilde{F}/\partial E)_0$, it is sufficient to differentiate f_1 or f_2 in the right-hand side of (7.13):
for even solutions

$$M_{\parallel} = \frac{\hbar^2}{2d^2}\left[\sin(ka)\sinh(\lambda b) f_2 \frac{\partial f_1}{\partial E}\right]_0,$$

for odd solutions:

$$M_{\parallel} = \frac{\hbar^2}{2d^2}\left[\sin(ka)\sinh(\lambda b) f_1 \frac{\partial f_2}{\partial E}\right]_0.$$

Finally, we come to the following expression for the longitudinal effective mass [7.1]

$$M_{\parallel} = \pm \tilde{h} \frac{m_A \sinh \lambda b}{k(a+b) C_\alpha^2}, \tag{7.21}$$

where

$$\tilde{h} = \eta\left(t_{22}^2 + \sigma^2 t_{12}^2\right) + \eta^{-1}\left[t_{11}^2 + (t_{21}/\sigma)^2\right] - 2\left(\sigma t_{11} t_{12} + \sigma^{-1} t_{21} t_{22}\right)\cot \lambda b,$$

$$C_a^2 = 2(a+b)\left\{a\left(1 \pm \frac{\sin ka}{ka}\right) + b\left(\frac{\sinh \lambda b}{\lambda b} \pm 1\right)\frac{1 \pm \cos ka}{\cosh \lambda b \pm 1}\left[t_{11} - t_{12}\rho\left(\tanh\frac{\lambda b}{2}\right)^{\pm 1}\right]\right\}^{-1}, \tag{7.22}$$

$\sigma = lk$, the upper and lower signs refer, respectively, to even and odd states, and the quantities k, λ, η are determined at $E = E_\nu^0$ (5.65). The coefficient C_a is contained in the expression for the Bloch function $U_0(z)$ in the form (5.64) for the even solutions and in the form

$$U_0(z) = \begin{cases} C_a \sin k\,(z - \bar{z}_a) & \text{in wells,} \\ C_b \sinh \lambda(z - \bar{z}_b) & \text{in barriers} \end{cases} \tag{7.23}$$

for the odd solutions. For the boundary conditions (5.5, 9), $t_{11} = t_{22} = 1, t_{12} = t_{21} = 0$ and the coefficient C_a^2 transforms to (5.67) for the even solutions, with $\tilde{h} \to \eta + \eta^{-1}$.

To analyze the spectrum near $Q_z = \pm\pi/d$, it may be recommended to use in place of (7.13) an equivalent equation

$$1 + \cos Q_z \mathrm{d} = \frac{1}{2}\sin(ka)\sinh(\lambda b) f_3 f_4, \tag{7.24}$$

where

$$f_3 = -\left(t_{11} - t_{12}\rho \coth\phi_b\right)\tan\phi_a + \eta\left(t_{22}\coth\phi_b - t_{21}\rho^{-1}\right),$$

$$f_4 = \eta^{-1}\left(t_{11} - t_{12}\rho\tanh\phi_g\right)\cot\phi_a + t_{22}\tanh\phi_b - t_{21}\rho^{-1}.$$

The equation $f_3 = 0$ determines the electron energy at Brillouin zone boundary in a state that is even under reflection through the center of the well, or odd, through that of the barrier. The equation $f_4 = 0$ corresponds to the states at Brillouin zone boundary of opposite parity.

To calculate $M_\perp$, we take into account that the electron energy $E_{\nu\mathbf{Q}}(V_0)$ in a SL with a potential barrier V_0 at the interfaces coincides with the energy

$$\frac{\hbar^2 Q_\perp^2}{2m_A} + E_{\nu\tilde{\mathbf{Q}}}(\tilde{V}_0),$$

where $\tilde{Q}_x = \tilde{Q}_y = 0$, $\tilde{Q}_z = Q_z$,

$$\tilde{V}_0 = V_0 + \frac{\hbar^2}{2}\left(\frac{1}{m_B} - \frac{1}{m_A}\right) Q_\perp^2.$$

Therefore,

$$\frac{\hbar^2 Q_\perp^2}{2M_\perp} = \frac{\hbar^2 Q_\perp^2}{2m_A} - \frac{\mathrm{d}E_\nu^0}{\mathrm{d}V_0}\frac{\hbar^2}{2}\left(\frac{1}{m_B} - \frac{1}{m_A}\right) Q_\perp^2$$

or

$$\frac{1}{M_\perp} = \frac{1}{m_A} + \left(\frac{1}{m_B} - \frac{1}{m_A}\right)\frac{\mathrm{d}E_\nu^0}{\mathrm{d}V_0}. \tag{7.25}$$

We come to the same expression if we take into account that $M_\perp^{-1}$ is obtained by averaging the reciprocal effective mass:

$$\begin{aligned}\frac{1}{M_\perp} &= \langle \nu, 0\,|m^{-1}|\,\nu, 0\rangle = \int U_0^2(z)\left[\frac{1}{m_A} + \left(\frac{1}{m_B} - \frac{1}{m_A}\right)\theta_B(z)\right]\mathrm{d}z \\ &= \frac{1}{m_A} + \left(\frac{1}{m_B} - \frac{1}{m_A}\right)\langle \nu, 0\,|\theta_B|\,\nu, 0\rangle .\end{aligned} \tag{7.26}$$

Here $\theta_B = 0$ in the wells and $\theta_B = 1$ in the barriers. By perturbation theory, a change of the potential barrier V_0 by δV_0 results in a change of the energy E_ν^0 in first order in δV_0 by an amount

$$\delta E_\nu^0 = \delta V_0 \langle \nu, 0\,|\theta_B|\,\nu, 0\rangle . \tag{7.27}$$

From (7.26, 27), one indeed obtains (7.25).

In the case of the boundary conditions (5.5, 9), one comes to the following expression for $\mathrm{d}E_\nu^0/\mathrm{d}V_0$:

$$\frac{\mathrm{d}E_\nu^0}{\mathrm{d}V_0} = \frac{1}{2}\left(\frac{\sinh \lambda b}{\lambda b} - (-1)^\nu\right)\frac{b}{a+b}C_b^2, \tag{7.28}$$

where

$$C_b = \begin{cases} \dfrac{\cos\phi_a}{\cosh\phi_b}C_a & \text{for even solutions,} \\ -\dfrac{\sin\phi_a}{\sinh\phi_b}C_a & \text{for odd solutions.}\end{cases} \tag{7.29}$$

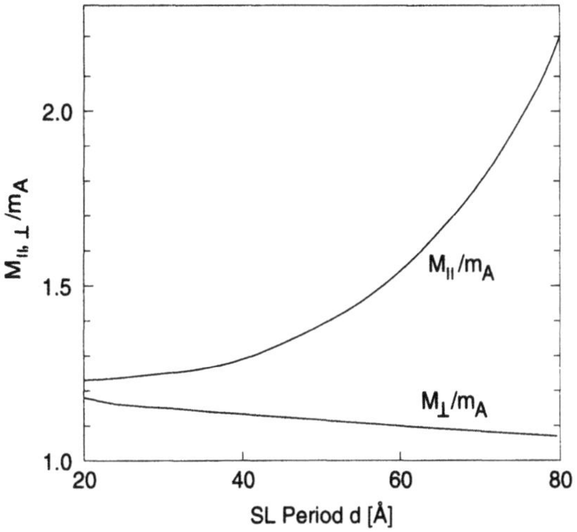

Fig. 7.1. Dependence of longitudinal and transverse electron effective masses at the bottom of the lowest e1 miniband in $GaAs/Al_{0.35}Ga_{0.65}As$ SL with equal well and barrier thicknesses on superlattice period. The parameters used in the calculation: $m_A = 0.067m$, $m_B/m_A = 1.43$, $V_0 = 249$ meV, with the boundary conditions (5.9)

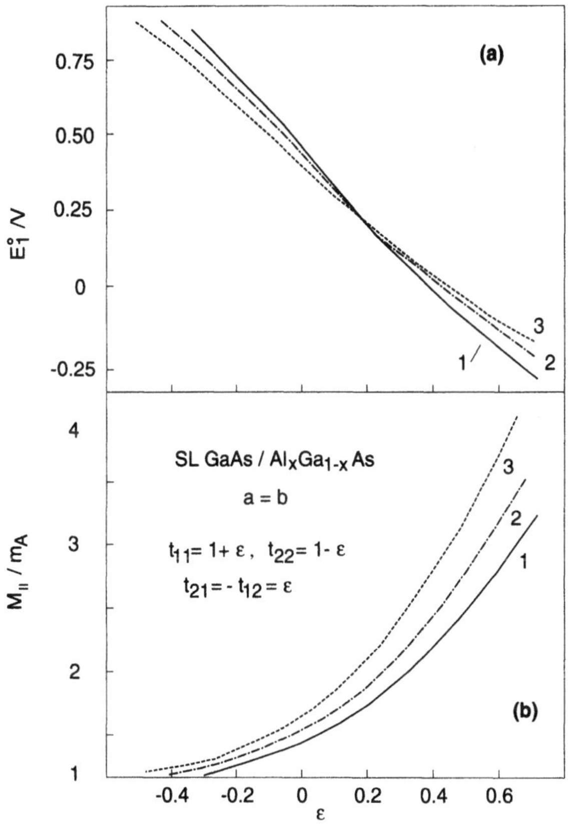

Fig. 7.2a,b. Dependence of **a** the energy E_1^0 and **b** longitudinal electron effective mass $M_{\parallel}$ at the bottom of the lowest miniband e1 on dimensionless parameter ε determining, by (7.30), the components of the matrix t_{ij} for boundary conditions (7.16). The calculation was performed for $GaAs/Al_{0.35}Ga_{0.65}$ As SL for $m_A = 0.067$ m, $m_B = 1.43m_A$, $V_0 = 249$ meV and for equal layer thicknesses $a = b = 20$ Å (curve 1), 25 Å (curve 2) and 30 Å (curve 3). The microscopic length l was set equal to $(\hbar^2/2m_A V_0)^{1/2}$ [7.1]

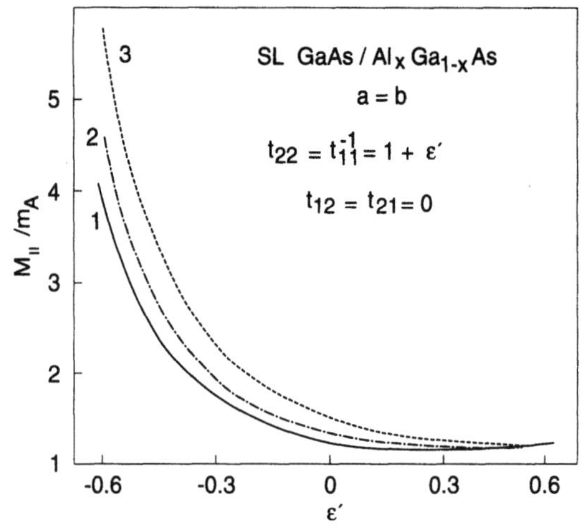

Fig. 7.3. Dependence of the longitudinal electron effective mass $M_{\parallel}$ at the bottom of the lowest miniband on dimensionless parameter ε' determining, by (7.31), the components of the matrix t_{ij} with the boundary conditions (7.16). The values of the SL parameters used in the calculation are given in the caption of Fig. 7.2 [7.1]

Since $\mathrm{d}E_{\nu}^{0}/\mathrm{d}V_0$ varies between 0 to 1, the values of the transverse mass lie between m_A and m_B.

Figure 7.1 presents in a graphical form the dependences $M_{\parallel}(d)$ and $M_{\perp}(d)$ for the GaAs/$Al_{0.35}Ga_{0.65}As$ superlattice with $a = b$. The effect of the boundary conditions on the subband structure is illustrated by Figs. 7.2 and 7.3. Two types of the single-parameter matrices $\hat{T}_{AB}$ were considered:

$$t_{11} = 1 + \varepsilon, \quad t_{12} = 1 - \varepsilon, \quad t_{21} = -t_{12} = \varepsilon \tag{7.30}$$

and

$$t_{22} = t_{11}^{-1} = 1 + \varepsilon', \quad t_{12} = t_{21} = 0. \tag{7.31}$$

Note that at a certain value $\varepsilon = \varepsilon_{\mathrm{cr}}$ the energy E_1^0 becomes zero. For $\varepsilon > \varepsilon_{\mathrm{cr}}$, this energy is negative and the quantity k imaginary. According to (7.18), $\varepsilon_{\mathrm{cr}}$ can be found from the condition

$$\left(t_{22} \tanh \phi_b - t_{21} \rho^{-1}\right)_{E=0} = 0, \tag{7.32}$$

whence we obtain

$$\varepsilon_{\mathrm{cr}} = \left[1 + \sqrt{\frac{m_B}{m_A}} \coth\left(\sqrt{\frac{m_B}{m_A}} \frac{b}{2l}\right)\right]^{-1}.$$

For a single quantum well ($b \to \infty$)

$$\varepsilon_{\mathrm{cr}} = \left(1 + \sqrt{m_B/m_A}\right)^{-1}.$$

For the diagonal boundary conditions (7.31) with $t_{11} \neq t_{22}$, (7.17) can be reduced to the dispersion equation corresponding to the boundary conditions (5.5, 9) through the substitution

$$m_B \to \tilde{m}_B = m_B t_{11}^4, \quad b \to \tilde{b} = b t_{11}^{-2},$$

$$Q_z \to \tilde{Q}_z = Q_z \frac{a+b}{a+\tilde{b}},$$

whence we obtain the following relations

$$E_1^0\left(\varepsilon', m_B, b\right) = E_1^0\left(0, \tilde{m}_B, \tilde{b}\right),$$

$$M_\parallel\left(\varepsilon', m_B, b\right) = M_\parallel\left(0, \tilde{m}_B, \tilde{b}\right).$$

7.1.2 Nonparabolicity of the Electron Miniband

For the boundary conditions (5.5, 9), the coefficients Λ_i of expansion (7.12) can be represented in the following form

$$\Lambda_1 = -\frac{\hbar^2}{8M_\parallel}\left(\frac{\tilde{F}_0''}{\tilde{F}_0'}\frac{\hbar^2}{M_\parallel} + \frac{d^2}{3}\right), \tag{7.33}$$

$$\Lambda_2 = \frac{\hbar^4}{8}\frac{d^2 E_\nu^2}{\mathrm{d}V_0^2}\left(\frac{1}{m_B} - \frac{1}{m_A}\right)^2,$$

$$\Lambda_3 = \frac{\hbar^4}{4}\left(\frac{1}{m_B} - \frac{1}{m_A}\right)\frac{\mathrm{d}}{\mathrm{d}V_0}\left(\frac{1}{M_\parallel}\right), \tag{7.34}$$

where $\tilde{F}$ is the right-hand part of (7.13), $\tilde{F}' = \partial\tilde{F}/\partial E$, $\tilde{F}'' = \partial^2\tilde{F}/\partial E^2$. Note that the coefficients Λ_2, Λ_3 are nonzero because m_A and m_B are not equal. For the type I superlattice GaAs/$Al_xGa_{1-x}As$, they are small compared to Λ_1.

The ratio of the derivatives in (7.33) can be transformed to

$$\frac{\tilde{F}_0''}{\tilde{F}_0'} = \left[\frac{(kf_1)''}{(kf_1)'} + 2\left(\frac{(\lambda f_2)'}{\lambda f_2} + \frac{(\sin ka)'}{\sin ka} + \frac{(\sinh\lambda b)'}{\sinh\lambda b} - \frac{k'}{k} - \frac{\lambda'}{\lambda}\right)\right]_0 \tag{7.35}$$

We are presenting below expressions for the derivatives of the functions contained in (7.35):

$$k' = k/2E, \quad \lambda' = -\lambda/\left[2\left(V_0 - E\right)\right],$$

$$(kf_1)' = \frac{m_A}{\hbar^2}\left(a\frac{1+\dfrac{\sin ka}{ka}}{1+\cos ka} + b\frac{1+\dfrac{\sinh\lambda b}{\lambda b}}{1+\cosh\lambda b}\right), \tag{7.36}$$

$$(\lambda f_2)' = -\frac{m_B}{\hbar^2}\left(a\frac{1-\dfrac{\sin ka}{ka}}{1-\cos ka} + b\frac{\dfrac{\sinh\lambda b}{\lambda b} - 1}{\cosh\lambda b - 1}\right), \tag{7.37}$$

$$(kf_1)'' = \frac{m_A}{\hbar^2}\left[k'aP\left(ka\right) + \lambda' bR\left(\lambda b\right)\right],$$

$$P(X) = \frac{1}{1+\cos X}\left[\frac{1}{X}\left(1 - \frac{\sin X}{X}\right) - \frac{\sin X}{1+\cos X}\right],$$

$$R(Y) = \frac{1}{1+\cosh Y}\left[\frac{1}{Y}\left(1-\frac{\sinh Y}{Y}\right)-\frac{\sinh Y}{1+\cosh Y}\right]. \tag{7.38}$$

7.1.3 Tight-Binding Approximation

According to (7.21), in a structure with thick barriers, such that

$$\exp(-\lambda b) \ll 1 \tag{7.39}$$

the dependence of $M_{\|}$ on b is described by the relation

$$M_{\|} \propto \frac{e^{\lambda b}}{(a+b)^2}.$$

Since the ratio $\tilde{F}''/\tilde{F}'$ in (7.33) remains finite with increasing b while $M_{\|}$ grows without limit, therefore in the case of sufficiently thick barriers

$$\Lambda_1 = -\frac{\hbar^2 d^2}{24 M_{\|}}. \tag{7.40}$$

In the limiting case (7.39), one can find from (7.13) not only the coefficient of Q_z^4 but the whole dependence $E(Q_z)$ as well. To do this, one has to leave in the expansion of the right-hand side of (7.13) in powers of $E - E^0$ the first term, and not expand in powers of $Q_z d$ the left-hand side of this equation. As a result, we obtain (3.163)

$$E_\nu\left(Q_z; \mathbf{Q}_\perp = 0\right) = E_\nu^0 + \frac{\hbar^2}{M_{\|} d^2}\left(1 - \cos Q_z \mathrm{d}\right), \tag{7.41}$$

with $M_{\|}$ defined by (7.20). Hence, in this case, the width of the miniband in the direction of Q_z is

$$\Delta = \frac{2\hbar^2}{\left|M_{\|}\right| d^2}. \tag{7.42}$$

The same result can be obtained in the tight binding approximation when the overlap of the single quantum well wave functions $f_\nu(z)$ and $f_\nu(z \pm d)$ in the neighboring wells is small, and the electron wave function in the SL can be written in the form

$$\chi_{\nu\mathbf{Q}}(\mathbf{x}) = \frac{e^{i\mathbf{Q}_\|\cdot\mathbf{x}_\|}}{\sqrt{NS}} \sum_l f_\nu(z - ld) e^{iQ_z l d}, \tag{7.43}$$

where N is the total number of periods in the structure. In the tight binding approximation, the miniband width Δ for equal masses m_A and m_B is determined by the expression

$$\Delta = 4V_0 \int f_\nu(z) f_\nu(z+d)\, \mathrm{d}z. \tag{7.44}$$

This yields for even solutions

$$\Delta = 4V_0 \left(\frac{C_a^2}{a+b}\right)_{b\to\infty} \cos\phi_a \mathrm{e}^{-\lambda(b+a/2)} \int\limits_{-a/2}^{a/2} \cos(kz)\mathrm{e}^{-\lambda z}\,\mathrm{d}z. \tag{7.45}$$

For k and λ related by

$$k\tan\phi_a = \lambda$$

the integral in (7.45) can be reduced to

$$\frac{2k\sin\phi_a}{\lambda^2+k2}\mathrm{e}^{\lambda a/2}$$

so that (7.45), indeed, transforms to (7.42), since for $m_A = m_B$ the sum $\lambda^2 + k^2 = 2m_A V_0/\hbar^2$ and $\eta + \eta^{-1} = 2/\sin ka$.

For $m_A \neq m_B$ one should use in place of (7.44) a more general expression (3.164, 165)

$$\Delta = 4\int \mathrm{d}z f_\nu(z)\left[V_0 + \frac{\hbar^2}{2}\left(\frac{1}{m_A} - \frac{1}{m_B}\right)\frac{\mathrm{d}}{\mathrm{d}z}\theta_A(z)\frac{\mathrm{d}}{\mathrm{d}z}\right] f_\nu(z+d) \tag{7.46}$$

since, in this case, the kinetic-energy operators for neighboring layers are different. Here $\theta_A = 1$ in the wells; $\theta_A = 0$ in the barriers.

Substituting in (7.11) for $E_\mathbf{k}$ the function

$$\frac{\hbar^2 k_x^2}{2M_\perp} + \frac{\Delta}{2}(1 - \cos k_z d),$$

we find the cyclotron mass in a magnetic field $\mathbf{B} \parallel y$

$$\begin{aligned} m_c(\varepsilon) &= \frac{\hbar}{\pi}\sqrt{2M_\perp} \int\limits_0^{Q_0} \frac{\mathrm{d}k_z}{\sqrt{\varepsilon - \frac{\Delta}{2}(1-\cos k_z\,\mathrm{d})}} \\ &= \frac{2\hbar}{\pi\,\mathrm{d}}\sqrt{\frac{2M_\perp}{\varepsilon}} F\left(\arcsin\sqrt{\frac{E}{\Delta}}, \sqrt{\frac{\Delta}{\varepsilon}}\right), \end{aligned} \tag{7.47}$$

where $\varepsilon = E - E_1^0$, $Q_0 = (2/d)\arcsin(\varepsilon/\Delta)^{1/2}$, $F(\varphi, k)$ is the elliptic integral of the first kind. For $\varepsilon/\Delta \ll 1$

$$F \approx \frac{\pi}{2}\sqrt{\frac{\varepsilon}{\Delta}}$$

and $m_{c,\perp} = (M_\parallel M_\perp)^{1/2}$.

Figure 7.4 illustrates the measurements of the longitudinal and transverse electron masses in a GaAs/Al$_x$Ga$_{1-x}$As SL with the layers of thickness $a = 80$ Å and $b \approx 20$ Å (the barriers were not strictly rectangular). The curves bounding the shaded areas represent the $M_\parallel(x)$ dependence calculated for an SL with rectangular wells and barriers and for potential jumps in the conduction

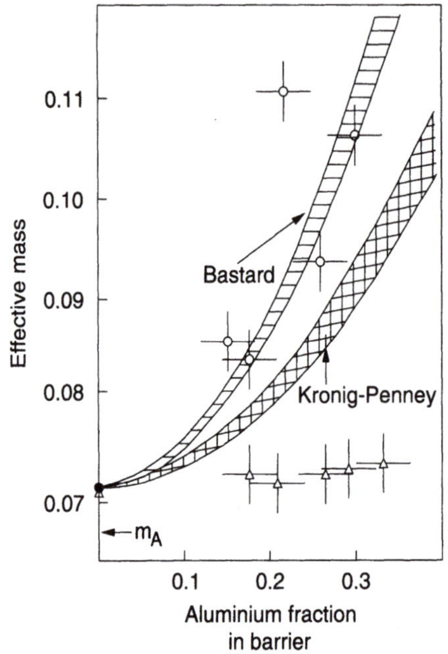

Fig. 7.4. Effective electron mass reduced to the free electron mass in the lowest miniband e1 in GaAs/$Al_xGa_{1-x}As$ SL with $a = 80$ Å, $b \approx 20$ Å vs. composition x. The experimental data were obtained from a cyclotron resonance study. Circles: longitudinal mass $M_\|$ (along the principal axis z), triangles: transverse mass $M_\perp$ (in x, y plane). The calculations performed for the boundary conditions (7.48) (Kronig-Penney) and (7.49) (Bastard) are displayed by shaded areas. The curves bounding these areas correspond to a potential jump in the conduction band, $V_0 = 0.65E_g$ (upper curves) and $V_0 = 0.61E_g$ (lower curves) [7.2]

band $V_0 = 0.65\Delta E_g$ (top curve) and $V_0 = 0.61\Delta E_g$ (bottom curve) using the following boundary conditions:
(i) Kronig-Penney:

$$\varphi_A = \varphi_B, \quad \left(\frac{\mathrm{d}\varphi}{\mathrm{d}z}\right)_A = \left(\frac{\mathrm{d}\varphi}{\mathrm{d}z}\right)_B, \tag{7.48}$$

(ii) Bastard:

$$\varphi_A = \varphi_B, \quad \frac{1}{m_A}\left(\frac{\mathrm{d}\varphi}{\mathrm{d}z}\right)_A = \frac{1}{m_B}\left(\frac{\mathrm{d}\varphi}{\mathrm{d}z}\right)_B. \tag{7.49}$$

7.2 Intersubband Absorption

In heterostructures, we find an additional mechanism of intraband absorption associated with direct optical transitions of free carriers between the subbands created by the super-structural potential. The calculation of the contribution to the dielectric constant coming from these transitions should include the partial filling of states in an expression of type (6.11). For a MQW structure, we have

$$\kappa_{\alpha\beta}(\omega) = \kappa_{\alpha\beta} + \frac{4\pi e^2}{\omega^2(a+b)S} \times \sum_{\substack{\nu s \nu' s' \\ \mathbf{k}_\perp}} \frac{v^{\alpha}_{\nu s,\nu' s'}(\mathbf{k}_\perp)\, v^{\beta}_{\nu' s',\nu s}(\mathbf{k}_\perp)}{E_{\nu' s' \mathbf{k}_\perp} - E_{\nu s \mathbf{k}_\perp} - \hbar\omega - \mathrm{i}\hbar\Gamma_{\nu'\nu}} \left(F_{\nu s \mathbf{k}_\perp} - F_{\nu' s' \mathbf{k}_\perp}\right), \tag{7.50}$$

where $\hat{\mathbf{v}} = \hat{\mathbf{p}}/m$ is the velocity operator, $F_{\nu s \mathbf{k}_\perp}$ the electron distribution function, ν and ν' are the subband numbers, s and s' the spin indices.

We will first consider a heterostructure consisting of semiconductor layers with a simple (isotropic and parabolic) conduction band. In this case

$$\mathbf{e} \cdot \mathbf{v}_{\nu' s', \nu s}(\mathbf{k}_\perp) = \delta_{ss'} \left[\hbar\,(\mathbf{k}_\perp \cdot \mathbf{e}_\perp)\,\langle \nu' \left| m^{-1}(z) \right| \nu \rangle + e_z \langle \nu' \left| \hat{v}_z \right| \nu \rangle \right], \tag{7.51}$$

where $\mathbf{e}$ is the light polarization vector, $\hat{v}_z$ the velocity operator in the effective-mass method,

$$\frac{1}{m(z)} = \theta_A(z)\frac{1}{m_A} + \theta_B(z)\frac{1}{m_B},$$

$$\langle \nu' \left| \mathcal{R} \right| \nu \rangle = \int f_{\nu' \mathbf{k}_\perp}(z)\, \mathcal{R} f_{\nu \mathbf{k}_\perp}(z)\, \mathrm{d}z. \tag{7.52}$$

In the limit of infinitely high barriers, (7.51) simplifies to

$$\mathbf{e} \cdot \mathbf{v}_{\nu' s', \nu s}(\mathbf{k}_\perp) = \delta_{ss'} \frac{\hbar}{m_A} \left[(\mathbf{k}_\perp \cdot \mathbf{e}_\perp)\, \delta_{\nu\nu'} + k_z^{(\nu',\nu)} e_z \right], \tag{7.53}$$

where

$$k_z^{(\nu',\nu)} = \mathrm{i}\frac{2}{a} \left[1 - (-1)^{\nu+\nu'} \right] \frac{\nu\nu'}{\nu'^2 - \nu^2}. \tag{7.54}$$

By (7.54), direct intersubband transitions $\nu \to \nu'$ are allowed in the $\mathbf{e} \parallel z$ polarization for states of opposite parity. The selection rules (7.53) hold for finite barriers as well, provided the effective masses m_A and m_B in the quantum well and barrier materials coincide. However, the expression for $k_z^{(\nu',\nu)}$ becomes now more complex

$$k_z^{(\nu',\nu)} = -k_z^{(\nu,\nu')} = \frac{\mathrm{i}}{2a} \bar{C}_{a\nu} \bar{C}_{a\nu'} \times \left[\left(g + \frac{h}{2} \right) \sin \phi_a^+ + \left(\frac{1}{g} + \frac{h}{2} \right) \sin \phi_a^- \right], \tag{7.55}$$

where ν and ν' are, respectively, the even and odd states,

$$\phi_a^\pm = (k_{\nu'} \pm k_\nu)\, a/2, \tag{7.56}$$

$$g = \frac{k_{\nu'} - k_\nu}{k_{\nu'} + k_\nu}, \quad h = \frac{\lambda_{\nu'} - \lambda_\nu}{\lambda_{\nu'} + \lambda_\nu}, \tag{7.57}$$

$$\bar{C}_{a\nu}^2 = 2 \left[a \left(1 + \frac{\sin k_\nu a}{k_\nu a} \right) + \frac{1}{\lambda_\nu} (1 + \cos k_\nu a) \right]^{-1},$$

$$\bar{C}_{a\nu'}^2 = 2 \left[a \left(1 - \frac{\sin k_{\nu'} a}{k_{\nu'} a} \right) + \frac{1}{\lambda_{\nu'}} (1 - \cos k_{\nu'} a) \right]^{-1}. \tag{7.58}$$

For $V_0 \to \infty$ we have $\lambda_{\nu,\nu'}^{-1} \to 0$, $k_\nu \to \pi\nu/a$, $\sin \phi_a^- \to -\sin \phi_a^+$ and (7.55) reduces to (7.54).

If the masses m_A and m_B are not equal, the matrix element $V^z_{\nu'\nu} = \langle \nu' \left| \hat{V}_z \right| \nu \rangle$ in (7.51) is, as before, nonzero only for states of opposite parity. When calculating it, one should take into account that the velocity operator in (7.51) is

$$\hat{v}_z = -\mathrm{i}\hbar \left[\frac{\theta_A}{m_A} + \frac{\theta_B}{m_B}, \frac{\partial}{\partial z} \right]_s , \tag{7.59}$$

where the square brackets denote the symmetrized product. As a result, one obtains for transitions at $\mathbf{k}_\perp = 0$ from an even state to an odd one

$$v^z_{\nu'\nu} = \mathrm{i}\frac{\hbar}{2a} \bar{C}_{a\nu} \bar{C}_{a\nu'} \left[\left(\frac{g}{m_A} + \frac{h}{2m_B} \right) \sin\phi_a^+ + \left(\frac{1}{gm_A} + \frac{h}{2m_B} \right) \sin\phi_a^- \right] . \tag{7.60}$$

For $m_A \neq m_B$, the matrix element $\langle \nu' | m^{-1} | \nu \rangle$ is nonzero not only for $\nu = \nu'$ but also for ν and ν' of the same parity as well.

Substituting (7.53, 54) in (7.50), we can find the resonant contribution to the dielectric tensor of the $\nu \rightarrow \nu'$ transitions involving a change of parity

$$\kappa^{(\mathrm{res})}_{\alpha\beta}(\omega) \Big|_{\omega \approx \omega_{\nu'\nu}} = \delta_{\alpha z} \delta_{\beta z} \frac{\Omega_{\nu'\nu}}{\omega_{\nu'\nu} - \omega - \mathrm{i}\Gamma_{\nu'\nu}} , \tag{7.61}$$

where the oscillator strength

$$\Omega_{\nu'\nu} = \frac{4\pi e^2}{\hbar(a+b)} \left(\frac{8}{\pi^2} \frac{\nu\nu'}{\left(\nu'^2 - \nu^2\right)^2} \right)^2 a^2 \left(\rho^s_\nu - \rho^s_{\nu'} \right) , \tag{7.62}$$

and ρ^s_ν is the two-dimensional concentration of electrons in subband ν, defined by

$$\rho^s_\nu = \frac{m_A}{\pi\hbar^2} k_{\mathrm{B}} T \ln\left(1 + \mathrm{e}^{\left(\mu - E^0_\nu\right)/k_{\mathrm{B}}T} \right) , \tag{7.63}$$

where μ is the Fermi energy, T is the temperature, $E^0_\nu \equiv E_{\nu,\mathbf{k}_\perp=0}$. In the derivation of (7.62) from (7.50), ω^2 was replaced by

$$\omega^2_{\nu'\nu} = \left[\frac{\hbar}{2m_A} \left(\frac{\pi}{a} \right)^2 \left(\nu'^2 - \nu^2 \right) \right]^2 .$$

Figure 7.5 illustrates the geometry of experiment on light transmission through a plane sample, with a sequence of quantum wells (shaded region) grown on one of its sides. Since in the $\mathbf{e} \perp z$ polarization intersubband transitions are forbidden, oblique incidence of p-polarized light is used. To exclude the losses associated with reflection from the outer boundaries of the sample, one may conveniently choose the light close to Brewster's angle. In this case we obtain the following expression for the transmission coefficient $T = |t|^2$:

$$-\ln T = \frac{\sin^2\varphi}{\cos\varphi} \frac{\omega}{c\sqrt{\kappa_{\mathrm{b}}}} \mathrm{Im}\left\{ \kappa_{zz} \right\} N(a+b)$$

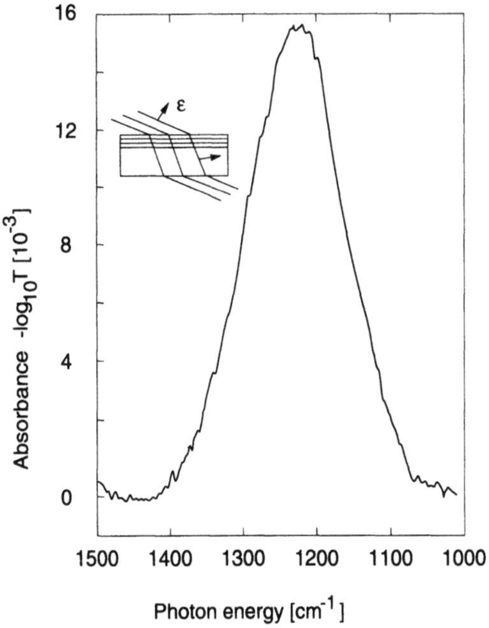

Fig. 7.5. Log transmission coefficient vs. photon energy for light incident at Brewster angle $\varphi_{Br} = 73°$ on a GaAs/AlGaAs MQW. On the top face of the sample are deposited 50 double layers with $a = 71 \pm 2$ Å, $b = 81 \pm 1$ Å, and a cap layer 0.5 μm thick [7.3]

$$= \frac{e^2}{\hbar c} \frac{4\pi N}{\kappa_b \sqrt{\kappa_b + 1}} \left(\frac{8}{\pi^2} \frac{\nu \nu'}{\left(\nu'^2 - \nu^2\right)^2} \right)^2$$

$$\times a^2 \left(\rho_\nu^s - \rho_{\nu'}^s\right) \frac{\omega_{\nu'\nu} \Gamma_{\nu'\nu}}{(\omega_{\nu'\nu} - \omega)^2 + \Gamma_{\nu'\nu}^2}, \tag{7.64}$$

where φ is the refraction angle, κ_b, the background dielectric constant, and N the number of quantum wells. It has been taken into consideration that in the Brewster geometry

$$\frac{\sin^2 \varphi}{\cos \varphi} = \frac{1}{\left[\kappa_b \left(\kappa_b + 1\right)\right]^{1/2}}.$$

For the $1 \to 2$ transitions

$$-\ln T = \frac{512}{27\pi} \frac{e^2}{\hbar c} \frac{N}{\kappa_b \sqrt{\kappa_b + 1}} \frac{\hbar}{m A} \left(\rho_1^s - \rho_2^s\right) \frac{\Gamma_{21}}{(\omega_{21} - \omega)^2 + \Gamma_{21}^2}. \tag{7.65}$$

Taking $\rho_1^s - \rho_2^s = 4 \times 10^{11}$ cm^{-2}, $\kappa_b = 12$, $\hbar\Gamma_{21} = 5$ meV, $m_A = 0.067$ m, $N = 50$, we obtain

$$\left(-\log_{10} T\right)_{\max} = -\log_{10} T_{\min} = 2 \times 10^{-2},$$

which coincides in order of magnitude with the observed value, $-\log_{10} T_{\min} = 1.5 \times 10^{-2}$ (Fig. 7.5). To increase the radiation energy absorbed in the sample, the side faces of the sample can be cleaved at an angle and the light directed through the side faces, thus exciting the waveguide modes (Fig. 7.6).

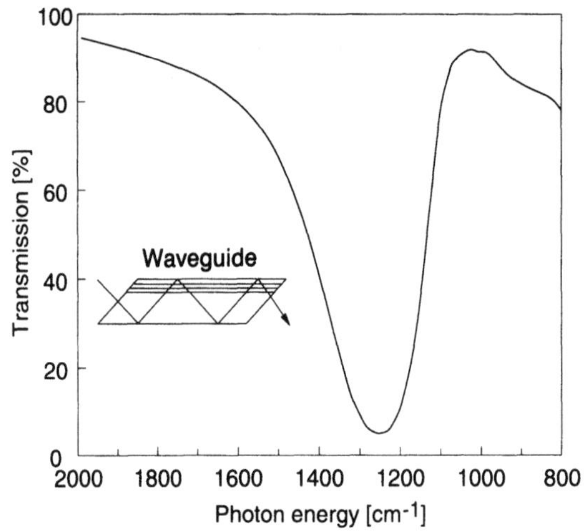

Fig. 7.6. Waveguide transmission vs. photon energy measured in the geometry shown schematically in the inset [7.3]

The quantity $-\log_{10} T$ increases, compared to the geometry shown in Fig. 7.5, both because of the increasing component $\mathcal{I}_z$ of the light wave electric field and due to multiple reflections from the outer boundaries (the transverse dimensions of the sample exceeding by far its thickness).

In contrast to a MQW structure, in a SL the electron energy depends on an additional quantum number, namely, the longitudinal component of the wave vector, k_z. After integration in k_z, the resonant features in (7.50) spread out, with root-type features appearing in the frequency dependence at the points

$$\hbar\omega_{\nu'\nu} = E^0_{\nu'\nu} \equiv E^0_{\nu'} - E^0_\nu \quad \text{and} \quad \hbar\omega'_{\nu'\nu} = E_{\nu'}(\pi/d) - E_\nu(\pi/d)\,.$$

For the contribution of the $1 \to 2$ transitions to the light absorption coefficient, one can write

$$K_{21}(\mathbf{e} \parallel z) = \frac{\omega}{c\sqrt{\kappa_b}} \mathrm{Im}\{\kappa_{zz}\} = \frac{4\pi^2 e^2}{\omega c\sqrt{\kappa_b}} \frac{2}{V} \times \sum_{\mathbf{Q}} \left|v^z_{21}(\mathbf{Q})\right|^2 \delta\left(E_{2\mathbf{Q}} - E_{1\mathbf{Q}} - \hbar\omega\right)\left(F_{1\mathbf{Q}} - F_{2\mathbf{Q}}\right). \tag{7.66}$$

In the particular case of $m_A = m_B$

$$E_{\nu\mathbf{Q}} = \frac{\hbar^2 Q_\perp^2}{2m_A} + E_\nu(Q_z)\,, \tag{7.67}$$

$$E_{2\mathbf{Q}} - E_{1\mathbf{Q}} = E_2(Q_z) - E_1(Q_z) \tag{7.68}$$

and

$$K_{21}(\mathbf{e} \parallel z) = \frac{4\pi e^2}{\hbar\omega c\sqrt{\kappa_b}} \times \left[\frac{1}{|v_2(Q_z) - v_1(Q_z)|} \frac{2}{S}\right.$$

$$\left. \sum_{\mathbf{Q}_\perp} \left| v_{21}^z(\mathbf{Q}) \right|^2 \left(F_{1\mathbf{Q}} - F_{2\mathbf{Q}} \right) \right]_{Q_z = Q_w} . \tag{7.69}$$

Here

$$v_\nu (Q_z) = \hbar^{-1} \partial E_\nu / \partial Q_z$$

and Q_ω satisfies the equation

$$E_2 (Q_\omega) - E_1 (Q_\omega) = \hbar\omega .$$

In the parabolic approximation, and assuming free carries to obey Boltzmann statistics, we come to

$$K_{21} (\mathbf{e} \parallel z) = \frac{4\pi e^2}{\hbar c \sqrt{\kappa_b}} \frac{\rho \left| v_{21}^z \right|^2}{\omega} \left(\frac{\pi \hbar^2 \mu_{21}^{\parallel}}{k_B T \left(E_{21}^0 - \hbar\omega \right) M_\parallel^{(1)}} \right)^{1/2}$$
$$\times \exp \left(- \frac{E_{21}^0 - \hbar\omega}{k_B T} \frac{\mu_{21}^{\parallel}}{M_\parallel^{(1)}} \right) , \tag{7.70}$$

where ρ is the volume concentration of the electrons,

$$\mu_{21}^{\parallel} = \frac{M_\parallel^{(1)} M_\parallel^{(2)}}{M_\parallel^{(1)} - M_\parallel^{(2)}}$$

$M_\parallel^{(1)}$ and $M_\parallel^{(2)}$ are the longitudinal electron effective masses in minibands 1 and 2, respectively (mass $M_\parallel^{(2)}$ is negative). The thermal energy $k_B T$ is assumed to be small compared to the width of the lower miniband. The interminiband matrix element of the velocity operator (for $m_A = m_B$) can be written as

$$v_{21}^z = -v_{12}^z = \frac{i\hbar}{2 m_A d} \left[C_{a1} C_{a2} \left(g \sin \phi_a^+ + g^{-1} \sin \phi_a^- \right) \right.$$
$$\left. + C_{b1} C_{b2} \left(h \sinh \phi_b^+ + h^{-1} \sinh \phi_b^- \right) \right] , \tag{7.71}$$

where the coefficients C_a, C_b are introduced in (5.64, 7.23) also (5.67, 7.21, 22), $\phi_b^\pm = (\lambda_2 \pm \lambda_1) b/2$, g and h are defined in (7.57).
In a wide-barrier SL (7.41)

$$E_2 (Q_z) = E_2^0 - \frac{1}{2} \Delta_2 (1 - \cos Q_z \mathrm{d}) ,$$
$$E_1 (Q_z) = E_1^0 + \frac{1}{2} \Delta_1 (1 - \cos Q_z \mathrm{d}) ,$$
$$|v_2 (Q_z) - v_1 (Q_z)| = (2\hbar)^{-1} (\Delta_1 + \Delta_2) \, \mathrm{d} \sin Q_z \mathrm{d} \tag{7.72}$$

and

$$Q_\omega = \frac{2}{d} \arcsin \left(\frac{E_{21}^0 - \hbar\omega}{\Delta_1 + \Delta_2} \right)^{1/2} ,$$

with the matrix element of the velocity operator for the $1 \to 2$ transitions defined by (7.60). The frequency dependence of the absorption coefficient has in this case the shape of a two-pronged fork

$$K_{21}(\mathbf{e} \parallel z) \propto \frac{1}{\hbar\omega} \frac{\exp\left(-\frac{\Delta_1}{\Delta_1+\Delta_2}\frac{E_{21}^0-\hbar\omega}{k_B T}\right)}{\left\{\left(E_{21}^0-\hbar\omega\right)\left[\hbar\omega-\left(E_{21}^0-\Delta_1-\Delta_2\right)\right]\right\}^{1/2}} \tag{7.73}$$

The relative intensity of the second peak decreases with decreasing temperature. For $m_A \neq m_B$ the transverse effective masses $M_\perp^{(1)}$ and $M_\perp^{(2)}$ are different. This difference may be neglected in (7.66) provided

$$E_{21}^0 - \hbar\omega \gg \frac{M_\perp^{(2)} - M_\perp^{(1)}}{M_\perp^{(1)}} k_B T .$$

In the opposite limiting case the root-type singularity $(E_{21}^0 - \hbar\omega)^{-1/2}$ is replaced by $(E_{21}^0 - \hbar\omega)^{1/2}$.

Intersubband transitions in the $\mathbf{e} \perp z$ polarization become allowed if the effective mass in the conduction band of a composite semiconductor is anisotropic and none of the principal axes of the reciprocal effective-mass tensor $m_{\alpha\beta}^{-1}$ coincides with the normal to the layer plane. Such a situation arises when layers of a many-valley semiconductor are grown in the z direction which does not coincide with the principal axis of the valley. This can be achieved, for instance, in a silicon MOS structure or a MQW structure Ge/Si_xGe_{1-x}.
If the electron energy in the principal axes 1, 2, 3 of the tensor $m_{\alpha\beta}^{-1}$

$$E_{\mathbf{k}} = \frac{\hbar^2}{2}\left(\frac{k_1^2+k_2^2}{m_t} + \frac{k_3^2}{m_l}\right), \tag{7.74}$$

then, in the coordinate frame x, y, z with $y \parallel 2$, it will be

$$E_{\mathbf{k}} = \frac{\hbar^2}{2}\left(\frac{k_x^2}{m_{xx}} + \frac{k_y^2}{m_{yy}} + \frac{k_z^2}{m_{zz}} + 2\frac{k_x k_z}{m_{xz}}\right), \tag{7.75}$$

where

$$\frac{1}{m_{xx}} = \frac{\cos^2\theta}{m_t} + \frac{\sin^2\theta}{m_l}, \quad \frac{1}{m_{zz}} = \frac{\sin^2\theta}{m_t} + \frac{\cos^2\theta}{m_l},$$
$$m_{yy} = m_t, \quad \frac{1}{m_{xz}} = \cos\theta\sin\theta\left(\frac{1}{m_t} - \frac{1}{m_l}\right), \tag{7.75a}$$

θ is the angle between the axes 3 and z. It is the presence of a mixed term $k_x k_z$ in (7.75) that produces the nonzero intersubband matrix element of the velocity operator

$$v_{\nu'\nu}^x = -\mathrm{i}\frac{\hbar}{m_{xz}}\left\langle \nu' \left| \frac{\partial}{\partial z} \right| \nu \right\rangle . \tag{7.76}$$

For the $1 \to 2$ transitions in a high-barrier quantum well, we have

$$v_{21}^x = \frac{\hbar k_z^{(2,1)}}{m_{xz}} = \mathrm{i}\frac{8\hbar}{3a}\cos\theta\sin\theta\left(\frac{1}{m_\mathrm{t}} - \frac{1}{m_\mathrm{l}}\right). \tag{7.77}$$

For the two lowest subbands in a MOS structure, one sometimes chooses trial functions in the form

$$\begin{aligned} f_1(z) &= C_1 z \mathrm{e}^{-\alpha_1 z}, \\ f_2(z) &= C_2 z\left(1 - \frac{\alpha_1 + \alpha_2}{3} z\right)\mathrm{e}^{-\alpha_2 z} \end{aligned} \tag{7.78}$$

with the normalizing factors

$$C_1 = 2\alpha_1^{3/2}, \quad C_2 = 2\left[\frac{3\alpha_2^5}{\left(\alpha_1^2 - \alpha_1\alpha_2 + \alpha_2^2\right)}\right]^{1/2}. \tag{7.78a}$$

In this case,

$$v_{21}^x = -2\mathrm{i}\frac{\hbar\alpha_2}{m_{xz}}\left(\frac{4t^3}{3\left(1+t^3\right)(1+t)^3}\right)^{1/2},$$

where $t = \alpha_2/\alpha_1$.

In a bulk semiconductor with a degenerate band Γ_8 (or Γ_8^+), direct optical transitions between the heavy and light-hole subbands are possible under arbitrary light polarization. In the isotropic approximation, $D = \sqrt{3}B$, for the coefficient of such an absorption we obtain

$$K = \frac{e^2 k}{\hbar c n_\omega}\left[F\left(E_\mathrm{hh}^*\right) - F\left(E_\mathrm{lh}^*\right)\right], \tag{7.79}$$

where n_ω is the refractive index at the frequency ω, E_hh^* and E_lh^* are the initial and final hole energies:

$$E_\mathrm{hh}^* = \frac{m_\mathrm{lh}}{m_\mathrm{hh} - m_\mathrm{lh}}\hbar\omega, \; E_\mathrm{lh}^* = \frac{m_\mathrm{hh}}{m_\mathrm{hh} - m_\mathrm{lh}}\hbar\omega = E_\mathrm{hh}^* + \hbar\omega,$$

and the wave vector of the holes involved in the transition is given by

$$\hbar k = \left(2m_\mathrm{hh}E_\mathrm{hh}^*\right)^{1/2} = \left(2m_\mathrm{lh}E_\mathrm{lh}^*\right)^{1/2} = \left(\frac{2m_\mathrm{lh}m_\mathrm{hh}}{m_\mathrm{hh} - m_\mathrm{lh}}\hbar\omega\right)^{1/2}.$$

In a MQW structure, transitions in the $\mathbf{e} \perp z$ polarization for $\mathbf{k}_\perp = 0$ are allowed between the νth heavy-hole subband and ν'th light-hole subband if the quantum numbers ν and ν' are of opposite parity, while for $\mathbf{k}_\perp \neq 0$ the inclusion of hybridization of the heavy and light-hole states makes possible transitions in the $\mathbf{e} \perp z$ polarization between subbands of the same series as well. In the effective-mass approximation, we can write the velocity operator in the form

$$v_\alpha = \hbar^{-1}\nabla_{k_\alpha}\mathcal{H}_{\Gamma_8}(\mathbf{k}) = Q_{\alpha\beta}k_\beta,$$

$$Q_{\alpha\beta} = \frac{2}{\hbar}\left[\left(A - \frac{5}{4}B + BJ_\alpha^2\right)\delta_{\alpha\beta} + \left(1 - \delta_{\alpha\beta}\right)\frac{D}{\sqrt{3}}\left[J_\alpha J_\beta\right]_s\right]. \tag{7.80}$$

Therefore, for the intersubband transitions $\nu j \to \nu' j'$ in a structure with $z \parallel [001]$, we obtain in the approximation of infinitely high barriers

$$\begin{aligned} \mathbf{e} \cdot \mathbf{v}_{\nu' j', \nu j}\,(\mathbf{k}_\perp = 0) &= \frac{2}{\hbar} k_z^{(\nu',\nu)} \\ &\times \left\{ (A \pm B) e_z \delta_{j'j} + \frac{D}{\sqrt{3}} \left[J_z, \mathbf{J}_\perp \cdot \mathbf{e}_\perp\right]_{j'j} \right\}. \end{aligned} \tag{7.81}$$

The effective Hamiltonian (3.44, 46) contains the matrices $[J_z J_x]_s$, $[J_z J_y]_s$ in combination with the factors $k_z k_x$, $k_z k_y$. Therefore, by Table 3.2, the matrix $[J_z, \mathbf{J}_\perp \cdot \mathbf{e}_\perp]_s$ does indeed have nonzero elements with $j = 3/2$, $j' = 1/2$, or $j = -3/2$, $j' = -1/2$.

7.3 Electron-Spin Resonance

The ESR frequency is determined by the magnitude of the electron g-factor. In the effective-mass method, the electron g-factor in a SL is obtained in an averaging procedure

$$g = g_A \langle \theta_A \rangle + g_B \langle \theta_B \rangle = g_A + (g_B - g_A)\,\langle \theta_B \rangle, \tag{7.82}$$

where $g_{A,B}$ is the g-factor in layer A or B, the angular brackets denoting the integral

$$\langle \theta_{A,B} \rangle = \int U_0^2(z)\theta_{A,B}(z)\,\mathrm{d}z. \tag{7.83}$$

Comparing (7.82) with (7.26), we come to the following relation between g and the mass $M_\perp$:

$$g = g_A + (g_B - g_A)\,\frac{M_\perp^{-1} - m_A^{-1}}{m_B^{-1} - m_A^{-1}}. \tag{7.84}$$

In semiconductors of T_d symmetry, one may conveniently represent the electron g-factor for the lowest conduction band c, Γ_6 as a sum

$$g = g_0 - \frac{4}{3}\sum_l \frac{p_{cl}^2}{m}\frac{\Delta_l}{\left(E_c^0 - E_{l\Gamma_8}^0\right)\left(E_c^0 - E_{l\Gamma_7}^0\right)}. \tag{7.85}$$

Here g_0 is the Landé factor for the free electron ($g_0 \approx 2$), and the summation is performed over all bands of symmetry Γ_5 split by spin-orbit coupling into Γ_8 and Γ_7 states, $\Delta_l = E_{l\Gamma_8}^0 - E_{l\Gamma_7}^0$, E_n^0 is the electron energy in band n for $k = 0$, the spin-orbit mixing of states in different bands is disregarded, the matrix element

$$p_{cl} = \mathrm{i}\langle S\,|p_x|\,X_l\rangle, \tag{7.86}$$

$X_l(\mathbf{x})$, $Y_l(\mathbf{x})$, $Z_l(\mathbf{x})$ are the Bloch functions in the band l for $k = 0$, which transform according to the representation Γ_5 as the coordinates x, y, z. The main contribution to the sum over l comes from the upper valence band Γ_5. The g-factor for the A_3B_5 and A_2B_6 semiconductors can be estimated by the expression (3.67)

$$g = 2\left(1 - \frac{2}{3}\frac{p_{\mathrm{cv}}^2}{m}\frac{\Delta}{E_{\mathrm{g}}\left(E_{\mathrm{g}} + \Delta\right)}\right), \tag{7.87}$$

where $p_{\mathrm{cv}} = \mathrm{i}\langle S\,|p_z|\,Z\rangle$ is the interband matrix element of the momentum operator. Using for GaAs the values $2p_{\mathrm{cv}}^2/m = 28.8$ eV, $E_{\mathrm{g}} = 1.52$ eV, $\Delta = 0.34$ eV [7.4], we obtain $g = -0.31$. Experiments yield $g_{\mathrm{exp}}(\mathrm{GaAs}) = -0.44$. We see that in GaAs both g_0 and the contribution of the valence band v, Γ_5 cancel to a considerable extent. The difference between the experimental value of the g-factor and the estimate made by (7.87) determines the contribution to (7.85) due to the higher bands with $l \neq v, \Gamma_5$.

In the solid solution $Al_xGa_{1-x}As$ the energy gap increases and the contribution of the valence band in (7.87) decreases with increasing x. Therefore, the electron g-factor vanishes for a certain composition $x_0 \simeq 0.12$ and becomes positive for $x > x_0$. In particular, $g(Al_{0.35}Ga_{0.65}As) \simeq 0.5$ [7.4]. Thus, in a $GaAs/Al_xGa_{1-x}As$ SL with $x \sim 0.35$, the g-factors g_A and g_B differ in sign, are close in absolute value, and their contributions to the resultant g-factor (7.82) nearly cancel out. In this case, the various corrections disregarded in (7.82) may play a noticeable role.

As the first correction, we will take into consideration the quadratic-in-k contribution to the electron g-factor in a bulk semiconductor. For crystals of symmetry T_d, the spin Hamiltonian with the quadratic terms included can be written as

$$\mathcal{H}_{\mathrm{sp}}(\mathbf{B}) = \frac{1}{2}\mu_{\mathrm{B}}\Big[\left(g + h_1k^2\right)\sigma\cdot\mathbf{B} + h_2(\mathbf{B}\cdot\mathbf{k})(\sigma\cdot\mathbf{k}) + h_3\sum_\alpha k_\alpha^2\sigma_\alpha B_\alpha\Big]. \tag{7.88}$$

For the coefficients h_i, one can derive expressions of type (7.85) using perturbation theory. One can obtain an expression for the effective electron Hamiltonian for the band c, Γ_6 in a magnetic field to within terms of fourth order in **k**-**p** interaction between the c, Γ_6 and v, $\Gamma_{7,8}$ bands, and of second order in interaction with the higher bands l:

$$\mathcal{H}_{ss'}(\mathbf{K}) = \left(E_{\mathrm{c}}^0 + \frac{\hbar^2K^2}{2m}\right)\delta_{ss'} + \sum_m \frac{\mathcal{H}_{sm}\mathcal{H}_{ms'}}{E_{\mathrm{c}}^0 - E_m^0} + \mathcal{H}'_{ss'}\left(1 - \sum_m \frac{|\mathcal{H}_{sm}|^2}{\left(E_{\mathrm{c}}^0 - E_m^0\right)^2}\right)$$

$$+\sum_{mm'}\frac{\mathcal{H}_{sm}\mathcal{H}'_{mm'}\mathcal{H}_{m's'}}{\left(E_{c}^{0}-E_{m}^{0}\right)\left(E_{c}^{0}-E_{m'}^{0}\right)}. \tag{7.89}$$

Here

$$\mathcal{H}_{sm}=\frac{\hbar}{m}\mathbf{K}\cdot\mathbf{p}_{sm}+\frac{1}{2}\left(\frac{\hbar}{m}\right)^{2}\sum_{l}(\mathbf{K}\cdot\mathbf{p}_{sl})(\mathbf{K}\cdot\mathbf{p}_{lm})$$
$$\times\left(\frac{1}{E_{c}^{0}-E_{l}^{0}}+\frac{1}{E_{m}^{0}-E_{l}^{0}}\right),\quad \mathbf{K}=-\mathrm{i}\frac{\partial}{\partial\mathbf{x}}-\frac{e}{c\hbar}\mathbf{A}(\mathbf{x}), \tag{7.90}$$

$\mathbf{A}(\mathbf{x})$ is the vector potential, the index m specifies the states in the upper valence bands Γ_8 and Γ_7, and

$$\mathcal{H}'_{ss'}=\sum_{l\neq v\Gamma_8,\Gamma_7}\frac{\mathcal{H}_{sl}\mathcal{H}_{ls'}}{E_{c}^{0}-E_{l}^{0}}+\frac{1}{2}g_{0}\mu_{\mathrm{B}}\sigma_{ss'}\cdot\mathbf{B}, \tag{7.91}$$

$$\mathcal{H}'_{mm'}=\frac{1}{2}\sum_{l}\mathcal{H}_{ml}\mathcal{H}_{lm'}\left(\frac{1}{E_{m}^{0}-E_{l}^{0}}+\frac{1}{E_{m'}^{0}-E_{l}^{0}}\right)$$
$$+\frac{1}{2}g_{0}\mu_{\mathrm{B}}\sigma_{mm'}\mathbf{B},$$
$$\sigma_{mm'}^{\alpha}=\int u_{m}^{(0)+}(\mathbf{x})\hat{\sigma}_{\alpha}u_{m'}^{(0)}(\mathbf{x})\,\mathrm{d}\mathbf{x}. \tag{7.92}$$

Neglecting the contribution of the upper bands, the coefficient h_3 of the anisotropic term in (7.88) is zero, and

$$h_{1}=h_{2}+\frac{4}{9}\frac{\hbar^{2}p_{cv}^{4}}{m^{3}}\frac{\Delta}{E_{g}\left(E_{g}+\Delta\right)}\left(\frac{4}{E_{g}^{2}}+\frac{2}{\left(E_{g}+\Delta\right)^{2}}+\frac{3}{E_{g}\left(E_{g}+\Delta\right)}\right),$$
$$h_{2}=-\frac{2}{9}g_{0}\frac{\hbar^{2}p_{cv}^{2}}{m^{2}}\frac{\Delta^{2}}{E_{g}^{2}\left(E_{g}+\Delta\right)^{2}}. \tag{7.93}$$

Since $p_{\mathrm{cv}}^2/m \gg E_{\mathrm{g}}, |h_2| \ll h_1$, and the anisotropy of the g-factor may be neglected.

One of the corrections to the g-factor that is linear in h_1 is obtained, similar to (7.82), by averaging the operator (7.88)

$$\Delta g=h_{1}^{A}k^{2}\langle\theta_{A}\rangle-h_{1}^{B}\lambda^{2}\langle\theta_{B}\rangle, \tag{7.94}$$

where h_1^A, h_1^B is the coefficient h_1 in a layer of type A or B. The second correction which is linear in h_1 is associated with the magnetic field-induced change of the effective mass of the spin-polarized electron by

$$\Delta m_{A,B}=-\frac{2m_{A,B}}{\hbar^{2}}(\mathbf{s}\cdot\mathbf{B})h_{1}\mu_{B}, \tag{7.95}$$

where $\mathbf{s}$ is the electron spin. If this change of the masses m_A and m_B is taken into account in the boundary condition (7.49) and, hence, in the equation

$$\frac{k}{m_A}\tan\frac{ka}{2} = \frac{\lambda}{m_B}\tanh\frac{\lambda b}{2}$$

for the energy E_1^0, we obtain instead of (7.94) the following contribution to the g-factor in the lowest miniband

$$\Delta g = \frac{h_1^A k^2 a v_- - h_1^B \lambda^2 b w_-}{a v_+ + b w_+}, \qquad (7.96)$$

where

$$v_\pm = \frac{1 \pm \dfrac{\sin ka}{ka}}{1+\cos ka}, \quad w_\pm = \frac{1 \pm \dfrac{\sinh \lambda b}{\lambda b}}{1+\cosh \lambda b}.$$

In the limit $a, b \to 0$, the contribution (7.96) vanishes, whereas the contribution (7.94) tends to

$$\frac{2ab}{(a+b)^2}\frac{1}{\hbar^2}V_0\left(h_1^A m_A - h_1^B m_B\right).$$

Taking into account the contribution to (7.88) proportional to h_1 makes the electron g-factor in heterostructures dependent on the angle between the magnetic field $\mathbf{B}$ and the axis z. We will show this to be true for a SQW structure assuming for the sake of simplicity that $h_1^A = h_1^B \equiv h_1$. Neglecting the magnetic field-induced mixing of states in different minibands, we obtain from (7.88):

$$g_N(\mathbf{B}) = g + h_1\left[2\frac{eB_\parallel}{\hbar c}\left(N+\frac{1}{2}\right) + \left(\frac{eB_\perp}{c\hbar}\right)^2 \langle z^2\rangle\right]$$

or

$$g_N(\mathbf{B}) = g + h_1\left[2\frac{eB}{\hbar c}\left(N+\frac{1}{2}\right)|\cos\theta| + \left(\frac{eB}{\hbar c}\right)^2 \langle z^2\rangle \sin^2\theta\right], \qquad (7.97)$$

where N is the Landau level number. This simple expression was used to explain the angular dependence of the electron g-factor in the GaAs/Al_xGa_{1-x}As heterostructure for the levels $N = 0, 1 \ldots 4$ [7.5]. The ESR was detected by measuring the magnetic-field dependence of the Hall signal at a fixed ac-field frequency.

Inclusion in (7.88) of terms proportional to h_2 and h_3 results in an anisotropy of the electron g-factor. If the principal axis of the structure is oriented along the [001] or [111] axis, then the spin Hamiltonian of the electron in a superlattice or a MQW structure is characterized by two constants

$$\mathcal{H}_{sp}(\mathbf{B}) = \frac{1}{2}\mu_B\left(g_\parallel \sigma_z B_z + g_\perp \boldsymbol{\sigma}_\perp \cdot \mathbf{B}_\perp\right). \qquad (7.98)$$

The ESR frequency in this case will be

$$\Omega_L = \left(g_\parallel^2 \cos^2\theta + g_\perp^2 \sin^2\theta\right)^{1/2} \mu_B B/\hbar. \tag{7.99}$$

The splitting of the heavy and light-hole states in the heterostructure valence band which is disregarded in the calculation of the corrections (7.94, 96) may contribute considerably to the electron g-factor anisotropy in the conduction band. Using (7.89) of the **k·p** method of perturbation theory for quasitwo-dimensional electrons in a SQW structure, we obtain in the second order of perturbation theory:

$$\begin{aligned} g_\perp - g_\parallel &= 2\frac{p_{cv}^2}{m}\sum_\nu \left(\frac{\Lambda_{c1,hh\nu}}{E_g + E_{c1}^0 + E_{hh\nu}^0} - \frac{\Lambda_{c1,lh\nu}}{E_g + E_{c1}^0 + E_{lh\nu}^0}\right) \\ &\approx 2\frac{p_{cv}^2}{m\left(E_g + E_{c1}^0\right)^2}\sum_\nu \left(E_{lh\nu}^0 \Lambda_{c1,lh\nu} - E_{hh\nu}^0 \Lambda_{c1,hh\nu}\right). \end{aligned} \tag{7.100}$$

The sum over ν can be readily evaluated by assuming the masses m_c, m_{hh} and m_{lh} to be independent of z. In this case, we will have

$$\begin{aligned} &\sum_\nu \left(E_{lh\nu}^0 \Lambda_{c1,lh\nu} - E_{hh\nu}^0 \Lambda_{c1,hh\nu}\right) \\ &= -\frac{\hbar^2}{2}\left(\frac{1}{m_{lh}} - \frac{1}{m_{hh}}\right)\int f_{e1}(z)\frac{d^2}{dz^2} f_{e1}(z) dz \\ &= m_c\left(\frac{1}{m_{lh}} - \frac{1}{m_{hh}}\right)\left(E_{e1}^0 - V_0\frac{dE_{e1}^0}{dV_0}\right), \end{aligned}$$

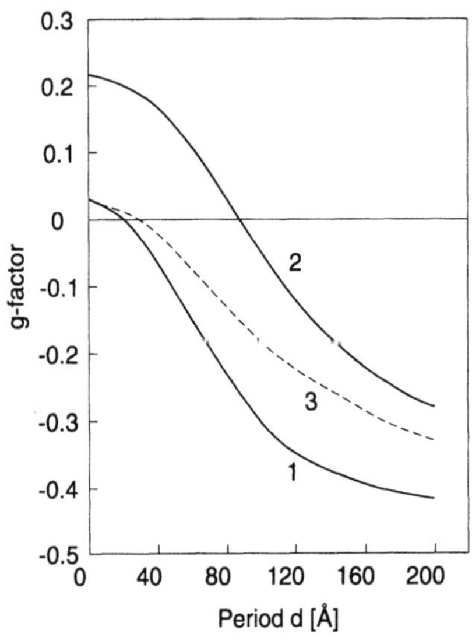

Fig. 7.7. Dependence of electron g factor in $GaAs/Al_{0.35}Ga_{0.65}As$ SL with $a = b$ on period $d = a + b$ without inclusion (curve 1) and with inclusion of the correction (7.94) (curve 2) or (7.96) (curve 3). The parameters used in the calculation: $g_A = -0.44$, $g_B = 0.5$, $h_1^A = 2.9 \times 10^{-14}\ \mathrm{cm}^2$, $h_1^B = 0.8 \times 10^{-14}\ \mathrm{cm}^2$ [7.6]

where $\mathrm{d}E_{e1}^0/\mathrm{d}V_0$ is obtained from (7.28) in the limit $b \to \infty$. For the boundary conditions (7.49)

$$\frac{\mathrm{d}E_{e1}^0}{\mathrm{d}V_0} = \left(1 + \frac{1 + \dfrac{\sin ka}{ka}}{1 + \cos ka} \right)^{-1} .$$

In a similar way, one can evaluate the electron g-factor anisotropy in a SL.

Figure 7.7 presents the dependence of g on the period of the GaAs/$Al_{0.35}Ga_{0.65}As$ SL with $a = b$, calculated by (7.84), as well as with inclusion of the corrections (7.94, 96). It is the correction (7.94) that is consistent with Kane's model for the boundary conditions (3.71) with $\alpha = 0$ [7.6].

7.4 IR Reflection in an Undoped Superlattice

For a crystal of cubic symmetry with two different atoms in the unit cell, the dielectric constant in the IR region can be represented in the form (4.65)

$$\kappa(\omega) = \kappa_\infty \frac{\omega_\mathrm{L}^2 - \omega^2 - 2\mathrm{i}\Gamma_\mathrm{L}\omega}{\omega_\mathrm{T}^2 - \omega^2 - 2\mathrm{i}\Gamma_\mathrm{T}\omega}, \tag{7.101}$$

where ω_T and ω_L are the resonance frequencies of the TO and LO phonon, respectively, κ_∞ is the hf dielectric constant, and $\Gamma_{\mathrm{L,T}}$ the phonon damping.

Consider a SL made up of alternating layers of type A and B with dielectric constants $\kappa^a(\omega)$ and $\kappa^b(\omega)$, described by expression (7.101) with a given set of parameters $\kappa_\infty^a, \omega_{\mathrm{T1}}, \omega_{\mathrm{L1}}$ for layer A, and $\kappa_\infty^b, \omega_{\mathrm{T2}}, \omega_{\mathrm{L2}}$ for layer B. For simplicity, in the main part of this section, we neglect the phonon damping. If the light wavelength is large compared to the period $d = a + b$, then, according to (6.9), the SL may be treated as a homogeneous anisotropic medium with the components of the dielectric constant tensor

$$\kappa_\perp(\omega) = \frac{a\kappa^a(\omega) + b\kappa^b(\omega)}{a+b}, \quad \kappa_\parallel(\omega) = \frac{(a+b)\kappa^a(\omega)\kappa^b(\omega)}{a\kappa^b(\omega) + b\kappa^a(\omega)}. \tag{7.102}$$

The function $\kappa_\perp(\omega)$ has poles at the points $\omega = \omega_{\mathrm{T1}}, \omega_{\mathrm{T2}}$ and the function $\kappa_\parallel(\omega)$ has zeroes at $\omega_{\mathrm{L1}}, \omega_{\mathrm{L2}}$. Denoting the zeroes of the function

$$a\kappa^a(\omega) + b\kappa^b(\omega)$$

by $\omega_{\perp,1}, \omega_{\perp,2}$ and those of the function

$$a\kappa^b(\omega) + b\kappa^a(\omega)$$

by $\omega_{\parallel,1}, \omega_{\parallel,2}$ we can represent the frequency dependence of $\kappa_\perp$ and $\kappa_\parallel$ in a factorized form

$$\kappa_\perp(\omega) = \kappa_\infty^\perp \frac{\left(\omega_{\perp,1}^2 - \omega^2\right)\left(\omega_{\perp,2}^2 - \omega^2\right)}{\left(\omega_{\mathrm{T1}}^2 - \omega^2\right)\left(\omega_{\mathrm{T2}}^2 - \omega^2\right)},$$

$$\kappa_{\parallel}(\omega) = \kappa_{\infty}^{\parallel} \frac{\left(\omega_{\mathrm{L1}}^2 - \omega^2\right)\left(\omega_{\mathrm{L2}}^2 - \omega^2\right)}{\left(\omega_{\parallel,1}^2 - \omega^2\right)\left(\omega_{\parallel,2}^2 - \omega^2\right)}, \tag{7.103}$$

where

$$\kappa_{\infty}^{\perp} = \frac{a\kappa_{\infty}^{a} + b\kappa_{\infty}^{b}}{a+b}, \quad \kappa_{\infty}^{\parallel} = \frac{(a+b)\kappa_{\infty}^{a}\kappa_{\infty}^{b}}{a\kappa_{\infty}^{b} + b\kappa_{\infty}^{a}}. \tag{7.104}$$

Note that in a SL with layers of equal thickness ($a = b$), the frequencies $\omega_{\perp,1}$ and $\omega_{\parallel,1}$ coincide.

For $\omega_{\mathrm{L1}} < \omega_{\mathrm{T2}}$, the frequencies $\omega_{\perp,1}$ and $\omega_{\parallel,1}$ satisfy the inequalities

$$\omega_{\mathrm{T1}} < \omega_{\perp,1}, \quad \omega_{\parallel,1} < \omega_{\mathrm{L1}}, \quad \omega_{\mathrm{T2}} < \omega_{\perp,2}, \quad \omega_{\parallel,2} < \omega_{\mathrm{L2}}. \tag{7.105}$$

If the longitudinal-transverse splittings $\omega_{\mathrm{LT}}^{(1)} = \omega_{\mathrm{L1}} - \omega_{\mathrm{T1}}$ are small compared to the separation between the resonances ω_{T1} and ω_{T2}, then the frequencies $\omega_{\perp,1}$ and $\omega_{\parallel,1}$ are determined by the relations

$$\omega_{\perp,1} - \omega_{\mathrm{T1}} = \frac{a}{a+b}\frac{\kappa_{\infty}^{a}}{\kappa_{\infty}^{\perp}}\omega_{\mathrm{LT}}^{(1)}, \quad \omega_{\parallel,1} - \omega_{\mathrm{T1}} = \frac{b}{a+b}\frac{\kappa_{\infty}^{\parallel}}{\kappa_{\infty}^{b}}\omega_{\mathrm{LT}}^{(1)}. \tag{7.106}$$

The expressions for $\omega_{\perp,2}$ and $\omega_{\parallel,2}$ can be obtained from (7.106) through the replacements $a \leftrightarrow b$, $1 \leftrightarrow 2$.

Consider the reflection of light at the boundary between vacuum and semi-infinite SL. Under normal incidence, or oblique incidence of s-polarized light, a transverse wave is excited in the SL with a dispersion described by the equation

$$(cq/\omega)^2 = \kappa_{\perp}(\omega). \tag{7.107}$$

The anisotropy of the dielectric constant of the SL manifests itself in the spectrum of obliquely reflected p-polarized light, when mixed, or TM waves are excited in the medium with the dispersion relation

$$\kappa_{\parallel} q_z^2 + \kappa_{\perp} q_{\perp}^2 = (\omega/c)^2 \kappa_{\parallel}\kappa_{\perp}. \tag{7.108}$$

For the reflection coefficients, we obtain

$$R = |r|^2, \quad r_{s,p} = \frac{1 - \bar{n}_{s,p}}{1 + \bar{n}_{s,p}},$$
$$\bar{n}_s = \frac{\left(\kappa_{\perp} - \sin^2\varphi_0\right)^{1/2}}{\cos\varphi_0}, \quad \bar{n}_p = \frac{1}{\cos\varphi_0}\frac{\left(\kappa_{\parallel} - \sin^2\varphi_0\right)^{1/2}}{\kappa_{\parallel}\kappa_{\perp}}. \tag{7.109}$$

At $\kappa_{\parallel} = \kappa_{\perp}$ these expressions transform to (6.82, 83).

In the case where a SL is grown on a substrate, one has to take into account the reflection of light at the SL-substrate boundary and interference of the light reflected from the outer and inner SL boundaries. If the substrate is thick enough that reflection from its back face may be disregarded, we obtain for the reflection coefficients r_s and r_p [7.7]

$$r_\mu = \frac{\left(1-\bar{n}_{2\mu}\right)\cos\Phi_\mu - \mathrm{i}\left(\dfrac{\bar{n}_{2\mu}}{\bar{n}_{1\mu}} - \bar{n}_{1\mu}\right)\sin\Phi_\mu}{\left(1+\bar{n}_{2\mu}\right)\cos\Phi_\mu - \mathrm{i}\left(\dfrac{\bar{n}_{2\mu}}{\bar{n}_{1\mu}} - \bar{n}_{1\mu}\right)\sin\Phi_\mu}. \tag{7.110}$$

Here $\bar{n}_{1s}$ and $\bar{n}_{1p}$ coincide with $\bar{n}_s$ and $\bar{n}_p$ in (7.109),

$$\Phi_s = \frac{\omega}{c}L\left(\kappa_\perp - \sin^2\varphi_0\right)^{1/2}, \quad \Phi_p = \frac{\omega}{c}L\left[\left(\kappa_\parallel - \sin^2\varphi_0\right)\frac{\kappa_\perp}{\kappa_\parallel}\right]^{1/2},$$

L is the SL thickness and κ_2 the substrate's dielectric constant. Reflectance spectra of the GaAs/AlAs SL grown on a GaAs substrate were measured and analyzed by *Lou* et al. [7.7].

In conclusion, we note that the solid-solution layer in a GaAs/$Al_xGa_{1-x}As$ SL with $x \neq 0, 1$ contains two optical modes associated with the vibrations of the Ga-As and Al-As atom pairs, and

$$\kappa^{\mathrm{b}}(\omega) = \kappa_\infty^{\mathrm{b}} \frac{\left(\omega_{\mathrm{L1}}^2 - \omega^2 - 2\mathrm{i}\Gamma_{\mathrm{L1}}\omega\right)\left(\omega_{\mathrm{L2}}^2 - \omega^2 - 2\mathrm{i}\Gamma_{\mathrm{L2}}\omega\right)}{\left(\omega_{\mathrm{T1}}^2 - \omega^2 - 2\mathrm{i}\Gamma_{\mathrm{T1}}\omega\right)\left(\omega_{\mathrm{T2}}^2 - \omega^2 - 2\mathrm{i}\Gamma_{\mathrm{T2}}\omega\right)}. \tag{7.111}$$

Here

$$\kappa_\infty^{\mathrm{b}} = (1-x)\kappa_\infty(\mathrm{GaAs}) + x\kappa_\infty(\mathrm{AlAs})$$

and ω_{T1} and ω_{T2} are close to the transverse optical vibrational frequencies 268 and 362 cm^{-1} in homogeneous GaAs and AlAs. For illustration, we present below the corresponding parameters for the $Al_{0.3}Ga_{0.7}As$ solid solution derived from a comparison of experimental and theoretical normal-reflectance spectra [7.8]:

$$\begin{aligned}
&\omega_{\mathrm{T1}} = 265.2\ \mathrm{cm}^{-1}, \quad \omega_{\mathrm{L1}} = 278.3\ \mathrm{cm}^{-1}\\
&\omega_{\mathrm{T2}} = 360.2\ \mathrm{cm}^{-1}, \quad \omega_{\mathrm{L2}} = 379.1\ \mathrm{cm}^{-1}\\
&\Gamma_{\mathrm{T1}} = 4.32\ \mathrm{cm}^{-1}, \quad \Gamma_{\mathrm{L1}} = 3.07\ \mathrm{cm}^{-1}\\
&\Gamma_{\mathrm{T2}} = 6.05\ \mathrm{cm}^{-1}, \quad \Gamma_{\mathrm{L2}} = 4.71\ \mathrm{cm}^{-1}\\
&\kappa_\infty^{\mathrm{b}} = 10.16.
\end{aligned}$$

8 Light Scattering

Under scattering of light one understands the appearance, in the medium illuminated by an external source, of new electromagnetic waves with frequencies and directions different from those of the initial wave. Light scattering and photoluminescence are aspects of a general phenomenon of secondary emission in matter. Moreover, under certain conditions (e.g., resonant fluorescence of atoms or exciton emission in a semiconductor under resonant pumping) to distinguish between these two phenomena becomes meaningless. In most cases, however, such a distinction can be justified; indeed, in light scattering defined in the traditional way the excited states of a system are virtual, whereas in photoluminescence the emission of the secondary photon is usually preceded by multiple transitions of the system between different real excited states.

The difference between the frequencies (or wave vectors) of the secondary and primary photons is directly determined by the frequency (or wave vector) of the particle or excitation participating in the scattering event. Therefore, scattering spectroscopy is used widely in studies of the energy spectra of quasiparticles in solids. After a general introductory section, we are going to consider successively light scattering from intersubband excitations in quantum-well structures and from acoustic and optical phonons, as well as that involving spin flip of carriers bound to charged impurity centers.

8.1 Theory of Light Scattering in Semiconductors

In the present section we are going to describe briefly the phenomenon of light scattering in a bulk semiconductor. We will use the notations ω_1, $\mathbf{q}_1$, $\mathbf{e}_1$ and ω_2, $\mathbf{q}_2$ and $\mathbf{e}_2$ for the frequency, wave vector and polarization vector of the initial and scattered electromagnetic waves, respectively. One can also introduce the transferred wave vector and frequency

$$\mathbf{q} = \mathbf{q}_1 - \mathbf{q}_2, \quad \omega = \omega_1 - \omega_2. \tag{8.1}$$

In the presence of absorption in the semiconductor at the frequency ω_1 or ω_2, the expression for $\mathbf{q}$ will contain the real parts of the wave vectors.

8.1.1 Scattering by Free Carriers

We introduce the differential light-scattering cross section defined as

$$\frac{d^2\sigma}{dw\, d\Omega} = \frac{\hbar\omega_2}{I_1 V} \frac{\Delta w}{\Delta\omega_2 \Delta\Omega_2}, \tag{8.2}$$

where V is the emitting volume, the energy-flux density of the primary radiation I_1 being related with the mean number of photons $\bar{N}_{\mathbf{q}_1}$ through

$$I_1 = \frac{c\hbar\omega}{V\sqrt{\kappa(\omega_1)}}\bar{N}_{\mathbf{q}1}, \tag{8.3}$$

Δw is the scattering rate for conduction electrons in the frequency region $\Delta\omega_2$ and within a solid angle $\Delta\Omega_2$. Note that the quantity $\mathrm{d}^2\sigma/\mathrm{d}\omega\,\mathrm{d}\Omega$ is defined in (8.2) as the specific differential cross section, i.e., the scattering cross section per unit volume and has the dimension $\mathrm{cm}^{-1}\mathrm{s}$ rather than $\mathrm{cm}^2\mathrm{s}$. With such a definition, the spectral intensity of the secondary radiation $J(\omega_2)$ propagating in vacuum in a unit solid angle is connected with the intensity I_1^0 incident on a semi-infinite crystal through the relation

$$\begin{aligned} J(\omega_2) &= (1-R)^2\frac{1}{\kappa(\omega_2)}\int_0^\infty I_1^0 \mathrm{e}^{-(K_1+K_2)z}\frac{\mathrm{d}^2\sigma}{\mathrm{d}\omega\,\mathrm{d}\Omega}\,\mathrm{d}z \\ &= \frac{(1-R)^2 I_1^0}{\kappa(\omega_2)(K_1+K_2)}\frac{\mathrm{d}^2\sigma}{\mathrm{d}\omega\,\mathrm{d}\Omega}. \end{aligned} \tag{8.4}$$

Here, K_i is the absorption coefficient for light of frequency $\omega_i (i=1,2)$ and R the reflection coefficient [we neglect the difference between $R(\omega_1)$ and $R(\omega_2)$]. For the sake of simplicity, we consider the geometry of backscattering under normal incidence of the primary wave. In writing (8.4), we have taken into account that for radiation backscattered perpendicular to the surface the ratio of the solid angles $\mathrm{d}\Omega_2^0/\mathrm{d}\Omega_2$ in vacuum and in the crystal is equal to the squared refractive index $\kappa(\omega_2)$.

In the limiting case of a rarefied plasma where the Coulomb interaction between electrons may be disregarded, we have

$$\begin{aligned} \Delta w = {} & \frac{V q_2^2 \Delta q_2 \Delta\Omega_2}{(2\pi)^3}\frac{2\pi}{\hbar}\frac{2\pi\hbar c^2}{\kappa(\omega_1)V\omega_1}\bar{N}_{\mathbf{q}_1}\frac{2\pi\hbar c^2}{\kappa(\omega_2)V\omega_2}r_0^2\left(\frac{m}{m_c}\right)^2 \\ & \times \left|\left(\mathbf{e}_1\cdot\mathbf{e}_2^*\right)\right|^2\sum_{\mathbf{k}s} F_{\mathbf{k}s}\left(1-F_{\mathbf{k}+\mathbf{q},s}\right)\delta\left(E_{\mathbf{k}+\mathbf{q}}-E_{\mathbf{k}}-\hbar\omega\right), \end{aligned} \tag{8.5}$$

where $r_0 = e^2/mc^2$ is the classical electron radius, m_c the effective mass, and $F_{\mathbf{k}s}$ the equilibrium electron distribution function. The electron-photon interaction operator used in (8.5) has the following form in the second-quantization representation

$$\mathcal{H}_{\text{el-phot}} = \frac{e^2}{m_c c^2} c_{\mathbf{q}_1} c_{\mathbf{q}_2}^{+}\left(\mathbf{A}_1\cdot\mathbf{A}_2^*\right)\sum_{\mathbf{k}s} a_{\mathbf{k}+\mathbf{q},s}^{+} a_{\mathbf{k}s}, \tag{8.6}$$

where

$$\mathbf{A}_i = \left(\frac{2\pi\hbar c^2}{\kappa(\omega_i)V\omega_i}\right)^{1/2}\mathbf{e}_i \quad (i=1,2),$$

$c_{\mathbf{q}}^{+}, c_{\mathbf{q}}$ are the creation and annihilation operators for the photons and $a_{\mathbf{k}s}^{+}, a_{\mathbf{k}s}$ those for the electrons.

Substituting (8.3, 5) in (8.2) and neglecting the dependence of $\kappa(\omega_i)$ on ω_i, we come to

$$\frac{\mathrm{d}^2\sigma}{\mathrm{d}\omega\,\mathrm{d}\Omega} = r_0^2\left(\frac{m}{m_c}\right)^2\left(\frac{\omega_2}{\omega_1}\right)^2 |\mathbf{e}_1\cdot\mathbf{e}_2|^2\hbar\frac{2}{V} \times \sum_{\mathbf{k}} F_{\mathbf{k}}\left(1-F_{\mathbf{k}+\mathbf{q}}\right)\delta\left(E_{\mathbf{k}+\mathbf{q}}-E_{\mathbf{k}}-\hbar\omega\right), \tag{8.7}$$

with the factor 2 accounting for spin degeneracy.

The scattering cross section (8.7) can be expressed in terms of the electronic susceptibility

$$\chi_{\mathrm{el}}(\omega,\mathbf{q}) = \frac{e^2}{q^2}\frac{2}{V}\sum_{\mathbf{k}}\frac{F_{\mathbf{k}}-F_{\mathbf{k}+\mathbf{q}}}{E_{\mathbf{k}+\mathbf{q}}-E_{\mathbf{k}}-\hbar(\omega+\mathrm{i}\Gamma)}. \tag{8.8}$$

Indeed, using the relations

$$F_{\mathbf{k}}-F_{\mathbf{k}+\mathbf{q}} = F_{\mathbf{k}}\left(1-F_{\mathbf{k}+\mathbf{q}}\right)\left(1-\mathrm{e}^{-\hbar\omega/k_{\mathrm{B}}T}\right)$$

and

$$\mathrm{Im}\{(\varepsilon-\mathrm{i}\hbar\Gamma)^{-1}\} = \mathrm{i}\pi\delta(\varepsilon)$$

we come, in place of (8.7), to

$$\frac{\mathrm{d}^2\sigma}{\mathrm{d}\omega\,\mathrm{d}\Omega} = r_0^2\left(\frac{m}{m_c}\right)^2\left(\frac{\omega_2}{\omega_1}\right)^2 |\mathbf{e}_1\cdot\mathbf{e}_2^*|^2\frac{\hbar q^2}{\pi e^2}(1+N_\omega)\,\mathrm{Im}\{\chi_{\mathrm{el}}(\omega,\mathbf{q})\}, \tag{8.9}$$

where $N_\omega = \left[\exp(\hbar\omega/k_{\mathrm{B}}T)-1\right]^{-1}$. In Stokes scattering, $\omega > 0$, $N_\omega > 0$ and, in the anti-Stokes case, $\omega < 0$, $N_\omega < 0$, and $1+N_\omega = -N_{|\omega|}$. Note also that

$$\mathrm{Im}\{\chi_{\mathrm{el}}(-\omega,\mathbf{q})\} = -\mathrm{Im}\{\chi_{\mathrm{el}}(\omega,\mathbf{q})\}.$$

Therefore, the relation

$$\frac{(1+N_{-\omega})\,\mathrm{Im}\{\chi_{\mathrm{el}}(-\omega,q)\}}{(1+N_\omega)\,\mathrm{Im}\{\chi_{\mathrm{el}}(\omega,q)\}}$$

defining the intensity ratio of the anti-Stokes to Stokes scattering lines is equal to $\exp(-\hbar\omega/k_{\mathrm{B}}T)$.

Since the operator (8.6) is proportional to the Fourier component of the electron-density operator

$$\rho_{\mathbf{q}} = \frac{1}{V}\sum_{\mathbf{k}s} a_{\mathbf{k}+\mathbf{q},s}^{+}a_{\mathbf{k}s}, \tag{8.10}$$

expression (8.9) defines the cross section of light scattering by charge-density fluctuations in a rarefield plasma. With inclusion of Coulomb correlations, the differential cross section of scattering by charge-density fluctuations takes on the form

$$\frac{d^2\sigma}{d\omega\, d\Omega} = r_0^2 \left(\frac{m}{m_c}\right)^2 \left(\frac{\omega_2}{\omega_1}\right)^2 |\mathbf{e}_1 \cdot \mathbf{e}_2^*|^2 \frac{\hbar q^2}{4\pi^2 e^2}$$

$$\times (1 + N_\omega)\, \kappa_\infty^2 \mathrm{Im}\left\{-\frac{1}{\kappa(\omega, \mathbf{q})}\right\}, \tag{8.11}$$

with the dielectric constant

$$\kappa(\omega, \mathbf{q}) = \kappa_\infty + \kappa_{\mathrm{el}}(\omega, \mathbf{q}),\ \kappa_{\mathrm{el}}(\omega, \mathbf{q}) = 4\pi \chi_{\mathrm{el}}(\omega, \mathbf{q}). \tag{8.12}$$

For simplicity, the contribution of optical phonons in (8.12) is neglected. The expressions (8.9, 11) differ in the factor $\kappa_\infty^2/|\kappa(\omega, \mathbf{q})|^2$ accounting for the screening of the charge fluctuations appearing in the system. For low-density plasmas where $|\kappa_{\mathrm{el}}| \ll \kappa_\infty$, this factor is close to unity. Equation (8.11) describes the scattering both from single-particle excitations with the transferred frequency $\omega = (E_{\mathbf{k}+\mathbf{q}} - E_{\mathbf{k}})/\hbar$ and plasma oscillations, or plasmons, whose frequency satisfies the equation $\kappa(\omega, \mathbf{q}) = 0$.

Expression (8.6) for the operator $\mathcal{H}_{\mathrm{el-phot}}$ describing light scattering by free electrons is valid in the case where the photon energy $\hbar\omega_i$ is small compared to the energy separation $E_c^0 - E_l^0$ from the other bands. If this condition is not met, one has to start from a more general expression

$$\mathcal{H}_{\mathrm{el-phot}} = r_0 c_{\mathbf{q}_1} c_{\mathbf{q}_2}^+ A_1 A_2 \sum_{\mathbf{k}ss'} \gamma_{s's} a_{\mathbf{k}+\mathbf{q},s'}^+ a_{\mathbf{k}s}, \tag{8.13}$$

where $A_i = |\mathbf{A}_i|$,

$$\gamma_{s's} = \left(\mathbf{e}_1 \cdot \mathbf{e}_2^*\right) \delta_{s's}$$

$$+ \frac{1}{m} \sum_l{}' \left(\frac{\left(\mathbf{e}_1 \cdot \mathbf{p}_{cs',l}\right)\left(\mathbf{e}_2^* \cdot \mathbf{p}_{l,cs}\right)}{E_c^0 - E_l^0 - \hbar\omega_1} + \frac{\left(\mathbf{e}_2 * \cdot \mathbf{p}_{cs',l}\right)}{E_c^0 - E_l^0 + \hbar\omega_2} \right). \tag{8.14}$$

In accordance with the definition of the reciprocal effective-mass tensor $m_{\alpha\beta}^{-1}$, for $\hbar\omega_i \ll |E_c^0 - E_l^0|$ we obtain

$$\gamma_{s's} = \delta_{s's} \sum_{\alpha\beta} \frac{m}{m_{\alpha\beta}} e_{1\alpha} e_{2\beta}^*.$$

In crystals of cubic symmetry, $m_{\alpha\beta} = m_c \delta_{\alpha\beta}$, $\gamma_{s's} = \delta_{s's}(m/m_c)(\mathbf{e}_1 \cdot \mathbf{e}_2^*)$ and (8.13) reduces to (8.6).

The matrix γ can be conveniently represented as a linear combination of four 2×2 matrices: $\hat{I}$, σ_x, σ_y and σ_z. For crystals of the GaAs type, this expansion has the form

$$\gamma = A\left(\mathbf{e}_1 \cdot \mathbf{e}_2^*\right)\hat{I} - \mathrm{i}B\left(\mathbf{e}_1 \times \mathbf{e}_2^*\right)\boldsymbol{\sigma}, \tag{8.15}$$

where

$$A = 1 + \frac{2p_{cv}^2}{3m}\left(\frac{2E_g}{E_g^2 - (\hbar\omega_1)^2} + \frac{E_g + \Delta}{\left(E_g + \Delta\right)^2 - (\hbar\omega_1)^2}\right), \tag{8.16}$$

$$B = \frac{2p_{cv}^2}{3m}\hbar\omega_l \left(\frac{1}{E_g^2 - (\hbar\omega_l)^2} - \frac{1}{\left(E_g + \Delta\right)^2 - (\hbar\omega_l)^2} \right). \tag{8.17}$$

In the calculations, we neglected the difference between the frequencies ω_l and ω_2, the sum over l in (8.14) including only the contributions of the upper valence bands Γ_8 and Γ_7. It was also assumed that the energies $|E_g - \hbar\omega_l|$, $|E_g + \Delta - \hbar\omega_l|$ exceed substantially the mean electron kinetic energy ($k_B T$ for the nondegenerate plasma). Substituting (8.15) in (8.13), we obtain in place of (8.13)

$$\mathcal{H}_{el-phot} = r_0 c_{\mathbf{q}_1} c_{\mathbf{q}_2}^+ A_1 A_2 V \left[A \left(\mathbf{e}_l \cdot \mathbf{e}_2^*\right) \rho_{\mathbf{q}} - 2\mathrm{i}B \left(\mathbf{e}_l \times \mathbf{e}_2^*\right) \mathbf{s}_{\mathbf{q}} \right], \tag{8.18}$$

where $\mathbf{s}_{\mathbf{q}}$ are the Fourier components of the electron-spin density operator:

$$\mathbf{s}_{\mathbf{q}} = \frac{1}{2V} \sum_{\mathbf{k}ss'} \sigma_{s's} a_{\mathbf{k}+\mathbf{q},s'}^+ a_{\mathbf{k}s}, \tag{8.19}$$

and the operator $\rho_{\mathbf{q}}$ is defined in accordance with (8.10). As follows from (8.18), light can be scattered not only by fluctuations of the electron density, but also by those of the spin density as well. The first contribution is described by (8.11), where the ratio m/m_c has to be replaced by the coefficient A. For the cross section of spin-dependent scattering, we obtain

$$\begin{aligned} \frac{\mathrm{d}^2\sigma}{\mathrm{d}\omega\,\mathrm{d}\Omega} &= r_0^2 B^2 \left(\frac{\omega_2}{\omega_l}\right)^2 |\mathbf{e}_l \times \mathbf{e}_2^*|^2 \hbar \frac{2}{V} \\ &\quad \times \sum_{\mathbf{k}} F_{\mathbf{k}} \left(1 - F_{\mathbf{k}+\mathbf{q}}\right) \delta\left(E_{\mathbf{k}+\mathbf{q}} - E_{\mathbf{k}} - \hbar\omega\right) \\ &= r_0^2 B^2 \left(\frac{\omega_2}{\omega_l}\right)^2 |\mathbf{e}_l \times \mathbf{e}_2^*|^2 \frac{\hbar q^2}{\pi e^2} (1 + N_\omega) \operatorname{Im}\{\chi_{el}(\omega, \mathbf{q})\}. \end{aligned} \tag{8.20}$$

Since the spin-density fluctuations are not accompanied by violation of neutrality and, therefore, not screened, the differential cross section (8.20) does not contain the factor $\kappa_\infty^2/|\kappa(\omega, \mathbf{q})|^2$. Note that (8.20) includes both the contribution due to spin-flip scattering $\mathbf{k}, 1/2 \to \mathbf{k}+\mathbf{q}, -1/2$ or $\mathbf{k}, -1/2 \to \mathbf{k}+\mathbf{q}, 1/2$ which is proportional to $|(\mathbf{e}_l \times \mathbf{e}_2^*)_+|^2$ and $|(\mathbf{e}_l \times \mathbf{e}_2^*)_-|^2$, respectively, and that coming from spin-dependent scattering without spin-flip, $\mathbf{k}, s \to \mathbf{k}+\mathbf{q}, s$ which is proportional to $|(\mathbf{e}_l \times \mathbf{e}_2^*)_z|^2$, where z is the spin quantization axis,

$$\left(\mathbf{e}_l \times \mathbf{e}_2^*\right)_\pm = \left(\mathbf{e}_l \times \mathbf{e}_2^*\right)_x \pm \mathrm{i} \left(\mathbf{e}_l \times \mathbf{e}_2^*\right)_y.$$

In a magnetic field $\mathbf{B} \parallel z$, the spin sublevels of the electron states split, the transferred frequency in spin-flip scattering $s \to -s$ being

$$\hbar\omega = E_{\mathbf{k}+\mathbf{q}} - E_{\mathbf{k}} - 2sg\mu_B B. \tag{8.21}$$

Here g is the electron g-factor, $E_{\mathbf{k}} = \hbar^2 k^2/2m_c$, and the magnetic field is assumed to be classical. In a quantizing field, contributions to light scattering arise not only from the spin-flip processes, but also from carrier transitions between the Landau levels.

Equations (8.11, 20) are valid provided $ql \gg 1$ (collisionless approximation), where l is the electron mean free path length. In the opposite limiting case $ql \ll 1$, light scattering by free carriers is collision-dominated, the expressions for the cross sections taking on a different form. In particular, the dependence of $\mathrm{d}^2\sigma/\mathrm{d}\omega\,\mathrm{d}\Omega$ on q for $ql \ll 1$ depends substantially on the particle diffusion coefficient in the case of scattering by charge fluctuations, or on the spin diffusion coefficient for scattering by spin-density fluctuations. It is instructive to follow the evolution of the spectra of light scattering from free carriers in n-GaAs or n-InP with increase of free-carrier concentration [8.1,2]. For low electron concentrations, $n \leqslant 10^{15}\ \mathrm{cm}^{-3}$, when q is large compared to the screening parameter $\lambda = (4\pi e^2 n/k_\mathrm{B}T)^{1/2}$, the scattering spectrum has a Gaussian shape. For higher concentrations, $n = (10^{16} - 3\times 10^{17})\ \mathrm{cm}^{-3}$, when $q/\lambda \ll 1$ and the condition of frequent collisions is met, $ql \ll 1$, the spectrum becomes Lorentzian with a halfwidth Dq^2, where D is the diffusion coefficient.

Besides these two mechanisms of electron light scattering in semiconductors, there are others, in particular, scattering by energy fluctuations (taking into account the nonparabolicity of the free carrier spectrum), by mass fluctuations (in a many-valley semiconductor with anisotropic effective electron masses), by collective electron-hole plasma oscillations, and scattering involving carrier transitions between different bands (e.g., between heavy and light-hole subbands).

8.1.2 Scattering by Phonons

The most substantial contribution to the phonon-assisted light scattering comes from the mechanism which involves interaction of light with the lattice not directly but rather through the electron subsystem. Lattice vibrations produce in the medium a transient optical superlattice, and it is from the latter that the scattering occurs. Therefore, the efficiency of scattering by acoustic phonons (Brillouin or Mandelshtam-Brillouin scattering) or optical phonons (Raman scattering) is connected intimately with the intensity of the corresponding fluctuations, $\delta\chi_{\alpha\beta}(\mathbf{x},t)$, of the dielectric constant of the medium. Here, the differential scattering cross section can be represented in the form

$$\frac{\mathrm{d}^2\sigma}{\mathrm{d}\omega\,\mathrm{d}\Omega} = \left(\frac{\omega_2}{c}\right)^4 V^2 \int \frac{\mathrm{d}t}{2\pi} \mathrm{e}^{\mathrm{i}\omega t} \langle \delta\chi^+(\mathbf{q},t)\delta\chi(\mathbf{q},0)\rangle, \tag{8.22}$$

where

$$\delta\chi(\mathbf{q},t) = \frac{1}{V} e^*_{2\alpha} e_{1\beta} \int \delta\chi_{\alpha\beta}(\mathbf{x},t)\mathbf{e}^{\mathrm{i}\mathbf{qx}}\,\mathbf{dx}. \tag{8.23}$$

In such a semiphenomenological description, the fluctuation $\delta\chi_{\alpha\beta}$ can be expanded in the normal coordinates of lattice vibrations which are written in the second-quantization representation. Therefore, $\delta\chi(\mathbf{q},t)$ is an operator acting on the wave function of the phonon subsystem, the angular brackets $\langle\ldots\rangle$ in (8.22) denoting the averaging of the operator product over the equilibrium phonon distribution.

The quantity $\delta\chi_{\alpha\beta}$ involved in the calculation of the cross section of scattering by acoustical phonons includes terms linear in the deformation tensor

$$\delta\chi_{\alpha\beta}(\mathbf{x},0)=\frac{\partial\chi_{\alpha\beta}}{\partial\varepsilon_{lm}}\varepsilon_{lm}, \tag{8.24}$$

where $\partial\chi_{\alpha\beta}/\partial\varepsilon_{lm}$ is the tensor of elasto-optical coefficients,

$$\varepsilon_{lm}=\frac{1}{2}\left(\frac{\partial u_l}{\partial x_m}+\frac{\partial u_m}{\partial x_l}\right),$$

the displacement vector

$$\mathbf{u}=\sum_{\mathbf{Q}\nu}\left(\frac{\hbar}{2\rho\omega_{\mathbf{Q}\nu}V}\right)^{1/2}\left(\mathbf{e}_{\mathbf{Q}\nu}\mathrm{e}^{\mathrm{i}\mathbf{Q}\mathbf{x}}b_{\mathbf{Q}\nu}+\mathbf{e}^*_{\mathbf{Q}\nu}\mathrm{e}^{-\mathrm{i}\mathbf{Q}\mathbf{x}}b^+_{\mathbf{Q}\nu}\right), \tag{8.25}$$

ρ is the density of material, $\omega_{\mathbf{Q}\nu}$ and $\mathbf{e}_{\mathbf{Q}\nu}$ are the frequency and polarization vector of the phonon of νth branch with a wave vector $\mathbf{Q}$, and $b^+_{\mathbf{Q}\nu}$ and $b_{\mathbf{Q}\nu}$ are the phonon creation and annihilation operators. In piezoelectrics, $\delta\chi_{\alpha\beta}$ includes, besides the deformation contribution, also an electro-optical term

$$\left(\partial\chi_{\alpha\beta}/\partial\mathcal{I}_n\right)_{\varepsilon_{lm}=0}\mathcal{I}_n,$$

where $\partial\chi_{\alpha\beta}/\partial\mathcal{I}_n$ is the electro-optical tensor and $\mathcal{I}_n$ the electric field induced by acoustic oscillations. In an intrinsic semiconductor which does not have free carriers, one can transfer from the operator $\delta\chi_{\alpha\beta}(\mathbf{x},0)$ to the Heisenberg operator $\delta\chi_{\alpha\beta}(\mathbf{x},t)$ (in the harmonic approximation) by replacing $\mathbf{Q}\cdot\mathbf{x}$ in (8.25) with $\mathbf{Q}\cdot\mathbf{x}-\omega_{\mathbf{Q}\nu}t$. Substituting the expression for $\delta\chi_{\alpha\beta}$ in (8.22) and averaging over the equilibrium phonon-density matrix, we obtain for the Brillouin scattering

$$\frac{\mathrm{d}^2\sigma}{\mathrm{d}\omega\,\mathrm{d}\Omega}=\left(\frac{\omega_2}{c}\right)^4\frac{\hbar q^2}{2\rho\omega_{\mathbf{q}\nu}}\left|\frac{\partial\chi_{\alpha\beta}}{\partial\varepsilon_{lm}}e^*_{2\alpha}e_{1\beta}e_{\mathbf{q}\nu,l}\frac{q_m}{q}\right|^2$$
$$\times\left[\left(N_{\mathbf{q}\nu}+1\right)\delta\left(\omega-\omega_{\mathbf{q}\nu}\right)+N_{\mathbf{q}\nu}\delta(\omega+\omega_{\mathbf{q}\nu})\right]. \tag{8.26}$$

Here, $N_{\mathbf{q}\nu}$ are the phonon occupation numbers, the vector $\mathbf{e}_{\mathbf{q}\nu}$ is for simplicity considered real, and $\partial\chi_{\alpha\beta}/\partial\varepsilon_{lm}$ includes the electro-optical contribution

$$\left(\frac{\partial\chi_{\alpha\beta}}{\partial\mathcal{I}_n}\right)_{\varepsilon_{lm}=0}\frac{\mathcal{I}_n(\mathbf{q})}{\varepsilon_{lm}(\mathbf{q})}.$$

Under ordinary conditions, $\hbar\omega_{\mathbf{q}\nu}\ll k_{\mathrm{B}}T$ and $N_{\mathbf{q}\nu},\ \left(N_{\mathbf{q}\nu}+1\right)\approx k_{\mathrm{B}}T/\hbar\omega_{\mathbf{q}\nu}$.

When describing light scattering by longitudinal optical phonons in a GaAs-type diatomic crystal, one uses the expansion

$$\delta\chi_{\alpha\beta}(\mathbf{x},0)=\frac{\partial\chi_{\alpha\beta}}{\partial u_l}u_l+\frac{\partial\chi_{\alpha\beta}}{\partial\mathcal{I}_l}\mathcal{I}_l, \tag{8.27}$$

where $\mathbf{u}$ is the displacement vector and $\mathcal{I}$ the electric field induced by such a displacement. The vector $\mathbf{u}$ is usually defined as the relative sublattice displacement $\mathbf{R}_1 - \mathbf{R}_2$ multiplied by $\sqrt{\bar{\rho}}$, where $\bar{\rho}$ is the reduced-mass density, i.e.,

$$\bar{\rho} = \rho \frac{M_1 M_2}{(M_1 + M_2)^2},$$

M_i is the mass of the atomic species, $i = 1, 2$. In the second-quantization representation, the operators $\mathbf{u}$ and $\mathcal{I}$ can be written (4.81):

$$\mathbf{u} = \frac{\beta}{\omega_{\mathrm{T}}^2 - \omega^2}\mathcal{I}, \quad \mathcal{I} = -\nabla\varphi,$$

$$\varphi = \mathrm{i}\left(\frac{2\pi\hbar\omega_{\mathrm{L}}}{V\kappa^*}\right)^{1/2} \sum_{\mathbf{Q}} \frac{1}{Q}\left(b_{\mathbf{Q}}\mathrm{e}^{\mathrm{i}\mathbf{Q}\cdot\mathbf{x}} - b_{\mathbf{Q}}^{+}\mathrm{e}^{-\mathrm{i}\mathbf{Q}\cdot\mathbf{x}}\right), \tag{8.28}$$

where

$$\beta = \left[(\kappa_0 - \kappa_\infty)\,\omega_{\mathrm{T}}^2/4\pi\right]^{1/2}, \tag{8.29}$$

$b_{\mathbf{Q}}^{+}$ and $b_{\mathbf{Q}}$ are the creation and annihilation operators for the longitudinal optical phonon. Using (8.22, 27-29) we come to

$$\frac{\mathrm{d}^2\sigma}{\mathrm{d}\omega\,\mathrm{d}\Omega} = \frac{2\pi\hbar\omega_{\mathrm{L}}}{\kappa^*}\left(\frac{\omega_2}{c}\right)^4 \left|\frac{\partial\chi_{\alpha\beta}}{\partial\mathcal{I}_l} e_{2\alpha}^* e_{1\beta}\frac{q_l}{q}\right|^2$$
$$\times\left(1 - \frac{C_1\kappa_\infty}{\kappa_0 - \kappa_\infty}\right)^2 \left[\left(N_{\omega_{\mathrm{L}}} + 1\right)\delta(\omega - \omega_{\mathrm{L}})\right.$$
$$\left. + N_{\omega_{\mathrm{L}}}\delta(\omega + \omega_{\mathrm{L}})\right], \tag{8.30}$$

where

$$C_1 = \frac{\beta}{\omega_{\mathrm{T}}^2}\left(\frac{\partial\chi_{\alpha\beta}}{\partial u_l} e_{2\alpha}^* e_{1\beta}\frac{q_l}{q}\right) \Big/ \left(\frac{\partial\chi_{\alpha\beta}}{\partial\mathcal{I}_l} e_{2\alpha}^* e_{1\beta}\frac{q_l}{q}\right).$$

In crystals of the class T_d

$$\frac{\partial\chi_{\alpha\beta}}{\partial\mathcal{I}_l} = \frac{\partial\chi_{xy}}{\partial\mathcal{I}_z}|\delta_{\alpha\beta l}|, \qquad \frac{\partial\chi_{\alpha\beta}}{\partial u_m} = \frac{\partial\chi_{xy}}{\partial u_z}|\delta_{\alpha\beta l}|,$$

where $\delta_{\alpha\beta l}$ is the unit antisymmetrical tensor of the third rank and

$$C_1 = \frac{\beta}{\omega_{\mathrm{T}}^2}\left(\frac{\partial\chi_{xy}}{\partial u_z}\Big/\frac{\partial\chi_{xy}}{\partial\mathcal{I}_z}\right). \tag{8.31}$$

Taking into account the phonon damping Γ determining the scattering line halfwidth, the function $\delta(\omega - \omega_{\mathrm{L}})$ in (8.30) should be replaced by

$$\frac{2}{\pi}\frac{\kappa^*}{\omega_{\mathrm{L}}}\mathrm{Im}\left\{-\frac{1}{\kappa(\omega)}\right\},$$

where $\kappa(\omega)$ is the dielectric constant (7.101):

$$\kappa(\omega) = \kappa_\infty + \kappa_{\mathrm{phon}}, \; \kappa_{\mathrm{phon}} = \frac{(\kappa_0 - \kappa_\infty)\,\omega_{\mathrm{T}}^2}{\omega_{\mathrm{T}}^2 - \omega^2 - 2\mathrm{i}\Gamma\omega}. \tag{8.32}$$

In a doped semiconductor with polar optical vibrations, the dielectric constant includes both the phonon and electron contributions

$$\kappa(\omega, q) = \kappa_\infty + \kappa_{\text{phon}} + \kappa_{\text{el}}. \tag{8.33}$$

In the region of the variables q and ω satisfying the inequality

$$qv \ll \omega, \tag{8.34}$$

where v is the rms electron velocity, the electron contribution can be written

$$\kappa_{\text{el}} = -\kappa_\infty \frac{\omega_{\text{pl}}^2}{\omega(\omega + \mathrm{i}\gamma)}, \tag{8.35}$$

where the plasma-oscillation frequency

$$\omega_{\text{pl}} = \left(\frac{4\pi e^2 n}{\kappa_\infty m_{\text{c}}}\right)^{1/2}. \tag{8.36}$$

The equation for longitudinal waves $\kappa(\omega) = 0$ has two solutions, ω_+ and ω_-, determining the frequencies of the mixed plasmon-phonon modes. The light-scattering cross section for each of these collective oscillations includes an electronic contribution determined by the interaction (8.6, 18), and a phonon contribution associated with the deformation and electro-optical mechanisms of modulation of the dielectric constant (8.27). These contributions are characterized by the polarization dependences

$$T_1 = \left|\left(\mathbf{e}_2^* \cdot \mathbf{e}_1\right)\right|^2 \quad \text{and} \quad T_2 = \left|\left|\delta_{\alpha\beta l}\right| e_{2\alpha}^* e_{1\beta} \frac{q_l}{q}\right|^2$$

respectively.

We present here also for completeness an expression for the cross section of scattering by transverse optical phonons in zinc-blende-type crystals

$$\frac{\mathrm{d}^2\sigma}{\mathrm{d}\omega\,\mathrm{d}\Omega} = \left(\frac{\omega_2}{c}\right)^4 \frac{\hbar}{2\omega_{\text{T}}} \left|\frac{\partial \chi_{xy}}{\partial u_z}\right|^2 \sum_{\nu=1,2} \left|\left|\delta_{\alpha\beta l}\right| e_{2\alpha}^* e_{1\beta} e_{\mathbf{q}\nu,l}\right|^2$$
$$\times \left[\left(N_{\omega_{\text{T}}} + 1\right)\delta\left(\omega - \omega_{\text{T}}\right) + N_{\omega_{\text{T}}}\delta\left(\omega + \omega_{\text{T}}\right)\right], \tag{8.37}$$

where the index ν identifies the states of transverse phonons ($\mathbf{e}_{\mathbf{q}\nu} \perp \mathbf{q}$). This relates to the limiting case, $q \gg \omega\sqrt{\kappa_\infty}/c$, which is usually realized in experiments on light scattering by phonons (with the exception of small-angle scattering where $q \ll q_{1,2}$). In this case, the excitation of a transverse optical phonon is not accompained by the creation of a substantial transverse electric field (which implies that the polariton effects are inessential), so that the electro-optical contribution in (8.37) can be disregarded.

8.1.3 Microscopic Theory

We will illustrate the microscopic calculation by the example of light scattering by acoustic phonons involving the deformation mechanism of electron-phonon interaction in a semiconductor with a nondegenerate band structure. According to the nonstationary perturbation theory, the rate of scattering by phonons into the interval $\Delta\omega_2 \Delta\Omega_2$ is determined by the expression

$$\Delta w = \frac{V q_2^2 \Delta q_2 \Delta\Omega_2}{(2\pi)^3} \frac{2\pi}{\hbar^2} \frac{2\pi\hbar}{\kappa(\omega_1) V \omega_1} \bar{N}_{\mathbf{q}_1} \frac{2\pi\hbar}{\kappa(\omega_2) V \omega_2}$$
$$\times \left| e_{2\alpha}^* R_{\alpha\beta} e_{1\beta} \right|^2 \left[\left(N_{\mathbf{q}\nu} + 1 \right) \delta\left(\omega - \omega_{\mathbf{q}\nu} \right) + N_{\mathbf{q}\nu} \delta\left(\omega + \omega_{\mathbf{q}\nu} \right) \right], \quad (8.38)$$

where the scattering tensor

$$R_{\alpha\beta}^{(\mathrm{res})} = \sum_{ff'} \frac{j_{o,f'}^{\alpha}(-\mathbf{q}_2) V_{f'f} j_{f,o}^{\beta}(\mathbf{q}_1)}{\left(E_{f'} - \hbar\omega_2\right)\left(E_f - \hbar\omega_1\right)} \quad (8.39)$$

and f, f' are the electron-hole excited states (6.31). For the sake of simplicity, we have included here only the resonant term increasing as $\hbar\omega_1 \to E_{\mathrm{g}}$. In the deformation-potential formalism, the matrix element of the electron-phonon interaction can be written

$$V_{f'f} = \left(\Xi_{\mathrm{c}} \langle f' \left| \mathrm{e}^{\mathrm{i}\mathbf{Q}\cdot\mathbf{x}_{\mathrm{e}}} \right| f \rangle - \Xi_{\mathrm{v}} \langle f' \left| \mathrm{e}^{\mathrm{i}\mathbf{Q}\cdot\mathbf{x}_{\mathrm{h}}} \right| f \rangle \right) \bar{\varepsilon}_{ll}, \quad (8.40)$$

where $\Xi_{\mathrm{c,v}}$ is the electron deformation potential constant in the conduction and valence bands, $\mathbf{x}_{\mathrm{e}}$, $\mathbf{x}_{\mathrm{h}}$ are the electron and hole coordinates,

$$\bar{\varepsilon}_{lm} = \left(\hbar / 2\rho\omega_{\mathbf{Q}\nu} V \right)^{1/2} \mathrm{i} e_{\mathbf{Q}\nu,l} Q_m, \quad (8.41)$$

$\mathbf{Q} = -\mathbf{q}$ for the phonon-absorption process ($\omega = -\omega_{\mathbf{q}\nu}$), and $\mathbf{Q} = \mathbf{q}$ for the case of phonon emission ($\omega = \omega_{\mathbf{q}\nu}$). Equation (8.40) was written taking into account that in an isotropic semiconductor with a nondegenerate band the deformation potential is determined by one constant (3.29)

$$V_D = \Xi \operatorname{div} \mathbf{u}. \quad (8.42)$$

In this model, the electrons and phonons interact only with longitudinal acoustical (LA) phonons since for TA phonons, $\bar{\varepsilon}_{ll} = 0$. In the zero order in $\mathbf{q}$

$$V_{f'f} = \left(\Xi_{\mathrm{c}} - \Xi_{\mathrm{v}} \right) \bar{\varepsilon}_{ll} \delta_{\nu,\mathrm{LA}}. \quad (8.43)$$

Substituting (8.43) in (8.39) and comparing the expression obtained with the resonant contribution (6.31) to the dielectric constant, we find that

$$R_{\alpha\beta}^{(\mathrm{res})} = -\left(\Xi_{\mathrm{c}} - \Xi_{\mathrm{v}} \right) \bar{\varepsilon}_{ll} V \omega_1^2 \frac{\partial \chi^{(\mathrm{res})}(\omega_1)}{\partial E_{\mathrm{g}}} \delta_{\alpha\beta} \delta_{\nu,\mathrm{LA}}$$
$$= -V\omega_1^2 \sum_{lm} \frac{\partial \chi_{\alpha\beta}^{(\mathrm{res})}(\omega_1)}{\partial \varepsilon_{lm}} \bar{\varepsilon}_{lm}, \quad (8.44)$$

where in the case under consideration

$$\frac{\partial \chi_{\alpha\beta}}{\partial \varepsilon_{lm}} = \delta_{lm} \delta_{\alpha\beta} \frac{\partial \chi}{\partial \varepsilon_{zz}}. \quad (8.45)$$

The difference between ω_1 and ω_2 was neglected in the derivation of (8.44). Therefore, expression (8.22) relating the scattering cross section with the spectrum of dielectric-constant fluctuations is valid in the frequency range where the difference between $\kappa(\omega_1)$ and $\kappa(\omega_2)$ is inessential. In the vicinity of the exciton resonances, the dielectric constant exhibits anomalous dispersion, so that scattering efficiency calculations should be based on the microscopic theory of the Brillouin resonant scattering of exciton polaritons. Using (8.2, 38, 41, 44) and setting in the common factors $\omega_1 = \omega_2$, one can show that the microscopic theory results in an expression for $\mathrm{d}^2\sigma/\mathrm{d}\omega\,\mathrm{d}\Omega$ coinciding with (8.26).

Let us discuss now the deformation mechanism of electron interaction with optical phonons and its contribution to the Raman light scattering. In a non-degenerate band, Γ_6 or Γ_7 , of GaAs type crystals the change in the electron energy E due to a relative sublattice displacement $\mathbf{R}_1 - \mathbf{R}_2$ is small since, according to (4.82), the expansion of the coefficient $\delta E/\delta u_l$ in powers of the wave vector starts with quadratic terms. In the degenerate valence band Γ_8, the deformation-interaction operator does not contain small factors and can be presented in the form (4.79). Therefore, for $\hbar\omega_1 \sim E_g$, the deformation contribution to the optical-phonon scattering tensor $R_{\alpha\beta}$ comes from $f \to f'$ processes involving scattering of the Γ_8 hole, so that $\partial\chi_{xy}/\partial u_z \propto d_o$.

It should be pointed out in conclusion that taking into account the Fröhlich interaction

$$-\, e\left[\varphi\left(\mathbf{x}_{\mathrm{e}}\right) - \varphi\left(\mathbf{x}_{\mathrm{h}}\right)\right] \tag{8.46}$$

of the electron-hole pair or exciton with LO phonons (the potential $\varphi(\mathbf{x})$ is defined in (8.28)) results only in linear-in-$\mathbf{q}$ corrections to the scattering tensor. However, under resonant excitation of a semiconductor with polar optical vibrations, it is this interaction that accounts for the formation of the series of equidistant emission lines $\omega_1 - n\omega_{\mathrm{LO}}$ with $n \geqslant 2$. Note that the 1LO line is much weaker than the 2LO. This secondary emission may be considered as n-phonon resonant Raman light scattering. Under certain conditions, it can also be interpreted as hot exciton photoluminescence.

8.2 Scattering by Intersubband Excitations

A new mechanism of light scattering from free carriers appears in heterostructures, accompanied by carrier transfer between the subbands (in a single heterojunction or QW structure) or minibands (in a SL). We present here an expression for the differential cross section of Raman scattering by intersubband electron excitations in the conduction band of a SQW structure:

$$\frac{\mathrm{d}^2\sigma}{\mathrm{d}\omega\,\mathrm{d}\Omega} \equiv \frac{\hbar\omega_2}{I_1 S}\frac{\Delta w}{\Delta\omega_2\Delta\Omega_2}$$

$$= \frac{\hbar}{c^4}\left(\frac{\omega_2}{\omega_1}\right)\frac{1}{S}\sum_{ff'}\left|e_{2\alpha}^{*}e_{1\beta}R_{\alpha\beta}(f', f)\right|^2$$

$$\times F_f \left(1 - F_{f'}\right) \delta \left(E_{f'} - E_f - \hbar\omega\right). \tag{8.47}$$

Here the indices f and f' denote the initial $(cs\nu\mathbf{k}_\perp)$ and final $(cs'\nu'\mathbf{k}'_\perp)$ electron states ($\mathbf{k}_\perp = \mathbf{k}'_\perp + \mathbf{q}_\perp$), the scattering cross section being reduced to unit area and $R_{\alpha\beta}$ being the scattering tensor. For $\hbar\omega_1 \sim E_g$, the main contribution comes from the resonant term

$$R^{(\mathrm{res})}_{\alpha\beta}(f', f) = \sum_{f''} \frac{j^{\alpha}_{f',\, f''}\,(-\mathbf{q}_2)\, j^{\beta}_{f'',f}\,(\mathbf{q}_1)}{E_{f''} - E_f - \hbar\omega_1}, \tag{8.48}$$

where f'' is the excited state of the crystal in which, besides the electron in state $f = (cs\nu\mathbf{k}_\perp)$, there is also an electron-hole pair (or exciton) created in the absorption of the $\hbar\omega_1$ photon. Neglecting Coulomb interaction between the carriers, (8.48) can be written

$$R^{(\mathrm{res})}_{\alpha\beta}(f', f) = \frac{e^2}{m^2} \sum_{\nu'' m} \frac{p^{\beta}_{\mathrm{c}s',\mathrm{v}m}\, p^{\alpha}_{\mathrm{v}m,\mathrm{c}s}}{E_{\mathrm{c}\nu'\mathbf{k}'} - E_{\mathrm{v}\nu''\mathbf{k}''} - \hbar\omega_1}$$
$$\times \langle e\nu'\mathbf{k}'|\mathrm{e}^{\mathrm{i}\mathbf{q}_1\cdot\mathbf{x}}|h\nu''\mathbf{k}''\rangle\langle h\nu''\mathbf{k}''|\mathrm{e}^{-\mathrm{i}\mathbf{q}_2\cdot\mathbf{x}}|e\nu\mathbf{k}\rangle, \tag{8.49}$$

where

$$\mathbf{k}'' = \mathbf{k} - \mathbf{q}_{2\perp} = \mathbf{k}' - \mathbf{q}_{1\perp},$$

and the index $\perp$ of the two-dimensional electron wave vector is dropped for the sake of simplicity. Neglecting the light wave vector compared to the reciprocal well width (the dipole approximation), we come to

$$e^*_{2\alpha} e_{1\beta} R^{(\mathrm{res})}_{\alpha\beta}(f', f) = \frac{e^2}{m}\gamma^{(\mathrm{res})}_{s's}(\nu', \nu), \tag{8.50}$$

$$\gamma^{(\mathrm{res})}_{s's}(\nu'\nu) = V_{s's} d_{\nu'\nu}\left(\mathbf{k}, \omega_1\right),$$
$$V_{s's} = m^{-1} \sum_j \left(\mathbf{e}_1 \cdot \mathbf{p}_{\mathrm{c}s',\mathrm{v}j}\right)\left(\mathbf{e}^*_2 \cdot \mathbf{p}_{\mathrm{v}j,\mathrm{c}s}\right), \tag{8.51}$$

$$d_{\nu'\nu}\left(\mathbf{k}, \omega_1\right) = \sum_{\nu''} \frac{\langle e\nu'\mathbf{k}|h\nu''\mathbf{k}\rangle\langle h\nu''\mathbf{k}|e\nu\mathbf{k}\rangle}{E_{c\nu'\mathbf{k}} - E_{v\nu''\mathbf{k}} - \hbar\omega_1}. \tag{8.52}$$

For the band pair c, Γ_6 and v, Γ_7 in a semiconductor of T_d symmetry, the components $V_{s's}$ may be presented in the following matrix form

$$\hat{V} = \frac{1}{3} p^2_{\mathrm{cv}} \left[\left(\mathbf{e}^*_2 \cdot \mathbf{e}_1\right) - \mathrm{i}\boldsymbol{\sigma} \cdot \left(\mathbf{e}^*_2 \times \mathbf{e}_1\right)\right]. \tag{8.53}$$

The contributions to $R_{\alpha\beta}$ of states in the heavy- and light-hole subbands of the valence band Γ_8 may be written in the simple form (8.53) only if the mixing of these states for $\mathbf{k} \neq 0$ is neglected. In this case, we will have

$$\hat{V}_{\mathrm{hh}} = \frac{1}{2} p^2_{\mathrm{cv}} \left[\left(\mathbf{e}^*_2 \cdot \mathbf{e}_1\right) - e^*_{2z} e_{1z} + \mathrm{i}\sigma_z \left(\mathbf{e}^*_2 \times \mathbf{e}_1\right)_z\right],$$

$$\hat{V}_{\mathrm{lh}} = \frac{1}{6} p^2_{\mathrm{cv}} \left[\left(\mathbf{e}^*_2 \cdot \mathbf{e}_1\right) + 3e^*_{2z} e_{1z} + 2\mathrm{i}\boldsymbol{\sigma} \cdot \left(\mathbf{e}^*_2 \times \mathbf{e}_1\right) - 3\mathrm{i}\sigma_z \left(\mathbf{e}^*_2 \times \mathbf{e}_1\right)_z\right].$$

The diagonal and nondiagonal components of the matrix $V_{s's}$ describe the scattering from intersubband excitations with and without electron spin-flip, respectively.

The selection rules for the $\nu \to \nu'$ intersubband absorption (sect. 7.2) and for light scattering involving the $\nu \to \nu'$ transition are different: namely, by (8.52), with the light scattered in a symmetric well the parities of the initial and final states are conserved. Note also that if the electron and hole parameters are identical, and the envelopes $f^{\mathrm{e}}_{\nu\mathbf{k}}(z)$ and $f^{\mathrm{h}}_{\nu\mathbf{k}}(z)$ coincide, the coefficients $d_{\nu'\nu}$ with $\nu \neq \nu'$ are zero. In real conditions, a deviation from the selection rules in parity may be due to an asymmetry of the heterostructure, additional scattering of the light-excited electron-hole pairs from static defects, and hybridization of the heavy and light hole states for $\mathbf{k} \neq 0$. Indeed, if hybridization is included, the coefficients $\gamma_{s's}$ (2,1) in (8.50) are proportional to the matrix element $k_z^{(2,1)}$ and represent linear combinations of quantities of the type

$$e^*_{2z}(\mathbf{k}\cdot\mathbf{e}_{1\perp}),\ e^*_{2z}(\mathbf{e}_{1\perp}\times\mathbf{k})_z\sigma_z,\ \left(\mathbf{e}^*_2\times\mathbf{e}_1\right)_z(\mathbf{k}\cdot\boldsymbol{\sigma}_\perp),\ \text{etc.},$$

transforming according to the representation B_2 of the D_{2d} group.

Figure 8.1 presents resonant light-scattering spectra measured in $z(yx)\bar{z}$ geometry (backscattering with crossed polarizer and analyzer) on three multilayer

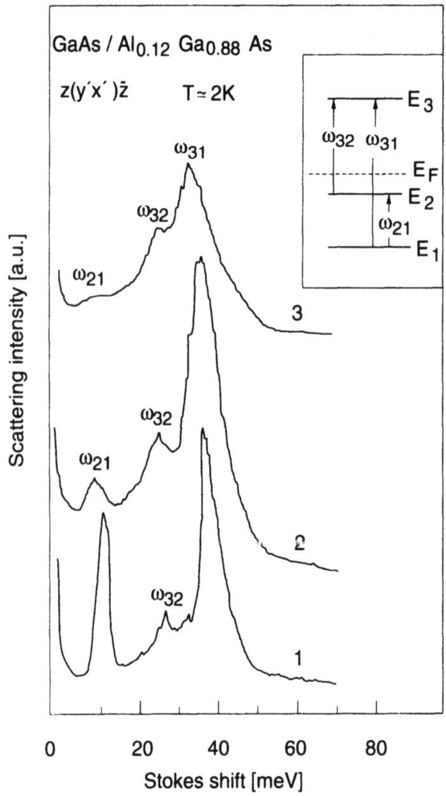

Fig. 8.1. Spectra of scattering by intersubband excitations in modulation-doped GaAs/AlGaAs heterostructures (well thickness $a \simeq 250$ Å). Energy $\hbar\omega_1 = 1.9$ eV. The inset explains the nomenclature of single-particle intersubband transitions. E_F - Fermi energy, $z \parallel [001]$, $x' \parallel [1\bar{1}0]$, $y' \parallel [110]$ [8.3]

structures GaAs/$Al_{0.12}Ga_{0.88}As$ excited with $\hbar\omega_1 \sim E^0_{c\Gamma_0} - E^0_{v\Gamma_7} = E_g + \Delta$. The strongest scattering line corresponds to the parity-conserving $1 \to 3$ transition. The efficiency of scattering involving the $1 \to 3$ and $1 \to 2$ transitions is shown in Figure 8.2.

Equation (8.47) was derived in the single-particle approximation. In this case, the change of the photon energy in scattering coincides with the difference between the single-particle energies $E_{f'f} = E_{f'} - E_f$. The inclusion of Coulomb interaction between electrons results in renormalization of the intersubband excitation energy. For the $1 \to 2$ transitions in a single quantum well, the renormalized intersubband excitation energies for the charge (E_{CD}) and spin (E_{SD}) density can be written as

$$E^2_{CD} = E^2_{21}(1 + \alpha_{21} - \beta_{21}),$$
$$E^2_{SD} = E^2_{21}(1 - \beta_{21}). \qquad (8.54)$$

Here the parameters α_{21} and β_{21} describe the depolarization and exchange-correlation (exciton) effects. For α_{21} one can obtain the following expression:

$$\alpha_{21} = 8\pi N_s e^2 E_{21}^{-1} L_{21} / \kappa(\omega_{CD}), \qquad (8.55)$$

where $N_s = N_s^1 - N_s^2$, $N_s^{(\nu)}$ is the two-dimensional electron density in subband ν, $\kappa(\omega)$ the dielectric constant (8.32), and

$$L_{21} = \int_{-\infty}^{+\infty} dz \left[\int_{-\infty}^{z} f_2(z') f_1(z') \, dz' \right]^2 . \qquad (8.56)$$

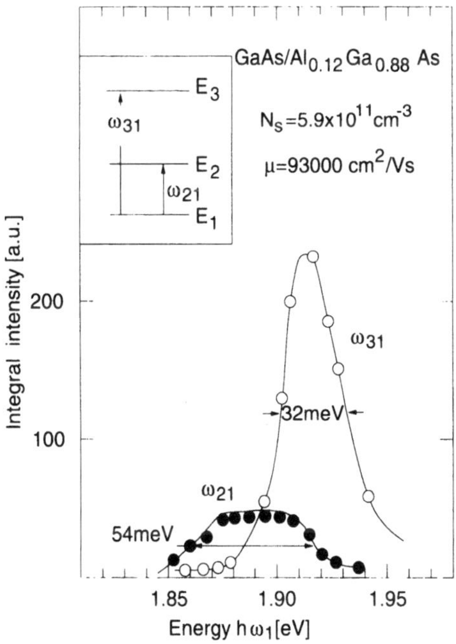

Fig. 8.2. Efficiency of light scattering by intersubband excitations vs. incident photon energy. The data relate to the sample whose scattering spectrum for $\hbar\omega_1 = 1.9$ eV is shown by curve 1 in Fig. 1 [8.3]

Including the dispersion $\kappa_{\text{phon}}(\omega)$ of the phonon contribution to the dielectric constant, we come to the dispersion equation for intersubband excitations of charge density (mixed excitation of LO phonon and intersubband plasmon)

$$\frac{\omega_{\text{L}}^2 - \omega^2}{\omega_{\text{T}}^2 - \omega^2} + \frac{\omega_{\text{pl}}^2(2,1)}{\omega_{21}^2 - \omega^2} = 0, \tag{8.57}$$

where

$$\omega_{\text{pl}}^2(2,1) = \frac{8\pi N_s e^2 \omega_{21} L_{21}}{\hbar \kappa_\infty}, \tag{8.58}$$

$\omega_{21} = E_{21}/\hbar$ and the exchange-correlation correction is neglected. Equation (8.57) has two solutions

$$\omega_\pm^2 = \frac{1}{2}\Bigg(\omega_{21}^2 + \omega_{\text{L}}^2 + \omega_{\text{pl}}^2(2,1) \pm \left\{\left[\omega_{21}^2 + \omega_{\text{L}}^2 + \omega_{\text{pl}}^2(2,1)\right]^2 - 4\left[\omega_{21}^2\omega_{\text{L}}^2 + \omega_{\text{pl}}^2(2,1)\omega_{\text{T}}^2\right]\right\}^{1/2}\Bigg). \tag{8.59}$$

Figure 8.3 presents spectra of light scattering by intersubband charge and spin density excitations measured on a single quantum well structure $GaAs/Al_{0.3}Ga_{0.7}As$ in parallel and crossed polarizations. The excitation-energy difference $E_{CD} - E_{SD}$ is seen to exceed 2 meV.

Not only inter- but also intra-subband excitations contribute to the light scattering from electrons in heterostructures. For two-dimensional electrons,

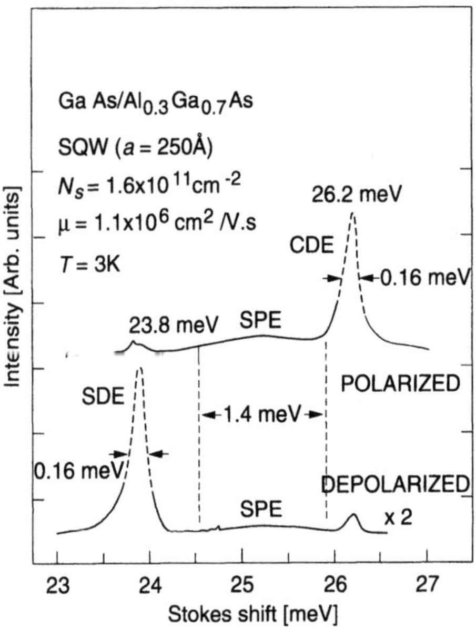

Fig. 8.3. Light scattering by collective intersubband excitations of charge density (CDE) and spin density (SDE), as well as by single-particle excitations (SPE) [8.4]

the dispersion of plasma oscillations exhibits a square-root behavior

$$\omega_{\rm pl}^{\rm 2D}(\mathbf{k}_\perp) = \left(\frac{2\pi e^2 N_s k_\perp}{\kappa_\infty m_{\rm c}}\right)^{1/2}, \tag{8.60}$$

where $\mathbf{k}_\perp$ is the two-dimensional plasmon wave vector.

In a periodic MQW structure, the plasmon frequency depends on the three-dimensional wave vector $\mathbf{Q}$ and differs from (8.60) in a structural factor

$$\omega_{\rm pl}^{SL}(\mathbf{Q}) = \omega_{\rm pl}^{\rm 2D}(Q_\perp)\, S^{1/2}(\mathbf{Q}),$$
$$S(\mathbf{Q}) = \frac{\sinh Q_\perp d}{\cosh Q_\perp d - \cos Q_z d}. \tag{8.61}$$

In the limit $Q_\perp d \gg 1$, the interaction between plasmons in different layers may be neglected, $S(\mathbf{Q}) \to 1$ and $\omega_{\rm pl}^{\rm SL} \to \omega_{\rm pl}^{\rm 2D}$. For $Q_\perp d \ll 1, |Q_z| d$ we have

$$S(\mathbf{Q}) \simeq \frac{Q_\perp d}{1 - \cos Q_z d}$$

and the plasmons are characterized by a linear dispersion law (acoustical plasmons)

$$\omega_{\rm SL}^{\rm pl}(\mathbf{Q}) = v(Q_z)\, Q_\perp,$$
$$v(Q_z) = \left(\frac{2\pi e^2 N_s}{\kappa_\infty m_{\rm c}} \frac{1}{1 - \cos Q_z d}\right)^{1/2}. \tag{8.62}$$

Of interest is also the limiting case $(Q_z d)^2 \ll (Q_\perp d)^2 \ll 1$, where

$$S(\mathbf{Q}) \simeq \frac{2}{Q_\perp d}, \quad \omega_{\rm pl}^{\rm SL} \simeq \left(\frac{4\pi^2 e^2 N_s}{\kappa_\infty m_{\rm c} d}\right)^{1/2}$$

and the plasma oscillations become quasi-three-dimensional.

Interaction of plasmons with polar optical phonons is included in (8.11, 12) by replacing κ_∞ with $\kappa_\infty + \kappa_{\rm phon}$. The dispersion equation thus obtained has two solutions (plasmon-phonon modes) which can be represented in the form (8.59) if one replaces $\omega_{\rm pl}$ (2,1) in (8.59) by the intrasubband plasmon frequency and sets $\omega_{21} = 0$.

8.3 Scattering by Acoustical Phonons with a Folded Dispersion Law

As pointed out in Chap. 4, each acoustic branch in a SL folds within the new Brillouin zone $|Q_z| \leqslant \pi/d$ into a series of mini-branches with forbidden mini-bands at the folding points $Q_z = 0, \pm\pi/d$. To describe light scattering from folded acoustical phonons in terms of the photoelastic mechanism, we expand the dielectric susceptibility fluctuation at frequency ω_1 in components of the deformation tensor

$$\delta\chi_{\alpha\beta}(\mathbf{x},t) = P_{\alpha\beta\gamma\delta}(z)\varepsilon_{\gamma\delta}(\mathbf{x},t). \tag{8.63}$$

In contrast to (8.24) derived for a homogeneous semiconductor, here the photoelastic coefficients $P_{\alpha\beta\gamma\delta} = \partial\chi_{\alpha\beta}/\partial\varepsilon_{\gamma\delta}$ depend on z and change stepwise at the SL heterojunctions. One can conveniently expand the tensor $\mathbf{P}$ in a Fourier series

$$\mathbf{P}(z) = \sum_m \mathbf{P}^{(m)} \mathrm{e}^{\mathrm{i}G_m z} \tag{8.64}$$

and the lattice displacement $\mathbf{u}$, in states of the folded acoustical phonons

$$\mathbf{u}_{\mathbf{Q}}^{(l)}(\mathbf{x}) = \mathrm{e}^{\mathrm{i}\mathbf{Q}\cdot\mathbf{x}} \sum_m \mathbf{u}_{\mathbf{Q}m}^{(l)} \mathrm{e}^{\mathrm{i}G_m z}, \tag{8.65}$$

where $G_m = 2\pi m/d (m = 0, \pm 1, \dots)$ and l is the phonon-branch index (in the notations of Fig. 4.1, $l = 0, \pm 1, \pm 2, \dots$). Using (8.64, 65), we obtain

$$\begin{aligned} \delta\chi_{\alpha\beta}(\mathbf{q}) &\equiv \frac{1}{V}\int \delta\chi_{\alpha\beta}(\mathbf{x})\mathrm{e}^{\mathrm{i}\mathbf{q}\cdot\mathbf{x}}\,\mathrm{d}\mathbf{x} \\ &= \sum_m P_{\alpha\beta\lambda\nu}^{-m} u_{\mathbf{q}m,\lambda} \mathrm{i}\,(q_\nu + G_m\delta_{\nu z})\,. \end{aligned} \tag{8.66}$$

For GaAs/AlGaAs-type superlattices, the modulation of photo-elastic coefficients plays a more essential role than the mixing in (8.65) of the various space harmonics. Therefore, the correlation of the fluctuations (8.22) can be evaluated by setting

$$\mathbf{u}_{\mathbf{Q}}^{(l)}(\mathbf{x}) = \left(\frac{\hbar}{2\rho\Omega_{\mathbf{Q}l}V}\right)^{1/2} \left(b_{\mathbf{Q}l} + b_{-\mathbf{Q}l}^{+}\right) \mathbf{e}_{\mathbf{Q}l}\mathrm{e}^{\mathrm{i}\mathbf{Q}\cdot\mathbf{x}+\mathrm{i}G_m z}, \tag{8.67}$$

where $m = l$ (sign Q_z), $\Omega_{\mathbf{Q}l}$ is the frequency and $\mathbf{e}_{\mathbf{Q}l}$ the phonon polarization. The intensity of the light scattered from the lth mode into the Stokes region of the spectrum can be written

$$J_l \propto \Omega_l \left|P^{(m)}\right|^2 (N_{\Omega_l} + 1)\delta(\omega - \Omega_l), \tag{8.68}$$

where the transferred frequency

$$\begin{aligned} \Omega_l &= v_{\mathrm{SL}}\,|\mathbf{G}_m + \mathbf{q}|\,, \\ G_{m,\nu} &= G_m\delta_{\nu z}, \quad v_{\mathrm{SL}} = d\left(\frac{a}{v_A} + \frac{b}{v_B}\right)^{-1}, \\ P^{(m)} &= P_{\alpha\beta\gamma z}^{(m)} e_{2\alpha}^{*} e_{1\beta} e_{\mathbf{Q}l,\gamma} \quad (\text{for } q \ll |G_m|)\,. \end{aligned} \tag{8.69, 8.70}$$

For the backscattering of the light propagating along the z-axis

$$\Omega_l = v_{SL}\left(\frac{2\pi|l|}{d} + \frac{l}{|l|}q\right). \tag{8.71}$$

Thus in the scattering spectrum, one should observe doublets $\pm|l|$ with an $|l|$ - independent splitting

$$\Delta\Omega = \Omega_{|l|} - \Omega_{-|l|} = 2v_{SL}q \simeq 2\frac{v_{SL}}{c}\omega_1\sqrt{\kappa(\omega_1)}. \tag{8.72}$$

For small q one should take into account the deviation from the linear dispersion law (8.71) and use a more general expression (4.39)

$$\Omega_l = \frac{v_{SL}}{d}\left\{2\pi|l| + \frac{l}{|l|}\left[\varepsilon^2\sin^2\left(\frac{\Omega_l^o}{v_A}a\right) + (qd)^2\right]^{1/2}\right\},$$

$$\varepsilon = \frac{\rho_B v_B - \rho_A v_A}{(\rho_A\rho_B v_A v_B)^{1/2}}, \quad \Omega_l^o = 2\pi|l|\frac{v_{SL}}{d}. \tag{8.73}$$

One can readily calculate the Fourier components of the coefficient P

$$P^{(o)} = d^{-1}(aP_A + bP_B),$$

$$P^{(m)} = (P_A - P_B)\frac{\sin(\pi ma/d)}{\pi m} \quad (m \neq 0), \tag{8.74}$$

where $P_{A,B}$ is the value of this coefficient in a layer of type A or B. We can now find the intensity ratio of the light scattered by phonons of the lth ($l \neq 0$) mode and of the lowest branch, $l = 0$ (the analog of the Brillouin scattering):

$$\frac{J_l}{J_o} = \gamma\left|\frac{P_A - P_B}{P^{(o)}}\right|^2 \frac{\sin^2(\pi la/d)}{(\pi l)^2},$$

where

$$\gamma = \frac{\Omega_l(N_{\Omega_l} + 1)}{\Omega_o(N_{\Omega_o} + 1)}.$$

Under ordinary conditions, $k_B T \gg \hbar\Omega_l$ and the coefficient $\gamma \simeq 1$.

For the light backscattered by longitudinal phonons with $\mathbf{e}_{Ql} \parallel z$ (folded LA phonons) in a GaAs/AlGaAs (001) superlattice, the coefficient $P^{(m)}$ in (8.68) can be represented in the form

$$P^{(m)} = P^{(m)}_{xxzz}(\mathbf{e}_2^* \cdot \mathbf{e}_1).$$

Figure 8.4 presents Raman spectra of scattering in a GaAs/$Al_{0.3}Ga_{0.7}As$ superlattice ($a = 42$ Å, $b = 8$ Å). In the $\mathbf{e}_2 \parallel \mathbf{e}_1$ geometry, one observes three doublets with $|l| = 1, 2, 3$. The arrows in Fig. 8.4 show the calculated line positions. In accordance with the selection rules, there is no scattering from folded LA phonons in the crossed geometry ($\mathbf{e}_2 \perp \mathbf{e}_1$). The peak at 160 cm^{-1} is attributed to 2TA scattering.

The dispersion of folded LA phonons in the vicinity of the first miniband determined from Raman spectra in a SL with $d = 25.5$ Å is depicted in Fig. 8.5.

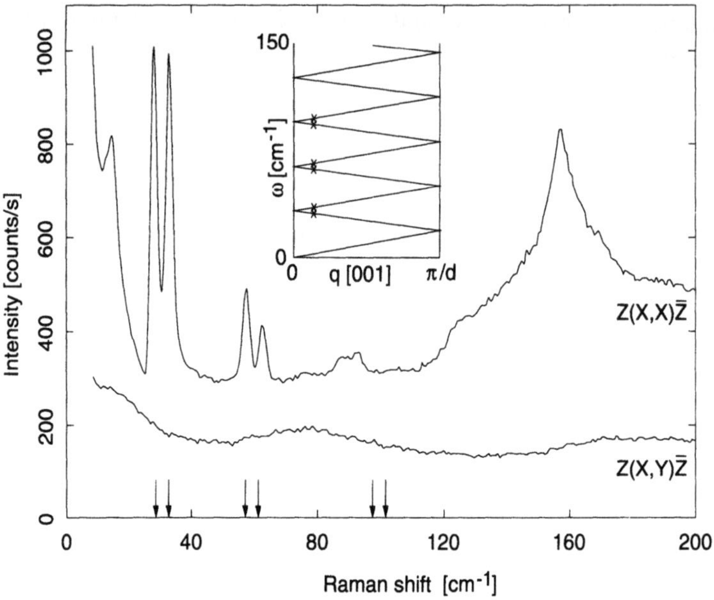

Fig. 8.4. Spectrum of light scattering by folded acoustic phonons in GaAs/$Al_{0.3}Ga_{0.7}As$ SL with $a = 42$ Å, $b = 8$ Å. Pump wavelength $\lambda_1 = 5145$ Å, $T = 300$ K. The inset shows phonon dispersion by Rytov's model. The arrows specify the peak positions derived from X-ray diffraction data [8.5]

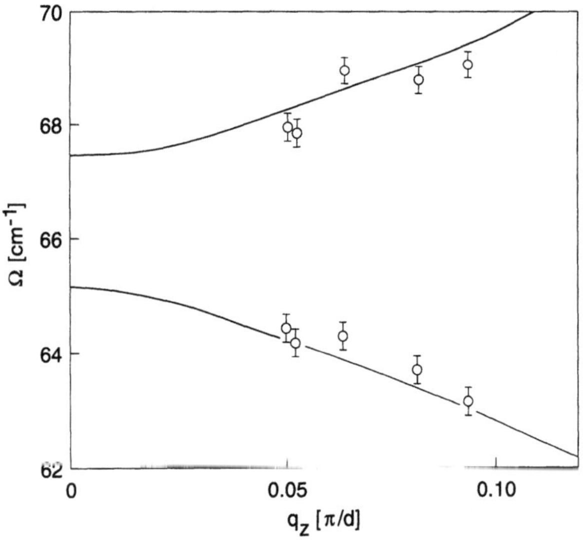

Fig. 8.5. Phonon dispersion near the first gap in the folded LA phonon spectrum. Points: experimental data for GaAs/AlAs SL with $a = 14$ Å, $b = 12$ Å; solid lines: linear chain model calculation [8.5]

The experimental data were obtained using laser lines in the range 4979 to 6794 Å. To measure phonon dispersion in the parabolic region ($qd < |\varepsilon|$), one has to use longer wavelength radiation or cross over from back to forward scattering geometry.

8.4 Scattering by Optical Phonons in Heterostructures

The linearly independent components of the Raman tensor $R_{\alpha\beta}$ can be found by the method of invariants. The principal rule here reads as follows: in light scattering by the long wavelength optical phonons (D_μ, i) the quantity $R(D_\mu, i) = e^*_{2\alpha} e_{1\beta} R_{\alpha\beta}(D_\mu, i)$ transforms under point-symmetry transformations as the i-th component of the basis of the representation D_μ. For instance, for phonons of symmetry A_1, B_2 and E in a type GaAs/AlAs(001) superlattice we have

$$\begin{aligned}
R(A_1) &= R_{1,\perp}\left(\mathbf{e}^*_{2\perp} \cdot \mathbf{e}_{1\perp}\right) + R_{1,\parallel} e^*_{2z} e_{1z},\\
R(B_2) &= R_2\left(e^*_{2x} e_{1y} + e^*_{2y} e_{1x}\right),\\
R(E,1) &= R_3 e^*_{2z} e_{1y} + R_4 e^*_{2y} e_{1z},\\
R(E,2) &= R_3 e^*_{2z} e_{1x} + R_4 e^*_{2x} e_{1z}.
\end{aligned} \tag{8.75}$$

To derive (8.75), we have to recall that as basis functions of the representations A_1, B_2, E of the group D_{2d} one may choose the functions $x^2 + y^2$ or z^2 (representation A_1), $xy(B_2)$, yz and $xz(E)$. According to (8.75), the tensors $R_{\alpha\beta}(D_\mu)$ can be represented in the following matrix form

$$\mathbf{R}(A_1) = \begin{bmatrix} R_{1,\perp} & 0 & 0 \\ 0 & R_{1,\perp} & 0 \\ 0 & 0 & R_{1,\parallel} \end{bmatrix}, \quad \mathbf{R}(B_2) = \begin{bmatrix} 0 & R_2 & 0 \\ R_2 & 0 & 0 \\ 0 & 0 & 0 \end{bmatrix},$$
$$\mathbf{R}(E,1) = \begin{bmatrix} 0 & 0 & 0 \\ 0 & 0 & R_4 \\ 0 & R_3 & 0 \end{bmatrix}, \quad \mathbf{R}(E,2) = \begin{bmatrix} 0 & 0 & R_4 \\ 0 & 0 & 0 \\ R_3 & 0 & 0 \end{bmatrix}. \tag{8.76}$$

For the sake of completeness, we write out also the tensors of scattering for phonons A_2 and B_1 in crystals of D_{2d} class

$$\mathbf{R}(A_2) = \begin{bmatrix} 0 & R_5 & 0 \\ -R_5 & 0 & 0 \\ 0 & 0 & 0 \end{bmatrix}, \quad \mathbf{R}(B_1) = \begin{bmatrix} R_6 & 0 & 0 \\ 0 & -R_6 & 0 \\ 0 & 0 & 0 \end{bmatrix}. \tag{8.77}$$

In bulk GaAs, long-wavelength optical vibrations are characterized by F_2 symmetry. This representation of the group T_d transforms with the symmetry lowered to D_{2d} to the representations $B_2 + E$. Therefore the scattering tensors for phonons of polarization $\mathbf{u} \parallel x, y, z$ in GaAs coincide with the tensors $\mathbf{R}(E,1)$, $\mathbf{R}(E,2)$, $\mathbf{R}(B_2)$, where one has to set $R_2 = R_3 = R_4$. Optical phonons confined in the quantum-well layer or barrier layer of the GaAs/AlAs(001) heterostructures are characterized by symmetry B_2, E for an even envelope, and A_1, E for an odd envelope of the lattice vibration amplitude.

In heterostructures grown along [111], the point symmetry lowers to C_{3v}. To find the Raman tensor in this case, one has to take into account that the basis for the irreducible representations of the group C_{3v} is formed by the following linear combinations of the products $e^*_{2\alpha}e_{1\beta}$:

$$e^*_{2z'}e_{1z'};\ e^*_{2x'}e_{1x'} + e^*_{2y'}e_{1y'} \quad \text{(representation } A_1\text{)},$$

$$e^*_{2x'}e_{1y'} - e^*_{2y'}e_{1x'} \quad \text{(representation } A_2\text{)},$$

$$\left.\begin{array}{l}\left(e^*_{2x'}e_{1z'},\, e^*_{2y'}e_{1z'}\right),\ \left(e^*_{2z'}e_{1x'},\, e^*_{2z'}e_{1y'}\right),\\ \left(-e^*_{2x'}e_{1x'} + e^*_{2y'}e_{1y'},\, e^*_{2x'}e_{1y'} + e^*_{2y'}e_{1x'}\right)\end{array}\right\}\text{(representation } E\text{)},$$

where a coordinate system with the axes $x' \parallel [1\bar{1}0]$, $y' \parallel [11\bar{2}]$, $z' \parallel [111]$ was chosen. In a similar way, one can find selection rules for heterostructures grown along [110] or [012] (classes C_{2v} and C_2, respectively).

In the microscopic description of scattering, we can use an expression for $R_{\alpha\beta}$ of the type (8.39), where $V_{f'f}$ is the matrix element of electron-hole pair interaction with the optical phonon. Two mechanisms contribute to $V_{f'f}$, one of them being the Fröhlich mechanism with the interaction operator

$$V = -e\left[\varphi\left(\mathbf{x}_{\mathrm{e}}\right) - \varphi\left(\mathbf{x}_{\mathrm{h}}\right)\right] \tag{8.78}$$

and the other, the deformation interaction (4.79) of the hole with the optical phonon. Here $\varphi(\mathbf{x})$ is the electric-field potential produced by polar lattice vibrations. As already pointed out in Sect. 8.1, for the Fröhlich mechanism the matrix element of single phonon light scattering in a homogeneous sample is linear in $\mathbf{q}$. Because of the heterostructure being spatially inhomogeneous in the direction of the z-axis, such scattering is allowed in the dipole approximation. Moreover, under resonance conditions, $\hbar\omega_1 \simeq E^0_{\mathrm{e}1} - E^0_{\mathrm{h}1}$, the main contribution to $z(xx)\bar{z}$ or $z(yy)\bar{z}$ scattering (the coefficient $R_{1,\perp}$ in (8.75)) comes from the Fröhlich electron-phonon interaction in the process $0 \to f \to f' \to 0$ with the same intermediate states, $f = f' = (\mathrm{e}1, \mathrm{h}1)$, and

$$V_{f'f} = -e\int\left[f^2_{\mathrm{e}1}(z) - f^2_{\mathrm{h}1}(z)\right]\varphi(z)\,\mathrm{d}z. \tag{8.79}$$

In GaAs/AlAs quantum-well structures this matrix is nonzero for confined LO phonons with an even function $\varphi(z)$ (mode A_1). Note that in the special case of the electron- and hole-band parameters being exactly equal, $f_{e1}(z) \equiv f_{h1}(z)$ and $V_{f'f} = 0$ for an arbitrary function $\varphi(z)$.

Deformation interaction (4.79) contributes to the Raman scattering by the phonon B_2 observed in the $z(xy)\bar{z}$ geometry (crossed polarizations) and described in (8.75) by the coefficient R_2. To check this, one can take out of all possible intermediate states f and f' those with $\mathbf{k}_{\mathrm{e}\perp} = \mathbf{k}_{\mathrm{h}\perp} = 0$. In this case, the matrix element $V_{f'f}$ will be nonzero only for transitions between the heavy and light hole subbands, i.e., $f = (e\nu s, \mathrm{hh}\nu m)$ and $f' = (e\nu s, \mathrm{lh}\nu m')$, or

$f = (e\nu, \mathrm{lh}\nu m)$ and $f' = (e\nu, \mathrm{hh}\nu m')$ with $|m - m'| = 2$. Taking the selection rules for interband optical transitions, one can readily show that

$$\sum_{\alpha,\beta=x,y,} e^*_{2\alpha} e_{1\beta} j^{\alpha}_{o,f'}(0) j^{\beta}_{f,o}(0) \propto e^*_{2+} e_{1+} \text{ or } e^*_{2-} e_{1-}$$

where $e_{1\pm} = e_{1x} \pm \mathrm{i} e_{1y}$, $e^*_{2\pm} = e^*_{2x} \pm \mathrm{i} e^*_{2y}$, and

$$e^*_{2\alpha} e_{1\beta} R_{\alpha\beta} \propto \left(e^*_{2+} e_{1+} - e^*_{2-} e_{1-} \right),$$

which is in agreement with (8.75).

Figure 8.6 shows Raman spectra for scattering from confined phonons measured on GaAs/AlAs superlattices made up of 400 double layers with

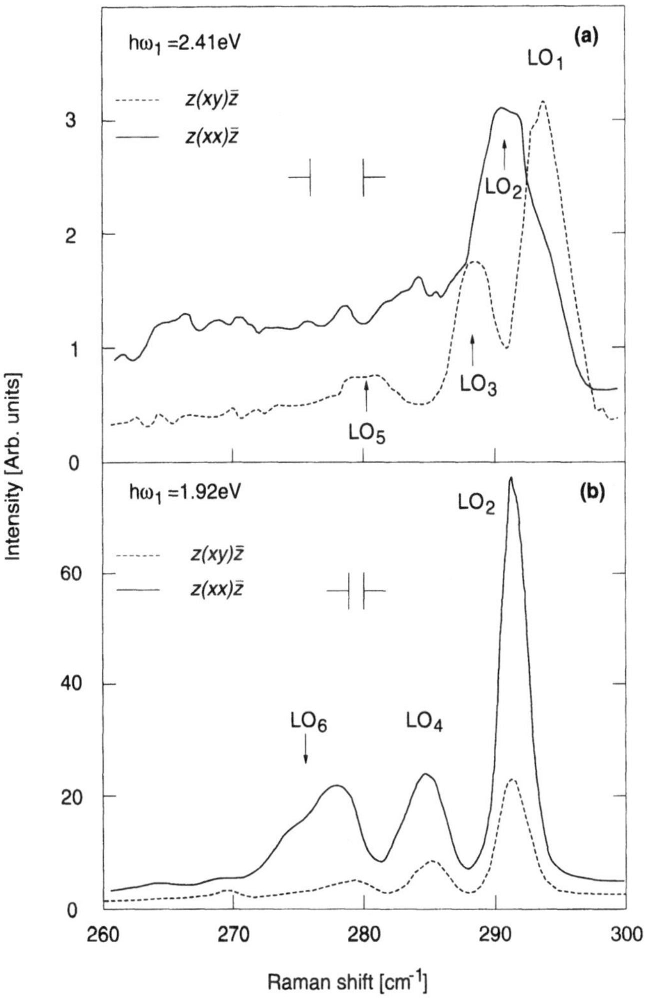

Fig. 8.6. a Nonresonant and **b** resonant Raman scattering spectra for GaAs/AlAs SL with $a = 20$ Å, $b = 60$ Å. The peak to the right of LO_6 phonon frequency is believed to be due to the interface mode [8.6]

$a = 20$ Å, $b = 60$ Å. When excited under nonresonant conditions, the cross sections of scattering by the phonons LO_{2l+1} (B_2 symmetry) and LO_{2l} (A_1 symmetry) observed in the $z(xy)\bar{z}$ and $z(xx)\bar{z}$ polarization, respectively, are comparable in order of magnitude. In accordance with theoretical predictions, under resonant excitation the Fröhlich mechanism dominates, and one observes scattering by LO_{2l} phonons. The presence of the same LO_{2l} lines (though of a considerably lower intensity) in crossed geometry, $z(xy)\bar{z}$, can be attributed to the influence of static defects on the Fröhlich interaction of carriers with optical phonons.

8.5 Acceptor Spin-Flip Raman Scattering

Sapega et al. [8.7] observed strong Raman scattering due to the spin-flip of holes bound to acceptors in p-type GaAs/AlGaAs (001) multiple quantum wells in a longitudinal magnetic field $\mathbf{B} \parallel z$ [8.8]. The scattering line polarization and the intensity ratio between the Stokes and anti-Stokes components depend strongly on the excitation frequency. Two *pure* limiting cases observed at the edges of the scattering excitation spectrum can be discriminated in experiment. When excited with circularly polarized light at the long-wavelength edge, the backscattering line, which is also completely polarized, is observed in the (σ_+, σ_-) or (σ_-, σ_+) configurations and is not seen in (σ_+, σ_+) or (σ_-, σ_-). Each of the (σ_+, σ_-) or (σ_-, σ_+) spectra contains either the Stokes or anti-Stokes component. Note that if the (σ_-, σ_+) geometry reveals a Stokes shift, then in the opposite geometry only the anti-Stokes component is present. It should be pointed out that the circular polarization $\sigma_\eta (\eta = \pm)$ of the initial light and the observed circular component $\sigma_\lambda (\lambda = \pm)$ of the backscattered light are denoted here by the sign of the photon angular-momentum projection onto the direction of the exciting light coinciding with the principal axis z of the structure. Using this nomenclature, the specularly reflected light is observed in the (σ_+, σ_+) or (σ_-, σ_-) configurations. In the opposite limiting case realized at the long-wavelength edge of the excitation specturm, one observes under circularly polarized excitation both the Stokes and anti-Stokes scattering lines simultaneously. These lines are circularly polarized (the polarization may be as high as 80%). The sign of the backscattered-light polarization coincides with that of specularly reflected light thus implying that the scattering occurs predominantly in the (σ_+, σ_+) and (σ_-, σ_-) configurations. The intensity ratio of the Stokes and anti-Stokes components can be approximated with good precision by $\exp(|\Delta E|/k_B T)$, where ΔE is the change of the photon energy in scattering proportional to the longitudinal magnetic-field component B_z. Since this scattering is observed to occur only in p-type samples, it is interpreted as a process accompanied by a $\pm 3/2 \to \mp 3/2$ spin flip of the hole bound to an acceptor resulting from its exchange interaction with the hole in a photo-excited localized exciton.

Let us consider now the possible mechanisms of this scattering process. First of all, we recall the selection rules for the optical excitation (and emission) of an electron-hole pair or exciton: transitions to the states $|-1/2, 3/2\rangle$ and $|1/2, -3/2\rangle$ are allowed, accordingly, in the σ_+ and σ_- polarizations, the states $|1/2, 3/2\rangle$, $|-1/2, -3/2\rangle$ being forbidden in the dipole approximation. Here in the notation $|s, m\rangle$ the first index specifies the electron-spin projection on the z-axis and the second, the hole-spin projection in terms of the Luttinger Hamiltonian.

The light scattering accompanied by a spin-flip $3/2 \rightarrow -3/2$ or $-3/2 \rightarrow 3/2$ of the acceptor-bound hole includes three stages:

(i) Initial photon absorption involving excitation of the exciton. The region of the scattering excitation spectrum studied by *Sapega* et al. [8.7] extends apparently over both the states of the bound exciton with a comparatively small localization radius (long-wavelength edge) and the weakly localized states, including the electron-hole pairs, modified by Coulomb interaction and localized in a quantizing magnetic field (short-wavelength edge).

(ii) Spin-flip of the acceptor-bound hole in the process:

$$|\pm 3/2\rangle_{\mathrm{A}} |\pm 1/2, \mp 3/2\rangle \rightarrow |\mp 3/2\rangle_{\mathrm{A}} |\mp 1/2, \pm 3/2\rangle \tag{8.80a}$$

or

$$|\pm 3/2\rangle_{\mathrm{A}} |s, m\rangle \rightarrow |\mp 3/2\rangle_{\mathrm{A}} |s, m\rangle, \tag{8.80b}$$

where $|j\rangle_{\mathrm{A}}$ is the state of the hole bound to the acceptor.

(iii) Secondary photon emission by the exciton. Process (b) represents actually hole–hole exchange interaction involving spin-flip of the acceptor-bound hole and preserving the spin of the exciton hole. Process (a) assumes the operation of three exchange-interaction mechanisms: flip-flop of the holes $|\pm 3/2\rangle_{\mathrm{A}} |s, \mp 3/2\rangle \rightarrow |\mp 3/2\rangle_{\mathrm{A}} |s, \pm 3/2\rangle$, spin flip of the electron and hole in the exciton $|\pm 1/2, \pm 3/2 \rightarrow |\mp 1/2, \mp 3/2\rangle$, and spin flip of the exciton-bound hole induced by the polarized hole at the acceptor. For these processes to occur, one should include in the exchange-interaction operator at least the terms

$$\begin{aligned} &\sigma_z^{\mathrm{h}} \left(\Delta_{1-}\sigma_+^{\mathrm{A}} + \Delta_{1+}\sigma_-^{\mathrm{A}}\right) + \sigma_z^{\mathrm{A}} \left(\Delta_{2-}\sigma_+^{\mathrm{h}} + \Delta_{2+}\sigma_-^{\mathrm{h}}\right) \\ &\quad + \Delta_\perp \left(\sigma_+^{\mathrm{h}}\sigma_-^{\mathrm{A}} + \sigma_-^{\mathrm{h}}\sigma_+^{\mathrm{A}}\right) + \Delta_\perp^{\mathrm{eh}} \left(\sigma_x^{\mathrm{e}}\sigma_x^{\mathrm{h}} - \sigma_y^{\mathrm{e}}\sigma_y^{\mathrm{h}}\right). \end{aligned} \tag{8.81}$$

Here $\Delta_{l-} = \Delta_{l+}^*$, $\sigma_\pm = (\sigma_x \pm \mathrm{i}\sigma_y)/2$, $\sigma_\alpha^{\mathrm{e}}$ are the Pauli matrices acting upon the electron spin states, and $\sigma_\alpha^{\mathrm{h}}$ and $\sigma_\alpha^{\mathrm{A}}$ are the analogs of the Pauli matrix for the hole in the exciton or at the acceptor in the basis $\alpha(x + \mathrm{i}y)$, $\beta(x - \mathrm{i}y)$.

If the inequalities

$$\left|\Delta_\perp^{\mathrm{eh}}\right| \gtrsim |\Delta_\perp| \gtrsim \left|g_{\mathrm{A,h}}\mu_{\mathrm{B}} B_z\right|$$

are met, where $g_{\mathrm{A,h}}$ is the g-factor of the hole at the acceptor and in the exciton, then the relative contribution to scattering due to processes (a) and (b) will be determined by the relative magnitude of the linearly independent coefficients $\Delta_1 = |\Delta_{1\pm}|$ and $\Delta_2 = |\Delta_{2\pm}|$. As the exciton localization radius increases, the constants $\Delta_\perp$, $\Delta_\perp^{\mathrm{eh}}$ decrease, the contribution of process (b) becoming predominant.

To evaluate the exchange constants $\Delta_{\perp}, \Delta_{1,2}$, we can represent the matrix element of hole–hole Coulomb interaction between the states $|s_1 j_1 j_1'\rangle = |j_1\rangle_A |s_1 j_1'\rangle$ and $|s_2 j_2 j_2'\rangle$ in the form of two terms [8.9]

$$\left\langle s_1 j_1 j_1' \left| \frac{e^2}{\kappa r_{12}} \right| s_2 j_2 j_2' \right\rangle = \left(V_{j_1 j_1', j_2 j_2'} + V^{\mathrm{exch}}_{j_1 j_1', j_2 j_2'} \right) \delta_{s_1 s_2}, \quad \text{where}$$

$$V_{j_1 j_1', j_2 j_2'} = \frac{e^2}{\kappa} \iint \frac{\mathrm{d}\mathbf{x}_1\, \mathrm{d}\mathbf{x}_2}{r_{12}} \psi_{\mathrm{A}}^{j_1^*}(\mathbf{x}_1)\, \psi_{\mathrm{A}}^{j_2}(\mathbf{x}_1) \int \mathrm{d}\mathbf{x}_{\mathrm{e}} \Psi_{\mathrm{exc}}^{s j_1'^*}(\mathbf{x}_{\mathrm{e}}, \mathbf{x}_2)$$
$$\times\, \Psi_{\mathrm{exc}}^{s j_2'}(\mathbf{x}_{\mathrm{e}}, \mathbf{x}_2)\,, \tag{8.82a}$$

$$V^{\mathrm{exch}}_{j_1 j_1', j_2 j_2'} = -\frac{e^2}{\kappa} \iint \frac{\mathrm{d}\mathbf{x}_1\, \mathrm{d}\mathbf{x}_2}{r_{12}} \psi_{\mathrm{A}}^{j_1^*}(\mathbf{x}_1)\, \psi_{\mathrm{A}}^{j_2}(\mathbf{x}_2) \int \mathrm{d}\mathbf{x}_{\mathrm{e}} \Psi_{\mathrm{exc}}^{s j_1'^*}(\mathbf{x}_{\mathrm{e}}, \mathbf{x}_2)$$
$$\times\, \Psi_{\mathrm{exc}}^{s j_2'}(\mathbf{x}_{\mathrm{e}}, \mathbf{x}_1)\,. \tag{8.82b}$$

Here $r_{12} = |\mathbf{x}_1 - \mathbf{x}_2|$, ψ_{A}^{j} is the wave function of the acceptor-bound hole, and Ψ_{exc}^{sj} is the exciton wave function in the two-particle representation. The first term describes the conventional Coulomb interaction which is compensated to a large extent by the hole interaction with the impurity ion and electron in the exciton. The second term describes the exchange interaction between the holes. The coefficients $\Delta_{l\pm}$, $\Delta_{\perp}$ introduced in (8.81) are connected with $V^{\mathrm{exch}}_{j_1 j_1, j_2 j_2'}$ through the relations

$$\Delta_{1-} = V^{\mathrm{exch}}_{\frac{3}{2}\frac{3}{2};-\frac{3}{2},\frac{3}{2}}; \quad \Delta_{2-} = V^{\mathrm{exch}}_{\frac{3}{2}\frac{3}{2};\frac{3}{2},-\frac{3}{2}};$$
$$\Delta_{\perp} = V^{\mathrm{exch}}_{-\frac{3}{2}\frac{3}{2};\frac{3}{2},-\frac{3}{2}}. \tag{8.83}$$

Neglecting the short-range correlations to the exchange interaction, we can express $V^{\mathrm{exch}}_{j_1 j_1', j_2 j_2'}$ in terms of the envelope wave function of the acceptor-bound hole (F_m^j) and of the exciton wave function ($G_m^{j'}$)

$$V^{\mathrm{exch}}_{j_1 j_1', j_2 j_2'} = -\frac{e^2}{\kappa} \sum_{m_1 m_2} \iint \frac{\mathrm{d}\mathbf{x}_1\, \mathrm{d}\mathbf{x}_2}{r_{12}} F_{m_1}^{j_1^*}(\mathbf{x}_1)$$
$$\times\, F_{m_2}^{j_2}(\mathbf{x}_2)\, T_{m_2 m_1}^{j_1' j_2'}(\mathbf{x}_1, \mathbf{x}_2)\,, \tag{8.84}$$

$$T_{m_1 m_2}^{j_1 j_2}(\mathbf{x}_1, \mathbf{x}_2) = \int \mathrm{d}\mathbf{x}_{\mathrm{e}} G_{m_1}^{j_1^*}(\mathbf{x}_{\mathrm{e}}, \mathbf{x}_2)\, G_{m_2}^{j_2}(\mathbf{x}_{\mathrm{e}}, \mathbf{x}_1)\,.$$

We will assume the localization radius a_B of the acceptor-bound hole in the (xy) plane to be small compared to that of the hole in a bound exciton, $a_{\mathrm{exc}}^{\mathrm{h}}$. In this case, (8.82a) is totally compensated for by the field generated by the impurity ion, yielding for (8.84)

$$V^{\mathrm{exch}}_{j_1 j_1', j_2 j_2'} = -\frac{e^2}{\kappa} \sum_{m_1 m_2} \iint \mathrm{d}z_1\, \mathrm{d}z_2\, Q_{m_1 m_2}^{j_1 j_2}(z_1, z_2)\, T_{m_2 m_1}^{j_1' j_2'}(\rho_{\mathrm{A}}; z_1, z_2)\,,$$

$$Q^{j_1 j_2}_{m_1 m_2}(z_1, z_2) = \iint \frac{\mathrm{d}\rho_1\,\mathrm{d}\rho_2}{r_{12}} F^{j_1^*}_{m_1}(\mathbf{x}_1)\, F^{j_2}_{m_2}(\mathbf{x}_2)\,, \tag{8.85}$$

where ρ is the component of vector $\mathbf{x}$ in the (x, y) plane, and ρ_A is the position of the impurity atom in this plane. Consider now the electron and the hole in the exciton to be localized in the (x, y) plane within an area δS. Assuming the inequalities

$$a_\mathrm{B} < a < \sqrt{\delta S}$$

to be satisfied, we obtain from (8.85),

$$V^{\mathrm{exch}} \sim \frac{e^2}{\kappa} \frac{a_\mathrm{B}^2}{a \delta S} \tag{8.86}$$

if the acceptor resides within the area of localization δS. We can now evaluate the quantity $|V^{\mathrm{exch}}|^2$ averaged over the acceptor impurity positions in a quantum well

$$\overline{|V^{\mathrm{exch}}|^2} \sim \left(\frac{e^2}{\kappa} \frac{a_\mathrm{B}^2}{a \delta S}\right)^2 N_\mathrm{A} \delta S, \tag{8.87}$$

where N_A is the two-dimensional acceptor density.

We can derive certain conclusions by setting a given symmetry of the perturbation $V(\mathbf{x}_\mathrm{e}, \mathbf{x}_\mathrm{h})$ responsible for the exciton localization. First of all, it should be stressed that the existence of a symmetry axis C_n of order $n = 2, 3, 4$ or 6 forbids light scattering involving spin flip of the acceptor-bound hole in a field $\mathbf{B} \parallel C_n$ in the absence of a perturbation $V(\mathbf{x}_\mathrm{e}, \mathbf{x}_\mathrm{h})$ lowering the symmetry. Indeed, in the presence of such a symmetry axis, the projection of the angular momentum should be preserved to within an integer multiple of n. In forward- or backscattering the photon-momentum projection changes by $\Delta M = \pm 2$ or not at all, whereas with a spin flip $|\pm 3/2\rangle_\mathrm{A} \to |\mp 3/2\rangle_\mathrm{A}$ the hole-momentum projection changes by $\Delta j = \pm 3$. Therefore, the conservation law $\Delta M - \Delta j = nN$ (N is any integer) cannot be satisfied.

It can be shown that if a perturbation $V(\mathbf{x}_\mathrm{e}, \mathbf{x}_\mathrm{h})$ possesses cylindrical symmetry $C_{\infty v}$, then the dependence of the coefficients $\Delta_{l\pm}$ on ρ_A can be represented in the form

$$\Delta^{(3)}_{l\pm} = f_l(\rho_\mathrm{A})\, \rho^3_{\mathrm{A}\pm}. \tag{8.88}$$

Here ρ_A is the acceptor position in the (x, y) plane reckoned from the center of cylindrical symmetry of the exciton state, $\rho_{\mathrm{A}\pm} = \rho_{\mathrm{A}x} \pm \mathrm{i}\rho_{\mathrm{A}y}$ and $f_l(\rho_\mathrm{A})$ is a function of $\rho_\mathrm{A} = |\rho_\mathrm{A}|$. Taking into account the tetragonal anisotropy in the (x, y) plane, the dependence $f_l(\rho_\mathrm{A})$ will contain a first-order spatial harmonic

$$\Delta^{(1)}_{l\pm} = \bar{f}_l(\rho_\mathrm{A})\, \rho_{\mathrm{A}\mp}. \tag{8.89}$$

The function $\bar{f}_l(\rho_\mathrm{A})$ tends identically to zero as $\gamma_2 - \gamma_3 \to 0$, where $\gamma_{2,3}$ are the parameters of the Luttinger Hamiltonian. Note that the estimate (8.86) relates to the coefficients $\Delta_\perp$, $\Delta_\parallel$. The presence of spatial harmonics in the expressions (8.88, 89) assumes the smallness of $\Delta_{l\pm}$ compared with $\Delta_{\perp,\parallel}$.

Let us neglect the effects associated with the absence of inversion center in bulk GaAs and assume the perturbation $V(\mathbf{x}_e, \mathbf{x}_h)$ to be characterized by the symmetry $D_{\infty h}$ or D_{4h}. Then the coefficients $\Delta_{l\pm}$ will vanish if the acceptor resides at the center of a quantum well. This means that the functions $f_l(\rho_A)$, $\bar{f}_l(\rho_A)$ in (8.88), (8.89) reverse their sign when the acceptor displacement z_i with respect to the well center changes its sign. In particular, in the vicinity of the point $z_i = 0$ these functions are proportional to z_i. As the impurity atom approaches the heterojunction, the overlap integral of the hole wave functions decreases, and the matrix elements (8.85) should decrease too. Therefore within the interval $(0, a/2)$ there are values of $|z_i|$ at which the coefficients reach a maximum.

We have considered here no-phonon light-scattering machanisms, involving acceptor-bound hole spin-flip, which are due to hole-hole exchange coupling. Theory gives an explanation for the experimentally observed scattering process (8.80b), and this explanation has no alternatives. As for the process of type (8.80a), the recent experiment [8.10] carried out in both normal and tilted magnetic fields has shown that the scattering is contributed by excitons bound to neutral acceptors, these complexes (A^oX) acting as resonant intermediate states, and occurs due to acoustic-phonon-assisted spin-flip of electrons in the complexes. As a result, the Raman shift is determined by both the hole and electron g-factors. In tilted magnetic fields, in addition to the acoustic-phonon-assisted Raman scattering, no-phonon bound-hole spin-flip becomes allowed since the electron in the A^oX complex can change its spin in Zeeman interaction with the in-plane magnetic field component. Therefore, tilted-field experiments provide a possibility of direct measurement of not only the acceptor-bound-hole, but also of the electron g-factor.

9 Polarized Luminescence in Quantum Wells and Superlattices

9.1 Luminescence as a Tool to Study Electronic Spectra and Kinetic Processes in Two-Dimensional Systems

Luminescence is an efficient tool for studying electronic spectra. Since the luminescence intensity is determined both by the density of electrons and the transition probabilities, and by the population of these states, luminescence can, in many cases, be used to advantage in investigating the fine structure of spectra, which does not show up in absorption, and provides the possibility to study the kinetics of population and depletion of these states. By studying the luminescence for different populations of the electronic states, one can determine the effect of electron–electron interaction on electronic spectra.

By the character of the recombination accompanied by light emission, one can distinguish between intrinsic, extrinsic and exciton luminescence. Intrinsic, or band-to-band luminescence is connected with the recombination of free electrons and holes. Exciton luminescence is created in the recombination of free, impurity-bound or localized excitons. If the energy of the free carriers or excitons involved in recombination exceeds substantially the thermal energy, such luminescence is called *hot*. In bulk-semiconductor solid solutions, exciton localization can result from fluctuations in composition accompanied by a change of the energy gap. Excitons are localized in regions with a lower exciton energy; for example, in CdSSe alloys these are regions with an enhanced Se content. In quantum wells and superlattices, localization is caused by fluctuations in the well and barrier thicknesses arising in the course of growth and resulting in a change of the energy of the excitons, which become localized in larger areas of wells. Localization produces a strong Stokes shift since excitons occupy low-energy states which do not contribute noticeably to the absorption spectra.

Extrinsic or impurity luminescence originates from the recombination of free electrons with the holes bound to acceptors, or of free holes with bound electrons, as well as from electron transitions between the levels of the same center or of neighboring centers. An example of the latter is radiative donor-acceptor recombination.

In the present chapter, the main emphasis will be placed on polarized luminescence appearing as a result of either optical spin orientation or thermal orientation of spins in a magnetic field. It is in these effects, to be considered in detail in the subsequent sections, that the symmetry of the structures under investigation manifests itself most directly. Prior to starting a comprehensive discussion of these topics, we will illustrate briefly the potential of the luminescence as a tool of investigation with a few examples appearing to be of considerable interest at present.

9.2 Luminescence in the Quantum Hall Regime, Quantum Beats

9.2.1 Electron-Electron Interaction in the Quantum Hall Effect

The integer quantum Hall effect was discovered by K. von Klitzing in 1980. This phenomenon is observed to occur in 2D-structures, namely, MOS structures, heterostructures and quantum wells in a strong magnetic field perpendicular to the well plane when the electron spectrum becomes discrete. In perfect structures, the electron spectrum in a magnetic field should consist of a series of Landau levels, and at each of them can reside $n_0 = (1/2\pi)(eB/c\hbar)$ electrons per unit area (we ignore here the possible spin or valley degeneracies).

In real structures, the field of the impurities or defects results in a broadening of each level into a band. Note that only the electrons close in energy to the center of each band are mobile. As a result, as the levels are populated, the non-diagonal component of conductivity σ_{xy} undergoes a jump when each subsequent level $[\nu]$ becomes half-filled, i.e., for half-integer values of the occupancy, $\nu = n/n_0$, where n is the surface concentration of electrons, reaching eventually a plateau with $\sigma_{xy} = e^2\nu/2\pi\hbar$. In the region of the plateau of σ_{xy}, the diagonal component of conductivity σ_{xx} is of the activation nature and reaches a maximum at half-integer values of ν.

In 1983, *Altukhov* et al. [9.1] revealed in the luminescence spectra of silicon MOS structures a new emission band, called δ line, whose position depends on gate bias. This band was attributed to the recombination of the electrons residing in the quantum well at the oxide-layer boundary with the non-equilibrium holes forming a space-charge layer of opposite sign located farther away from the surface. The features observed in this and later studies implied that under the experimental conditions chosen the holes are free and form, together with the electrons, a two-dimensional e-h plasma.

In 1984, *Kukushkin* and *Timofeev* observed in MOS structures a 2D emission line originating from electron recombination with the holes bound to the acceptors located near a quantum well filled by electrons [9.2]. Lineshape measurements performed later [9.3,4] permitted a direct determination of the single-electron density of states. It turned out that the widths Γ of the Landau levels depend substantially on the population of the uppermost filled level. When this level is only partially filled, the electrons occupying the lower states screen the impurity field partially, thus reducing the level width Γ. When all levels are filled completely, there is no screening and Γ increases. Figure 9.1 presents the dependence of the width Γ_i on ν when the third Landau level, $N = 2$, is being filled, with the levels $N = 0, 1$ fully occupied. Each of these levels is fourfold

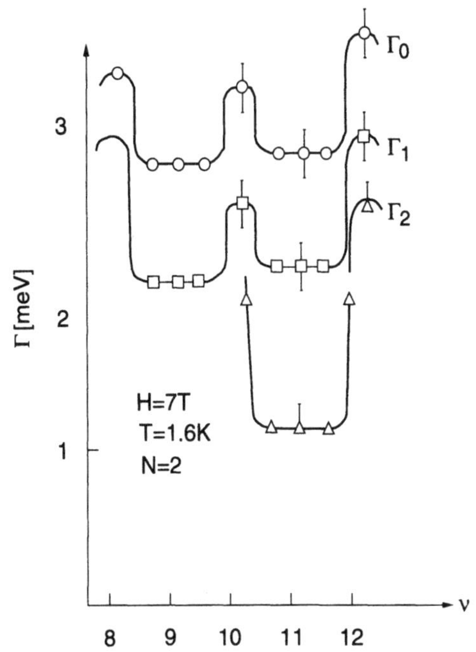

Fig. 9.1. Width Γ_N of Landau level ($N = 2$) vs. population factor ν for $T = 1.6$ K, $B = 7$T (after [9.3])

degenerate in spin and two valleys with $\mathbf{k}_0^+ = (O, O, \Delta)$, $\mathbf{k}_0^- = (O, O, -\Delta)$. The maximum Γ_i for $\nu = 10$ corresponds to a total occupation of the lowest spin level. Each of these spin levels splits into two terms corresponding to the symmetric and antisymmetric valley functions $\Psi^\pm = (\Psi_{\mathbf{k}_0^+} \pm \Psi_{\mathbf{k}_0^-})/\sqrt{2}$. This splitting is an analog of the orbit-valley splitting of a donor center in silicon strained along [001] when only these valleys contribute to the donor wave functions, see [9.5].

It turned out that both spin- and orbit-valley splittings depend substantially on the population of the corresponding states. The reason for this lies in the exchange interaction which lowers the energy of the electrons filling the same states. Indeed, in the structures studied, the magnitude of the orbit-valley splitting ΔE_v for fully occupied states Ψ^+ or empty ones Ψ^-, i.e., for even ν, did not exceed 0.4 meV for $\nu = 10$ and increased proportionately with ν. For odd ν this splitting increased dramatically, so that, for instance, for $\nu = 3$ it reached 1.4 meV [9.3,4]. Similarly, the electron g-factor determining the spin splitting was close to the free electron g-factor in Si, $g_0 = 2$, when all the four states corresponding to the given N became fully populated, whereas when these states were only half-filled, it increased rapidly approaching, for instance, 9 for $\nu = 2$. As ν increases, the exchange contribution to the orbit-valley and spin splittings decreases [9.3,4].

For non-integer values of $\nu = p/q$, and with odd q, one also observes in high-mobility samples at low temperatures and in strong magnetic fields additional plateaus in the dependence $\sigma_{xy}(\nu)$ and minima of σ_{xx}. This effect, called fractional quantum Hall effect, is explaned for the formation of an incompressible quantum liquid at $\nu = p/q$, as a result of electron–electron interaction. Elementary excitations of such a liquid are quasiholes and quasielectrons with the charges $\pm e^* = e/q$. The energy Δ of *electron-hole* pair formation, called gap width, can be derived from the activation relation $\sigma_{xx} \propto \mathrm{e}^{-\Delta/k_\mathrm{B}T}$ for $\nu = p/q$. An experiment [9.6] performed on Si MOS structures revealed, simultaneously with a decrease of σ_{xx}, also a nonmonotonic shift of the position $\bar{\omega}$ of the 2D line maximum for the same values of ν (Fig. 9.2). Note that within experimental error the shift $\delta\bar{\omega}$ for $\nu = 7/3$ and $\nu = 8/3$ coincided with the value of 3Δ. This was accounted for by the fact that for $\nu < \nu_o = p/q$ recombination of one electron involved generation of q "holes", and for $\nu > \nu_o$, the disappearance of *quasielectrons*, and therefore the difference between the energies of a recombining electron for $\nu > \nu_o$ and $\nu < \nu_o$ is exactly $q\Delta$.

Subsequent experiments carried out on GaAs/GaAlAs heterostructures showed, however, [9.7, 8, 9] that such a direct relation between $\delta\bar{\omega}$ and the gap width Δ is not universal. As follows from the calculations of *Apalkov* and *Rashba* [9.10], the value of $\delta\bar{\omega}$ depends considerably on the actual character of interaction of the 2D-layer electrons with the center to which the recombining hole is bound. If this center is far enough from the 2D-layer so that its interaction with electrons may be neglected both in the initial state when the acceptor is neutral and in the final state when it is charged, then at the point $\nu = \nu_o$ not the quantity $\bar{\omega}$ but rather its derivative undergoes a discontinuity, and $\delta(\mathrm{d}\bar{\omega}/\mathrm{d}\nu)_{\nu=\nu_o} = 2q\Delta/\nu_o = 2q^2\Delta/p$. The value of Δ derived by this expression from the jumps of $d\bar{\omega}/d\nu$ in a δ-doped GaAs/GaAlAs heterostructure turned out [9.11] to be about three times greater than the gap obtained from the activation dependence of σ_{xx}, and approaches the value of Δ calculated theoretically.

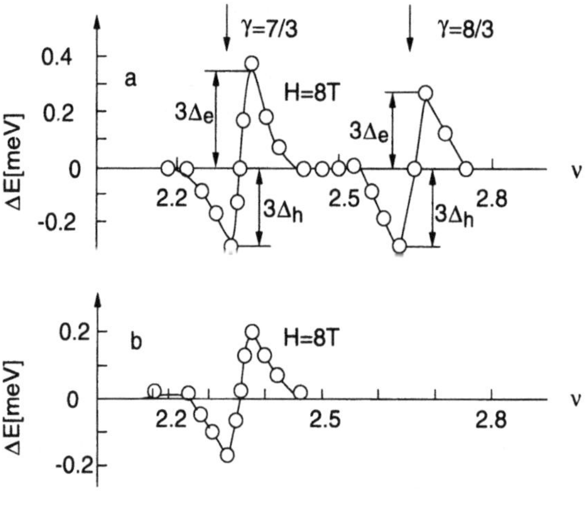

Fig. 9.2. Nonmonotonic shift of the maximum of emission line 2D vs. population factor ν, $T = 1.5$ K, $B = 8$T (after [9.6])

9.2.2 Exciton Effects

The position of the 2D line can vary substantially as a result of interaction of the hole and electron from the 2D layer which leads to exciton formation. As shown by *Averkiev* and *Pikus* [9.12], the exciton binding energy remains positive despite the strong screening of the Coulomb field by free electrons, provided the electron concentration is not too high. *Altuhov* et al. [9.13] revealed manifestations of exciton effects in luminescence spectra. In a strong magnetic field, the exciton binding energy increases, thus increasing the role of the exciton effects. As shown by *Apalkov* and *Rashba* [9.14], the presence of the incompressible 2D liquid under conditions of the fractal Hall effect leads to a dramatic suppression of exciton dispersion, the shape of the optical spectra depending substantially on that of the well.

Turberfield et al. [9.8] observed strong luminescence due to the recombination of electrons from the upper subband $1(e_1)$ with holes while the contribution of electrons from subband $0(e_0)$ was insignificant. At the same time, electric measurements showed subband 1 to be practically empty. It was shown [9.12] that the binding energy of the exciton $e_1 - h$ can exceed by far under these conditions that of the $e_0 - h$ exciton, since in the first case the condition of orthogonality of the wave functions of the electron in the exciton and of the free electrons is automatically met.

9.2.3 The Wigner Crystal

Theoretical calculations suggest that the ground state of 2D electrons in a magnetic field for small $\nu \lesssim 1/5$ should be the Wigner crystal rather than the

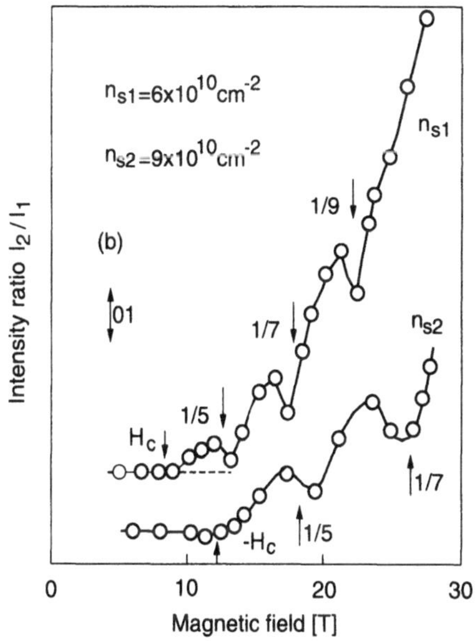

Fig. 9.3. Line intensity ratio I_2/I_1 vs. population factor ν for T = 0.4 K (after [9.11]). The line I_2 is assumed to be due to luminescence of the electrons forming the Wigner crystal

quantum liquid. *Jiang* et al. [9.15] were the first to eliminate the strong 2D-electron pinning which occurs usually at small ν and reveal the anomalously strong increase of ρ_{xx} at $\nu \simeq 0.21$. The authors consider this effect to be proof of a liquid – crystal transition taking place at this value of ν.

Measurements of 2D electron-luminescence spectra in the GaAs/GaAlAs heterostructures showed that, besides line 1 seen in strong magnetic fields at low temperatures, one observes at $\nu \leqslant 1/5$ a second line 2 shifted to longer wavelengths [9.11]. The intensity ratio I_2/I_1 exhibits minima at $\nu = 1/5$, $1/7$, $1/9$ and falls off rapidly with increasing temperature (Fig. 9.3). At $\nu = 1/5$ the corresponding critical temperature $T_c = 1.4$ K. Both this critical behavior of $I_2(T)$ and the other features observed by *Buhman* et al. [9.11] suggest that line 2 originates from the luminescence of the electrons making up the Wigner crystal. It is assumed that for $\nu = p/q$ the lowest-energy stable phase should be the quantum liquid, the ground state far from these values of ν for $\nu \lesssim 1/5$ being the Wigner crystal.

9.2.4 Transient Luminescence and Quantum Beats

Progress in picosecond spectroscopy opened up the possibility of a direct measurement of the times of energy relaxation, carrier trapping in the quantum wells and by impurities, recombination, spin relaxation, etc. Of particular interest is the direct measurement of tunneling times in asymmetric double quantum wells.

These experiments are illustrated schematically in Fig. 9.4. In the asymmetric quantum-well structure, one can measure directly the radiation due to pair or exciton recombination in the narrow and wide wells differing in the photon energy. Under pulse excitation, the reciprocal radiation decay time in the narrow well 2 is $1/\tau_2 = 1/\tau_T + 1/\tau_R^{(2)}$, where $\tau_R^{(i)}$ is the recombination time

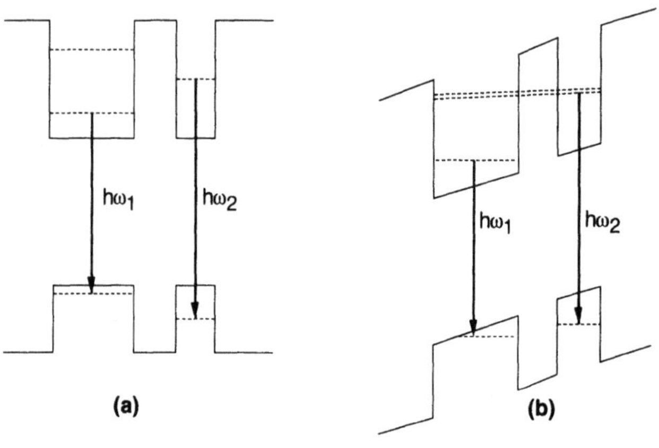

Fig. 9.4a,b. Scheme of electron tunneling time determination from measurement of transient luminescence in double asymmetric quantum wells (after [9.16])

in the ith well, and τ_T is the time of carrier transfer from well 2 to well 1. In the broad well 1 the contribution to radiation comes from the electrons created in well 1 for which $\tau_1 = \tau_R^{(1)}$, as well as from the electrons tunneling from well 2 whose concentration varies as $n(t) \sim (\tau_T - \tau_1)^{-1}(e^{-t/\tau_T} - e^{-t/\tau_1})$. In the absence of an electric field, the time τ_T is determined by nonresonant tunneling through the barrier

$$\tau_T^{-1} \simeq \left(E_2/2m_A^*\right)^{1/2} a \exp\left[-\left(8m_B^*\right)^{1/2}(V_o - E_2)^{1/2}\, b/\hbar\right].$$

Here m_A^* and m_B^* are the effective masses in the wells and the barriers, E_2 is the electron energy in well 2, a is the width of this well, V_0 and b are the height and thickness of the barrier.

The times τ_T measured by *Alexander* et al. [9.16] turned out to be in agreement with calculations. By applying an electric field, one can bring the first level of well 2 into alignment with the second level of well 1. This results in the formation of a symmetrized and an antisymmetrized states, $\Psi_\pm = (\Psi^{(2)} \pm \Psi^{(1)})/\sqrt{2}$, split by a value ΔE determined primarily by the barrier transmission with the wave packet oscillating between wells 1 and 2 with a period $\Delta T_c = 2\pi\hbar/\Delta E$. The transfer time τ_T will be determined in this case by the time $\tau_c = \Delta T_c/2$ and the time of transition from the second to the first level in well 1 and, as is usual in such cases, is equal to the largest of these times. The experimentally revealed decrease [9.16] of the time at resonance and the strong dependence of τ_T on barrier thickness at $b = 6$ nm and 8 nm suggest that it is the time τ_c that is dominant.

Similar measurements were carried out [9.17] on the structures displayed in Fig. 9.5. The electron-hole pairs were created in these experiments by light in the wide well 1. At a bias large enough to raise the hole level in well 2 above that in well 1, the luminescence spectra revealed two lines corresponding to the recombination of the direct and indirect excitons formed by electrons of well 1

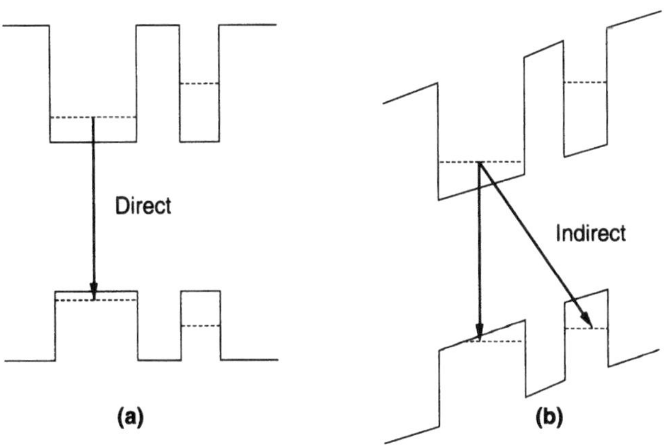

Fig. 9.5a,b. Scheme of hole tunneling time determination from measurement of transient luminescence in double asymmetric quantum wells (after [9.17])

and holes of wells 1 and 2, respectively. Note that the line corresponding to the indirect exciton recombination is red-shifted with increasing bias. A strong decrease of the hole transfer time from well 1 to well 2 was also observed to occur at resonance when the first level of well 1 and the second level of well 2 coincide.

Progress in picosecond techniques permitted one also to observe quantum beats appearing under simultaneous excitation by a short pulse of two closely lying levels 1 and 2. For quantum beats to be observed, the spacing between these levels, ΔE, should be less than the width of the coherent exciting beam, i.e., its duration should not exceed $\hbar/\Delta E$ and, at the same time, should be less than $\hbar/\tau_\varphi$, where τ_φ is the phase relaxation time. Excitons in quantum wells turned out to be convenient subjects for the observation of such beats. At the closely lying levels, one can employ light and heavy exciton levels [9.18, 19] and excitonic levels in well areas differing in width by one lattice constant [9.18, 20].

These studies used for the observation of quantum beats the two-pulse self-diffracted transient grating technique. In a study [9.21] whose results will be discussed in more detail in Sect. 9.3 a simpler technique was employed. Two split states, ψ_1 and ψ_2, optically active in the polarizations $\mathbf{e}\|x'$ and $\mathbf{e}\|y'$, accordingly, were excited simultaneously by a $\mathbf{e}\|x$ light beam whose plane of polarization made an angle θ with the x' axis. The excited state, $\psi = \psi_1 \cos\theta + \psi_2 \sin\theta$, decayed in time as $\psi(t) = \psi_1 \cos\theta \exp(-\mathrm{i}E_1 t/\hbar) + \psi_2 \sin\theta \exp(-\mathrm{i}E_2 t/\hbar)$. One measured the radiation intensity in two polarizations, namely, $\tilde{\mathbf{e}}\|x$ and $\tilde{\mathbf{e}}\|y$. For the degree of polarization we have

$$P_{\text{lin}} = \frac{I_x - I_y}{I_x + I_y} = \sin^2 2\theta \cos\left[(E_1 - E_2)\,t/\hbar\right] \mathrm{e}^{-t/\tau_\varphi} + \cos^2 2\theta . \tag{9.1}$$

We readily see that the oscillating contribution in P_{lin} has a maximum at $\theta = \pi/4$, its measurement providing a direct determination of the magnitude of $\Delta E = E_1 - E_2$.

9.3 Optical Spin Orientation and Alignment of Electron Momenta

The orientation of electron spins under interband excitation by circularly polarized light and electron alignment in momentum space under excitation by linearly polarized light originate from the degeneracy and spin-orbit splitting of the valence band. The first of these effects manifests itself in the circular polarization of the recombination radiation of thermalized or hot electrons, and the second, in the linear polarization of the radiation of the hot electrons, i.e., electrons which preserved, completely or partially, the initially anisotropic distribution in momentum.

The mechanism of electron-spin orientation is illustrated by a simple diagram of transitions from states with moments $J = 3/2$ ($J_z = \pm 3/2, \pm 1/2$)

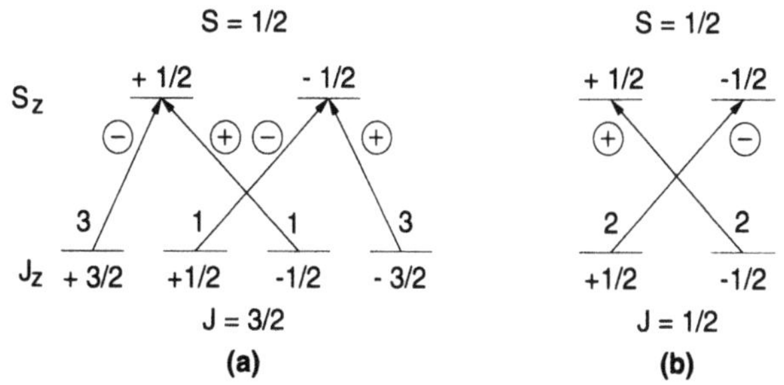

Fig. 9.6a,b. Diagram of optical transitions between P and S levels in an atom

and $J = 1/2$ ($J_z = \pm 1/2$) created in the spin-orbit splitting of the P level with an orbital moment $J = 1$ ($J_z = 0, \pm 1$) to S-states ($s_z = \pm 1/2$). This transition diagram relates to atoms with a filled P-shell. The symbols σ_+ and σ_- in Fig. 9.6 denote the transitions allowed in the corresponding polarization, and the numbers, the relative transition intensity. It is assumed that the axis of quantization z coincides with the direction of light propagation.

We see that for the σ_+ -induced transition from $J = 3/2$ levels with the photon angular momentum $m_z = +1$, the degree of s-electron orientation

$$2s = \frac{n_+ - n_-}{n_+ + n_-} = -\frac{1}{2}. \tag{9.2}$$

Here s is the mean value of the excited electron spin s_z, n_+ and n_- being, respectively, the numbers of electrons with the spins $s_z = +1/2$ and $s_z = -1/2$.

Similarly, in the reverse transitions from the $s_z = -1/2$ state, the relative emission probabilities for photons with moments $m_z = +1$ and $m_z = -1$ are 3 and 1, respectively, and those from the $s_z = +1/2$ state, 1 and 3. Therefore, the polarization of the radiation propagating in the direction of the exciting beam (or in the reverse direction)

$$P_{\text{circ}} = \frac{I_+ - I_-}{I_+ + I_-} = \frac{1}{4}. \tag{9.3}$$

Here I_+ and I_- are the radiation intensities for the σ_+ and σ_- polarizations, accordingly. When excited from the $J = 1/2$ states, the degree of electron orientation $2s = 1$, the relative intensity of such transitions being one half that for the excitation from states with $J = 3/2$. Therefore in the absence of spin-orbit splitting electrons do not become oriented.

9.3.1 Electron Orientation in Cubic A_3B_5 Crystals

Calculations of the degree of electron-spin orientation in crystals should take into account the relation between the electron spin and momentum directions in the valence band.

The corresponding calculations were first done by *Dyakonov* and *Perel* [9.22] who showed that the mean spin of the electrons excited by circularly polarized light from the band Γ_8, is $-1/4$, just as is the case with transitions from the $J = 3/2$ level in an atom, the degree of the luminescence polarization for thermal electrons in the absence of spin relaxation being $+1/4$.

As we intend to discuss later the electron-spin orientation and momentum alignment in (001) quantum wells, we will calculate these effects in bulk crystals in the basis (3.109) limiting ourselves to the case of light propagating along the [001] or [111] axes. In the isotropic approximation, i.e., for $\gamma_2 = \gamma_3 \equiv \bar{\gamma}$, the expressions thus obtained will be valid for any direction of light propagation.

The electron (or hole) spin state is described by the density matrix $\rho_{nn'}$, the rate of variation of the $\rho_{nn'}$ component in the course of carrier generation being determined by the generation matrix $G_{nn'}$. The generation matrix G for the electrons excited from the branch l of the valence band is determined by the expression

$$G^l_{nn'}(\mathbf{e}, \mathbf{k}) = \sum_m (\mathbf{e} \cdot \mathbf{v}^l_{nm})(\mathbf{e} \cdot \mathbf{v}^l_{n'm})^* \tag{9.4}$$

to within a constant proportional to the light intensity. Here $\mathbf{e}$ is the exciting-light polarization vector, n and n' are the spin indices of the conduction-band electron, m are the indices of degenerate states of the valence-band branch

$$\mathbf{v}_{nm} = \hbar^{-1} \nabla_{\mathbf{k}} \mathcal{H}_{nm}, \tag{9.5}$$

where $\mathcal{H}_{nm}$ are the corresponding interband Hamiltonian components. It is assumed here that the valence band is practically filled, i.e., the corresponding density matrix $\rho_{mm'} = \delta_{mm'}$. The energy of the excited electrons depends on the light frequency and direction of $\mathbf{k}$ and is determined, as usual, by the conditions of energy and momentum conservation.

The recombination-radiation polarization determined by the recombination of conduction electrons and holes of the valence branch l is given by the matrix $d^l_{\alpha\beta} = \langle \tilde{e}_\alpha \tilde{e}^*_\beta \rangle$ where $\tilde{e}$ is the polarization vector of the emitted light, which can be written, to within a constant as

$$d^l_{\alpha\beta} = \sum_{mnn'k} v^{\alpha *}_{nm} v^{\beta}_{n'm} \rho_{nn'} \rho_{mm}, \tag{9.6}$$

where ρ is the electron density matrix. The relation between ρ and the generation matrix G is determined by the solution of the corresponding kinetic equation, the radiation frequency depending, in the general case, both on the energy and direction $\mathbf{k}$ of the electron. Equation (9.6) assumes that the hole density matrix is diagonal and depends only on the hole energy: $\rho_{mm'} = \delta_{mm'} \rho^l(\mathbf{k})$. One can conveniently describe the polarization dependence of the radiation by the matrix Q^l:

$$Q^l(\tilde{\mathbf{e}}, \mathbf{e}) = \sum_{\alpha\beta} d^l_{\alpha\beta} \tilde{e}^*_\alpha \tilde{e}_\beta, \tag{9.7}$$

which, according to (9.4, 6, 7), can be written, to within a constant, as

$$Q^l = \sum_{nn'k} G^{l*}_{nn'}(\tilde{\mathbf{e}}, \mathbf{k}) \rho_{nn'}(\mathbf{k}) \rho^l(\mathbf{k}). \tag{9.8}$$

If the function, corresponding to the state m of the valence-band electron, is

$$\varphi_m = \sum_l C_l^m \varphi_l, \tag{9.9}$$

where φ_l ($l = 1$–4) are the valence-band basis functions, then

$$\mathbf{e} \cdot \mathbf{v}_{nm} = \sum_l C_l^m (\mathbf{e} \cdot \mathbf{v}_{nl}). \tag{9.10}$$

The components of the matrix $\mathbf{v}_{nl}$ can be readily obtained from Table 3.7 by using (9.5).

Table 9.1 presents components of the matrix $(\mathbf{e} \cdot \mathbf{v}_{nm})$ in the basis (3.109). It has been taken into account that the basis functions of the valence-band electron, φ^{e}, are related with those of the holes, φ^{h}, through $\varphi^{\mathrm{e}} = -K\varphi^{\mathrm{h}}$, where K is the time inversion operation, and are chosen in the form $\tilde{\varphi}^{\mathrm{e}}_{1,2} = \tilde{\varphi}^{\mathrm{h}}_{1,2}$, $\tilde{\varphi}^{\mathrm{e}}_{3,4} = -\tilde{\varphi}^{\mathrm{h}}_{3,4}$. The common factor $s = i/m \langle S|p_z|Z\rangle$ is dropped.

Using Table 9.1, one can readily show that for transitions from the heavy ($l = 2$) and light ($l = 1$) hole branches whose basis functions are determined by (3.108a):

$$\begin{aligned}
G_+^l &= \frac{1}{2}\left(G^l_{1/2,1/2} + G^l_{-1/2,-1/2}\right) \\
&= \frac{1}{3}\left(2 - \frac{G-F}{E_l - E_j}\right) + \frac{2}{\sqrt{3}\left(E_l - E_j\right)}\left(e_+ e_-^* I + e_+^* e_- I^*\right), \\
G_-^l &= \frac{1}{2}\left(G^l_{1/2,1/2} - G^l_{-1/2,-1/2}\right) \\
&= \frac{1}{3}\left(1 - 2\frac{G-F}{E_l - E_j}\right)\left(|e_+|^2 - |e_-|^2\right), \\
G^l_{1/2,-1/2} &= G^{l*}_{-1/2,1/2} = \frac{2}{\sqrt{3}}\frac{H}{E_l - E_j}\left(|e_+|^2 - |e_-|^2\right).
\end{aligned} \tag{9.11}$$

Here E_l is the energy of holes in branch l ($j \neq l$).

The same expressions are applicable also to the case where the quantization axis z is directed along [111], and G, F, H and I are determined by (3.115). To calculate the mean value of the components of G, one has to integrate (9.8) over all values of $\mathbf{k}$ taking into account that for a given frequency of

Table 9.1. Components of the $(\mathbf{e} \cdot \mathbf{v})$ matrix in the basis of the hole functions $\tilde{\varphi}_i$ (3.109) and electronic functions $\pm 1/2$

$n \backslash m$	1	2	3	4
1/2	$e_+ e^{i\varphi}$	$-\frac{1}{\sqrt{3}} e_- e^{-i\eta}$	$-\frac{1}{\sqrt{3}} e_- e^{-i\eta}$	$-e_+ e^{i\varphi}$
−1/2	$e_- e^{-i\varphi}$	$\frac{1}{\sqrt{3}} e_+ e^{i\eta}$	$\frac{1}{\sqrt{3}} e_+ e^{-i\eta}$	$e_- e^{-i\varphi}$

excitation the value of **k** depends on its direction. By virtue of cubic symmetry, the mean value $\langle k_i^2 \rangle$ does not depend on the index $i = x, y, z$, and therefore $\langle G - F \rangle = 2\gamma_2 \langle 3k_3^2 - k^2 \rangle = 0$, just as $\langle H \rangle_l = \langle I \rangle = 0$, and, consequently, for the quantities G_+^l and G_-^l averaged over all directions **k**, one can write

$$\bar{G}_+^l = \frac{2}{3}, \quad \bar{G}_-^l = \frac{1}{3} \left(|e_+|^2 - |e_-|^2 \right). \tag{9.12}$$

This means that the degree of orientation of the electrons excited from the light and heavy-hole branches is the same and for the σ_+light($|e_+|^2 = 0, |e_-|^2 = 1$) $2s = \bar{G}_- / \bar{G}_+ = -1/2$.

If during the lifetime τ_o, the electrons become completely isotropic in momentum, and the spin relaxation time τ_s exceeds substantially τ_o, then the degree of circular polarization

$$P_{\text{circ}} = P_{\text{circ}}^0 = \left(\bar{G}_- / \bar{G}_+ \right)^2 = 1/4. \tag{9.13}$$

If, however, the times τ_o and τ_s are comparable, then

$$\rho_+ = \rho_{1/2,1/2} + \rho_{-1/2,-1/2} = \bar{G}_+ / \tau_o,$$

$$\rho_- = \rho_{1/2,1/2} - \rho_{-1/2,-1/2} = \bar{G}_- / T,$$

where

$$T^{-1} = \tau_o^{-1} + \tau_s^{-1}$$

and

$$P_{\text{circ}} = P_{\text{circ}}^o \frac{T}{\tau_o}. \tag{9.14}$$

The time T can be directly determined using the Hanle effect, i.e., by measuring the radiation depolarization caused by spin precession in a transverse magnetic field B,

$$P_{\text{circ}}(B_\perp) = \frac{P_{\text{circ}}(0)}{1 + (\Omega_B T)^2}. \tag{9.15}$$

Here $\Omega_B = g_e \mu_B B / \hbar$ is the Larmor frequency, where g_e the electron g-factor and μ_B the Bohr magneton.

In the case of recombination of the hot electrons excited from the heavy-hole subband and preserving the initial momentum with the thermalized holes of this subband, the matrix Q is determined by the expression

$$Q^{(2,2)}(\tilde{\mathbf{e}}, \mathbf{e}) = \langle G_+^{(2)*}(\tilde{\mathbf{e}}) G_+^{(2)}(\mathbf{e}) + G_-^{(2)*}(\tilde{\mathbf{e}}) G_-^{(2)}(\mathbf{e}) \rangle + \text{Re} \left\{ \langle G_{1/2,-1/2}^{(2)}(\tilde{\mathbf{e}}) G_{-1/2,1/2}^{(2)}(\mathbf{e}) \rangle \right\}. \tag{9.16}$$

Here the brackets $\langle \rangle$ denote, as earlier, averaging over all values of **k** corresponding to the given pump frequency ω.

If we neglect the warping of the isoenergetic surfaces for holes, then the averaging, $\langle \rangle$, reduces to averaging over all directions of **k**. Substituting the components G^l from (9.11), we find for (001) wells in this approximation for the light propagating along [001]:

$$Q^{(2,2)}(\tilde{\mathbf{e}}, \mathbf{e}) = \frac{1}{3\bar{\gamma}^2}\left[\left(20\bar{\gamma}^2 + \gamma_2^2\right) + 3\left(3\gamma_2^2 + 2\gamma_3^2\right)\right.$$
$$\left. \times \left(|\tilde{e}_+|^2 - |\tilde{e}_-|^2\right)\left(|e_+|^2 - |e_-|^2\right)\right]. \qquad (9.17)$$

In the isotropic approximation, i.e., for $\gamma_2 = \gamma_3 = \bar{\gamma}$, the circular polarization of the hot luminescence under excitation by circularly polarized light $P_{\text{circ}} = 5/7$.

In a longitudinal magnetic field, the Lorentz force rotates the component $\mathbf{k}_\perp$ perpendicular to the z axis and, as a result, the component $\rho_{1/2,-1/2}$ proportional to H, i.e., to $k_z(k_x - \mathrm{i}k_y)$, decreases, while the components $\rho_{1/2,1/2}$ and $\rho_{-1/2,-1/2}$ depending only on k_z^2 and $k_\perp^2$ do not change. Therefore, the degree of circular polarization falls off as

$$P_{\text{circ}}\left(B_\parallel\right) = \frac{9\gamma_2^2 + 6\gamma_3^2\left(1 + \omega_c^2\tau_p^2\right)^{-1}}{20\bar{\gamma}^2 + \gamma_2^2}, \qquad (9.18)$$

where ω_c is the cyclotron frequency and τ_p the relaxation time of the second polynomial in the momentum distribution function.

Similarly, in the case of light propagation along [111], we have, in accordance with (9.8, 13, 3.115)

$$P_{\text{circ}}\left(B_\parallel\right) = \frac{9\gamma_3^2 + 2(2\gamma_2^2 + \gamma_3^2)(1 + \omega_c^2\tau_p^2)^{-1}}{20\bar{\gamma}^2 + \gamma_3^2}. \qquad (9.19)$$

Dymnikov et al. [9.23] were the first to calculate the degree of circular polarization of the hot luminescence in A_3B_5 crystals.

Electron-spin orientation under excitation by circularly polarized light was first detected experimentally by *Lampel* [9.24] in silicon by means of nonoptical methods. Circular polarization of the luminescence of oriented electrons and its suppression by a transverse magnetic field were first observed by *Parsons* and *Revzett* [9.25] on GaAs. The circular polarization of hot luminescence of oriented electrons and its depolarization in a longitudinal magnetic field was revealed by *Karlik* et al. [9.26]. These effects were studied subsequently in detail in many works summarized in reviews [9.27–30].

9.3.2 Optical Alignment of Electron Momenta in Cubic Crystals

As follows from (9.8), in the case of excitation by linearly polarized light, the generation matrix G can be written

$$G_+^l = G_{1/2,1/2}^l = G_{-1/2,-1/2}^l = \frac{1}{3}\left(2 - \frac{G-F}{E_l - E_j}\right) + \frac{2}{\sqrt{3}}\frac{1}{E_l - E_j}$$
$$\left[\left(e_x^2 - e_y^2\right)\operatorname{Re}\{I\} - 2e_xe_y\operatorname{Im}\{I\}\right]. \qquad (9.20)$$

In the isotropic approximation

$$G_+^{(1,2)} = \frac{2}{3} \mp \left(\frac{k_z^2}{k^2} - \frac{2}{3} + \frac{(\mathbf{k}_\perp \mathbf{e})^2}{k^2} \right). \tag{9.21}$$

We see that the electron distribution in momentum under excitation by linearly polarized light is anisotropic, as this was shown, for instance, by *Bir* and *Pikus* ([9.5] p 412). Recombination of hot electrons retaining an anisotropic distribution also produces linearly polarized radiation [9.23]. In accordance with (9.16), in the case of recombination of hot electrons excited from the subband l and retaining their momentum with heavy holes, the matrix Q can be written

$$Q^{(2,l)}(\tilde{\mathbf{e}}, \mathbf{e}) = \langle G_+^2(\tilde{\mathbf{e}}) G_+^l(\mathbf{e}) \rangle. \tag{9.22}$$

According to (9.21), in the isotropic approximation

$$Q^{(2,2)} = \left[6 + 2(\tilde{\mathbf{e}} \cdot \mathbf{e})^2\right]/15, \quad Q^{(2,1)} = (4/45)\left[11 - 3(\tilde{\mathbf{e}} \cdot \mathbf{e})^2\right]. \tag{9.23}$$

Therefore, the degree of linear polarization under excitation by light polarized along the x axis ($e_x = 1$) from the heavy-hole subband

$$P_{\text{lin}}^{(2,2)} = \frac{I_x - I_y}{I_x + I_y} = \frac{1}{7}. \tag{9.24a}$$

In the case of excitation from the light-hole subband,

$$P_{\text{lin}}^{(2,1)} = -\frac{3}{19}. \tag{9.24b}$$

The inclusion of anisotropy makes the degree of polarization dependent both on the direction of propagation of the exciting light and on the position of its polarization plane. Neglecting the hole spectrum anisotropy we obtain, in accordance with (9.20, 22), for the case of light propagating along [001]

$$\begin{aligned} Q^{(2,2)}(\tilde{\mathbf{e}}, \mathbf{e}) = \frac{1}{45\bar{\gamma}^2} &\left[20\bar{\gamma}^2 + \gamma_2^2 + 3\gamma_2^2 \left(\tilde{e}_x^2 - \tilde{e}_y^2\right)\left(e_x^2 - e_y^2\right)\right. \\ &\left. + 12\gamma_3^2 \tilde{e}_x \tilde{e}_y e_x e_y\right]. \end{aligned} \tag{9.25}$$

We see that the degree of linear polarization of the luminescence varies from $3\gamma_2^2/(20\bar{\gamma}^2 + \gamma_2^2)$ for $\mathbf{e}||x$ or $\mathbf{e}||y$ to $3\gamma_3^2/(20\bar{\gamma}^2 + \gamma_2^2)$ when $\mathbf{e}$ makes an angle of $\pm 45°$ to these axes.

If the light propagates along [111], the degree of polarization does not depend on the position of the plane of polarization and, in accordance with (9.20, 22, 3.115), will be $\left(\gamma_2^2 + 2\gamma_3^2\right) / \left(20\bar{\gamma}^2 + \gamma_3^2\right)$ if the hole-spectrum anisotropy is neglected. Linearly polarized hot luminescence of momentum-aligned electrons was first observed on GaAs [9.26]. The results of this and subsequent studies are discussed in detail in the reviews [9.28]. The luminescence observed under usual experimental conditions is due to electron recombination with holes at acceptors. Under these conditions, the degree of polarization depends on the acceptor wave function anisotropy and, as a rule, exceeds the above values.

Momentum relaxation reduces the degree of polarization. Since at the same time the electron energy decreases too, this results in a fast falloff of the degree of polarization with decreasing radiation frequency. In a longitudinal magnetic field, as a result of the bending of electron trajectories by the Lorentz force, the polarization decreases as

$$P_{\mathrm{lin}}\left(B_{\|}\right) = \frac{P_{\mathrm{lin}}(0)}{1+(2\omega_{\mathrm{c}}\tau_{\mathrm{p}})^2}. \tag{9.26}$$

At the same time the plane of polarization rotates by an angle $\varphi = (1/2)$ arctan $(2\omega_{\mathrm{c}}\tau_{\mathrm{p}})$.

9.3.3 Electron-Spin Orientation in Quantum Wells and Superlattices

As already pointed out, in GaAs/AlGaAs quantum wells the barrier for holes may be considered to be infinitely high because of the large hole effective mass. The wave function of the valence-band electrons or holes is defined in this case by (3.109), (3.110), and their spectrum, by (3.111). From the condition of the wave functions vanishing at the well boundaries at $x = 0$ and $x = a$, one can derive the following expressions for the coefficients C_i^m of the function F_I:

$$C_1^{(1)} = -C_2^{(1)*} = C_o \det\begin{bmatrix} -R_1^* & R_2 & -R_2^* \\ E-F_1 & E-F_2 & E-F_2 \\ (E-F_1)\,\mathrm{e}^{-\mathrm{i}k_1a} & (E-F_2)\,\mathrm{e}^{\mathrm{i}k_2a} & (E-F_2)\,\mathrm{e}^{-\mathrm{i}k_2a} \end{bmatrix}$$

$$C_1^{(2)} = -C_2^{(2)*} = C_o \det\begin{bmatrix} R_1^* & -R_1 & R_2^* \\ E-F_1 & E-F_1 & E-F_2 \\ (E-F_1)\,\mathrm{e}^{-\mathrm{i}k_1a} & (E-F_1)\,\mathrm{e}^{\mathrm{i}k_1a} & (E-F_2)\,\mathrm{e}^{\mathrm{i}k_2a} \end{bmatrix}. \tag{9.27}$$

The real constant C_o is determined from the condition of normalization. For the function F_{II}, the coefficients $\tilde{C}_l^{(m)}$ differ from $C_l^{(m)}$ in R_i being replaced by R_i^*. The matrix elements of the operator (**ev**) for transitions from states I and II to electron states $s_z = \pm 1/2$ calculated using Table 9.1 are presented in Table 9.2. Just as in the case of Table 9.1, the common factor s is dropped. Here

$$A_1 = \mathrm{Re}\left\{C_1^{(1)}R_1J_1 + C_1^{(2)}R_2J_2\right\}, \tag{9.28}$$

$$B_1 = 3^{-1/2}\left[(E-F_1)\,\mathrm{Im}\left\{C_1^{(1)}J_1\right\} + (E-F_2)\,\mathrm{Im}\left\{C_1^{(2)}J_2\right\}\right],$$

Table 9.2. Components of the $(\mathbf{e}\cdot\mathbf{v})$ matrix in the basis of the hole functions I, II defined by (9.24) and electronic functions $\pm 1/2$

$n \backslash m$	I	II
1/2	$-A_1\mathrm{e}^{i\varphi}e_+ + iB_1\mathrm{e}^{-i\eta}e_-$	$-A_2\mathrm{e}^{i\varphi}e_+ - iB_2\mathrm{e}^{-i\eta}e_-$
−1/2	$-A_1\mathrm{e}^{-i\varphi}e_- iB_1\mathrm{e}^{i\eta}e_+$	$A_2e_{-i\varphi}e_- - iB_2\mathrm{e}^{i\eta}e_+$

$$J_i = \int_0^a \mathrm{e}^{\mathrm{i}k_i z} \varphi_{\mathrm{e}}(z)\,\mathrm{d}z, \tag{9.29}$$

where $\varphi_{\mathrm{e}}(z)$ is the electron wave function defined by (3.83, 89). (One has to take into consideration that in (3.81) the origin is set at the well's center, and in (9.29), at the well-barrier boundary). The expressions for A_2 and B_2 differ from (9.28) in R_i being replaced by R_i^*. Recalling that $C_i\left(R_i^*, k_i\right) = C_i^*\left(R_i, -k_i\right)$, one can readily show that $A_2^2 = A_1^2$, $B_2^2 = B_1^2$, and $A_2 B_2 = -A_1 B_1$.

Accordingly, for the generation function, we have

$$\begin{aligned} G_+ &= A_1^2 + B_1^2 - 4A_1 B_1 \mathrm{Im}\left\{e_+ e_-^* \mathrm{e}^{\mathrm{i}(\varphi+\eta)}\right\}, \\ G_- &= \left(A_1^2 - B_1^2\right)\left(|e_+|^2 - |e_-|^2\right), \\ G_{+-} &= G_{-+} = 0. \end{aligned} \tag{9.30}$$

Note that for symmetrical wells, i.e., for the equivalence of the z and $-z$ directions, the nondiagonal components of the matrix G are always zero.

When exciting from the lowest heavy-hole level near the threshold, i.e., for $k_\perp = 0$, according to (3.112), we obtain $E = F_2, k_2 = \pi/a$, and under excitation by anticlockwise polarized light ($e_- = 1$) the spin of the excited electrons $s = -1/2$, that is, they are fully oriented. Accordingly, in the absence of spin relaxation, the degree of polarization of the radiation created in recombination with heavy holes near the bottom of the band is 100%. When excited from the lowest light-hole level at $k_\perp = 0$ we have, by (3.112), $E = G_1, k_1 = \pi/a$, the spin of the created electrons $s = +1/2$, and the polarization of radiation produced in recombination with heavy holes is -100%. If, however, the electrons recombine with acceptor-bound holes, then the degree of polarization will depend on the actual relation between the Bohr radius of the acceptor a_B and its distance to the well boundary z_0 or well thickness a. For $a_B < z_0$ the acceptor may be considered 3D, the degree of polarization P_{circ}, just as in the case of a bulk crystal, being $-s$. If, however, $a_B > z_0$, then the ground state of the acceptor splits, the recombination with the holes at the lowest level yielding $P_{\mathrm{circ}} = -2s$. As the excitation frequency increases, i.e., as $k_\perp$ increases, the degree of electron orientation decreases. *Twardowski* and *Hermann* [9.31], and *Merkulov* et al. [9.32] were the first to calculate the electron-spin orientation and radiation polarization in quantum wells. *Merkulov* et al. [9.32] showed that in the isotropic approximation and for infinitely high barriers, both for the holes and for electrons, the degree of electron orientation depends only on the ratio $t_{\mathrm{h}} = k_\perp^2/\left(k_2^2 + k_\perp^2\right)$ or, equivalently, on the ratio $t_{\mathrm{l}} = k_\perp^2/\left(k_1^2 + k_\perp^2\right) = m_{\mathrm{h}}/m_{\mathrm{l}} t_{\mathrm{h}}$, where m_{l} and m_{h} are the effective masses of the holes, defined by (3.51). Figure 9.7 presents the dependence of mean spin on the quantity t_{h} for transitions to the first electronic level from the lowest levels of the heavy (curve 1) and light (curve 2) holes for $m_{\mathrm{l}}/m_{\mathrm{h}} = 0.18$. The reversal of the sign of s with increasing t_{h} revealed by curve 1 is due to the anticrossing of the heavy- and light-hole levels at comparatively small $k_\perp$, so

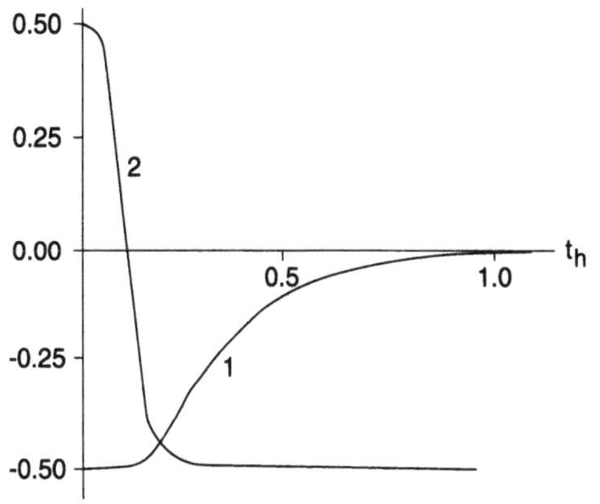

Fig. 9.7. Dependence of the mean spin of excited electrons on the quantity $t_h = k_\perp^2/(k_\perp^2 + k_z^2)$ for transitions to the first electronic level from the lowest levels of heavy (curve 1) and light (curve 2) holes (after [9.32])

that for large $k_\perp$ the dominant contribution to the heavy-hole wave function comes from states with $J_z = \pm 1/2$ rather than with $J_z = \pm 3/2$ as is the case with small $k_\perp$. For infinitely high barriers for the holes and electrons and for $k_\perp = 0$, only transitions from the valence band states hhν and lhν to the electronic state $\psi_{\nu'}^{e} = \sqrt{2/a}\sin(\pi\nu' z/a)$ with $\nu' = \nu$ are allowed. As $k_\perp$ increases, transitions between states with $\nu' \neq \nu$, in particular to states with opposite parity, also become allowed. Electron orientation in GaAs/GaAlAs quantum wells was observed to occur by *Weisbuch* et al. [9.33], and, in the GaAs/GaAlAs superlattice, by *Ivchenko* et al. [9.34]. In quantum wells the degree of circular polarization of the luminescence under excitation at the absorption edge was found to be in excess of 20%, the sign of polarization undergoing reversal with increasing excitation frequency in full agreement with theoretical calculations presented in Fig. 9.7. Measurements of the Hanle effect in superlattices [9.34] showed two groups of electrons to contribute to the luminescence, namely, bound (with a long lifetime) and free (short lifetime) carriers. *Ivchenko* et al. [9.34] observed also a substantial increase of the degree of polarization in a longitudinal magnetic field due to a reduced spin relaxation rate.

When the spin-relaxation time of holes in bulk crystals practically coincides with the momentum-relaxation time, then hole-spin orientation cannot be observed. If the degeneracy is removed, the spin-relaxation time increases [9.35], and it is this that makes it possible to observe the optical orientation of holes in quantum wells or uniaxaly deformed cubic crystals. This optical orientation of holes in quantum wells was first observed by *Uenoyama* and *Sham* [9.41,42].

Spin relaxation in quantum wells. The main mechanisms of spin relaxation accompanying optical orientation of electrons in bulk crystals are as follows:

Elliott-Yafet (EY) *mechanism* involves the mixing of opposite-spin wave functions with $k \neq 0$ as a result of **k-p** interaction with other bands.

Dyakonov-Perel (DP) *mechanism* is caused by spin splitting of the conduction band in off-centrosymmetric crystals with $\mathbf{k} \neq 0$ [9.36,37].

Bir-Aronov-Pikus (BAP) *mechanism* involves electron scattering on free or bound holes with spin flip [9.38]. Scattering from paramagnetic impurities can play a similar role.

In bulk crystals the EY mechanism can be essential only in narrow-gap materials with a large spin-orbit valence-band splitting, e.g. in InSb. The BAP mechanism predominates at low temperatures and sufficiently high hole concentrations, and the DP mechanism, at high temperatures [9.39].

In quantum wells and superlattices the EY mechanism is predominant for holes. In [001] wells the $|3/2\rangle_1$ ground state, as seen from (3.46), mixes with the first light hole level $|-1/2\rangle_1$, and the latter, with the second $|-3/2\rangle_2$ level. Similarly, the $|-3/2\rangle_1$ state mixes with $|1/2\rangle_1$ and $|3/2\rangle_2$. This makes possible heavy-hole scattering from impurities or static defects with spin flip. Since the matrix element I in (3.46) which mixes the states $|3/2\rangle_1$ and $|-1/2\rangle_1$ is proportional to $k^2 = k_x^2 + k_y^2$, and the matrix element H mixing $|-1/2\rangle_1$ with $|-3/2\rangle_2$ is proportional to k, when the above scattering mechanisms dominate at $ka < 1$, we have

$$\frac{1}{\tau_{\mathrm{s}}^{\mathrm{h}}} \sim \frac{1}{\tau_{\mathrm{p}}}(ka)^6,$$

where a is the well thickness.

Ferreira and *Bastard* [9.40] carried out comprehensive quantitative calculations of the spin-relaxation rate for this case. In scattering from acoustic phonons it is sufficient to take into account only the mixing of $|3/2\rangle_1$ and $|-1/2\rangle_1$ states, or of $|-3/2\rangle_1$ with $|1/2\rangle_1$, since vibrations causing ε_{xz} or ε_{yz} deformation, as seen from (3.77), make possible $|-1/2\rangle_1 \rightarrow |-3/2\rangle_1$ or $|1/2\rangle_1 \rightarrow |3/2\rangle_1$ transitions. Therefore, for this mechanism considered by *Uenoyama* and *Sham* [9.41,42] we have

$$\frac{1}{\tau_{\mathrm{s}}^{\mathrm{h}}} \sim \frac{1}{\tau_{\mathrm{p}}}(ka)^4.$$

We see that scattering from phonons at $ka \ll 1$ can provide a dominant contribution to $(\tau_{\mathrm{s}}^{\mathrm{h}})^{-1}$ even in cases where the main contribution to τ_{p}^{-1} comes from scattering on impurities or defects.

The major mechanism of spin relaxation for electrons in samples with a moderate hole concentration is that due to DP.

The Hamiltonian $\mathcal{H}_{\mathrm{s}}(k)$ including both linear- and cubic-in-k terms and determining spin splitting in [001] quantum wells can be written, according to (3.38, 101), in thc form

$$\mathcal{H}_{\mathrm{s}}(\mathbf{k}) = \frac{1}{2}\sigma\,(\Omega_1 + \Omega_3)\,, \tag{9.31a}$$

where

$$\Omega_1 = \Omega_1(-\cos\varphi, \sin\varphi, 0), \quad \Omega_3 = \Omega_3(-\cos 3\varphi, -\sin 3\varphi, 0),$$

$$\hbar\Omega_1 = 2\gamma k\left(\langle k_z^2\rangle - \frac{1}{4}k^2\right), \quad \hbar\Omega_3 = \frac{1}{2}\gamma k^3,$$

and φ is the angle between k and the x axis. Iteration of the kinetic equation to second order in the parameter $\Omega_n\tau_n$ yields the following values of the spin relaxation rate [9.36–38]:

$$\frac{\partial s_i}{\partial t} = -\frac{S_i}{\tau^{\mathrm{e}}_{\mathrm{s}ij}},$$

$$\frac{1}{\tau^{\mathrm{e}}_{\mathrm{s}zz}} = \frac{2}{\tau^{\mathrm{e}}_{\mathrm{s}xx}} = \frac{2}{\tau^{\mathrm{e}}_{\mathrm{s}yy}} = \Omega_1^2\tau_1 + \Omega_3^2\tau_3; \quad \frac{1}{\tau^{\mathrm{e}}_{\mathrm{s}ij}} = 0 (i \neq j). \tag{9.31b}$$

Here τ_n is the relaxation rate for the corresponding harmonic of the distribution function:

$$\frac{1}{\tau_n} = \int \sigma(\theta)(1 - \cos n\theta)\mathrm{d}\theta,$$

where $\sigma(\theta)$ is the scattering cross section. Note that $\tau_1 \equiv \tau_p$.

A characteristic feature of the DP mechanism is a strong decrease of the spin relaxation rate in a magnetic field $\mathbf{B} \parallel [001]$ because of the cyclotron rotation [9.42]. Inclusion into (9.1a) of only linear-in-k terms yields

$$\frac{1}{\tau^{\mathrm{e}}_{\mathrm{s}zz}(B)} = \frac{\Omega_1^2\tau_\mathrm{p}}{1+(\omega_\mathrm{c}\tau_\mathrm{p})^2}, \tag{9.31c}$$

where $\omega_c = eB/mc$ is the cyclotron frequency.

The BAP mechanism in quantum wells is most efficient when electrons are scattered from bound holes under the conditions where the splitting of the $\pm 3/2$ and $\pm 1/2$ acceptor terms is less than the electron kinetic energy, thus allowing transitions between all four hole states.

Using the general relations derived in [9.43], one can readily show the electron spin relaxation rate in this case to be determined by

$$\frac{1}{\tau^{\mathrm{e}}_{\mathrm{s}}} = \frac{1}{8}\Delta^2_{\mathrm{ex}}\frac{m^*}{\hbar^3}\left[\int \mathrm{d}z|f_\mathrm{e}(z)f_\mathrm{h}(z)|^2\right]^2 \frac{|\varphi^{2\mathrm{D}}(0)|^4}{|f^{3\mathrm{D}}(0)|^4}N_\mathrm{h}.$$

Here, Δ_{ex} is the spin splitting of the ± 2 and ± 1 exciton levels, $f^{3\mathrm{D}}(0)$ is the value of the wave function $f(r)$ at $r = r_\mathrm{e} - r_\mathrm{h} = 0$ for the 3D exciton, $f_\mathrm{e}(z)$ and $f_\mathrm{h}(z)$ are the corresponding wave functions of the 2D electron and hole, and N_h is the hole concentration.

The efficiency of the BAP mechanism drops substantially if the splitting of the heavy and light hole levels exceeds the electron kinetic energy. *Wagner* et al. [9.44] revealed that, in δ-doped quantum wells at the acceptor concentration $N_\mathrm{h} = 8 \times 10^{12}$ cm^{-2} and $T = 6$ K, $\tau^{\mathrm{e}}_{\mathrm{s}}$ is two orders of magnitude larger than that observed under similar conditions at homogeneous doping. The increase of $\tau^{\mathrm{e}}_{\mathrm{s}}$ is attributed to a smaller overlap of the $f_\mathrm{e}(z)$ and $f_\mathrm{h}(z)$ wave functions under δ-doping because of the electrons being repelled from the δ-layer by negatively charged acceptors. These experiments reveal that under these conditions and for homogeneous doping the BAP mechanism is predominant, at least at acceptor concentrations $N_\mathrm{h} > 10^{11}$ cm^{-2}.

9.3.4 Electron-Nuclear Coupling Under Optical Orientation

Due to the exchange coupling between the electrons and nuclei, the orientation of electrons results in that of the nuclei. The orientation of nuclei occurs practically only in interaction with bound electrons whose time of interaction with the nuclei is determined by their lifetime at the center or the spin relaxation time dominated by exchange with free electrons. For the latter, the interaction time is equal to the transit time, $\hbar/E_{\mathbf{k}}$, where $E_{\mathbf{k}}$ is the electron energy, and is very small. The orientation of nuclei occurs only in an external magnetic field **B** which eliminates the dipole–dipole nuclear interaction characterized by the effective field B_L. Note that the nuclear spin is directed along or opposite to the magnetic field, its sign determined by the electron-spin orientation and the sign of the exchange-interaction constant. Nuclear orientation produces an effective magnetic field $\mathbf{B}_N$ acting upon the electrons. For $B \gg B_L$, the field $\mathbf{B}_N \sim (\mathbf{S} \cdot \mathbf{B})\mathbf{B}/B^2$. The orientation of electrons, in turn, results in the appearance of an effective field $\mathbf{B}_e$ acting upon the nuclei. These fields are revealed particularly clearly when studying the Hanle effect in an oblique field. If the external field **B** makes an angle α to the pump beam, then the effective transverse field acting upon the electron $B_{\text{eff}} = (B + B_N)\sin\alpha$, the component of **s** perpendicular to the field **B** decreasing, in accordance with (9.15), with $\hbar\Omega = g\mu_B B_{\text{eff}}$, while the longitudinal component remains unchanged

$$S(\mathbf{B}) = S(0)\left(\frac{\sin^2\alpha}{1+(\Omega T)^2} + \cos^2\alpha\right). \tag{9.32}$$

For $B_N \neq 0$ the maximum of the Hanle curve shifts to the point $B = -B_N$.

The corresponding effects in bulk crystals were studied by many researchers [9.27-30]. A similar investigation involving electron orientation in GaAs/AlAs quantum wells was first carried out by *Kalevich* et al. [9.45]. Curve 1 in Figure 9.8 is the conventional Hanle curve obtained at $\alpha = 85°$ with unoriented nuclei. These measurements were done by varying the sign of circular polarization with a period $T = 3\times10^{-5}$s, which is much shorter than the nuclear

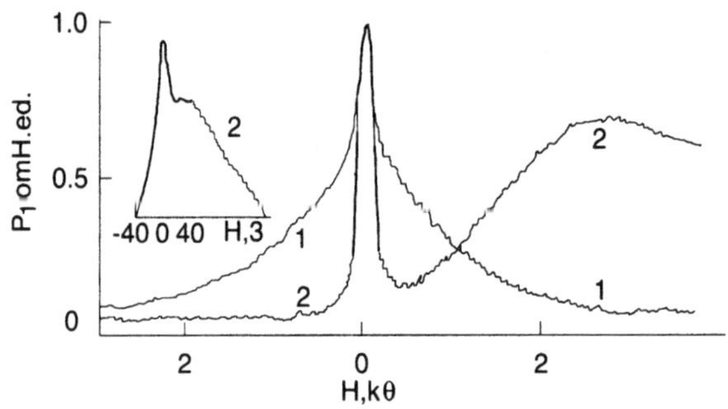

Fig. 9.8. Variation of circular polarization of luminescence in a quantum well in an oblique magnetic field ($\alpha = 85°$): curve 1 – under periodic reversal of the sign of pump light circular polarization; curve 2 – for fixed sign of circular polarization (after [9.45])

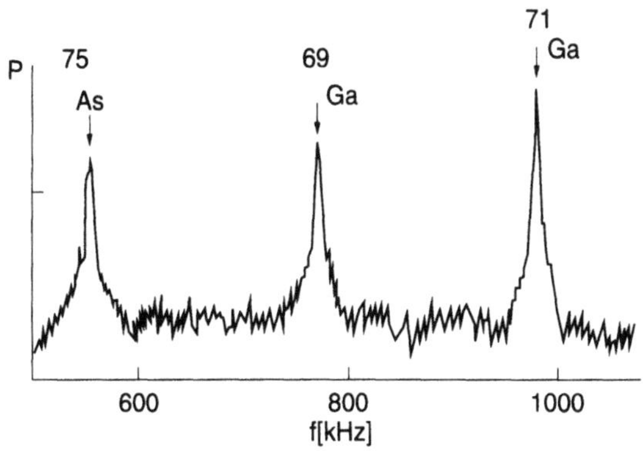

Fig. 9.9. Variation of luminescence polarization in an oblique magnetic field when an ac field at nuclear resonance frequency is turned on (after [9.45])

orientation time. Curve 2 was measured at a fixed sign of the circular polarization. The function $S(\mathbf{B})$ reaches a maximum at the point where $B = -B_N$, i.e., $B_{eff} = 0$. The sharp peak at low B is accounted for by the fact that nuclear orientation does not occur at $|\mathbf{B} + \mathbf{B}_e| < B_L$, $B_N = 0$, and B_{eff} is also, accordingly, low. Application of an ac field at the nuclear resonance frequency of the ^{69}Ga, ^{71}Ga or ^{75}As nuclei results in depolarization of the corresponding nuclei and a decrease of B_N. Depending on the actual sign and magnitude of $\mathbf{B}$, this decrease of B_N may result either in an increase or in a decrease of the degree of the radiation polarization, as is seen from Figure 9.9.

A similar effect is observed at a modulation of the pump-light intensity or at a circular polarization large enough to result in the appearance of a high-frequency component of B_E at the NMR frequency.

9.3.5 Alignment of Electron Momenta in Quantum Wells

According to (9.30), for excitation by linearly polarized light

$$G_+ = G_{1/2,1/2} = G_{-1/2,-1/2} = A_l^2 + B_l^2 - 2A_l B_l \sin(2\psi + \varphi + \eta) \quad (9.33)$$

where ψ is the angle between $\mathbf{e}$ and the x-axis and $e^{i(\varphi+\eta)} = -iI/|I|$. In the isotropic approximation $\varphi + \eta = -2\chi - (\pi/2)$, where χ is the angle between $\mathbf{k}_\perp$ and the [001] axis and, accordingly, $\sin(2\psi + \varphi + \eta) = -\cos 2(\chi - \psi)$. As already pointed out, for excitation from heavy-hole levels near the threshold, i.e., for $k_\perp = 0$, $B_l = 0$, while in the case of excitation from the light-hole levels $A_l = 0$. Therefore, practically no alignment occurs near the threshold. The degree of alignment increases with $k_\perp$. As shown by *Merkulov* et al. [9.32], in the isotropic approximation and for infinitely high barriers for both the electrons and holes, the degree of alignment (just as that of spin orientation) depends only on the ratio $t_h = k_\perp^2/(k_\perp^2 + k_2^2)$ or $t_l = k_\perp^2/(k_\perp^2 + k_1^2)$. In this approximation and when excitation occurs from the ith level,

$$G_+^{(l)} = G_o\left\{1 + \alpha_0^{(l)}\left[2\,(\mathbf{e}\cdot\mathbf{k}_\perp)^2 - 1\right]\right\}. \quad (9.34)$$

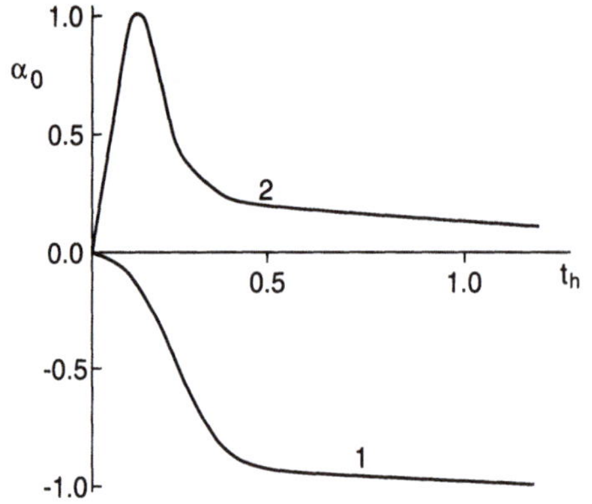

Fig. 9.10. Dependence of electron momentum alignment parameter α_c on the quantity $t_h = k_\perp^2/(k_z^2+k_\perp^2)$ under excitation from the lowest level of heavy (curve 1) and light (curve 2) holes (after [9.32])

Figure 9.10 displays the dependence of the alignment parameter $\alpha^{(i)}$ on t_h for $m_l/m_h = 0.18$ for the case of excitation to the first electronic level from the heavy hole (curve 1) and light hole (curve 2) levels. The degree of hot-luminescence polarization in the recombination of electrons with heavy holes in the absence of momentum relaxation is determined by (9.18). In the isotropic approximation

$$Q^{(l,2)}(\tilde{\mathbf{e}}, \mathbf{e}) = Q_o \left\{ 1 + \frac{1}{2}\alpha_o^{(l)}\alpha_o^{(2)} \left[2(\tilde{\mathbf{e}} \cdot \mathbf{e})^2 - 1 \right] \right\}. \tag{9.35}$$

When anisotropy is included, the degree of polarization depends on the direction of light propagation and orientation of e with respect to the principal axes of the crystal. As follows from symmetry considerations, for light propagating along [001] $Q^{(l,2)}$ is determined by an expression similar to (9.25)

$$Q^{(l,2)} = Q_0 \left[1 + \beta^{(l)} \left(\tilde{e}_x^2 - \tilde{e}_y^2 \right) \left(e_x^2 - e_y^2 \right) + 4\gamma^{(l)} \tilde{e}_x \tilde{e}_y e_x e_y \right], \tag{9.36}$$

where, in contrast to the case of bulk crystals, $\beta^{(l)}$ and $\gamma^{(l)}$ depend on the pump frequency. When exciting near the threshold, A_1 and B_1 in (9.33), just as the hole energy $E(\mathbf{k}_\perp)$, do not depend on the direction of $\mathbf{k}_\perp$. In this case

$$\beta^{(l)} = \frac{6A_1^{(l)}A_1^{(2)}B_1^{(l)}B_1^{(2)}\gamma_2^2}{\left(A_1^{(l)2} + B_1^{(l)2}\right)\left(A_1^{(2)2} + B_1^{(2)2}\right)}, \quad \gamma^{(l)} = \frac{\gamma_3^2}{\gamma_2^2}\beta^{(l)}. \tag{9.37}$$

At high pump frequencies, the anisotropies of the coefficients A_1, B_1 of the hole spectrum play a substantial role. As a result, the energy of the excited electrons will also depend on the direction of $\mathbf{k}_\perp$, and the polarization at the short-wavelength edge will be determined by electrons with the maximum energy; these are electrons with $\mathbf{k}_\perp \parallel [110]$ or $\mathbf{k}_\perp \parallel [1\bar{1}0]$, since for these directions of $\mathbf{k}_\perp$ the hole energy is the lowest. Therefore, at such frequencies, the degree of polarization is the highest for $\mathbf{e}$ parallel to [110] or $[1\bar{1}0]$ and is substantially smaller for $\mathbf{e}$ parallel to [100] or [010]. In the case of electrons recombining with acceptor-bound holes, the degree of polarization depends markedly on the anisotropy of the acceptor wave function as well.

Figure 9.11 presents the angular indicatrices of the luminescence linear polarization calculated by *Portnoi* [9.46] for the case of excitation from the first heavy-hole level E_{21} and of transitions back to this level for the GaAs/AlGaAs well at high excitation energies $\hbar\omega - E_g \gg E_{21}$. The figure presents the dependence of the degree of polarization ξ on the orientation of **e**. Curve 1 shows the value of ξ for the recombination due to radiation of the most energetic electrons, curve 2 corresponding to the mean value of ξ for all excited electrons. Electron alignment in GaAs/AlGaAs quantum wells was observed by *Zakharchenya* [9.47] and *Kop'ev* et al. [9.48]. Figure 9.12 shows the angular indicatrix of the linear polarization of luminescence at the high frequency edge of the emission line for two values of the excited electron energy E_0. We readily see that both the degree of polarization and anisotropy increase with increasing E_0, namely, the ratio $\xi_{max}(\mathbf{e}\| \; [110]) \; / \; \xi_{min}(\mathbf{e}\| \; [100])$ grows with E_0 from 3.0 to 4.15. According to (9.36, 37), for low excitation energies E_0, this ratio for

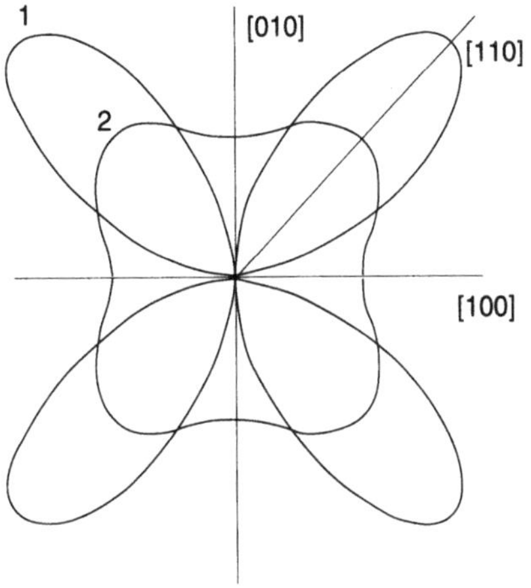

Fig. 9.11. Angular indicatrix of linear luminescence polarization for excitation and recombination to the lowest heavy hole level. Curve 1 – low temperatures (radiation is due to electrons with the highest energy); curve 2 – high temperatures (radiation comes from all excited electrons) (after [9.46])

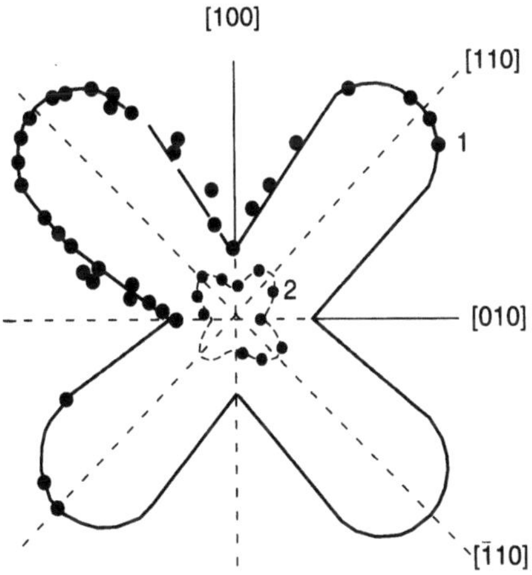

Fig. 9.12. Angular indicatrix of linear polarization of luminescence in GaAs/AlGaAs superlattice at the high frequency edge of emission line. Curve 1, $E_0 = 215$ meV; curve 2, $E_0 = 60$ meV (after [9.48])

GaAs is $\gamma_3^2/\gamma_2^2 = 2$. *Kop'ev* et al. [9.48] investigated also the depolarization of radiation in a longitudinal magnetic field. Just as in bulk crystals, the depolarization here is caused by the change of electron momentum direction induced by the Lorentz force and is described by (9.26). Such experiments provide a possibility of directly measuring the hot electron momentum relaxation time τ_p. For the samples studied it turned out to be 1.6×10^{-12}s.

9.4 Optical Orientation and Alignment of Excitons

In contrast to electrons and holes, excitons possess an integer rather than half-integer spin. Therefore for excitons, just as for atoms, it is possible to achieve not only spin-orientation under excitation by circularly polarized light but also alignment of the dipole moments under linearly polarized pumping, an effect that reveals itself in the linear polarization of the luminescence. This dipole-moment alignment is in no way related to the exciton alignment in momentum **K**, which under resonant excitation of direct excitons is always equal to the photon momentum **q** and, in two-dimensional systems, to its transverse component $\mathbf{q}_\perp$.

In contrast to the electrons, the alignment and orientation of excitons requires neither spin-orbit interaction nor the corresponding valence band splitting. Under resonant excitation the exciton angular momentum or dipole moment is equal to the corresponding momentum of the absorbed photon. The optical orientation of excitons was first observed by *Gross* et al. [9.49] in 1971 with CdSe. The possibility of exciton alignment under linearly polarized excitation was predicted by *Bir* and *Pikus* [9.50,51] and first observed in 1974 also with CdSe [9.52].

Comprehensive references to later studies can be found in the reviews [9.53-55]. Exciton orientation in quantum wells was first observed by *Masselink* et al. [9.56] and in superlattices, by *Uraltsev* et al. [9.57]. *Stolz* et al. [9.58] and Ivchenko et al. [9.59, 60] observed exciton alignment in quantum wells.

9.4.1 Exchange Interaction

Taking into account the spin states of the electron and the hole, the exciton ground state is always degenerate. In the effective-mass approximation, the ground-state degeneracy of the exciton at the point $K = 0$ or of the charged center-bound exciton is equal to the product of degeneracies of the conduction and valence bands at the extremum point, its wave function transforming according to the representation $D_{\text{ex}} = D_{\text{c}} \times D_{\text{v}}$, where D_{c} and D_{v} are the representations according to which the corresponding electron and hole wave functions transform. The representation D_{ex} is at least fourfold degenerate, i.e., it is reducible and expandable into irreducible representations. As pointed out in

Chap. 5, exchange interaction between the electron and the hole lifts the degeneracy and splits the exciton ground state into the terms corresponding to these irreducible representations. The mechanism of the exchange interaction involves short-range Coulomb interaction of the conduction and valence band electrons. The long-range (annihilation) interaction, which is of the exchange type for the electron-positron pair, results for the electron-hole pair in the dipole–dipole interaction, making the energy of the exciton dependent on the mutual orientation of its dipole moment and momentum. For bound excitons this interaction provides an additional contribution to exchange interaction constants. According to (5.142), for the cubic A_3B_5 crystals, the Hamiltonian $\mathcal{H}_{ex}$ describing the exchange interaction of the Γ_6 electron and Γ_8 hole can be written in the form

$$\mathcal{H}_{ex} = \frac{\Delta_0}{4}(\sigma \cdot \mathbf{J}) + \tilde{\Delta}_1 \sum_i \sigma_i J_i^3. \tag{9.38}$$

The constant $\tilde{\Delta}_1$ is due to the mixing of the Γ_{15} and Γ_{12} states by spin-orbit coupling and is usually small. In accordance with (9.38), in (001) quantum wells, the ground state of the exciton formed with a heavy hole splits into three terms

$$\begin{aligned}
\Phi_1 &= |-1/2, 3/2\rangle, \\
\Phi_{-1} &= |1/2, -3/2\rangle \quad (\text{representation } \Gamma_5(E), \text{energy} E = -\Delta), \\
\Phi_3 &= (\Phi_2 - \Phi_{-2})/\sqrt{2} \quad (\text{representation } \Gamma_1(A_1), E = \Delta - \Delta_1), \\
\Phi_4 &= (\Phi_2 + \Phi_{-2})/\sqrt{2} \quad (\text{representation } \Gamma_2(A_2), E = \Delta + \Delta_1). \quad (9.38a)
\end{aligned}$$

Here $\Phi_2 = |1/2, 3/2\rangle$, $\Phi_{-2} = |-1/2, -3/2\rangle$, the brackets specifying the representations of the point group D_{2d}.

The ground state of the exciton containing the light hole also splits into three terms; similar to (5.143)

$$\begin{aligned}
\Phi_1 &= |1/2, 1/2\rangle, \quad \Phi_{-1} = |-1/2, -1/2\rangle \\
&\quad (\text{representation } \Gamma_5(E), \text{ energy } E = \Delta), \\
\Phi_- &= (|1/2, -1/2\rangle - |-1/2, 1/2\rangle)/\sqrt{2} \\
&\quad (\text{representation } \Gamma_2(B_2), E = -5\Delta), \\
\Phi_+ &= (|1/2, -1/2\rangle + |-1/2, 1/2\rangle)/\sqrt{2} \\
&\quad (\text{representation} \Gamma_1(A_1), E = +3\Delta). \quad (9.38b)
\end{aligned}$$

The constants Δ and Δ_1 in (9.38a) are related with Δ_0 and $\tilde{\Delta}_1$ in (9.38) for the bulk exciton through the expression

$$\Delta = \frac{3}{8}(\Delta_0 + 9\tilde{\Delta}_1)\theta, \quad \Delta_1 = \frac{3}{2}\tilde{\Delta}_1\theta,$$

where θ is the ratio of the overlap integrals for the two- and three-dimensional excitons

$$\theta = \frac{a_B^3}{\tilde{a}_B^2} \int |\psi_e(z)|^2 |\psi_h(z)|^2 dz. \tag{9.38c}$$

Here a_B and $\tilde{a}_B$ are the Bohr radii of 3D and 2D excitons, $\psi_e(z)$ and $\psi_h(z)$ being the electron and hole wave functions, accordingly, for $k_\perp = 0$. For a well of width a with an infinitely high barrier, the integral in (9.38c) is equal to $\frac{3}{2}a^{-1}$. For the light exciton the constant Δ in (9.38b)

$$\Delta = \Delta_0 \theta / 8 .$$

9.4.2 Exchange-Deformation Splitting

A specific feature of excitons as integer-spin particles is the possibility to remove their degeneracy completely, whereas for particles with a half-integer spin, in accordance with the requirement of time inversion invariance, degeneracy always exists. Therefore, uniaxial deformation of crystals may result in a splitting of the exciton states even when the initial states of the holes and electrons do not split. What causes the lifting of degeneracy is the change in the exchange coupling, an effect first observed on A_2B_6 crystals [9.61-63]. A similar effect can be revealed in quantum wells and superlattices [9.64, 65]. Thus, for instance, in (001) wells deformed along [100] or [110], partial mixing of the heavy and light hole states occurs. By (3.77), in deformed crystals, the heavy-hole wave function is determined for $k_\perp = 0$ by the expressions

$$\tilde{\psi}_{3/2} = \psi_{3/2} + \frac{i}{\Delta E}\psi_{-1/2}, \quad \tilde{\psi}_{-3/2} = \psi_{-3/2} + \frac{i}{\Delta E}\psi_{1/2},$$

where i is defined by (3.77a), and ΔE is the spacing between the lowest levels of the light and heavy holes. Accordingly, the exchange interaction Hamiltonian in the basis of the function ($i = \pm 1, \pm 2$) is described by Table 9.3 and can be written in the form

$$\mathcal{H}'_{ex} = -\frac{2}{3}\left[\tilde{\delta}'\left(V_x\sigma_x - V_y\sigma_y\right) - \tilde{\delta}\left(V_x\sigma_y - V_y\sigma_x\right)\right] \tag{9.39}$$

Table 9.3. Exchange-interaction Hamiltonian in the basis of the excitonic functions Φ_i ($\delta = \tilde{\delta}' + i\tilde{\delta}$)

	1	−1	2	−2
1	$-\Delta$	δ	0	0
−1	δ^*	$-\Delta$	0	0
2	0	0	Δ	Δ_1
−2	0	0	Δ_1	Δ

where

$$\tilde{\delta}' = 2b\left(\Delta - \frac{1}{2}\Delta_1\right)\left(\varepsilon_{xx} - \varepsilon_{yy}\right)/\Delta E,$$

$$\tilde{\delta} = -\frac{4}{\sqrt{3}}d\left(\Delta - \frac{1}{2}\Delta_1\right)\varepsilon_{xy}/\Delta E.$$

As a result, the term $\Phi_\pm$ splits into two levels with energies $E_{\pm 1} = -\Delta \pm |\tilde{\delta}|$, and for each of them the time inversion operation sends the wave function into itself.

9.4.3 Exchange Splitting of Bound Excitons

As shown by *Aleiner* and *Ivchenko* [9.66] (see also [9.67]), the lower symmetry of the well boundary may produce a similar, additional exchange splitting of bound excitons localized near it. Indeed, in the GaAs/AlAs superlattice [001] the symmetry group for each of the boundaries is C_{2v} while for the well as a whole it is D_{2d}.

As a result, at the boundary between layers A and B, mixing of the heavy- and light-hole wave functions occurs, which may be described by the boundary conditions

$$\varphi_j^A = \varphi_j^B, \quad \nabla_z^j \varphi_j^A = \nabla_z^j \varphi_j^B - \frac{2}{\sqrt{3}} t_{\mathrm{lh}} \sum_{j'} \left[J_x J_y\right]_{s,jj'} \varphi_{j'}^B, \tag{9.40a}$$

where

$$\nabla_z^{\pm 3/2} = a_0 \frac{m}{m_{\mathrm{hh}}}\frac{\partial}{\partial z}, \quad \nabla_z^{\pm 1/2} = a_0 \frac{m}{m_{\mathrm{lh}}}\frac{\partial}{\partial z}.$$

However, in contrast to strain that mixes the light-and heavy-hole functions of the same parity, in this case the symmetric function of the heavy holes and the antisymmetric one of the light holes are mixed. Thus we obtain for the heavy-hole ground state

$$\psi_{3/2}^{\pm} = C(Z)|\pm 3/2\rangle \mp \mathrm{i}\beta S(Z)|\mp 1/2\rangle. \tag{9.40b}$$

Here, $C(z)$ and $S(z)$ are the normalized even and odd functions. The mixing constant β is proportional to t_{lh} and depends on the well thickness and barrier height. For free excitons this mixing does not produce additional exchange splitting of the radiative doublet, since it does not destroy the well symmetry as a whole. However, for the exciton localized at one of the boundaries, and whose electron wave function $\varphi_{\mathrm{e}}(z)$ does not possess a definite symmetry, the constant $\tilde{\delta}$ in (9.39) is nonzero

$$\tilde{\delta} = -\frac{4}{\sqrt{3}}\beta\left(\Delta - \frac{1}{2}\Delta_1\right)\int C(z)S(z)|\varphi_{\mathrm{e}}(z)|^2 \mathrm{d}z \left[\int C^2(z)|\varphi_{\mathrm{e}}(z)|^2 \mathrm{d}z\right]^{-1}. \tag{9.40c}$$

Note that the values of $\tilde{\delta}$ for excitons localized at opposite boundaries differ in sign.

9.4.4 Long-Range Exchange Interaction

As pointed out in Sect. 6.3, Long-Range (LR) exchange interaction produces longitudinal-transverse splitting of exciton terms. In a cubic crystal the energy of the longitudinal exciton exceeds that of the transverse one by the amount Δ_{LT}^{3D} which, for the ground state of the $\Gamma_6 \times \Gamma_8$ exciton, is

$$\Delta_{LT}^{3D} = \frac{16}{3}\pi \frac{e^2|P|^2}{\kappa E_g}|f^{3D}(0)|^2, \tag{9.41a}$$

where

$$P = \mathrm{i}\frac{\hbar}{m}\langle S|P_z|Z\rangle .$$

The matrix $\mathcal{H}_{ex}^{LR}$ of the long-range exchange interaction for excitons in a quantum well can be calculated using the general expression for the corresponding operator ΔV obtained by *Bir* and *Pikus* [9.5] (see also [9.68]).

According to *Pikus* and *Bir* [9.69]

$$\Delta V_{\substack{m'n'\\mn}}\begin{pmatrix} \mathbf{r}_1' & \mathbf{r}_2' \\ \mathbf{r}_1 & \mathbf{r}_2 \end{pmatrix} = -\sum_{\alpha\beta} Q^{\alpha\beta}_{\substack{m'n'\\mn}} \frac{\partial^2 V(\mathbf{r}_1 - \mathbf{r}_2')}{\partial r_{1\alpha}\partial r_{1\beta}} \delta(\mathbf{r}_1 - \mathbf{r}_2)\,\delta\left(\mathbf{r}_1' - \mathbf{r}_2'\right). \tag{9.41b}$$

where mn and $m'n'$ are the electron- and hole-spin indices, $\mathbf{r}_1$, $\mathbf{r}_2$ and $\mathbf{r}_1'$, $\mathbf{r}_2'$ are the coordinates, $V(\mathbf{r})$ is the Coulomb interaction energy, and

$$Q^{\alpha\beta}_{\substack{m'n'\\mn}} = \frac{\hbar^2}{m^2 E_g} p^{\alpha}_{m',Kn'} p^{\beta}_{Kn,m} . \tag{9.41c}$$

This expression takes into account that the Bloch functions of the valence-band electrons φ_n^e are related to the corresponding wave functions of holes through $\varphi_n^e = -K\varphi_n^h$, K being the time-inversion operator.

Introducing the coordinates of relative motion $\mathbf{r}$ and of the center-of-gravity motion $\mathbf{R}$ in the well plane, $\mathbf{r} = \mathbf{r}_1 - \mathbf{r}_2$, $\mathbf{R} = a\mathbf{r}_1 + b\mathbf{r}_2$ with $a + b = 1$, we can write the quantity $V\left(\mathbf{r}_1 - \mathbf{r}_2'\right)$ for $\mathbf{r}_1 = \mathbf{r}_2$, $\mathbf{r}_1' = \mathbf{r}_2'$ in the form

$$V\left(\mathbf{r}_1 - \mathbf{r}_2'\right) = \frac{4\pi e^2}{\kappa}\sum_{\mathbf{K}}\int \frac{dk_z}{2\pi}\frac{1}{\mathcal{K}^2 + k_z^2} \exp\left[\mathrm{i}\mathcal{K}\cdot(\mathbf{R} - \mathbf{R}') + \mathrm{i}k_z\left(z_1 - z_2'\right)\right]. \tag{9.41d}$$

Since

$$\int_{-\infty}^{+\infty} \frac{dk_z}{2\pi}\frac{k_z^p}{\mathcal{K}^2 + k_z^2}\mathrm{e}^{\mathrm{i}k_z z} = \frac{1}{2\mathcal{K}}\mathrm{e}^{-\mathcal{K}|z|}(\mathrm{i}\mathcal{K})^p\left(\frac{z}{|z|}\right)^p$$

we obtain for the matrix element of the operator (9.41b) in the basis of the functions

$$\Psi^{\mathcal{K}}_{mn}(r, \mathbf{R}, z_1, z_2) = \varphi_m^e \varphi_n^h f^{2D}(r)\mathrm{e}^{\mathrm{i}\mathcal{K}\cdot\mathbf{R}}\psi_e(z_1)\,\psi_h(z_2) \tag{9.42a}$$

the following expression

$$\Delta V^{\mathcal{K}\mathcal{K}'}_{\substack{m'n'\\mn}} = \frac{2\pi e^2}{\kappa \mathcal{K}} |f^{2D}(O)|^2 \delta_{\mathcal{K}\mathcal{K}'} \left\{ \sum_{\alpha'\beta'} Q^{\alpha'\beta'}_{\substack{m'n'\\mn}} \mathcal{K}_{\alpha'} \mathcal{K}_{\beta'} J_0 \right.$$

$$\left. + \sum_{\alpha'} \left(Q^{\alpha' z}_{\substack{m'n'\\mn}} + Q^{z\alpha'}_{\substack{m'n'\\mn}} \right) \mathcal{K}_{\alpha'} \mathcal{K} J_1 + Q^{zz}_{\substack{m'n'\\mn}} \mathcal{K}^2 J_2 \right\}.$$

Here α', $\beta' = x, y$, and

$$J_p = \int_{-\infty}^{\infty} \mathrm{d}z \int_{-\infty}^{\infty} \mathrm{d}z' \psi_e(z') \psi_h^*(z') \psi_e^*(z) \psi_h(z) \mathrm{e}^{-\mathcal{K}|z-z'|} \mathrm{i}^p \left(\frac{z-z'}{|z-z'|} \right)^p .$$

We see that $J_p = J_p^*$. Therefore, for the real functions $\psi_e(z)$ and $\psi_h(z)$ we have $J_1 = 0$, while for the even (or odd) functions $\psi_e(z)$ and $\psi_h(z)$

$$J_0(\mathcal{K}) = -J_2(\mathcal{K}) = \int_{-\infty}^{+\infty} \mathrm{d}z \psi_e(z) \psi_h(z) \int_{-\infty}^{z} \mathrm{d}z' \psi_e(z') \psi_h(z') \mathrm{e}^{-\mathcal{K}|z-z'|}.$$

For infinitely high barriers for electrons and holes

$$\psi_e(z) = \psi_h(z) = \sqrt{\frac{2}{a}} \sin \frac{\pi z}{a}.$$

In this case

$$J_0\left(\frac{\pi\kappa}{a}\right) = \frac{32\left(\pi\kappa - 1 + \mathrm{e}^{-\pi\kappa}\right) + \pi\kappa^3\left(20 + 3\kappa^2\right)}{\left[\pi\kappa\left(\kappa^2 + 4\right)\right]^2}.$$

For $\kappa = \mathcal{K}a/\pi \ll 1$

$$J_0(\mathcal{K}) = 1 - \frac{1}{3}\mathcal{K}a\left(1 - \frac{15}{4\pi^2}\right).$$

For heavy $\pm 3/2$ holes only the matrix elements $Q_{\substack{m'n'\\mn}}$ between the optically active states $(-1/2, 3/2)$, $(1/2, -3/2)$ are nonzero. In this basis, according to (3.43)

$$\Delta V^{hh}_{\mathcal{K}\mathcal{K}'} = \frac{1}{2} \Delta^{2D}_{LT}(\mathcal{K}) D_{hh} \delta_{\mathcal{K}\mathcal{K}'} \tag{9.42b}$$

and

$$\Delta^{2D}_{LT} = \frac{3}{8} \Delta^{3D}_{LT} \left| \frac{f^{2D}(0)}{f^{3D}(0)} \right|^2 \mathcal{K} J_0(\mathcal{K}),$$

$$D_{hh} = \begin{bmatrix} 1 & -\mathrm{e}^{-2i\varphi} \\ -\mathrm{e}^{2i\varphi} & 1 \end{bmatrix}$$

where φ is the angle between $\mathcal{K}$ and the x axis.

For light holes in the basis (1/2, -1/2), (-1/2, 1/2), (1/2, 1/2), (-1/2, -1/2) we obtain according to (3.43)

$$\Delta V_{\mathcal{K}\mathcal{K}'} = \frac{1}{6}\Delta_{\mathrm{LT}}^{\mathrm{2D}}(\mathcal{K})D_{\mathrm{lh}}\delta_{\mathcal{K}\mathcal{K}'}, \tag{9.42c}$$

where

$$D_{\mathrm{lh}} = \begin{bmatrix} -4 & 4 & 0 & 0 \\ 4 & -4 & 0 & 0 \\ 0 & 0 & 1 & \mathrm{e}^{2\mathrm{i}\varphi} \\ 0 & 0 & \mathrm{e}^{-2\mathrm{i}\varphi} & 1 \end{bmatrix}.$$

9.4.5 Selection Rules

The probability of exciton excitation in the state m, n, whose wave function is described by (9.42b), is proportional to (see (6.18)),

$$|(\mathbf{e}\cdot\mathbf{v}_{mn})|^2\left|f^{\mathrm{2D}}(0)\right|^2\left|\int\psi_{\mathrm{e}}^*(z)\psi_{\mathrm{h}}(z)\mathrm{d}z\right|^2,$$

where $\mathbf{e}$ is the polarization vector,

$$\mathbf{v}_{mn} = -\langle\varphi_m^{\mathrm{e}}\left|\hat{\mathbf{v}}\right|K\varphi_n^{\mathrm{h}}\rangle.$$

Light propagating perpendicular to the well plane excites only the states $\Phi_{\pm 1}$ of the heavy or light exciton with the moment $M = \pm 1$. Depending on the sign of the polarization, circularly polarized light excites only one of these states. Light propagating in the well plane with $\mathbf{e}\| z$ excites the light exciton state Φ_-. The corresponding matrix elements of the operator $(\mathbf{e}\cdot\mathbf{v})$ are given in Table 9.4 (with the constant factor dropped).

9.4.6 Density Matrix

The optical orientation of excitons is described in terms of the phenomenological method of the density matrix. The method is applicable to excitons bound to charged centers or isoelectronic traps. It is applicable to free excitons if the density matrix can be represented as a product of a matrix ρ depending only on the electron- and hole-spin indices and a distribution function $f(\mathbf{K})$

Table 9.4. Components of the $(\mathbf{e}\cdot\mathbf{v})$ matrix in the basis of the excitonic functions defined by (9.39) and (9.40)

	Heavy excitons		Light excitons		
	Φ_1	Φ_{-1}	Φ_1	Φ_{-1}	Φ_-
$(\mathbf{ev})$	e_-	e_+	$\frac{1}{\sqrt{3}}e_-$	$\frac{1}{\sqrt{3}}e_+$	$\frac{2}{\sqrt{3}}e_z$

depending only on the exciton wave vector. Such a representation is valid if the momentum relaxation time τ_p is much shorter than the exciton lifetime τ_0 and the spin relaxation time τ_s. For excitons bound to neutral donors or acceptors, one has to take into account the exchange interaction of a pair of electrons or holes, which is much stronger than that of the electron and the hole. Therefore, such pairs usually form a complex with a zero net spin, only the remaining hole or electron undergoing orientation. The character of orientation of the unpaired electron or hole depends on the mechanism of the complex formation which may involve sequential capture of free particles or capture of a free exciton.

If the exciton state is described by a wave function $\Phi = \Sigma_m c_m \Phi_m$, where Φ_m is one of the basis functions, then the density matrix will be $\rho_{mm'} = c_m c^*_{m'}$. For a mixed state, the quantity $c_m c^*_{m'}$ is averaged over the statistical ensemble. The diagonal components ρ_{mm} determine the probability to find an exciton in the state m, and the nondiagonal ones, $\rho_{mm'}$ with $m' \neq m$, the correlation between the m' and m states. For a pure state $|\rho_{mm'}|^2 = \rho_{mm}\rho_{m'm'}$. The relation between the matrix $d_{\alpha\beta}$ determining the polarization of the secondary radiation and the density matrix is described by an expression similar to (9.6):

$$d_{\alpha\beta} = \sum_{mm'} v^{\alpha *}_m v^{\beta}_{m'} \rho_{mm'} \tag{9.43}$$

and the matrix $Q(\tilde{\mathbf{e}}, \mathbf{e})$, by (9.7). The relation between the generation matrix G and density matrix ρ is described by the kinetic equation

$$\frac{\partial \rho}{\partial t} = \left(\frac{\partial \rho}{\partial t}\right)_{\mathrm{rec}} + \left(\frac{\partial \rho}{\partial t}\right)_{\mathrm{sr}} - \frac{\mathrm{i}}{\hbar}\left[\mathcal{H}'_{\mathrm{ex}}\rho\right] - \frac{\mathrm{i}}{\hbar}\left[\mathcal{H}'\rho\right] + G. \tag{9.44}$$

The first two terms in the right-hand part of (9.44) describe the change in the exciton density matrix as a result of recombination (or dissociation into a free pair) and spin relaxation, the operator $\mathcal{H}'$ covering the effect of external perturbations, e.g., of the magnetic field.

The generation matrix G depends on the conditions of exciton formation. In the case of resonance excitation, it is defined by an expression similar to (9.4)

$$G_{mm'} = \sum_{\alpha\beta} v^{\alpha}_m v^{\beta *}_{m'} d^0_{\alpha\beta}, \tag{9.45}$$

where

$$d^0_{\alpha\beta} = \langle e_\alpha e^*_\beta \rangle .$$

If, however, the exciton is formed by binding of pairs, then $G \sim \rho_e \rho_h$ where the matrices ρ_e and ρ_h describe the spin states of the electron and hole at the moment of binding. In the latter case one may observe only exciton orientation whereas the alignment may occur only under the action of a magnetic field or deformation. Under steady-state excitation, $\partial\rho/\partial t = 0$ in (9.44). Equation (9.44) assumes that the splittings of the terms caused both by exchange interaction and by external fields are small compared to $k_B T$, and therefore does not include the terms describing energy relaxation.

In the general case there are two recombination channels, a radiative and a nonradiative (including dissociation) ones, characterized for the state m by the times $\tau_0^{(m)}$ and $\tau_i^{(m)}$. Accordingly,

$$\left(\frac{\partial \rho_{mm'}}{\partial t}\right)_{\text{rec}} = -\frac{\rho_{mm'}}{\tau_{mm'}}, \tag{9.46}$$

where

$$\frac{1}{\tau_{mm'}} = \frac{1}{2}\left(\frac{1}{\tau^{(m)}} + \frac{1}{\tau^{(m')}}\right),$$

for $m = m'$, $\tau_{mm} = \tau^{(m)}$.

9.4.7 Exciton Spin Relaxation

The exciton spin state can change as a result of either the electron or hole, or the exciton spin flip.

If spin relaxation occurs as a result of independent relaxation of the electron and hole spins characterized by the times τ_s^e and τ_s^h, and the exciton states m and m' correspond to the spin projections s and s' for the electron, and j and j' for the hole, i.e., $\rho_{mm'} \equiv \rho_{sj,s'j'}$, then

$$\left(\frac{\partial \rho_{sj,s'j'}}{\partial t}\right)_{\text{sr}} = -\frac{1}{\tau_s^e}\left(\rho_{sj,s'j'} - \frac{1}{2}\delta_{ss'}\sum_{s''}\rho_{s''j,s''j'}\right) - \frac{1}{\tau_s^h}\left(\rho_{sj,s'j'} - \frac{1}{2}\delta_{jj'}\sum_{j''}\rho_{sj'',s'j''}\right). \tag{9.47a}$$

One can readily check that this expression meets the condition of conservation of the total number of excitons, i.e.,

$$\sum_{sj}\left(\frac{\partial \rho_{sj,sj}}{\partial t}\right)_{\text{sr}} = 0.$$

If the state m is a superposition of electron and hole states with different s and j, i.e.,

$$\phi_m = \sum_{sj} C_{sj}^m \phi_{sj}$$

then the expression for $(\partial \rho_{mm'}/\partial t)_{\text{sr}}$ can be derived from (9.47a) by means of the relation

$$\rho_{mm'} = \sum_{sjs'j'} C_{sj}^m C_{s'j'}^{m'*} \rho_{sj,s'j'}. \tag{9.47b}$$

Under optical orientation of the exciton involving a change of only the electron- or hole-spin state the degree of circular polarization of the radiation does not change, since the exciton transfers here from optically active states with the moment $m = \pm 1$ to inactive states with $m = \pm 2$. Accordingly, the radiation depolarization rate is dominated by the larger of the times τ_s^e and τ_s^h.

In the optical alignment, however, a change in the spin of one of the particles destroys the nondiagonal component $\rho_{1,-1}$ and, therefore, the depolarization rate is determined by the smaller of the times τ_s^e and τ_s^h.

The main processes involved in the hole and electron spin relaxation in the exciton are the Elliott-Yafet and Dyakonov-Perel mechanisms, respectively. As shown by *Bir* and *Pikus* [9.69], short-range exchange interaction at $\tau_s^h \ll \tau_s^e$ can also produce spin relaxation of the exciton-bound electron. However, in contrast to the three-dimensional exciton, the electron- and heavy-hole spins can change simultaneously only in transitions between optically inactive $m = \pm 2$ states, the corresponding time of these transitions being proportional to $\hbar^2/\Delta_1^2\tau_s$.

Long-range exchange interaction involved in exciton scattering changes its spin state as a whole. The corresponding Hamiltonian (9.42b) written in the basis of $m = \pm 1$ functions is similar to (9.31a)

$$\mathcal{H}^L = \frac{1}{2}(\sigma \cdot \boldsymbol{\Omega}_L), \tag{9.48a}$$

where

$$\boldsymbol{\Omega}_L = -\Omega_o(\cos 2\varphi, \sin 2\varphi, 0), \quad \hbar\Omega_o = \Delta_{LT}^{2D}.$$

According to (9.31b), we have for $\Omega_o\tau_2 \ll 1$

$$\frac{1}{\tau_{zz}^s} = \frac{2}{\tau_{xx}^s} = \frac{2}{\tau_{yy}^s} = \Omega_o^2\tau_2. \tag{9.48b}$$

The longitudinal, $T_1 = \tau_{zz}^s$, and transverse, $T_2 = \tau_{xx}^s = \tau_{yy}^s$ relaxation times determine, respectively, the rates of decrease of the circular polarization, i.e. of the $\rho_{11} - \rho_{-1,-1}$ component, and of the linear polarization characterized by the $\rho_{1,-1} = \rho_{-1,1}^*$ component. As pointed out in [9.68], this mechanism is characterized by a strong decrease of the spin relaxation rate in a transverse electric field E_z due to a reduced overlap of the electron, $f_e(z)$, and hole, $f_h(z)$, wave functions that determine the magnitude of Ω_0. A transverse magnetic field B_z splitting the $m = 1$ and $m = -1$ states likewise increases the T_1 and T_2 times.

9.4.8 Exciton Orientation and Alignment in Type II GaAs/AlAs Superlattices

As pointed out in Chap. 3, in GaAs/AlAs (001) superlattices the holes are localized in GaAs wells, and the electrons, primarily in AlAs barriers, which results in an enhanced lifetime which depends substantially on the width of the wells and barriers. The pattern of the exciton ground state exchange splitting in superlattices is the same as that in quantum wells. As already pointed out in Chap. 5, studies [9.21, 9.77], which will be discussed in more detail in the next section, revealed in these superlattices the existence of exchange splitting similar to that produced by deformation along [110] or [1$\bar{1}$0]. Note that the same sample has regions with splittings equal in magnitude and opposite in sign, so that the numbers of excitons of both types are equal. As pointed out by *Gourdon* and *Lavallard* [9.65], the value of $|\tilde{\delta}|$ depends substantially on well and barrier thicknesses, increasing from 0.2 μeV in 23/41 structures to 9.4 μeV in 17/11 structures. The Hamiltonian $\mathcal{H}_{\text{ex}}$ for these excitons is specified by Table 9.3 with an imaginary constant $\delta = \mathrm{i}\tilde{\delta}$, which corresponds to the deformation ε_{xy}. Although the states $\Phi_{\pm 2}$ are optically inactive, they can be filled also under resonant pumping as a result of spin relaxation. The lifetime in these states is τ_o and, in the optically active states $\Phi_{\pm 1}$, it is $\tau_{11} = (\tau_o^{-1} + \tau_i^{-1})^{-1}$. Using (9.44), we can now calculate the function $Q(\mathbf{e}, \tilde{\mathbf{e}})$ determining the luminescence polarization under the following simplifying assumptions:

(i) Exchange splitting $\Delta \gg \hbar/T$, where T is the exciton lifetime in a given spin state. Therefore, the nondiagonal elements of the density matrix ρ_{ij} with $\mathrm{i} = \pm 1$, $j = \pm 2$, or vice versa, are zero.

(ii) Exchange splitting Δ_1 in Table 9.3 is much smaller than Δ.

(iii) The main contribution to the spin relaxation comes from the electron-spin relaxation.

In this approximation

$$\begin{aligned}
\left(\frac{\partial \rho_{11}}{\partial t}\right)_{\text{sr}} &= -\left(\frac{\partial \rho_{22}}{\partial t}\right)_{\text{sr}} = -\frac{\rho_{11} - \rho_{22}}{\tau_{\text{s}}}, \\
\left(\frac{\partial \rho_{-1,-1}}{\partial t}\right)_{\text{sr}} &= -\left(\frac{\partial \rho_{-2,-2}}{\partial t}\right)_{\text{sr}} = -\frac{\rho_{-1,-1} - \rho_{-2,-2}}{\tau_{\text{s}}}, \\
\left(\frac{\partial \rho_{ij}}{\partial t}\right)_{\text{sr}} &= -\frac{\rho_{ij}}{\tau_{\text{s}}} \quad (i \neq j),
\end{aligned} \tag{9.49}$$

where $\tau_{\text{s}} = \tau_{\text{s}}^{\text{e}}$.

Using the coupled equations (9.44) and the selection rules given in Table 9.4, one can readily show that

$$\begin{aligned}
Q(\tilde{\mathbf{e}}, \mathbf{e}) = \frac{1}{\tau_i} \Big(T' &+ \left(1 + \tilde{\omega}^2 T T'\right)^{-1} \left\{ T' \left(|\tilde{e}_+|^2 - |\tilde{e}_-|^2\right) \right. \\
&\times \left[|e_+|^2 - |e_-|^2 + \tilde{\omega} T \left(|e_x|^2 - |e_y|^2\right) \right] \\
&\left. + T \left(|\tilde{e}_x|^2 - |\tilde{e}_y|^2\right) \left[|e_x|^2 - |e_y|^2 - \tilde{\omega} T' \left(|e_+|^2 - |e_-|^2\right) \right] \right\} \\
&+ T \left(|\tilde{e}_{x'}|^2 - |\tilde{e}_{y'}|^2\right) \left(|e_{x'}|^2 - |e_{y'}|^2\right) \Big) .
\end{aligned} \tag{9.50}$$

Here

$$T = \left(\tau_{11}^{-1} + \tau_s^{-1}\right)^{-1}, \quad T' = \left[\tau_{11}^{-1} + (\tau_o + 2\tau_s)^{-1}\right]^{-1},$$
$$\tilde{\omega} = (2/\hbar)\mathrm{Im}\{\delta\} = 2\tilde{\delta}/\hbar$$

with the x and y axes along [100] and [010], and x' and y', along [110] and [1$\bar{1}$0]. It was found [9.59, 60, 70] that polarized luminescence appeared in the GaAs/AlAs superlattices only under excitation by light polarized along the x' or y' axis. This suggests that in these structures $\tilde{\omega}^2 TT' \gg 1$.

The degree of linear polarization under excitation by light polarized in the plane making an angle φ to the [110] axis was studied [9.21] on 22 Å GaAs/41 Å AlAs multiple quantum wells. In this case, $e_{x'} = \cos\varphi$, $e_{y'} = \sin\varphi$, $e_x = \cos(\pi/4 - \varphi)$, $e_y = \cos(\pi/4 + \varphi)$, and for the degree of linear polarization we obtain, in accordance with (9.50), for $\tilde{e}_1 \| e, \tilde{e}_2 \perp e$:

$$P_{\mathrm{lin}}(\varphi) = \frac{I(\tilde{\mathbf{e}}_1) - I(\tilde{\mathbf{e}}_2)}{I(\tilde{\mathbf{e}}_1) + I(\tilde{\mathbf{e}}_2)} = \frac{T}{T'}\left(\cos^2 2\varphi + \frac{\sin^2 2\varphi}{1 + \tilde{\omega}^2 TT'}\right).$$

The dependence $P_{\mathrm{lin}}(\varphi)$ displayed in Fig. 9.13 shows that for these samples $\tilde{\omega}^2 TT' \gg 1$ as well. Direct measurement of the photoluminescence decay time showed that $T' = 25$ ms and $T = 6$ ms. A similar dependence $P_{\mathrm{lin}}(\varphi)$ is given by *Permogorov* et al. [9.70]. *Van der Poel* et al. [9.21] also observed quantum beats. The method used was described in Sect. 9.1. When excitons were excited by an x-polarized, short light pulse, with $\mathbf{e}\|[100]$, the emission intensities in the x and y polarizations, I_x and I_y, oscillated, in accordance with (9.1), in opposite phase with a frequency $2\omega = 8 \times 10^7$ Hz, corresponding to $\Delta E = 2\hbar\tilde{\omega} = 0.3\mu$eV, and a beat decay time 2×10^{-8}s. For such a value of $\tilde{\omega}$ and $T \approx T' = 2.5 \times 10^{-5}$s [9.21], we obtain the product $\tilde{\omega}T = 6 \times 10^3$.

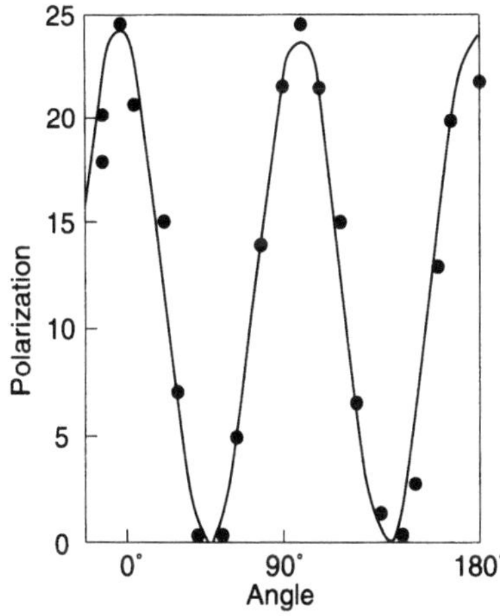

Fig. 9.13. Linear polarization of exciton luminescence in GaAs/AlAs quantum wells vs. angle φ between polarization vector and [110] axis (after [9.21])

9.4.9 Longitudinal Magnetic Field

In accordance with (3.19), Hamiltonian $\mathcal{H}_B$ determining the excitation-term splitting in a magnetic field can be written[1]

$$\mathcal{H}_B = \frac{1}{2}\mu_B \left[g_{e\parallel}\sigma_z B_z + g_{e\perp}\left(\boldsymbol{\sigma}_\perp \cdot \mathbf{B}_\perp\right)\right] + g_o\mu_B \left[\mathcal{K}(\mathbf{J}\cdot\mathbf{B}) + q\sum_i J_i^3 B_i\right]. \tag{9.51}$$

Note that the constants $\mathcal{K}$ and q for the exciton differ from the corresponding constants κ_0, q_0 for the free holes and depend, besides $\mathcal{K}_0$ and q_0, also on the ratio of the constants γ_1, γ_2 and γ_3 that determine the spectrum of the holes.

For the electron-heavy-hole exciton in a longitudinal magnetic field, one can write the non-zero matrix elements of the Hamiltonian in the basis $\Phi_{\pm1}$, $\Phi_{\pm2}$ as

$$\mathcal{H}^B_{1,1} = -\mathcal{H}^B_{-1,-1} = \frac{\hbar\omega_B}{2} = \left[\frac{3}{2}\left(\mathcal{K}+\frac{9}{4}q\right)g_o - \frac{1}{2}g_{e\parallel}\right]\mu_B B_z,$$
$$\mathcal{H}^B_{2,2} = -\mathcal{H}^B_{-2,-2} = \frac{\hbar\omega'_B}{2} = \left[\frac{3}{2}\left(\mathcal{K}+\frac{9}{4}q\right)g_o + \frac{1}{2}g_{e\parallel}\right]\mu_B B_z. \tag{9.51a}$$

If $\hbar\omega_B$ and $\hbar\omega'_B$ are small compared to Δ, then it is sufficient to take into account in (9.44) only the splitting of levels with $M_z = \pm 1$, i.e., the components $\mathcal{H}_{1,1}$ and $\mathcal{H}_{-1,-1}$. In this approximation, and if the above conditions (i)–(iii) are met, the degree of polarization of the radiation under corresponding excitation will be determined by Table 9.5. Here $P_{\rm lin}$ is the degree of linear polarization along the x, y axes, and $P'_{\rm lin}$, that along the x', y' axes. We see that in sufficiently strong fields, i.e., for $|\omega_B| \gg |\tilde{\omega}|$, it becomes possible to observe circularly

Table 9.5. Polarization of radiation in a longitudinal magnetic field for various excitation conditions

Rad. \ Excit.	$P_{\rm circ} = 1$	$P_{\rm lin} = 1$, $e \parallel x$	$P'_{\rm lin} = 1$, $e \parallel x'$
$P_{\rm circ}$	$(1+\omega_B^2 T^2)R$	$\omega T R$	$\omega_B \omega T^2 R$
$P_{\rm lin}$	$-\omega T R$	$\frac{T}{T'}R$	$\frac{\omega_B T^2}{T'}R$
$P'_{\rm lin}$	$\omega\omega_B T^2 R$	$\omega T R$	$\frac{T}{T'}(1+\omega^2 T T')R$

Here $R = (1+\omega^2 T T' + \omega_B^2 T^2)^{-1}$.

[1] Note that, when transforming the Hamiltonian $\mathcal{H}_{\rm v}$ for the valence-band electron into the Hamiltonian $\mathcal{H}_{\rm h}$ for the holes, one has, apart from substituting $-\mathcal{H}_{\rm v}$ for $\mathcal{H}_{\rm v}$, also to replace J_i by $-J_i$ and k_i by $-k_i$ since the electron and hole wave functions are related through the time inversion operation. Therefore, the Hamiltonian $\mathcal{H}_B$ has the same sign for the electrons and the holes.

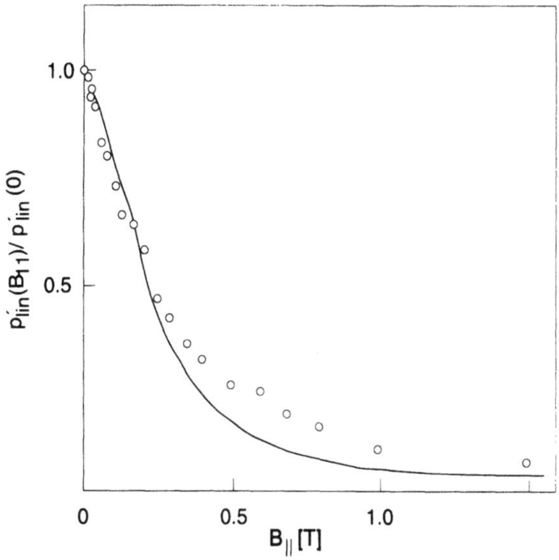

Fig. 9.14. Variation of the linear polarization of exciton luminescence P'_{lin} in GaAs/AlAs quantum wells in a longitudinal magnetic field $B_{\|}$ (after [9.59])

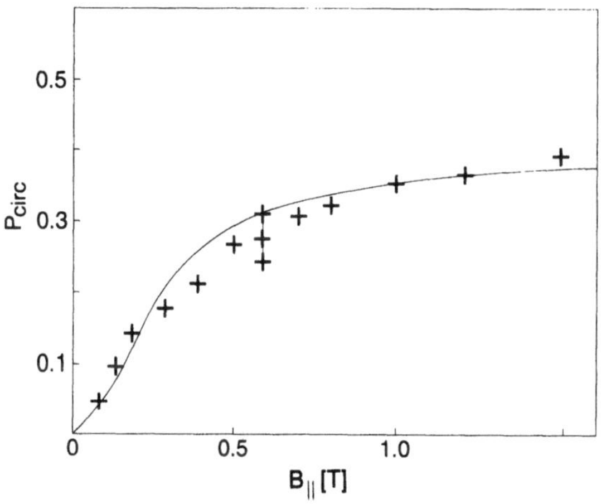

Fig. 9.15. Variation of the circular polarization of exciton luminescence P_{circ} in GaAs/AlAs quantum wells in a longitudinal magnetic field $B_{\|}$ (after [9.59])

polarized radiation under excitation by circularly polarized light. The degree of linear polarization P'_{lin}, conversely, decreases with increasing $|\omega_B|$. Figures 9.14 and 9.15 present the dependences $P'_{\text{lin}}(B_{\|})$ and $P_{\text{circ}}(B_{\|})$ obtained by *Ivchenko* et al. [9.59]. As seen also from Table 9.5, for a fixed sign of $\tilde{\omega}$ in a magnetic field it becomes possible to observe circularly polarized luminescence under excitation by linearly polarized light with $|e_{x'}|^2 - |e_{y'}|^2 \neq 0$ and, conversely, linearly polarized luminescence P'_{lin} under excitation by circularly polarized light. The degree of corresponding polarization reaches a maximum for $\omega_{\text{B}}^2 = \tilde{\omega}^2(T'/T)$ and is equal to $1/2(T/T')^{1/2} = 1/2(P'_{\text{lin}}(0))^{1/2}$. The absence of this effect in the experiment of Ivchenko et al. [9.59] suggests that the mean value $\bar{\tilde{\omega}}$ is close to zero.

9.4.10 Transverse Magnetic Field

Heavy-hole level splitting in a transverse field is determined by the constant q in (9.51). Usually $q \ll 1$. Lowering the symmetry from D_{2d} to C_{2v} mixes the light- and heavy-hole states and results in the appearance in the Hamiltonian $\mathcal{H}_B^{\mathrm{hh}}$ of additional terms [9.71]. Based on general symmetry considerations, the Hamiltonian $\mathcal{H}_B^{\mathrm{hh}}$ in a transverse magnetic field can be written

$$\mathcal{H}_B^{\mathrm{hh}} = \mu_B g_o \left[q_1 \left(J_x^3 B_x + J_y^3 B_y\right) + q_2 \left(J_x^3 B_y + J_y^3 B_x\right)\right]. \tag{9.53a}$$

If the mixing of the heavy and light hole wave functions is due to a strain ε_{xy}, then $q_1 = q, q_2$ being determined by an expression similar to (9.39)

$$q_2^\varepsilon = \frac{4}{\sqrt{3}} \mathcal{K} \frac{d\varepsilon_{xy}}{\Delta E}, \tag{9.53b}$$

which means that q_2^ε and $\tilde{\delta}$ are related through

$$q_2^\varepsilon = -\mathcal{K} \frac{\tilde{\delta}}{\Delta - \frac{1}{2}\Delta_1}. \tag{9.53c}$$

If, however, the mixing is caused by a lower boundary symmetry, then, according to (9.40b), an additional contribution to q_1 appears

$$q_1 = q + q', \quad q' = \frac{4}{3}\mathcal{K}\beta^2. \tag{9.53d}$$

Exciton localization at the boundary results in a partial admixture to the symmetric function $C(z)$ of the heavy hole of an antisymmetric one, $S(z)$, so that for a localized exciton

$$f_{nn}(z) = C(z) + \alpha S(z).$$

While hardly affecting the value of $\tilde{\delta}$, this mixing produces a contribution to q_2 which may be comparable to q_1'. We can write [9.71]

$$q_2 = \frac{4}{\sqrt{3}} \mathcal{K} \alpha \beta. \tag{9.53e}$$

In contrast to (9.53c), the sign of the constant q_2 is the same for the right-hand and left-hand polarized excitons for which the constants α and β differ in sign. Thus in the model of [9.66] the constants q_1' and q_2 are quadratic in the mixing constants and, thus, are substantially smaller than q_2.[2]

In a transverse magnetic field $\mathbf{B} \perp z$, the matrix $\mathcal{H}_B$ in the basis of the functions $\phi_i (i = \pm 1, \pm 2)$, in accordance with (9.51), has the form presented in Table 9.6. When only the main contribution to the exchange interaction is included, i.e., when the terms containing $\tilde{\delta}$, Δ_1, δ_B', q_1 and q_2 are neglected,

[2] The data presented below [9.78] yield for a 25 Å GaAs/25 Å AlAs sample $\Delta = -0.85$ μeV, $\Delta_1 = 0.1$ μeV, $\tilde{\delta} = 0.4$ μeV, $\mathcal{K} = -0.4$, and $q < 3 \times 10^{-3}$. Calculation with (9.53c) yields $q_2 = 0.16$, whereas the values of q_1' and q_2 calculated using (9.53d and e) do not exceed 2×10^{-3}.

Table 9.6. Hamiltonian $\mathcal{H}_B$ in a transverse magnetic field in the basis of the functions $\Phi_{\pm 1}, \Phi_{\pm 2}$, see (9.39)

	2	−2	1	−1
2	0	0	δ_B^*	δ_B'
−2	0	0	$\delta_B'^*$	δ_B
1	δ_B	δ_B'	0	0
−1	$\delta_B'^*$	δ_B^*	0	0

Here

$$\delta_B = \frac{1}{\sqrt{2}} g_{e\perp}\mu_B B_+, \; B_\pm = (B_x \pm i B_y)/\sqrt{2},$$

$$\delta_B' = \frac{3\sqrt{2}}{4} g_0 \mu_B (q_1 B_+ + i q_2 B_-)$$

the spectrum remains degenerate, and the energies

$$E = \pm E_o, \tag{9.54a}$$

where

$$E_o = \left(\Delta^2 + |\delta_B|^2\right)^{1/2}$$

and the eigenfunctions are

$$\left.\begin{aligned} \Psi_1 &= C_1 \left[(E_o + \Delta)\,\phi_1 - \delta_B^* \phi_2\right], \\ \Psi_2 &= C_1 \left[(E_o + \Delta)\,\phi_{-1} - \delta_B \phi_{-2}\right], \end{aligned}\right\} E = -E_o$$

$$\left.\begin{aligned} \Psi_3 &= C_2 \left[(E_o + \Delta)\,\phi_1 + \delta_B \phi_2\right], \\ \Psi_4 &= C_2 \left[(E_o + \Delta)\,\phi_{-1} + \delta_B^* \phi_{-2}\right], \end{aligned}\right\} E = E_o \tag{9.54b}$$

with

$$C_1 = [2E_o (E_o + \Delta)]^{-1/2}, \quad C_2 = [2E_o (E_o - \Delta)]^{-1/2}.$$

For $\Delta > 0$ and $B_\perp = 0$, the optically active states are $\psi_{\pm 1}$ and we will keep this in mind in what follows. For $\Delta < 0$, one can conveniently set $E_0 = -(\Delta^2 + |\delta_B|^2)^{1/2}$.

In the basis of the functions (9.54) the Hamiltonian $\mathcal{H}'$ taking into account small exchange and magnetic splittings, has the following non-zero matrix elements:

$$\mathcal{H}_{1,-1} = \mathcal{H}^*_{-1,1} = \frac{\hbar}{2}\left[-\left(\omega_3 + \omega_3'\right) + \mathrm{i}\left(\omega_1 + \omega_1' + \omega_1''\right)\right],$$

$$\mathcal{H}_{2,-2} = \mathcal{H}^*_{-2,2} = \frac{\hbar}{2}\left[\omega_3 - \omega_3' + \mathrm{i}\left(\omega_2 - \omega_1' + \omega_2''\right)\right], \tag{9.55}$$

where

$$\omega_1 = \frac{1}{2}\tilde{\omega}\left(1 + \frac{\Delta}{E_o}\right), \quad \omega_2 = \frac{1}{2}\tilde{\omega}\left(1 - \frac{\Delta}{E_o}\right),$$

$$\hbar\omega_3 = \frac{3}{4E_o} q \mu_{\mathrm{B}}^2 g_o g_{\mathrm{e}} \left(B_x^2 - B_y^2\right),$$

$$\hbar\omega_1' = \frac{3}{2E_o}\mu_B^2 g_o g_e \left(q_1 B_x B_y + \frac{1}{2} q_2 B_\perp^2\right),$$

$$\hbar\omega_3' = \left(1 - \frac{\Delta}{E_o}\right)\Delta_1 \frac{B_x^2 - B_y^2}{B^2},$$

$$\hbar\omega_3'' = \left(1 + \frac{\Delta}{E_o}\right)\Delta_1 \frac{B_x^2 - B_y^2}{B^2},$$

$$\hbar\omega_1'' = 2\left(1 - \frac{\Delta}{E_o}\right)\Delta_1 \frac{B_x B_y}{B^2},$$

$$\hbar\omega_2'' = 2\left(1 + \frac{\Delta}{E_o}\right)\Delta_1 \frac{B_x B_y}{B^2}.$$

The matrix elements $\mathcal{H}_{ij}$ with $i = \pm 1$, $j = \pm 2$, or vice versa, may be disregarded. Using (9.48), one can show that the matrix $(\partial\rho/\partial t)_{sr}$ in the basis of the functions (9.54) has the same form (9.48) as that in the basis of the functions Φ_i.

In a transverse magnetic field, the states $\Psi_{\pm 2}$ also become optically active. The generation matrix and the reciprocal time of radiative recombination for the functions $\Psi_{\pm 1}$ contain, compared with the functions $\Phi_{\pm 1}$, also an additional factor $1/2(1 + \Delta/E_0)$, and for the functions $\Psi_{\pm 2}$, a factor $1/2(1 - \Delta/E_0)$.

Rather than presenting here fairly bulky expressions for the density matrix components determining the degree of polarization of the radiation, we will dwell on the main results.

(i) In a strong transverse field, i.e., for $|\delta_B| \geqslant |\Delta|$, the radiation intensity I decreases. The ratio of I^∞ (for $|E_0| \gg |\Delta|$) to I^0 (for $E_0 = \Delta$) can be written

$$\frac{I^\infty}{I^o} = 1 - \frac{2\tau_i \tau_s}{(\tau_o + 2\tau_s)(\tau_o + 2\tau_i)}. \tag{9.56}$$

For $\tau_s \to 0$, $I^\infty \to I^0$, since spin relaxation results, in this case, in the same filling of the $\Phi_{\pm 2}$ states at $B_\perp = 0$, as does a strong magnetic field which mixes them with $\Phi_{\pm 1}$.

(ii) For $\Delta_1 = 0$ and $q = 0$, linearly polarized luminescence appears only for **e** parallel to [110] or [1$\bar{1}$0], since the increase of oscillator strength of the terms $\psi_{\pm 2}$ is accompanied by the splitting of these terms; indeed, for $|E_0| \gg |\Delta|$ the splitting of both terms is $|\tilde{\delta}|$, which is one half that of the $\Phi_{\pm 1}$ splitting for $B_\perp = 0$. The degree of linear polarization P'_{lin} does not depend on the direction of $B_\perp$ and decreases with increasing $B_\perp$ from

$$P'^{o}_{\text{lin}} = 1 - \frac{\tau_o \tau_i (\tau_o + \tau_s)}{(\tau_s \tau_o + \tau_s \tau_i + \tau_o \tau_i)(2\tau_s + \tau_o)}$$

for $E_o = |\Delta|$ down to

$$P'^{\infty}_{\text{lin}} = 1 - \frac{2\tau_o \tau_i}{\tau_s \tau_o + 2\tau_s \tau_i + 2\tau_o \tau_i} \tag{9.57a}$$

for $E_o \gg |\Delta|$. For $\tau_i \ll \tau_o$ we obtain

$$\frac{P'^{\infty}_{\text{lin}}}{P'^{o}_{\text{lin}}} = 1 - \frac{\tau_i}{\tau_s + 2\tau_i} \tag{9.57b}$$

and, for $\tau_o \ll \tau_i$,

$$\frac{P'^{\infty}_{\text{lin}}}{P'^{o}_{\text{lin}}} = 1 - \frac{\tau_o}{2(\tau_s + \tau_o)}. \tag{9.57c}$$

The effect of transverse magnetic field on polarized luminescence in GaAs/AlAs superlattices was studied by *Ivchenko* et al. [9.59]. Figure 9.16 presents the magnetic field dependences of the radiation intensity and degree of polarization P'_{lin}. In accordance with (9.56, 57a) the intensity I is seen to decrease in a strong field by 25%, and P'_{lin}, by 30%.

(iii) If for $\mathbf{e}\|[100]$, $[010]$ the field $B_\perp$ is also directed along [100], [010], then linearly polarized luminescence appears for $|\omega'_3| \simeq |\omega''_3| \geqslant |\tilde{\omega}|$, the degree of polarization for $|\tilde{\omega}|T \gg 1$ being defined by an expression differing from (9.57a) in the factor $4\omega'^2_3/\left(\tilde{\omega}^2 + 4\omega'^2_3\right)$. A magnetic field directed along [110] or $[1\bar{1}0]$ does not affect P'_{lin}, while a field along [100] or [010] suppresses this polarization for $|\omega'_3|T \gg 1$, therefore

$$P'_{\text{lin}}(B) = P'^{\infty}_{\text{lin}}(0)\frac{\tilde{\omega}^2}{\tilde{\omega}^2 + 4\omega'^2_3}, \tag{9.58}$$

where $P'_{\text{lin}}(0)$ is determined by (9.57a)

(iv) In the range of intermediate magnetic fields with $\mathbf{B}$ parallel to [110] or $[1\bar{1}0]$ there are several narrow regions where the spacings between the two

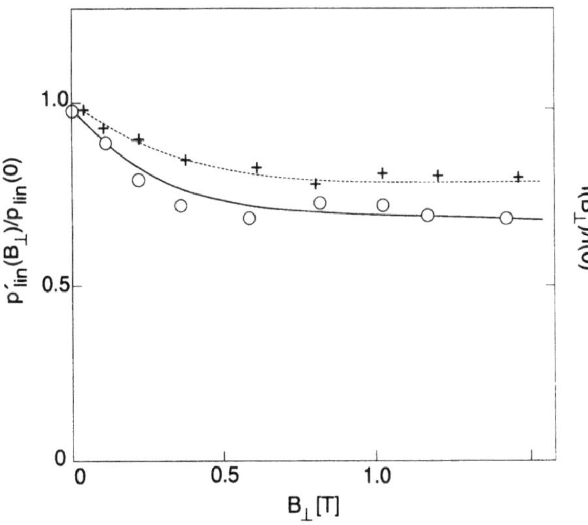

Fig. 9.16. Variation of linear polarization P'_{lin} and of exciton luminescence intensity I in GaAs/AlAs quantum wells in a transverse magnetic field $B_\perp$ (after [9.59])

nearest terms become less than $(\hbar^2/TT')^{1/2}$. In such fields observation of linear polarization P_{lin} with the pump light polarized along x or y becomes possible.

As shown in [9.72], for $\mathbf{B} \parallel [110]$ or $\mathbf{B} \parallel [1\bar{1}0]$ the 4×4 Hamiltonian $\mathcal{H} = \mathcal{H}_{\text{ex}} + \mathcal{H}'_{\text{ex}} + \mathcal{H}_B$ defined by (9.38, 39, and 53a) in the basis of functions

$$\Psi_1 = \frac{1}{\sqrt{2}}(\phi_1 + \mathrm{i}\phi_{-1}),$$

$$\Psi_2 = \frac{1}{\sqrt{2}}(\phi_2 - \gamma\phi_{-2}),$$

$$\Psi_3 = \frac{1}{\sqrt{2}}(\mathrm{i}\phi_1 + \phi_{-1}),$$

$$\Psi_4 = \frac{1}{\sqrt{2}}(\gamma\phi_2 + \phi_{-2}),$$

where $\gamma = 2B_xB_y/B^2 = \pm 1$, splits into two 2×2 submatrices, $\mathcal{H}_{12}$ and $\mathcal{H}_{34}$, with the corresponding eigenenergies

$$E_{1,2} = \frac{1}{2}\left(\tilde{\delta} - \gamma\Delta_1\right) \pm \left\{\left[-\Delta + \frac{1}{2}\left(\tilde{\delta} + \gamma\Delta_1\right)\right]^2 + |\delta_B|^2(1-\alpha)^2\right\}^{1/2},$$

$$E_{3,4} = -\frac{1}{2}\left(\tilde{\delta} - \gamma\Delta_1\right) \pm \left\{\left[-\Delta - \frac{1}{2}\left(\tilde{\delta} + \gamma\Delta_1\right)\right]^2 + |\delta_B|^2(1-\alpha)^2\right\}^{1/2}, \tag{9.59}$$

where $\alpha = (3/2)g_0/g_e(\gamma q_1 + q_2)$. As B increases from 0 to ∞, see (9.59), the difference $E_1 - E_3$ changes sign at $\tilde{\delta}(\gamma q_1 + q_2) > 0$, i.e., for both directions of B there exists one point where these levels cross for one of the signs of $\tilde{\delta}$. The difference $E_2 - E_4$ changes its sign at $\Delta_1(q_1 + \gamma q_2) > 0$, i.e., for $|q_2| \gg |q_1|$ these levels must cross. For $|q_1| > |q_2|$ these levels must cross for $\Delta_1 q_1 > 0$, and do not cross, or cross an even number of times, for $\Delta_1 q_1 < 0$.

Levels belonging to different pairs 1, 3 and 2, 4, do not cross for $|\Delta| \gg |\tilde{\delta}| + |\Delta_1|$. Equation (9.59) implies that, when this inequality is met for $|q_1|$, $|q_2| \ll 1$, levels 2 and 4 cross at the point where

$$\frac{\Delta}{E_0} = -\frac{\tilde{\delta} - \gamma\Delta_1}{\delta + \gamma\Delta_1},$$

At this point the degree of linear polarization

$$P_{\text{lin}} \simeq \frac{1}{2}\frac{\Delta_1}{|\Delta_1| + |\tilde{\delta}|}\frac{T}{T'}.$$

At small q_1, q_2 the other crossing points lie in the region of stronger magnetic fields where the oscillator strength for all four states is practically the same and, hence, at these points

$$P_{\text{lin}} \simeq \frac{1}{4}\frac{T}{T'}.$$

9.4.11 Hot-Exciton Orientation and Alignment

Excitation by light with a frequency above the exciton resonance creates, in addition to free electron-hole pairs, hot excitons. Since the momentum of such excitons exceeds substantially that of the photon, their excitation, just as radiative recombination, involves usually the emission of one or more optical phonons. Accordingly, the emission spectrum reveals a series of narrow lines shifted along with the pump-light frequency by 2, 3, ... frequencies of the longitudinal optical phonons ω_{LO}. Such spectra were observed on many A_2B_6 crystals [9.53, 54]. These lines are polarized circularly or linearly in accordance with the pump-light polarization. Note that the degree of polarization decreases with increasing number of emitted phonons, which is accounted for by the loss of orientation of the holes in scattering because of a correlation between the directions of their spin and momentum [9.53]. As a result of a strong coupling of hot excitons with the LO phonons, the time of phonon emission is very small, so that no depolarization of the radiation was observed in the magnetic fields used. It should be stressed that the linewidth does not depend on this time and is determined only by the lifetime of the LO phonon. In a consistent quantum-mechanical description, this secondary radiation should be considered as a result of resonant light scattering involving emission of 2, 3,... optical phonons. Both excitons and electron-hole pairs can serve as intermediate states. The persisting high degree of linear polarization suggests that the exciton contribution is predominant. Such luminescence was not observed earlier in A_3B_5 crystals under excitation above the resonance frequency, which can be accounted for by the smaller constant of interaction between the electrons and phonons in these crystals.

A similar effect of hot-exciton alignment has recently been observed on GaAs/AlGaAs quantum wells [9.73]. This effect was called by *Kop'ev* et al. [9.73] *geminate recombination* since in this case, in contrast to the conventional recombination, a photo-electron annihilates with a photo-hole created in the same event of light absorption. It turned out that the intensity of the corresponding radiation grows rapidly with increasing magnetic field, polarization becoming noticeable only in fields in excess of 4T. The radiation was observed at frequencies shifted from the pump frequency by 1, 2, 3 LO phonon frequencies in GaAs, as well as by 1, 2 LO phonon frequencies in AlAs. In a strong magnetic field, one observes an intense emission band close to the pump frequency. Similar emission was seen to occur in magnetic fields above 2T when excited by light of a frequency such that LO phonon emission cannot occur (Fig. 9.17). All this implies a substantial contribution of scattering by acoustical phonons and impurity centers or defects. The role of the magnetic field can be explained

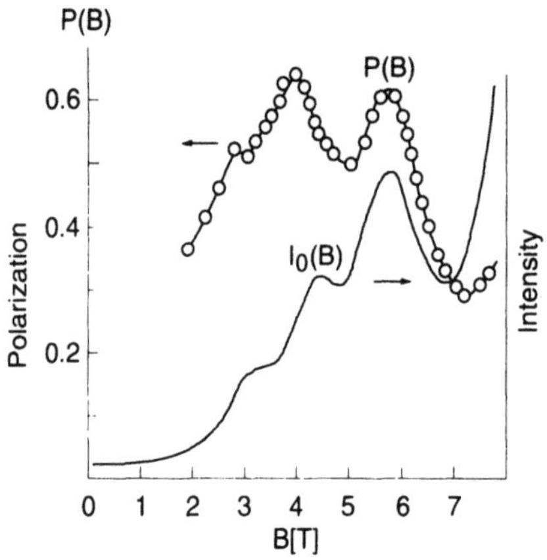

Fig. 9.17. Dependence of linear polarization P_{lin} and of hot exciton luminescence intensity in GaAs/AlAs quantum wells on magnetic field for pump energy $\hbar\omega = 1.65$ eV [9.73]

thus: by increasing the binding energy of the excitons it increases their contribution as intermediate states in scattering. In terms of classical theory, the explanation lies in that, by bending the trajectories of electrons and holes, the magnetic field increases the probability of their repeated encounters [9.74].

9.5 Polarized Luminescence of Excitons and Impurities in an External Magnetic Field

The magnetic-field-induced splitting of the spin levels of impurity centers, free carriers or excitons results in a predominant population of lower-energy spin levels. Therefore, even when carriers are excited by unpolarized light, in a magnetic field the luminescence becomes polarized. The population of the corresponding levels depends not only on their energy but on the relative magnitude of the carrier lifetime τ_o for the given level and the spin relaxation time τ_s as well. Note that while in optical orientation experiments the maximum polarization is reached for $\tau_s \gg \tau_o$, the magnetic-field-induced polarization, conversely, is the largest for $\tau_o \gg \tau_s$ when the population reaches equilibrium corresponding to the given temperature T. If the time of carrier trapping by the centers or of exciton formation is comparable with the carrier lifetime in the bound state, then specific orientation in the magnetic field may also occur in the course of binding.

9.5.1 Exciton Orientation in a Magnetic Field and Optically Detectable Magnetic Resonance (ODMR)

Exciton radiation polarization in type II GaAs/AlAs superlattices in a constant magnetic field B and its variation in a resonant microwave field b were investigated in a number of publications [9.75–78]. For the ODMR to be observable, the reciprocal spin lifetime $T^{-1} = \tau_o^{-1} + \tau_s^{-1}$ must be less than or comparable with the transition rate in the microwave field and must at least be substantially less than the field frequency ω. This condition is met in type-II superlattices, where the exciton lifetime is of the order of 10^{-6} s due to the weak overlap of the electron and hole wave functions. In the absence of magnetic field, the exciton ground state includes four levels. In accordance with Table 9.3 and (9.39), for imaginary $\delta = \mathrm{i}\tilde{\delta}$ the functions corresponding to these levels are as follows

$$\Psi_{1,2} = (\Phi_1 \pm \mathrm{i}\Phi_{-1})(E = -\Delta \mp \tilde{\delta}),$$
$$\Psi_{3,4} = (\Phi_2 \pm \Phi_{-2})(E = \Delta \pm \Delta_1). \tag{9.60}$$

As this will be shown later in detail, experiments reveal [9.61–64] that for the structures studied $\Delta < 0$, i.e., the lowest states are $\Psi_{3,4}$.

As follows from the selection rules (Table 9.4), the optically active terms are Ψ_1 in the x' polarization and Ψ_2 in the y' polarization, where $x' \parallel [110]$, $y' \parallel [1\bar{1}0]$. Figure 9.18 shows the splitting scheme for the two signs of $\tilde{\delta}$. For equal numbers of excitons with both signs of $\tilde{\delta}$ the radiation will be unpolarized. Linearly polarized ac magnetic field $\mathbf{b}_\perp$ causes the transitions depicted in Fig 9.18. This field acts only on the electron spins.

The corresponding Hamiltonian

$$\tilde{\mathcal{H}} = \frac{1}{2} g_{\mathrm{e}\perp}\mu_{\mathrm{B}} (\boldsymbol{\sigma}_\perp \cdot \mathbf{b}_\perp) = \frac{1}{\sqrt{2}} g_{\mathrm{e}\perp}\mu_{\mathrm{B}} (\sigma_+ b_- + \sigma_- b_+), \tag{9.61}$$

where

$$\sigma_\pm = \left(\sigma_x \pm \mathrm{i}\sigma_y\right)/2, \quad b_\pm = \left(b_x \pm \mathrm{i}b_y\right)/\sqrt{2}.$$

We readily see that in the basis of the functions Ψ_i

$$\tilde{\mathcal{H}}_{31} = \tilde{\mathcal{H}}_{42} \sim b_{y'}, \quad \tilde{\mathcal{H}}_{32} = \tilde{\mathcal{H}}_{41} \sim b_{x'}.$$

For a fixed sign of $\tilde{\delta}$, in the x' (or y') polarization of the microwave field one should observe two ODMR spectral lines that correspond to an increase of the Ψ_1 (or Ψ_2) population and, hence, to an increase of the radiation intensity $I_{x'}$, (or $I_{y'}$,).

Experiments carried out on 25 Å GaAs/25 Å AlAs structures revealed four lines with different signs of the difference $\Delta I = I_{x'}, - I_{y'},$. The frequency order of the signs shows that for these structures $\Delta_1 > 0$ and $|\tilde{\delta}| > \Delta_1$. The

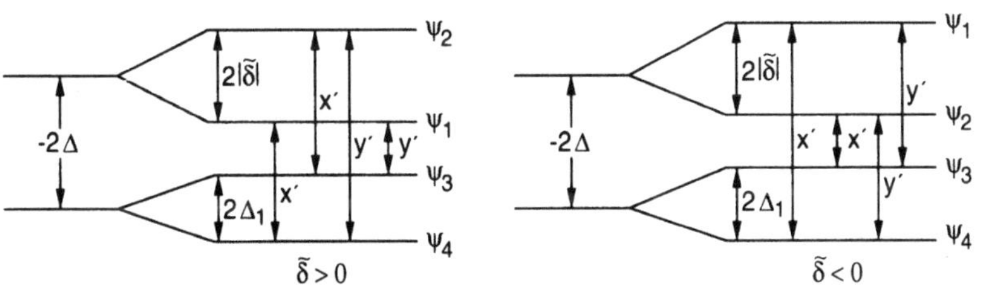

Fig. 9.18. Diagram of exciton levels in GaAs/AlAs quantum wells

values of these constants were found to be $\tilde{\delta} = \pm 0.4\ \mu$eV, $\Delta_1 = 0.2\ \mu$eV for $\Delta = -0.82\ \mu$eV.

9.5.2 Longitudinal Magnetic Field

In accordance with (9.51) in a field $B_z \parallel [001]$ there must be four terms with the energies

$$E_{\pm 1} = -\Delta \mp \left[(\hbar\omega_1)^2 + \tilde{\delta}^2\right]^{1/2}, \quad E_{\pm 2} = \Delta \pm \left[(\hbar\omega_2)^2 + \Delta_1^2\right]^{1/2}, \quad (9.62)$$

where

$$\hbar\omega_{1,2} = \frac{1}{2}\left[\mp 3\left(\mathcal{K} + \frac{9}{4}q\right) g_o + g_e\right] \mu_B B_z.$$

Since $\mathcal{K} < 0$ and $g_e < 6|\mathcal{K}|$, the lowest state for $\hbar|\omega_{1,2}| \gg \tilde{\delta}, \Delta_1$ is Φ_1 with $E_{-1} = -E_1$. The lowest of the terms $\Phi_{\pm 2}$ is Φ_2 with the energy $E_2 \simeq \Delta - \hbar\omega_2$, and the higher one is Φ_{-2} with $E_{-2} = \Delta + \hbar\omega_2$. The microwave field induces transitions between the states Φ_1 and Φ_2 by filling the state Φ_1 at the expense of the inactive state Φ_2 and, thus, increases the intensity I_+, and those between the states Φ_{-2} and Φ_{-1}, by increasing the population of the state Φ_{-1}, i.e., the intensity I_- (Fig. 9.19). For a fixed microwave frequency ω, the first of these transitions occurs at $g_e \mu_B B_z^{(1)} = \hbar\omega - 2\Delta$ and the second, at $g_e \mu_B B_z^{(2)} = \hbar\omega + 2\Delta$. The fact that the $\Phi_{-2} \to \Phi_{-1}$ transition occurs at a lower field B_z shows that $\Delta < 0$, and from the difference $B_z^{(1)} - B_z^{(2)}$ it was found that $\Delta = -0.85\mu$eV, which agrees within experimental accuracy with the value obtained in the absence of magnetic field.

9.5.3 Transverse Magnetic Field

In accordance with (9.54) and Table 9.4, in a field $B_x \parallel [110]$ there are four levels with the wave functions

$$\tilde{\Phi}_1 = (\Psi_1 + \mathrm{i}\Psi_{-1})/\sqrt{2} \quad E = E_0 - \kappa_1,$$

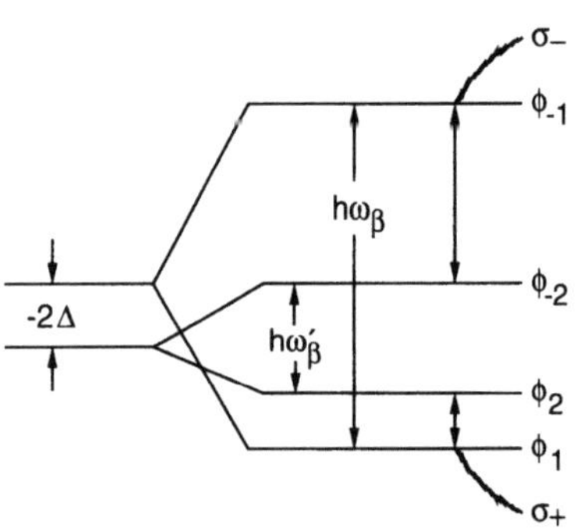

Fig. 9.19. Diagram of exciton levels in GaAs/AlAs quantum wells in a longitudinal magnetic field

$$\tilde{\Phi}_2 = (\Psi_1 - \mathrm{i}\Psi_{-1})/\sqrt{2} \quad E = E_0 + \kappa_1,$$
$$\tilde{\Phi}_3 = (\Psi_2 + \mathrm{i}\Psi_{-2})/\sqrt{2} \quad E = -E_0 - \kappa_1,$$
$$\tilde{\Phi}_4 = (\Psi_2 - \mathrm{i}\Psi_{-2})/\sqrt{2} \quad E = -E_0 + \kappa_1. \tag{9.63}$$

Here

$$\kappa_1 = (\tilde{\delta} - \Delta_1)/2, \quad E_0 = g_{\mathrm{e}\perp}\mu_{\mathrm{B}}B/2,$$
$$\Psi_1 = \left(\Phi_1 + c^*\Phi_2\right)/\sqrt{2}, \quad \Psi_{-1} = \left(\Phi_{-1} + c\Phi_{-2}\right)/\sqrt{2},$$
$$\Psi_2 = \left(\Phi_1 - c^*\Phi_2\right)/\sqrt{2}, \quad \Psi_{-2} = \left(\Phi_{-1} - c\Phi_{-2}\right)/\sqrt{2}, \tag{9.63a}$$

for $g_e > 0$, $c = (1+\mathrm{i})/\sqrt{2}$.

It is assumed that $E_0 \gg |\Delta|$, $\Delta < 0$. We see the lowest states to be $\tilde{\Phi}_{3,4}$. When corrections of the order of Δ/E_0 are included, these states are found to have the longest lifetime, with $1/\tau_i \sim 1/2(1 + \Delta/E_0)$. Therefore, they are more populated. The interlevel transitions 1–4 and 2–3 are possible only in a transverse microwave field, $\mathbf{b} \perp \mathbf{B}$, i.e., for $\mathbf{b} \parallel y'$. The probabilities of these transitions are the same, the transitions 1–3, 2–4, as well as 1–2, 3–4, being forbidden. In a strong magnetic field all the states are optically active. As follows from the selection rules, the recombination of the $\tilde{\Phi}_1$ and $\tilde{\Phi}_3$ excitons is accompanied by x'-polarized radiation, and that of the $\tilde{\Phi}_2$ and $\tilde{\Phi}_4$ excitons, by radiation in the y' polarization (Fig. 9.20). Therefore in the $3 \rightarrow 2$ transitions resulting, in the presence of a microwave field, in a decrease of the Φ_3 population and an increase of the $\tilde{\Phi}_2$ population, the change of the degree of polarization, $\Delta I = I_{x'} - I_{y'}$, should be negative, while for the 4–1 transitions, ΔI is positive. These selection rules remain valid also for $\mathbf{B} \parallel y'$, where transitions in the microwave field occur for $\mathbf{b} \parallel x'$. Note that in this case $c = -(1-\mathrm{i})/\sqrt{2}$ in (9.63a) and, accordingly, κ_1 in (9.63) is replaced by $\kappa_2 = (\tilde{\delta} + \Delta_1)/2$. For excitons with the opposite sign of $\tilde{\delta}$, the order of the 3,4 and 1,2 levels is reversed. Therefore, for each orientation of the magnetic field, i.e., for $\mathbf{B} \parallel x'$ and $\mathbf{B} \parallel y'$, each dependence $\Delta I(B) = I_{x'} - I_{y'}$ should exhibit four lines.

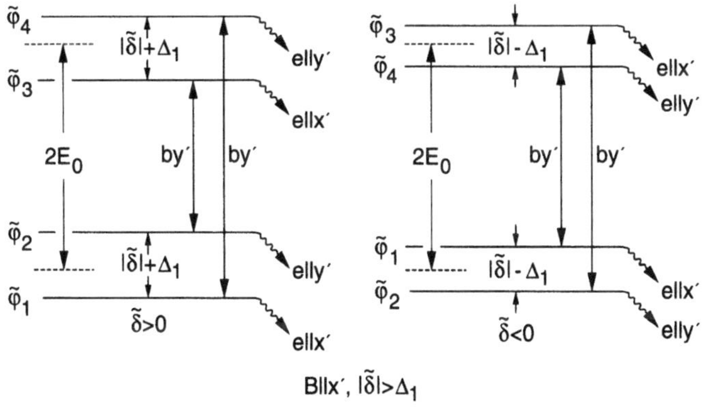

Fig. 9.20. Diagram of exciton levels in GaAs/AlAs quantum wells in a transverse magnetic field

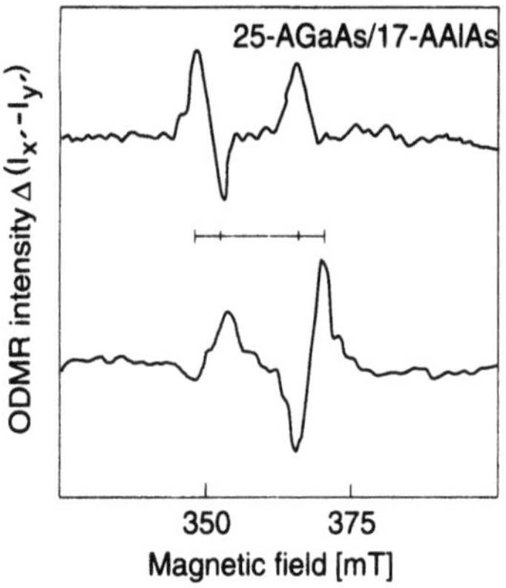

Fig. 9.21. Intensity variation of the linearly polarized component of exciton luminescence $I_{x'} - I_{y'}$ in GaAs/AlAs quantum wells in microwave field (after [9.78])

Table 9.7. Variation of polarization, $I_{x'} - I_{y'}$, in a microwave field

Transition frequencies $\hbar\omega$	$\Delta I = I_{x'} - I_{y'}$	
	$\mathbf{B} \parallel x'$	$\mathbf{B} \parallel y'$
$2E_0 + \tilde{\delta} + \Delta_1$	> 0	< 0
$2E_0 + \tilde{\delta} - \Delta_1$	< 0	> 0
$2E_0 - \tilde{\delta} + \Delta_1$	> 0	< 0
$2E_0 - \tilde{\delta} - \Delta_1$	< 0	> 0

The corresponding resonance transition frequencies and the sign of $I_{x'} - I_{y'}$ are presented in Table 9.7. On the experimental curves of Fig. 9.21, to a higher frequency ω corresponds a lower resonance field B. This order of the maxima and minima in the $\Delta I(B)$ dependence confirms that $\Delta_1 > 0$ and $|\delta| > \Delta_1$.

The experimental curves of Fig. 9.21 [9.78] were used to obtain the values $\tilde{\delta} = 0.4\ \mu\text{eV}$, $\Delta_1 = 0.1\ \mu\text{eV}$. Apart from this, the positions of the resonance maxima in the longitudinal and transverse fields permitted one to establish that the g-factors of the electrons in the X_z-valley for a 25 Å GaAs/25 Å AlAs structure are $g_{e\parallel} = 1.888$, $g_{e\perp} = 1.975$ for the electrons and for the holes $\mathcal{K} = -0.4$, $|q| < 3 \times 10^{-3}$. Similar values, $g_{e\parallel} = 1.898$, $g_{e\perp} = 1.969$, $\mathcal{K} = -0.5$, $|q| < 2 \times 10^{-2}$ were obtained by means of a comparable technique for superlattices with the wells and barriers containing five GaAs and AlAs layers each [9.78].

9.5.4 Polarized Luminescence Caused by the Orientation of Shallow Acceptor-Bound Holes

The luminescence spectra produced by electrons recombining with acceptor-bound holes were found [9.79, 80] to represent a broad band whose shape depends on the excitation level. As shown in Chap. 5, this is accounted for by the hole binding energy depending on the acceptor position in the well. In

$GaAs/Al_{0.4}Ga_{0.6}As$ quantum wells with $a = 120$ Å the dominant transitions at low excitation levels are those to the A_c-acceptors lying close to the well's center with a binding energy of 41 meV, and at high excitation levels, to the A_i-acceptors located close to or at the well boundary. The binding energy of these acceptors is 22.5 meV, the concentration of the A_i acceptors, as shown by experiments, exceeding considerably that of the A_c-acceptors. The degree of the luminescence polarization in a longitudinal magnetic field also varies along the line profile and depends on the excitation level.

The acceptor-level splitting in a magnetic field is described by an expression similar to (9.51), the constants $\mathcal{K}$ and q differing from the corresponding constants for free excitons and holes and depending on the acceptor binding energy and position.

In accordance with the selection rules (Table 9.4), in a longitudinal field B_z the polarization of the radiation due to recombination of the holes with unoriented electrons can be written

$$P_{\mathrm{circ}} = \frac{\tau_{\mathrm{h}}}{\tau_{\mathrm{h}} + \tau_{\mathrm{s}}} \Phi(B), \tag{9.64}$$

where τ_{h} and τ_{s} are the hole lifetime and spin relaxation time, respectively,

$$\Phi(B) = \frac{3 \sinh x_1 + \sinh x_2 \mathrm{e}^{-\Delta/k_B T}}{3 \cosh x_1 + \cosh x_2 \mathrm{e}^{-\Delta/k_B T}}. \tag{9.65}$$

Here

$$x_1 = \frac{3}{2} g_o \left(\mathcal{K} + \frac{9}{4} q\right) \frac{\mu_{\mathrm{B}} B_z}{k_{\mathrm{B}} T}, \quad x_2 = \frac{1}{2} g_o \left(\mathcal{K} + \frac{1}{4} q\right) \frac{\mu_{\mathrm{B}} B_z}{k_{\mathrm{B}} T}$$

and Δ is the splitting of the acceptor-bound light and heavy-hole levels.

In a strong magnetic field, $\Phi(B) \to 1$, the degree of polarization P_{circ} depends only on the ratio $\tau_{\mathrm{s}}/\tau_{\mathrm{h}}$. Measurements of P_{circ} in different spectral regions showed that the lifetime ratio of the A_i- and A_c-acceptors $\tau_{\mathrm{hi}}/\tau_{\mathrm{hc}} = 3$ and does not practically depend on excitation intensity, the lifetimes τ_{hi} and τ_{hc} proper decreasing with increasing pump level. This suggests that the electron trapping cross section for the A_i-acceptors is one third that of the A_c-acceptors. Having found the ratio $\tau_{\mathrm{s}}/\tau_{\mathrm{h}}$, one could measure P_{circ} (B) in the weak-field domain where

$$P_{\mathrm{circ}}(B) = \frac{1}{2} \frac{\tau_{\mathrm{h}}}{\tau_{\mathrm{h}} + \tau_{\mathrm{s}}} (g_{\mathrm{eff}} \mu_{\mathrm{B}} B_z / k_{\mathrm{B}} T) \tag{9.66}$$

and determine the effective acceptor g-factor

$$g_{\mathrm{eff}} = g_o \frac{\left[3\left(\mathcal{K} + \frac{9}{4} q\right) + \frac{1}{3}\left(\mathcal{K} + \frac{1}{4} q\right) \mathrm{e}^{-\Delta/k_{\mathrm{B}} T}\right]}{\left(1 + \frac{1}{3} \mathrm{e}^{-\Delta/k_{\mathrm{B}} T}\right)}.$$

For the A_c-acceptors, this quantity was found to be $|g_c| = 1.65 \pm 0.15$ and, for the A_i-acceptors, $|g_i| = 0.66 \pm 0.15$. This decrease of g_{eff} for the A_i-acceptors is connected with a change in the constants $\mathcal{K}$ and q.

Simultaneous measurement of the luminescence polarization under electron photoexcitation by circularly polarized light has offered a possibility [9.81] to determine the electron lifetime at low excitation levels, τ_{ec}, when the A_c-acceptors are practically fully populated and the A_i-acceptors empty, and at high excitations, τ_{ei}, when the A_i-acceptors are populated. The lifetimes were derived from the value of P_{circ}, and its variation in a transverse magnetic field, from (9.11,12). It was assumed that $P^0_{circ} = 0.25\%$. It was found that $P^c_{circ} = 2\%$, $P^i_{circ} = 7\%$, and $\tau_{ec}/\tau_{ei} = 5.7$, whence, recalling that $\tau_{hi}/\tau_{hc} = 3$, one could determine that the concentration of the A_i-acceptors is about 20 times that of the A_c. These experiments reveal the possibility of determining the acceptor distribution in a well and establish the dependence of the hole trapping cross sections and g-factor on acceptor position by simultaneously measuring the degree of radiation polarization under optical carrier orientation and thermal orientation in an external magnetic field.

10 Nonlinear Optics

Many nonlinear phenomena can be conveniently described phenomenologically by expanding the electric polarization **P** in powers of the light-wave electric field

$$\begin{aligned} P_\alpha = &\ \chi^{(1)}_{\alpha\beta}(\omega) E_\beta(\omega, \mathbf{k}) \exp(\mathrm{i}\mathbf{k}\cdot\mathbf{x} - \mathrm{i}\omega t) \\ &+ \chi^{(2)}_{\alpha\beta\gamma}(\omega_1, \omega_2)\, E_\beta(\omega_1, \mathbf{k}_1)\, E_\gamma(\omega_2, \mathbf{k}_2) \\ &\times \exp[\mathrm{i}(\mathbf{k}_1 + \mathbf{k}_2)\cdot\mathbf{x} - \mathrm{i}(\omega_1 + \omega_2)t] \\ &+ \chi^{(3)}_{\alpha\beta\gamma\delta}(\omega_1, \omega_2, \omega_3)\, E_\beta(\omega_1, \mathbf{k}_1)\, E_\gamma(\omega_2, \mathbf{k}_2)\, E_\delta(\omega_3, \mathbf{k}_3) \\ &\times \exp[\mathrm{i}(\mathbf{k}_1 + \mathbf{k}_2 + \mathbf{k}_3)\cdot\mathbf{x} - \mathrm{i}(\omega_1 + \omega_2 + \omega_3)t]\,. \end{aligned} \qquad (10.1)$$

Here ω_j is the frequency, $\mathbf{k}_j$ the wave vector, $\mathbf{E}(\omega_j, \mathbf{k}_j)$ the amplitude of the jth component of the electromagnetic field; for single crystals or short-period SLs, $\mathbf{x}$ is the three-dimensional radius-vector and, in quantum-well structures, $\mathbf{x} = (x, y, z_l)$, where z_l is the position of the lth well, for instance, the coordinate of its center (the selection of point z_l within a well is arbitrary if its width a is small compared with the light wavelength). The frequencies in (10.1) may assume both positive and negative values, the corresponding amplitudes being related through

$$\mathbf{E}^*(\omega, \mathbf{k}) = \mathbf{E}(-\omega, -\mathbf{k}).$$

For the sake of simplicity, we neglect in (10.1) the spatial dispersion of the susceptibility, i.e., the dependence of $\chi^{(n)}$ on $\mathbf{k}, \mathbf{k}_1, \mathbf{k}_2 \dots$

The susceptibility $\chi^{(1)}$ describes the conventional linear response, and $\chi^{(2)}$ – generation of optical harmonics at the sum or difference frequency (three-wave interaction), in particular, at $\omega_1 = \omega_2$, second-harmonic generation. The third-order susceptibility $\chi^{(3)}$ describes a variety of phenomena, namely, third-harmonic generation (at $\omega_1 = \omega_2 = \omega_3$), two-photon absorption of a monochromatic light wave (two of the three values of ω_j coincide, the third one differing from them in sign, e.g., $\omega_2 = \omega_3 = -\omega_1$) or two monochromatic waves (two of the three frequencies differ in sign and do not coincide with the third, e.g., $\omega_1 = -\omega_2 \neq \pm\omega_3$), variation of the absorption or reflection coefficient of a probe beam linear in pump-light intensity (photoabsorption or photoreflection), photoinduced gyrotropy or birefringence, four-wave mixing of two light waves with frequencies ω_1 and ω_2 accompanied by generation of a new harmonic at the frequency $2\omega_1 - \omega_2$ (for $\mathbf{k}_1 \parallel \mathbf{k}_2$) or of a new spatial harmonic $2\mathbf{k}_1 - \mathbf{k}_2$ (for $\omega_1 = \omega_2$).

One should discriminate among the various nonlinear optical phenomena in heterostructures those that are allowed by the symmetry of homogeneous compositional materials and those that originate from the lowering of symmetry due to the transition from homogeneous material to a heterostructure. We will place particular emphasis on this in the sections dealing with photogalvanic effects.

10.1 Two-Photon Absorption

One can write the following expression for the interband electron transition rate in a unit volume under two-photon excitation of a semiconductor by monochromatic light

$$w^{(2)}(\mathbf{e},\omega) = \frac{2\pi}{\hbar}\left(\frac{eA}{cm}\right)^4 \sum_{\mathbf{k}}\sum_{cv} \left|M_{cv}^{(2)}(\mathbf{k},\mathbf{e}|\omega)\right|^2 \times \delta\left(E_{c\mathbf{k}} - E_{v\mathbf{k}} - 2\hbar\omega\right). \qquad (10.2)$$

Here, A is the amplitude, $\mathbf{e}$ the polarization vector, ω the light frequency, c and v are the conduction and valence band indices for which the argument of the δ-function can vanish, and the compound matrix element of the two-photon transition is defined as

$$M_{cv}^{(2)}(\mathbf{k},\mathbf{e}|\omega) = \sum_r \frac{\left(\mathbf{e}\cdot\mathbf{p}_{c\mathbf{k},r\mathbf{k}}\right)\left(\mathbf{e}\cdot\mathbf{p}_{r\mathbf{k},v\mathbf{k}}\right)}{E_{r\mathbf{k}} - E_{v\mathbf{k}} - \hbar\omega}. \qquad (10.3)$$

The summation here is performed over all possible states, both filled and unoccupied. In this case, the expressions for $M_{cv}^{(2)}$ obtained when using the perturbation operator in the form

$$\frac{e}{c}\frac{\mathbf{p}\mathbf{A}}{m} \quad \text{or} \quad e\mathbf{x}\cdot\mathcal{I}$$

coincide ($\mathcal{I}$ is the electric field of the light wave). For the field of a plane monochromatic wave, we use the representation

$$\mathbf{A}(\mathbf{x},t) = \mathbf{A}(\omega)\mathrm{e}^{\mathrm{i}\mathbf{k}\cdot\mathbf{x}-\mathrm{i}\omega t} + \mathbf{A}^*(\omega)\mathrm{e}^{-\mathrm{i}\mathbf{k}\cdot\mathbf{x}+\mathrm{i}\omega t}. \qquad (10.4)$$

The amplitude $A = |\mathbf{A}(\omega)|$ or $\mathcal{I} = |\mathcal{I}(\omega)|$ is connected with the light intensity I (energy flux through unit surface) through the relation

$$A^2 = \left(\frac{c}{\omega}\right)^2 \mathcal{I}^2 = \frac{2\pi c}{n_\omega \omega^2} I, \qquad (10.5)$$

where n_ω is the refractive index at frequency ω. This permits us to rewrite the factor in front of the sum in (10.2) in the form

$$\left(\frac{eA}{cm}\right)^4 = \left(\frac{2\pi e^2}{m^2\omega^2 c n_\omega}\right)^2 I^2. \qquad (10.6)$$

If the light wave is quasimonochromatic, the probability $w^{(2)}$ should be multiplied by the correlation factor $g_2 = \langle I^2 \rangle / I^2$, where the angular brackets denote averaging over the intensity distribution (with $I \equiv \langle I \rangle$). For multimode radiation with a random phase distribution $g_2 = 2$, and for a biharmonic wave

$$g_2 = 1 + 2I_1 I_2 / I^2$$

where I_1 and I_2 are the intensities of the components, the total intensity being $I = I_1 + I_2$. The two-photon absorption coefficient $K^{(2)}$ is related to $w^{(2)}$ through

$$K^{(2)} = \hbar\omega \frac{\mathrm{d}w^{(2)}}{\mathrm{d}I} = 2\hbar\omega \frac{w^{(2)}}{I} \tag{10.7}$$

and with the nonlinear susceptibility, through the expression

$$K^{(2)}(\omega, \mathbf{e}) \propto \mathrm{Im}\left\{ \chi^{(3)}_{\alpha\beta\gamma\delta}(-\omega, \omega, \omega) e^*_\alpha e^*_\beta e_\gamma e_\delta \right\} I.$$

10.1.1 Two-Photon $v\Gamma_8, \Gamma_7 \to c\Gamma_6$ Transitions

Consider first two-photon transitions of electrons between nondegenerate bands v and c in homogeneous semiconductors, e.g., transitions to the conduction band Γ_6 from the valence band Γ_7 in GaAs. If

$$2\hbar\omega - (E_\mathrm{g} + \Delta) \ll \hbar\omega, \Delta,$$

then the electron energy spectrum in the bands Γ_6 and Γ_7 can be considered parabolic (and spherical). With the same accuracy, one can neglect in the expansion of $M^{(2)}_{\mathrm{cv}}$ in powers of $\mathbf{k}$ terms of higher than first order

$$M^{(2)}_{\mathrm{cs,v}\Gamma_{7j}}(\mathbf{k}, \mathbf{e}|\omega) = \sum_r \frac{\left(\mathbf{e} \cdot \mathbf{p}_{\mathrm{cs},r;\mathbf{k}}\right)\left(\mathbf{e} \cdot \mathbf{p}_{r,\mathrm{v}\Gamma_{7j};\mathbf{k}}\right)}{E^o_r - E^o_{\mathrm{v}\Gamma_7} - \hbar\omega}. \tag{10.8}$$

Here E^o_r is the electron energy in band r for $k = 0$, s and j are the electron spin indices for the bands Γ_6 and Γ_7, respectively, and $\mathbf{p}_{r,r';\mathbf{k}}$ is the momentum-operator matrix in the multiband model in the linear-in-$\mathbf{k}$ approximation:

$$p^\alpha_{r,r';\mathbf{k}} = p^\alpha_{r,r'} + \hbar k_\alpha \delta_{rr'} + \frac{\hbar}{m} \sum_{r_1} \left(\frac{p^\alpha_{rr_1}\left(\mathbf{k} \cdot \mathbf{p}_{r_1 r'}\right)}{E^o_{r'} - E^o_{r_1}} + \frac{\left(\mathbf{k} \cdot \mathbf{p}_{rr_1}\right) p^\alpha_{r_1 r'}}{E^o_r - E^o_{r_1}} \right). \tag{10.9}$$

For transitions from the degenerate valence band Γ_8 and for $2\hbar\omega - E_\mathrm{g} \ll \hbar\omega, \Delta$ we have in place of (10.8)

$$M^{(2)}_{\mathrm{cs,v}\Gamma_{8j}}(\mathbf{k}, \mathbf{e}|\omega) = \sum_m V_{\mathrm{cs,v}\Gamma_8 m}(\mathbf{k}, \mathbf{e}|\omega) F_{m\mathbf{k},j}, \tag{10.10}$$

where the four-component column $\hat{F}_{m\mathbf{k}}$ is the eigenvector of the effective electron Hamiltonian in the valence band Γ_8 (3.49),

$$V^{(2)}_{cs,\mathrm{v}\Gamma_8 m}(\mathbf{k},\mathbf{e}|\omega) = \sum_r \frac{\left(\mathbf{e}\cdot\mathbf{p}_{cs,r;\mathbf{k}}\right)\left(\mathbf{e}\cdot\mathbf{p}_{r,\mathrm{v}\Gamma_8 m;\mathbf{k}}\right)}{E^o_r - E^o_{\mathrm{v}\Gamma_8} - \hbar\omega}. \tag{10.11}$$

For $k = 0$, the contribution to $V^{(2)}_{\mathrm{cv}}$ comes from two-photon allowed-allowed transitions $\mathrm{v} \to r \to \mathrm{c}$ for which both matrix elements, $\mathbf{e}\cdot\mathbf{p}_{cr}(\mathbf{k})$ and $\mathbf{e}\cdot\mathbf{p}_{r\mathrm{v}}(\mathbf{k})$, are nonzero as $k \to 0$. The linear-in-k contribution to $V^{(2)}_{\mathrm{cv}}$ is determined by allowed – forbidden transitions for which one of the above matrix elements is proportional to k. In noncentrosymmetric crystals, symmetry allows the presence in $M^{(2)}_{\mathrm{cv}}$ and $V^{(2)}_{\mathrm{cv}}$ of both k-independent and linear-in-k terms. However, in A_3B_5 and A_2B_6 semiconductors the allowed–allowed transitions provide, as a rule, a negligible contribution to two-photon absorption so that in (10.8, 11) one may restrict oneself to including only the linear-in-k contribution.

The direct product $\Gamma_6 \times \Gamma_8$ contains the representations $\mu = E, F_1, F_2$. We choose the bases of these representations and introduce the matrices

$$T^{\mu\alpha}_{sm} = \frac{\langle cs|V^\mu_\alpha|\mathrm{v}m\rangle}{\left[\sum\limits_{s'm'} |\langle cs'|V^\mu_\alpha|\mathrm{v}m'\rangle|^2\right]^{1/2}}, \tag{10.12}$$

where V^μ_α are operators transforming according to the irreducible representation μ. For fixed bases of the irreducible representations, these matrices do not depend on the actual choice of the operators V^μ_α; indeed, the form of these operators affects only the common phase factor which does not depend on α. The matrices $\hat{T}^{\mu\alpha}$ form a complete set in the 2×4 matrix space

$$\mathrm{Tr}\left(\hat{T}^{\mu\alpha+}\hat{T}^{\mu'\beta}\right) = \delta_{\mu\mu'}\delta_{\alpha\beta}. \tag{10.13}$$

When choosing the basis of Γ_6 states in the form $\alpha S, \beta S$, and the Γ_8 basis in the form (3.43), these matrices can be written in the following compact representation:

$$\begin{gathered}
\sum_\alpha \hat{T}^{F_2\alpha} f_\alpha = \frac{1}{\sqrt{2}}\begin{bmatrix} -\frac{\sqrt{3}}{2}f_+ & f_z & \frac{1}{2}f_- & 0 \\ 0 & -\frac{1}{2}f_+ & f_z & \frac{\sqrt{3}}{2}f_- \end{bmatrix}, \\
\sum_\beta \hat{T}^{F_1\beta} g_\beta = \frac{1}{\sqrt{2}}\begin{bmatrix} -\frac{1}{2}g_- & 0 & \frac{\sqrt{3}}{2}g_+ & -g_z \\ -g_z & \frac{\sqrt{3}}{2}g_- & 0 & \frac{1}{2}g_+ \end{bmatrix}, \\
\hat{T}^{E1}h_1 + \hat{T}^{E2}h_2 = \frac{1}{\sqrt{2}}\begin{bmatrix} 0 & ih_1 & 0 & -ih_2 \\ ih_2 & 0 & -ih_1 & 0 \end{bmatrix}.
\end{gathered} \tag{10.14}$$

To construct the matrices $\hat{T}^{F_2\alpha}$, it is sufficient to calculate the interband matrix elements of the momentum operator $p^\alpha_{cs,\mathrm{v}\Gamma_8 m}$. To find the matrices $\hat{T}^{F_1\beta}$ and

$\hat{T}^{E\gamma}$, one should proceed in the following way. First, one uses the components of the vectors **A** and **B** to construct the combinations

$$B_{\beta+1}A_\beta A_{\beta+1} - B_{\beta+2}A_\beta A_{\beta+2}$$

and

$$\sqrt{3}\left(B_x A_z A_y - B_y A_x A_z\right), \quad 2B_z A_x A_y - B_x A_y A_z - B_y A_z A_x$$

and replaces B_α in these combinations with the matrices $\hat{T}^{F_2\alpha}$, and $A_\alpha A_\beta$, with the 4×4 matrices $[J_\alpha J_\beta]_s$. After this, one multiplies these matrices in the required order and normalizes the resultant matrices in accordance with (10.12).

Using the matrices $\hat{T}^{\mu\alpha}$, the linear-in-k contribution in (10.11) can be represented in an invariant form

$$\hat{V}^{(2)} = \frac{2}{\sqrt{3}}\sum_\alpha \left(\sum_{p=1}^{3} a_p R_\alpha^{(p)}\right)\hat{T}^{F_2\alpha} + \sum_\beta \left(\sum_{p=4,5} a_p R_\beta^{(p)}\right)\hat{T}^{F_1\beta} + a_6 \sum_{\gamma=1,2} R_\gamma^{(6)}\hat{T}^{E\gamma}. \quad (10.15)$$

Here $R_\alpha^{(p)}$ are combinations composed of the products $e_\beta e_\gamma k_\delta$ and transforming according to the irreducible representations of the group T_d:

(a) representation F_2

$$\mathbf{R}^{(1)} = (\mathbf{e}\cdot\mathbf{e})\mathbf{k}, \quad \mathbf{R}^{(2)} = (\mathbf{e}\cdot\mathbf{k})\mathbf{e}, \quad R_\alpha^{(3)} = e_\alpha^2 k_\alpha, \quad (10.16a)$$

(b) representation F_1

$$R_\beta^{(4)} = k_\beta\left(e_{\beta+1}^2 - e_{\beta+2}^2\right), \quad R_\beta^{(5)} = e_\beta\left(k_{\beta+1}e_{\beta+1} - k_{\beta+2}e_{\beta+2}\right), \quad (10.16b)$$

(c) representation E

$$R_1^{(6)} = \sqrt{3}e_z\left(e_y k_x - e_x k_y\right), \quad R_2^{(6)} = 2k_z e_x e_y - e_z\left(e_y k_x + e_x k_y\right) \quad (10.16c)$$

the factor $2/\sqrt{3}$ in the first term being isolated for the sake of convenience. If the common phase factors in the $\hat{T}^{\mu\alpha}$ matrices are chosen in the form (10.14), the coefficients a_p are real.

In Kane's model, which includes only the **k-p** interaction between the bands $c\Gamma_6$, $v\Gamma_8$ and $v\Gamma_7$, and assumes only these three bands to be virtual states of two-photon transitions, we have [10.1]

$$a_1 = \mathcal{L}\left(2 - \frac{2}{3}q - \frac{1}{3}s\right), \quad a_2 = \mathcal{L}\left(6 - \frac{4}{3}\eta - \frac{1}{3}s\right),$$

$$a_3 = 0, \quad a_5 = a_6 = -a_4 = \frac{2}{3}\mathcal{L}s, \quad (10.17)$$

where

$$\mathcal{L} = \frac{\hbar}{m}\frac{p_{cv}^2}{E_g^2}, \quad \eta = \frac{\Delta}{E_g + \Delta},$$

$$q = \frac{2\Delta}{E_g + 2\Delta}, \quad s = q\frac{3E_g + 2\Delta}{2\left(E_g + \Delta\right)} \tag{10.18}$$

and the relation $\hbar\omega \simeq E_g/2$ is used. In the same model, the effective masses of the electron in band Γ_6 and of the light hole in Γ_8 are defined by the relations (3.67)

$$\frac{m}{m_c} = 1 + \frac{2}{3}\frac{P_{cv}^2}{mE_g}\frac{3E_g + 2\Delta}{E_g + \Delta},$$

$$\frac{m}{m_{lh}} = -1 + \frac{4}{3}\frac{P_{cv}^2}{mE_g}. \tag{10.19}$$

In the multiband model, the coefficients a_p vary, first, due to the contribution of the upper bands to m_c^{-1} and to the parameters A, B, D in the **k-p** method of perturbation theory, which modify the intraband matrix elements

$$\mathbf{e}\mathbf{p}_{cc'}(\mathbf{k}) = \delta_{cc'} m\hbar^{-1}\mathbf{e}\nabla_\mathbf{k} E_{c\mathbf{k}},$$

$$\mathbf{e}\mathbf{p}_{v\Gamma_{8j},v\Gamma_{8j'}}(\mathbf{k}) = \delta_{jj'} m\hbar^{-1}\mathbf{e}\nabla_\mathbf{k} E_{v\Gamma_8 j\mathbf{k}} \; (|j| = |j'|)$$

and the intersubband matrix elements $\mathbf{e}\cdot\mathbf{p}_{v\Gamma_{8j},v\Gamma_{8j'}}(\mathbf{k})$ (for $|j| \neq |j'|$), $\mathbf{e}\cdot\mathbf{p}_{v\Gamma_{7m},v\Gamma_{8j}}(\mathbf{k})$ and, second, due to the contribution of the upper bands $r \neq c\Gamma_6, v\Gamma_8, v\Gamma_7$ to the compound matrix element (10.3). Since the two contributions of the upper bands to a_p are comparable, then, when one of them is taken into consideration, so should the other. For the same reason, the inclusion in the energy denominators $E_{r\mathbf{k}} - E_{v\mathbf{k}} - \hbar\omega$ of the dependence on $\mathbf{k}$ (in place of using their values $E_r^0 - E_v^0 - \hbar\omega$ at $k = 0$) requires simultaneous inclusion of the nonlinear-in-k contribution to the matrix elements of the momentum operator. Indeed, according to (3.66), the inclusion of non-parabolicity $\Delta E \propto k^4$ changes the reciprocal electron effective mass in the band Γ_6 by

$$\delta m_c^{-1} = -\theta\frac{\hbar^2 k^2}{2m_c^2 E_g}, \quad \theta = \frac{3 - 2\eta + \eta^2}{3 - \eta}. \tag{10.20}$$

The relative change of the intraband matrix element is

$$-\theta\frac{\hbar^2 k^2}{m_c E_g},$$

which coincides in order of magnitude with the relative value of the correction

$$(E_{r\mathbf{k}} - E_{r'\mathbf{k}} - \hbar\omega)^{-1} - \left(E_r^0 - E_{r'}^0 - \hbar\omega\right)^{-1}$$

for $r = c\Gamma_6, r' = v\Gamma_8 \pm 3/2$.

10.1.2 Two-Photon Transitions in QW Structures

Equations (10.2, 3) can be used to describe two-photon absorption in heterostructures as well, while bearing in mind that the indices c, r, and v should include the numbering of the subbands or minibands appearing as one crosses over from a homogeneous semiconductor to a layered structure. In the case of a single quantum well, the summation in (10.2) is performed over the two-dimensional wave vector $\mathbf{k}_\perp$. Note that $w_{\mathrm{SQW}}^{(2)}$ is the two-dimensional electron generation rate, i.e., the number of electrons created per unit quantum-layer area per unit time. In a thick barrier SL, the two-photon absorption coefficient is connected with $w_{\mathrm{SQW}}^{(2)}$ through the relation

$$K^{(2)} = 2\hbar\omega \frac{w_{\mathrm{SQW}}^{(2)}}{I(a+b)}. \tag{10.21}$$

If the confinement energy

$$\frac{1}{2m_{\mathrm{c,v}}}\left(\hbar\frac{\pi}{a}\right)^2$$

is small compared to E_{g}, then the calculation of $M_{\mathrm{cv}}^{(2)}$ does not require a special summation over virtual states, so that one can use the results of the calculation of the matrix $M_{\mathrm{cv}}^{(2)}$ for homogeneous semiconductors. Thus, for instance, for two-photon transitions $\mathrm{v}\Gamma_8 \to \mathrm{c}\Gamma_6$ in a single GaAs/AlGaAs quantum well, we will have

$$M_{\mathrm{c}\nu's,\mathrm{v}\nu j}^{(2)}(\mathbf{k}_\perp, \mathbf{e}|\omega) = \sum_m \int f_{\nu'\mathbf{k}_\perp}^{c*}(z) V_{cs,\mathrm{v}\Gamma_8 m}^{(2)}\left(\mathbf{k}_\perp, \hat{k}_z, \mathbf{e}|\omega\right) f_{\nu\mathbf{k}_\perp m}^{v}(z)\mathrm{d}z, \tag{10.22}$$

where $\hat{k}_z = -\mathrm{i}\partial/\partial z$, the matrix $V_{\mathrm{cv}}^{(2)}$ was introduced in (10.10), and $f_{\nu\mathbf{k}}^{l}$ is the electron envelope wave function in subband ν in a state with the two-dimensional wave vector $\mathbf{k}_\perp$ (3.83). A more exact calculation of $M_{\mathrm{cv}}^{(2)}$, in which the energy denominator $E_{r\mathbf{k}} - E_{\nu\mathbf{k}} - \hbar\omega$ is not replaced by $E_r^0 - E_{\mathrm{v}}^0 - \hbar\omega$, requires simultaneous inclusion of corrections of the same order in the matrix elements of the momentum operator. This comment is equivalent to the one concerning the inclusion of the dependence of $E_{r\mathbf{k}} - E_{\nu\mathbf{k}} - \hbar\omega$ on $\mathbf{k}$ in three-dimensional crystals.

Let us calculate $w_{\mathrm{SQW}}^{(2)}(\mathbf{e}|\omega)$ for the transitions $\mathrm{hh}\nu \to \mathrm{e}\nu'$ and $\mathrm{lh}\nu \to \mathrm{e}\nu' (\nu, \nu' = 1, 2)$ for the GaAs/AlGaAs structure in the infinite-barrier approximation neglecting the light- and heavy-hole state mixing, when the selection rules (7.53) are valid. In this case,

$$M_{c\nu's,\nu\nu j}^{(2)}(\mathbf{k}_\perp, \mathbf{e}|\omega) = V_{cs,\mathrm{v}\Gamma_8 j}^{(2)}\left(\mathbf{k}_\perp\delta_{\nu\nu'}, k_z^{(\nu',\nu)}, \mathbf{e}|\omega\right). \tag{10.23}$$

Therefore, two-photon interband transitions which do or do not involve a change of the quantum number are related, respectively, with a linear-in-k_z or linear-in-$\mathbf{k}_\perp$ contribution to $V_{\mathrm{cv}}^{(2)}$.

Let us evaluate $K^{(2)}$ for a thick-barrier SL using Kane's three-band model, in which the coefficients a_p in expansion (10.15) are defined in accordance with (10.17). Leaving out intermediate algebra, we are giving here the final result presenting the absorption coefficient in the form

$$K^{(2)}_{cv',vv}(\mathbf{e}) = 2\hbar\omega \frac{2\pi}{\hbar}\left(\frac{2\pi e^2}{m^2\omega^2 c n_\omega}\right)^2 I \frac{\mu_{cv}}{2\pi\hbar^2}\frac{\mathcal{L}^2}{a^2(a+b)}\zeta_{\nu'\nu}(\mathbf{e}), \tag{10.24}$$

where v is the heavy- or light-hole subband index and μ_{cv} the reduced effective mass of an electron-hole pair. The dimensionless factor $\zeta_{\nu'\nu}$ depends on the indices of the subbands involved in the transition, and for the three different light-polarization states, namely, linear ($\mathbf{e} \perp z$ and $\mathbf{e} \parallel z$) and circular ($\mathbf{e} \perp z$, $e_y = \pm \mathrm{i} e_x$), assumes the values

(a) transition hh1 → e1

$$\begin{aligned} \zeta_{\text{lin}}(\mathbf{e} \perp z) &= \frac{1}{2}\zeta_1 \left[(\bar{a}_1 + \bar{a}_2)^2 + \left(\bar{a}_1 + \frac{1}{2}\bar{a}_4\right)^2\right], \\ \zeta_{\text{lin}}(\mathbf{e} \parallel z) &= \zeta_1 \left(\bar{a}_1 - \frac{1}{2}\bar{a}_4\right)^2, \\ \zeta_{\text{circ}}(\mathbf{e} \perp z) &= \frac{1}{2}\zeta_1 \left(\bar{a}_2 + \frac{1}{2}\bar{a}_5\right)^2; \end{aligned} \tag{10.25a}$$

(b) transition hh1 → e2, or hh2 → e1

$$\begin{aligned} \zeta_{\text{lin}}(\mathbf{e} \perp z) &= \zeta_2 \bar{a}_4^2, \\ \zeta_{\text{lin}}(\mathbf{e} \parallel z) &= 0, \\ \zeta_{\text{circ}}(\mathbf{e} \perp z) &= \zeta_2 \left(\bar{a}_4 - \bar{a}_6\right)^2; \end{aligned} \tag{10.25b}$$

(c) transition lh1 → e1

$$\begin{aligned} \zeta_{\text{lin}}(\mathbf{e} \perp z) &= \frac{1}{6}\zeta_1 \left[(\bar{a}_1 + \bar{a}_2)^2 + \left(\bar{a}_1 - \frac{3}{2}\bar{a}_4\right)^2\right], \\ \zeta_{\text{lin}}(\mathbf{e} \parallel z) &= \frac{1}{3}\zeta_1 \left(\bar{a}_1 + \frac{3}{2}\bar{a}_4\right)^2, \\ \zeta_{\text{circ}}(\mathbf{e} \perp z) &= \frac{1}{6}\zeta_1 \left[\left(\bar{a}_2 + \frac{3}{2}\bar{a}_4\right)^2 + \frac{9}{4}\bar{a}_4^2\right]; \end{aligned} \tag{10.25c}$$

(d) transition lh1 → e2, or lh2 → e1

$$\begin{aligned} \zeta_{\text{lin}}(\mathbf{e} \perp z) &= \frac{4}{3}\zeta_2 \bar{a}_1^2, \\ \zeta_{\text{lin}}(\mathbf{e} \parallel z) &= \frac{4}{3}\zeta_2 \left(\bar{a}_1 + \bar{a}_2\right)^2, \\ \zeta_{\text{circ}}(\mathbf{e} \perp z) &= \zeta_2 \bar{a}_6^2, \end{aligned} \tag{10.25d}$$

where $\bar{a}_p = a_p/\mathcal{L}$, the coefficients a_p are defined in (10.17),

$$\zeta_1 = \frac{2\mu_{cv}}{\hbar^2}\left(2\hbar\omega - E^o_{cv'} + E^o_{vv}\right), \quad \zeta_2 = a^2 |k_z^{(1,2)}|^2. \tag{10.26}$$

Expressions (10.25a–d) were obtained neglecting the Coulomb interaction between the electron and the hole. The polarization dependences of two-photon excitation of the excitons e1 – v2$(1s)$, e2 – v1$(1s)$ and e1 – v1$(2p_{x,y})$ are described by similar expressions differing from (10.25a–d) in the common factors ζ_1 and ζ_2. The two-photon transitions involving excitation of the excitons e1 – v1$(1s)$ or e1 – v2$(2p)$, e2 – v1$(2p)$ are too weak because of the smallness of the contribution of allowed – allowed transitions to the compound matrix element $M_{\mathrm{cv}}^{(2)}$ in a homogeneous material and are not observed experimentally.

According to (10.17), the coefficient a_2 exceeds by far the other coefficients a_p. Therefore, two-photon absorption is easier to reveal in the geometry in which the coefficient a_2 contributes to $\zeta_{\nu'\nu}$. Figures 10.1a and b present two-photon excitation spectra of the photoluminescence due to radiative recombination of the exciton e1-hh1 (1s) in a thick-barrier GaAs/$Al_{0.35}Ga_{0.65}As$ superlattice for two polarization states, $\mathbf{e} \perp z$ and $\mathbf{e} \parallel z$. At the same time, Fig. 10.1a shows the frequency dependence of one-photon photocurrent excitation with a clearly pronounced feature near the resonance frequency of the 1s-exciton e1-hh1. In accordance with the above selection rules, two-photon absorption does not manifest itself in this part of the spectrum. The other features in the spectrum likewise are in accord with theory, namely, two-photon absorption peaks at the resonance frequency of the excitons e2-lh1 (1s) and e1-lh2 (1s) in the $\mathbf{e} \parallel z$ polarization, the presence of absorption in the resonance region of the exciton e1-hh1 (2p) for $\mathbf{e} \perp z$, linear growth of photoluminescence signal with increasing $2\hbar\omega$ for $2\hbar\omega > E^0_{\mathrm{c}\nu'} - E^0_{\mathrm{v}\nu}$ for transitions with $\nu' = \nu$ and the absence of such a behavior for the $|\nu' - \nu| = 1$ transitions.

The polarization dependence $K^{(2)}(\mathbf{e})$ for $\mathbf{e} \perp z$ is characterized by three linearly independent coefficients

$$K^{(2)}(\mathbf{e}) = K_1^{(2)} + K_2^{(2)}|\mathbf{e}\cdot\mathbf{e}|^2 + K_3^{(2)}\left(|e_x|^4 + |e_y|^4\right). \tag{10.27}$$

This expression can be derived based on a group-theoretical analysis for the point symmetry D_{2d} (one-photon absorption for $\mathbf{e} \perp z$ independent of the polarization state). In Kane's model, used to calculate a_p, there is no anisotropy of two-photon absorption in the (x, y) plane since in this model the electron-energy spectrum in a SQW or SL possesses cylindrical symmetry. In the multiband model $a_3 \neq 0$, and the relations $a_4 = -a_5 = -a_6$ break down, as a result of which the coefficient $K_3^{(2)}$ of the anisotropic term in (10.27) is nonzero.

Two-photon absorption is different for the linear and circular polarizations even in the isotropic model: indeed, according to (10.25a), one can write the expression for the Linear-Circular Dichroism (LCD) for the transitions hh1 → e1 in the following way:

$$\Lambda^{(2)} \equiv \frac{K_{\mathrm{lin}}^{(2)}(\mathbf{e} \perp z)}{K_{\mathrm{circ}}^{(2)}(\mathbf{e} \perp z)} = \frac{(\bar{a}_1 + \bar{a}_2)^2 + \left(\bar{a}_1 + \frac{1}{2}\bar{a}_4\right)^2}{\left(\bar{a}_1 - \frac{1}{2}\bar{a}_4\right)^2}. \tag{10.28}$$

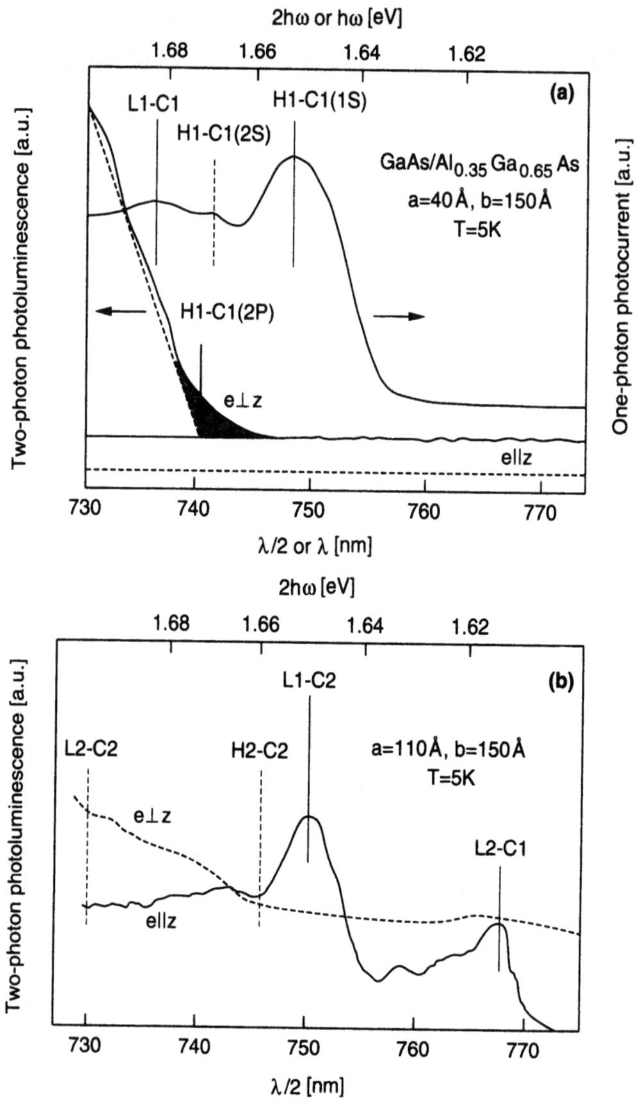

Fig. 10.1. **a** Single-photon (right scale) and two-photon (left scale) light absorption spectra in $GaAs/Al_{0.35}Ga_{0.65}As$ MQW structure. Single photon spectrum was measured at normal incidence ($\mathbf{e} \perp z$). The dotted line below shows that for the sample with $a = 40$ Å there is no two-photon absorption in $\mathbf{e} \| z$ polarization within the wavelength range studied. Vertical lines indicate the predicted positions of exciton resonances. The vertical arrow indicates the threshold of $hh1 \to e1$ intersubband transitions. **b** Two-photon light absorption spectra for a similar structure but with wider quantum wells ($a = 110$ Å). The solid and dotted lines were measured in $\mathbf{e} \| z$ and $\mathbf{e} \perp z$ polarizations, respectively. The solid and dotted vertical lines indicate predicted positions of exciton resonances which are active, accordingly, in two-photon allowed-forbidden and one photon allowed transitions [10.2]

For $\Delta/E_g = 0.3$, we obtain for LCD $\Lambda^{(2)} \simeq 1.6$, which differs noticeably from the value $\Lambda^{(2)} \simeq 1.2$ calculated for bulk GaAs.

10.1.3 Two-Photon Absorption in a Magnetic Field

In a quantizing magnetic field $\mathbf{B} \parallel z$, the spectrum of interband two-photon absorption in a quantum-well structure undergoes rearrangement and becomes a series of narrow lines

$$w_{\mathrm{cv}}^{(2)}(\mathbf{e}_1, \mathbf{e}_2) \propto I_1 I_2 \left| M^{(2)}_{\mathrm{c}\nu' N', \mathrm{v}\nu N}(\mathbf{e}_1, \mathbf{e}_2) \right|^2 \times \delta\left(E_{c\nu' N'} - E_{v\nu N} - \hbar\omega_1 - \hbar\omega_2\right). \quad (10.29)$$

Here N, N' are the Landau-level numbers; for generality, we consider two-photon absorption in the two-beam technique, indices 1 and 2 identifying the parameters of the first and second light waves. In experiments the frequency of the stronger radiation, say, ω_1, usually satisfies the inequality $2\hbar\omega_1 < E_{\mathrm{c}\nu'} - E^0_{\mathrm{v}\nu}$, so that two-photon transitions involving absorption of two photons $\hbar\omega_1$ are forbidden by energy conservation. When describing real spectra, the δ-function in (10.29) should be replaced by a function taking into account homogeneous and inhomogeneous broadening of the levels. Recalling that in bulk material the main contribution to $M^{(2)}_{\mathrm{cv}}$ comes from allowed – forbidden transitions, we obtain the following selection rules:

$N_{\mathrm{c}} - N_{\mathrm{v}} = \pm 1$ for two-photon transitions h1 $\to$ e1,

$N_{\mathrm{c}} = N_{\mathrm{v}}$ for two-photon transitions h1 $\to$ e2 or h2 $\to$ e1.

It should be stressed that for one-photon allowed transitions the selection rules are different:

$N_{\mathrm{c}} = N_{\mathrm{v}}$ for one-photon transitions h1 $\to$ e1,

$N_{\mathrm{c}} = N_{\mathrm{v}} \pm 1$ for one-photon transitions h1 $\to$ e2 or h2 $\to$ e1.

Fröhlich et al. [10.3] studied two-photon magnetoabsorption in GaAs/AlGaAs quantum-well structures using as an intense radiation source a CO_2 laser with $\hbar\omega_1 = 0.117$ eV ($\hbar\omega_1 \ll E_g$). In these conditions, the transitions playing the dominant role in two-photon absorption are

$$\mathrm{v} \xrightarrow{\hbar\omega_2} \mathrm{c} \xrightarrow{\hbar\omega_1} \mathrm{c},$$

in which case the photon $\hbar\omega_2$ is absorbed in the first step, and the intermediate states are in the conduction band, and

$$\mathrm{v} \xrightarrow{\hbar\omega_1} \mathrm{v} \xrightarrow{\hbar\omega_2} \mathrm{c},$$

where the interband transition is also initiated by the photon $\hbar\omega_2$, and the intermediate states lie in the valence band. As a result, the principal contribution to the compound matrix element comes from the term proportional to $\mathbf{R}^{(2)} = (\mathbf{e}_1 \cdot \mathbf{k}_\perp)\mathbf{e}_2$. Therefore, we have the following selection rules for the transitions h1 → e1 for circularly polarized CO_2 radiation:

$N_c - N_v = 1$ for σ_+ polarization, and

$N_c - N_v = -1$ for σ_- polarization.

10.2 Photoreflectance

In bulk materials, photoreflectance, i.e., photoinduced change of reflectance, may be considered a contactless method of electromodulation. Indeed, the electron-hole pairs created by a strong source of radiation (pumping) modulate the built-in surface electric field and affect the reflection of the probe beam. Investigation of the photoreflectance caused by this and other mechanisms is a simple and reliable modulation method of determining the energy spacing between the subbands in structures with quantum wells and minibands in a SL. Interpretation of photoreflectance spectra needed for correct assignment of the electronic subbands and minibands does not, as a rule, require the knowledge of the nature of formation of the photoreflectance signal. At the same time, investigation of the mechanisms of photoreflectance or photoabsorption provides valuable information on spectral line broadening and on the character of exciton interaction with defects and photocarriers in heterostructures.

Spectra of resonant photoreflection from a heterostructure can be described within the approach discussed in sect. 6.5 assuming that pumping results in a change of one of the parameters determining the dielectric constant (6.99, 71), namely, the resonance frequency, damping, or oscillator strength. According to (6.80, 84), the exciton contribution to the coefficient of reflection from a quantum-well structure is proportional to the function

$$f(x, \Phi) = \frac{\tilde{\omega}}{\Gamma} \frac{\sin \Phi + x \cos \Phi}{1 + x^2}, \tag{10.30}$$

where $x = (\omega - \omega_0)/\Gamma$. For small variations of this function, the photoreflectance spectrum is determined by the relation

$$\Delta R \propto \frac{\partial f}{\partial \omega_0} \Delta\omega_0 + \frac{\partial f}{\partial \tilde{\omega}} \Delta\tilde{\omega} + \frac{\partial f}{\partial \Gamma} \Delta\Gamma, \tag{10.31}$$

where $\Delta\omega_0$, $\Delta\tilde{\omega}$ and $\Delta\Gamma$ are photoinduced deviations from the dark (equilibrium) values ω_0, $\tilde{\omega}$, and Γ. The spectral response of each of the three derivatives in (10.31) is totally determined by the value of the phase Φ and is shown in Fig. 10.2 for (a) $\Phi = 2\pi N$, and (b) $\Phi = 2\pi N + \pi/2$, where N is an integer. We recall that this phase can be derived from an experimental reflectance spectrum in the absence of illumination. Thus, by comparing the shape of the measured spectrum $\Delta R(\omega)$ with those of the three derivatives of the function $f(x, \Phi)$, one can determine the dominant mechanism of photomodulation.

Extremely high photoreflectance from some GaAs/AlGaAs structures with a single quantum well surrounded by two equivalent short-period SL was found

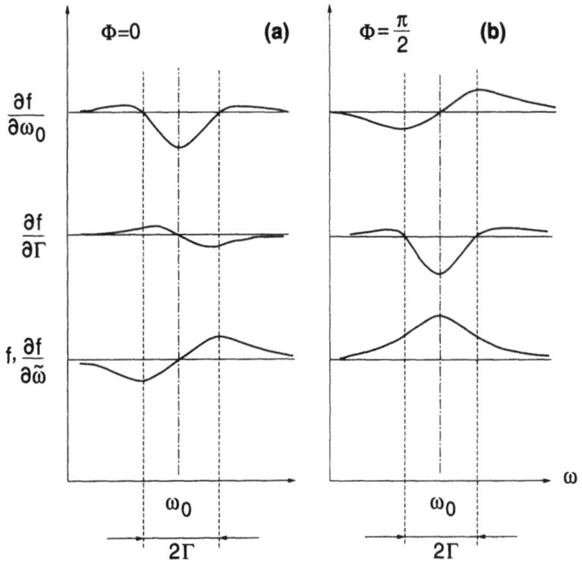

Fig. 10.2a,b. Frequency dependence of the derivatives of the function $f(x, \Phi)$ with respect to ω_0, Γ, and $\tilde{\omega}$ (see (10.30) and (10.31)) for **a** $\Phi = 0$ and **b** $\Phi = \pi/2$

to exist [10.4] under excitation well above the band gap. The dotted and solid lines in Fig. 10.3 show experimental and calculated spectra of oblique reflection in equilibrium and when illuminated by He-Ne laser radiation. The relative $\Delta R/R$ signal is seen to be over 10% at the maximum, which is much greater than the values $\Delta R/R \sim 10^{-4} - 10^{-3}$ observed in MQW structures at normal incidence. The large amplitude of the photoreflectance signal makes possible its measurement under steady-state illumination without modulating the pump intensity.

The best agreement between theory and experiment is obtained under the assumption of constant $\omega_{\rm LT}^{\sigma}$ and ω_0^{σ}, and of decreasing exciton damping Γ under illumination, from 2.4 to 1.5 meV, and from 3.5 to 2.9 meV, respectively, for the heavy- (σ = hh) and light- (σ = lh) hole exciton. This effect exhibits a threshold nature with respect to the pump photon energy $\hbar\omega_{\rm p}$; namely, under the conditions where carrier photoexcitation occurs only inside the quantum well, no photoreflection is observed within experimental error. Conversely, if $\hbar\omega_{\rm p} > E_{\rm g}({\rm Al}_x{\rm Ga}_{1-x}{\rm As})$, steady-state illumination of very low intensity produces a strong variation of the resonant reflectance. The signal rise time $\tau_{\rm R}$ is short ($\tau_{\rm R} < 1$ ms for pump intensity 1 mW/cm^2). After turning off the illumination, the signal dies out in a nonexponential way with an initial decay time

$$\tau_{\rm d} = \frac{\Delta R}{{\rm d}\Delta R/{\rm d}t} \sim 1.4 \text{ ms}.$$

The short rise time and long decay time of the photoreflectance signal, as well as the existence of a threshold frequency, can be interpreted by assuming

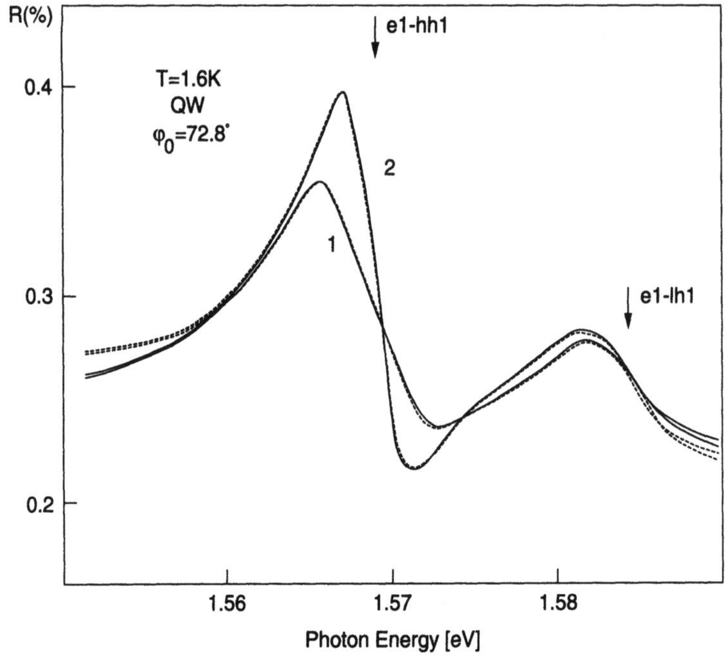

Fig. 10.3. Experimental reflectance spectra (solid curves) measured in GaAs/AlGaAs SQW structure under oblique incidence in (1) equilibrium conditions and (2) under additional illumination. Theoretical (dotted) curves were calculated by (6.88)–(6.90) for the same parameters $\kappa_1 = 12.5$, $\hbar\omega_0^h = 1.568$ eV, $\hbar\omega_0^l = 1.583$ eV, $\hbar\omega_{LT}^h = 0.48$ meV, $\hbar\omega_{LT}^l = 0.25$ meV but for different values of damping: $\hbar\Gamma_h = 2.4$ meV, $\hbar\Gamma_l = 3.5$ meV for curve 1, and $\hbar\Gamma_h = 1.5$ meV, $\hbar\Gamma_l = 2.9$ meV for curve 2 [10.4]

the effect to originate from photo-induced neutralization of impurity centers (or defects) located in the superlattice barriers of the structure in question and creating there deep levels for free carriers. At equilibrium these centers are charged and produce in the quantum well a random Coulomb potential acting upon the free carriers in the well. The scattering of excitons from this random potential contributes to their damping. When excited above the barrier, electron-hole pairs are created not only in a single quantum well but in the adjacent SLs also. Photoexcited carriers are trapped at the deep centers and neutralize them, thus excluding their contribution to the exciton scattering. As a result, the damping Γ decreases noticeably and this is what is actually observed in experiment [10.4].

10.3 Diffraction from a Light-Induced Spatial Grating

The pronounced decrease of the exciton damping under excitation well above the barrier described in the preceding section can be used to obtain photoinduced spatial modulation of Γ. Let a QW structure be illuminated by two coherent beams falling on the sample surface at the angles φ_1 and $-\varphi_2$, the light incidence plane coinciding with the (xz) plane. Interference modulates the pump-light intensity along the x-axis

$$J(x) = J_0 + J_1 \cos(2\pi x/\Lambda) \tag{10.32}$$

with a period

$$\Lambda = \lambda |\sin\varphi_1 + \sin\varphi_2|^{-1}, \tag{10.33}$$

where Λ is assumed to be considerably longer than the light wavelength and the diffusion length of the photocarriers participating in the impurity neutralization discussed in Sect. 10.2. Then a grating of neutralized centers will form near the quantum well, the exciton damping Γ varying periodically along the x-axis. The existence of the dependence $\Gamma(x)$ can be visualized in experiments on resonant diffraction of a probe light wave.

The diffraction efficiency can be evaluated using the relation between the electric fields of the reflected and incident waves (6.63, 69)

$$\mathcal{I}(x) = \left(r_{01} + \frac{t_{01} t_{10} \mathrm{i}\tilde{\omega} \mathrm{e}^{\mathrm{i}\Phi}}{\omega_0 - \omega - \mathrm{i}\Gamma(x,T)} \right) \mathcal{I}_0(x), \tag{10.34}$$

where T is the duration of optical excitation. In the simplest model of deep level filling by photocarriers, the spatial dependence of the damping can be represented in the form

$$\Gamma(x,T) = \Gamma_0 + \Delta\Gamma \exp[-\xi J(x) T], \tag{10.35}$$

where ξ is a constant, $J(x)$ is defined in (10.32), and $\Delta\Gamma$ is the difference between the equilibrium and steady-state damping values. The angle φ_{d}, at which the first-order diffraction maximum is observed, is connected with φ_1, φ_2 and the probe beam incidence angle φ_0 through the relation

$$\sin\varphi_{\mathrm{d}} = \sin\varphi_0 \pm \frac{\lambda_{\mathrm{probe}}}{\lambda_{\mathrm{pump}}} (\sin\varphi_1 + \sin\varphi_2)\,. \tag{10.36}$$

Assuming for simplicity the oscillating component $\delta\Gamma$ to be small compared to Γ_0, we obtain for the diffraction efficiency

$$R_{\pm 1} = \left| \frac{\mathcal{I}_{\pm 1}}{\mathcal{I}_0} \right|^2 = \left[t_{01} t_{10} I_1(u_1)\, \mathrm{e}^{-u_0} \frac{\Delta\Gamma \tilde{\omega}}{(\omega_0 - \omega)^2 + \Gamma_0^2} \right]^2 . \tag{10.37}$$

Here $u_i = \xi J_i T\, (i = 0, 1)$, the intensities J_0 and J_1 were introduced in (10.32), $I_\nu(z)$ is the modified Bessel function of νth order. Note that, in contrast to the reflection coefficient R in (6.80), the spectral response $R_{\pm 1}(\omega)$ does not depend on the phase Φ and has the shape of a symmetrical maximum with an effective FWHM of $\Gamma_0/\sqrt{2}$. In GaAs/AlGaAs heterostructures, this spectrum has two peaks corresponding to the e1-hh1 and e1-lh1 1s-excitons with different values of $\Delta\Gamma$.

Figure 10.4 shows an experimental first-order diffraction spectrum obtained in backscattering geometry. Two coherent pump beams and one probe beam are

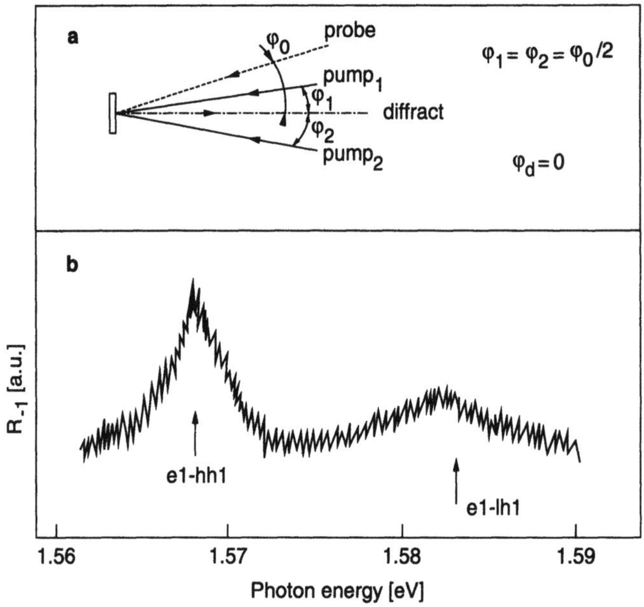

Fig. 10.4. **a** Schematic representation of geometry of diffraction from the light-induced lateral lattice. **b** First order diffraction spectrum in the region of e1-hh1 and e1-lh1 exciton resonances. Measurements were performed on the sample whose reflectance spectrum is shown in Fig. 10.3 [10.5]

polarized perpendicular to the common incidence plane. To prevent the photoluminescence from contributing to the measured signal, the recording was done with a delay which, while exceeding the luminescence decay time, was nevertheless small compared to the photoreflectance fall-off time. In accordance with theoretical predictions, the signal is observed to exist only close to the exciton resonance frequencies. As follows from the data on the exciton parameters given in the legend of Fig. 10.3,

$$(\Delta\Gamma_{\rm l}\tilde{\omega}_{\rm l})^2 \ll (\Delta\Gamma_{\rm h}\tilde{\omega}_{\rm h})^2 \,.$$

This explains why the e1-hh1 exciton peak manifests itself stronger in experiment. The diffraction signal reduced to the reflection coefficient R is 10^{-3}, which is a typical value for diffraction from light-induced gratings in bulk crystals. Taking into account the microscopic size of the single quantum well in question ($\sim$ 100 Å), we come to the conclusion that the observed diffraction is extremely efficient.

10.4 Third-Harmonic Generation

Consider the nonlinear interaction of conduction electrons with light assuming the photon energy to be high compared to $\hbar/\tau_p$, where τ_p is the momentum relaxation time and is small compared with the semiconductor gap width or with distance to the higher conduction bands. In this case, the efficiency of third-harmonic generation is intimately connected with the nonparabolicity of the electron energy spectrum, namely, the third-order polarization $\mathbf{P}^{(3)}$ or current density $\mathbf{j}^{(3)}$ is determined by the cubic-in-field correction to the electron velocity:

$$\mathbf{j}^{(3)} = \frac{\mathrm{d}\mathbf{P}^{(3)}}{\mathrm{d}t} = -2e \sum_{\mathbf{k}} \mathbf{v}^{(3)} F_{\mathbf{k}}. \tag{10.38}$$

Here $F_{\mathbf{k}}$ is the equilibrium distribution function,

$$v_{\alpha}^{(3)} = \frac{1}{6\hbar^4} \sum_{\beta\gamma\delta} \frac{\partial^4 E_k}{\partial k_\alpha \partial k_\beta \partial k_\gamma \partial k_\delta} \left(\frac{e}{c}\right)^3 A_\beta(\mathbf{x}, t) A_\gamma(\mathbf{x}, t) A_\delta(\mathbf{x}, t), \tag{10.39}$$

where $\mathbf{A}(\mathbf{x}, t)$ is the light-wave vector potential. Since the electrons are usually located close to the band bottom, one may restrict oneself, when calculating (10.38) in the expansion of energy $E_{\mathbf{k}}$ in powers of $\mathbf{k}$, to fourth-order terms. In crystals of cubic symmetry, these terms are characterized by two linearly independent constants

$$\Delta E_{\mathrm{c}\mathbf{k}} = \Lambda^{(1)} k^4 + \Lambda^{(2)} \left(k_x^4 + k_y^4 + k_z^4\right), \tag{10.40}$$

where x, y, z are the crystal principal axes. In Kane's model, where $E_{\mathrm{c}\mathbf{k}}$ satisfies (3.64), we have

$$\Lambda^{(1)} = -\theta \frac{\hbar^4}{4m_{\mathrm{c}}^2 E_{\mathrm{g}}}, \quad \Lambda^{(2)} = 0 \tag{10.41}$$

with the coefficient θ introduced in (10.20). Recalling that

$$\partial k^4 / \partial k_\alpha = 4k^2 k_\alpha,$$

we obtain from (10.38)–(10.41)

$$\mathbf{j}^{(3)} = \theta \frac{e^4 n}{c^3 m_{\mathrm{c}}^2 E_{\mathrm{g}}} \left[\mathbf{A}(\mathbf{x}, t)\right]^2 \mathbf{A}(\mathbf{x}, t), \tag{10.42}$$

where n is the electron concentration. Representing the electric field of a plane monochromatic wave in the form

$$\mathcal{I}(\mathbf{x}, t) = \mathcal{I}(\omega) \mathrm{e}^{\mathrm{i}\mathbf{k}\cdot\mathbf{x} - \mathrm{i}\omega t} + \mathcal{I}^*(\omega) \mathrm{e}^{-\mathrm{i}\mathbf{k}\cdot\mathbf{x} + \mathrm{i}\omega t},$$

we find the polarization at the triple frequency

$$\mathbf{P}(3\omega) = -\frac{\theta}{3} \frac{e^4 n}{\omega^4 m_{\mathrm{c}}^2 E_{\mathrm{g}}} [\mathcal{I}(\omega) \cdot \mathcal{I}(\omega)] \mathcal{I}(\omega). \tag{10.43}$$

In a semiconductor SL, the nonparabolicity of the electron minibands is more strongly pronounced than that of the conduction band in a homogeneous compositional material. Therefore, using a SL for nonlinear optical conversion may turn out to be advantageous under certain conditions [10.6, 7]. Equations (10.42, 43) can be used to describe third-harmonic generation in a SL provided $\hbar\omega$ is small compared with the separation between the first and second subbands. Neglecting the contribution of the terms $\Lambda_2 Q_\perp^4$ and $\Lambda_3 Q_z^2 Q_\perp^2$ in expansion (7.12), we obtain

$$P_z(3\omega) = \frac{4}{3}\frac{e^4 n}{(\hbar\omega)^4}\Lambda_1 \mathcal{I}_z^3(\omega). \tag{10.44}$$

According to (7.40, 42), in the tight-binding approximation

$$\Lambda_1 = -\frac{\hbar^4}{12 M_\parallel^2 \Delta_1}, \tag{10.45}$$

where Δ_1 is the width of the lowest miniband. Comparing (10.41) with (10.45), we come to the conclusion that in this approximation the ratio of the coefficients in the nonlinear equations (10.44) and (10.43) can be written in the form

$$\frac{1}{3\theta}\left(\frac{m_c}{M_\parallel}\right)^2 \frac{E_g}{\Delta_1}.$$

In a thin barrier SL, one has to use the more precise expression (7.33) in the calculations of Λ_1.

10.5 Linear and Circular Photogalvanic (Photovoltaic) Effects

In accordance with the general expression (10.1), when the frequencies ω_1 and $-\omega_2$ are equal, illumination of off-centrosymmetric crystals will produce, besides polarization P at the double frequency, also a steady-state effect, i.e., polarization at the difference frequency $\omega_1 + \omega_2 = 0$. This effect was called optical rectification or d–c effect. It can be observed by measuring the current $j = \mathrm{d}p/\mathrm{d}t$ at the moments when the illumination is turned on or off. The d-c effect was observed to occur in a number of crystals [10.8, 9]. Apart from this, another nonlinear phenomenon can be revealed, namely, the appearance of a steady-state current. In contrast to the d-c effect, this current flows during the time the illumination is on and appears only when light is absorbed.[1] Based on general symmetry considerations, one can write the expression for the photogalvanic effect (PGE) in the form

$$j_\alpha = I\left\{\chi_{\alpha\beta\delta}\mathrm{Re}\left\{e_\beta e_\delta^*\right\} + \gamma_{\alpha\beta}\mathrm{i}\left[\mathbf{e}\times\mathbf{e}^*\right]_\beta\right\}, \tag{10.46}$$

where I is the light intensity and $\mathbf{e}$ the polarization vector. The vector $\mathrm{i}\left[\mathbf{e}\times\mathbf{e}^*\right]$ determines the degree of circular polarization P_{circ}, and for a transverse light wave it is $(\mathbf{q}/q)P_{\mathrm{circ}}$, where $\mathbf{q}$ is the wave vector.

The tensor χ in (10.46) is nonzero only for piezoelectric crystals and transforms as the piezoelectric tensor. In nonpyroelectric crystals, the corresponding

[1] This effect can be observed also for $\omega_1 \neq -\omega_2$. In the case when $\omega = \omega_1 + \omega_2 < \tau_p^{-1}$, where τ_p is the momentum relaxation time, the current at the frequency ω is determined by the same expressions as the steady-state current at $\omega_1 = -\omega_2$. At $\omega \gtrsim \tau_p^{-1}$ the current vanishes rapidly.

current can be excited only by linearly polarized light. For instance, in crystals of the class T_d, the tensor χ has one linearly independent component $\chi_{xyz} = \chi_{xzy}$. Therefore, the current described by the first term in (10.46) was named the linear PGE current. In pyroelectrics, linear PGE can be observed also when illuminated by unpolarized or circularly polarized light.

The tensor γ in (10.46) is nonzero only in optically active (gyrotropic) crystals and transforms as the gyration tensor. The current corresponding to the second term in (10.46) is called the circular PGE current. It appears only under illumination with circularly polarized light and reverses direction when the sign of circular polarization is changed.

The linear PGE was observed in some insulators as early as the 1950s, and possibly even earlier, but was correctly identified as a new phenomenon only in 1974–75 [10.10, 11] (for more details see [10.12, 13]). In semiconductors, the linear PGE was revealed already in 1972 in tellurium [10.14, 15] and studied most comprehensively on p-GaAs [10.16].

The circular PGE was theoretically predicted in [10.17, 18] and observed for the first time in tellurium [10.19], where the tensor γ has three nonzero components γ_{zz}, $\gamma_{xx} = \gamma_{yy}$; however, later its existence was revealed in many crystals. It was most extensively studied on tellurium [10.20]. As seen from (10.46), under time inversion the second term in (10.46), just as the current j_α, reverses the sign, whereas the first term does not change it. Violation of time inversion invariance does not naturally imply the vanishing of the corresponding coefficient $\chi_{\alpha\beta\delta}$ (the standard expression for current $\mathbf{j} = \sigma \mathcal{I}$ is also not invariant under time inversion); however, it results in a radical difference between the mechanisms of linear and circular PGE.

The linear PGE current was shown [10.21–25] to have two contributions. The first of them is ballistic and is described by the conventional expression:

$$\mathbf{j} = e \sum_{n,n'} W_{n'n} \left(\mathbf{v}_n \tau_{pn} - \mathbf{v}_{n'} \tau_{pn'} \right) . \tag{10.47}$$

Here $W_{n'n}$ is the probability of transition from the state n to n':

$$W_{n'n} = \frac{2\pi}{\hbar} |M_{n'n}|^2 \left(f_n - f_{n'} \right) \delta \left(E_{n'} - E_n \right) , \tag{10.48}$$

$M_{n'n}$ is the transition matrix element, $\mathbf{v}_n$ and τ_{pn} are the electron velocity and momentum relaxation time in the state n, f_n is the distribution function, i.e., the occupancy of the state n. The energy E_n includes the phonon or photon energy in the initial or final state. Equation (10.47) is a contribution to the general expression for the current

$$\mathbf{j} = -e \sum_{nn'} \rho_{nn'} \mathbf{v}_{n'n} \tag{10.49}$$

of diagonal components $\rho_{nn} \equiv f_n$ of the density matrix and of the velocity $\mathbf{v}_{nn} = \mathbf{v}_n$. The ballistic current is nonzero only if one includes in $M_{nn'}$ carrier interaction, not only with the photon but also with the phonons or impurities as well, or the interaction of the electron and hole created by light with one another.

The second contribution to the linear PGE current comes from the inclusion in (10.49) of the nondiagonal components $\rho_{nn'}$ and $\mathbf{v}_{nn'}$ with $n' \neq n$. This current was shown [10.24] to originate from the displacement of the wave packet's center of mass in quantum transitions and can be written

$$\mathbf{j} = -e \sum_{nn'} W_{n'n} \mathbf{R}_{n'n}. \tag{10.50}$$

For the displacement $\mathbf{R}$ we have

$$\mathbf{R}_{n'n} = -(\nabla_{\mathbf{k}} + \nabla_{\mathbf{k}'})\,\varphi_{n'n} + \boldsymbol{\Omega}_{n'} - \boldsymbol{\Omega}_n, \tag{10.51}$$

where $\varphi_{n'n}$ is the phase of the transition matrix element, $\mathbf{k}$ and $\mathbf{k}'$ are the wave vectors in the states n and n', and $\boldsymbol{\Omega}_n$ is the diagonal matrix element of the coordinate

$$\boldsymbol{\Omega}_n = \mathrm{i} \int u_n^* \nabla_{\mathbf{k}} u_n \mathrm{d}\mathbf{x}.$$

In steady state, when the processes of generation, scattering and recombination are taken into consideration, the contributions associated with $\boldsymbol{\Omega}_n$ vanish since they describe the charge redistribution, and in steady state this redistribution is fixed.

The first term in (10.51) can be rewritten

$$\begin{aligned}\mathbf{R}^{(1)}_{n'n} = &-\big[\mathrm{Re}\,\{M_{n'n}\}\,(\nabla_{\mathbf{k}} + \nabla_{\mathbf{k}'})\,\mathrm{Im}\,\{M_{n'n}\} \\ &- \mathrm{Im}\,\{M_{n'n}\}\,(\nabla_{\mathbf{k}} + \nabla_{\mathbf{k}'})\,\mathrm{Re}\,\{M_{n'n}\}\big]\,|M_{n'n}|^{-2}.\end{aligned} \tag{10.52}$$

In optical transitions where the momentum $\hbar\mathbf{k}$ is preserved the sum $\nabla_{\mathbf{k}} + \nabla_{\mathbf{k}'}$ in (10.51,52) is replaced by $\nabla_{\mathbf{k}}$. The contribution to the current described by (10.50) is called the shift term. The circular PGE has only the ballistic contribution. In contrast to the linear PGE, this contribution can be nonzero in direct optical transitions as well. We consider these phenomena in more detail in the following subsection.

10.5.1 Photogalvanic Effect in an Inversion Channel on the Vicinal Face of Silicon

Linear PGE was observed to occur in optical transitions between quantum levels in the inversion layer of a MOS transistor prepared on the silicon surface, which was set at an angle $\theta = 9.50°$ from the (001) plane about the [110] direction [10.39]. The existence of a nonsymmetrical well and of a deviation from the (001) plane lowers the symmetry from D_{4h} to C_s; this is what makes the appearance of the PGE current possible. A theory of linear and circular PGE for this structure has been developed in [10.26].

This effect was observed to occur [10.27] in transitions between the 1st and 2nd quantum levels near the resonance frequency $\omega = E_{21}/\hbar$, where $E_{21} = E_2 - E_1$, E_i is the energy of the ith level at the temperature $T = 4.2$ K when all electrons reside in the 1st level. Levels 1 and 2 are electron levels in the two of the six silicon conduction band valleys, $\Delta(001)$ and $\Delta(00\bar{1})$, for which the effective mass m_{zz}, where z is the surface normal, is the largest. The extrema of these valleys lie at the points $k_x^0 = \pm\Delta \sin\theta$. In the coordinate frame x, y, z where $y \parallel [110]$, $x \perp y$, z the electron spectrum in these valleys can be written

$$E(\mathbf{k}) = \frac{\hbar^2}{2}\left(\frac{k_x^2}{m_{xx}} + \frac{k_y^2}{m_{yy}} + \frac{k_z^2}{m_{zz}} + 2\frac{k_x k_z}{m_{xz}}\right), \tag{10.53}$$

where the masses m_{ii}, m_{xz} were introduced in (7.75 a). By virtue of (3.95, 96), we can write the wave function in the form

$$F(\mathbf{x}) = \exp\left[\mathrm{i}\left(k_x x + k_y y\right)\right]\varphi(z), \tag{10.54}$$

where

$$\varphi(z) = \exp\left(-\mathrm{i}\frac{m_{zz}}{m_{xz}}k_x z\right) f(z) \tag{10.55}$$

and $f(z)$ is determined by the equation

$$\left[\frac{\hbar^2\hat{k}_z^2}{2m_{zz}} + V(z) - E_\parallel\right] f(z) = 0. \tag{10.56}$$

For the total energy, we obtain

$$E = E_\parallel + \frac{\hbar^2 k_x^2}{2\tilde{m}_{xx}} + \frac{\hbar^2 k_y^2}{2\tilde{m}_{yy}}, \tag{10.57}$$

where, according to (7.75 a),

$$\tilde{m}_{xx}^{-1} = m_{xx}^{-1} - m_{zz}/m_{xz}^2 = m_{zz}/m_\parallel m_\perp .$$

The potential $V(z)$ is produced by the external electric field. For the field at the boundary, we have

$$\mathcal{I}(0) = \frac{4\pi e^2}{\kappa}\left(N_s + N_a\right), \tag{10.58}$$

where N_s is the electron concentration in the well, N_a the number of charged acceptors in the inversion layer (per 1 cm^2 of surface area) and κ the dielectric constant.

10.5.2 Ballistic Contribution

As already mentioned, ballistic current calculations should include simultaneously the interaction with photons and the scattering processes. The contribution to the current comes only from the terms in $|M|^2$ associated with the interference of two compound matrix elements of second order, or of the first and third orders. Next, it is assumed that the scattering is only elastic, both intra- and intersubband, its cross section being independent of the scattering angle. The corresponding transition matrix elements will be denoted by v_{11}, v_{22}, v_{12}

and assumed to be real. The calculation should include both optical transitions between levels and transitions within one branch. The corresponding matrix elements (7.76):

$$M_{ij} = -\frac{e\hbar A_x}{c\, m_{xz}} k_z^{(i,j)}, \tag{10.59}$$

$$\mu_{ii} = \frac{e\hbar}{c}\left(\frac{k_x A_x}{\tilde{m}_{xx}} + \frac{k_y A_y}{m_{yy}}\right). \tag{10.60}$$

Here **A** is the vector potential.

Figure 10.5 shows the transitions contributing to the current. The interference between the third-order matrix elements and the first-order matrix element corresponding to scattering does not contribute to the current under the conditions of Fermi degeneracy and elastic scattering, since in this case $f_{n'} - f_n = 0$ in (10.48). When summing over intermediate states, the energy denominator in the composite matrix element is transformed using the identity

$$\left.\frac{1}{E_i - E_k + \mathrm{i}\gamma}\right|_{\gamma\to+0} = \mathcal{P}\frac{1}{E_i - E_k} - \mathrm{i}\pi\delta\left(E_i - E_k\right), \tag{10.61}$$

where the symbol $\mathcal{P}$ denotes the principal value of the integral. Only the terms containing the product of the real term for one of the denominators in (10.61) by the imaginary term for the other denominator contribute to the current. Therefore, the effect is nonzero solely for the transitions where the energy conservation law holds not only for the initial and final states but also for one of the intermediate states as well.

Taking into account the compound matrix elements shown in Fig. 10.5, calculation by (10.47,48) yields the following expression for the current

$$j_x = \frac{eKI}{\hbar\omega} L|e_x|^2, \quad j_y = \frac{eKI}{\hbar\omega} L\frac{m_{\parallel}}{m_{zz}} \mathrm{Re}\left\{e_x e_y^*\right\}. \tag{10.62}$$

Here I is the light intensity related to A through (10.5)

$$I = \frac{\omega^2 n_\omega}{2\pi c}|A|^2 \tag{10.63}$$

where ω is the frequency, c the velocity of light, n_ω the refractive index and K the absorption coefficient. It is assumed that the dominant contribution to absorption comes from direct transitions between levels 1 and 2. In this case

$$K = \frac{4\pi^2 e^2 N_s m_{zz}}{n_\omega c m_{xz}^2} Z_{21}^2 \omega^2 |e_x|^2 \delta\left(E_{21} - \hbar\omega\right). \tag{10.64}$$

This expression was obtained taking into account that $k_z = -\mathrm{i}m_{zz}[zH]/\hbar^2$ and, accordingly,

$$k_z^{(ij)} = \frac{\mathrm{i}m_{zz}Z_{ij}}{\hbar^2}\left(E_i - E_j\right), \tag{10.65}$$

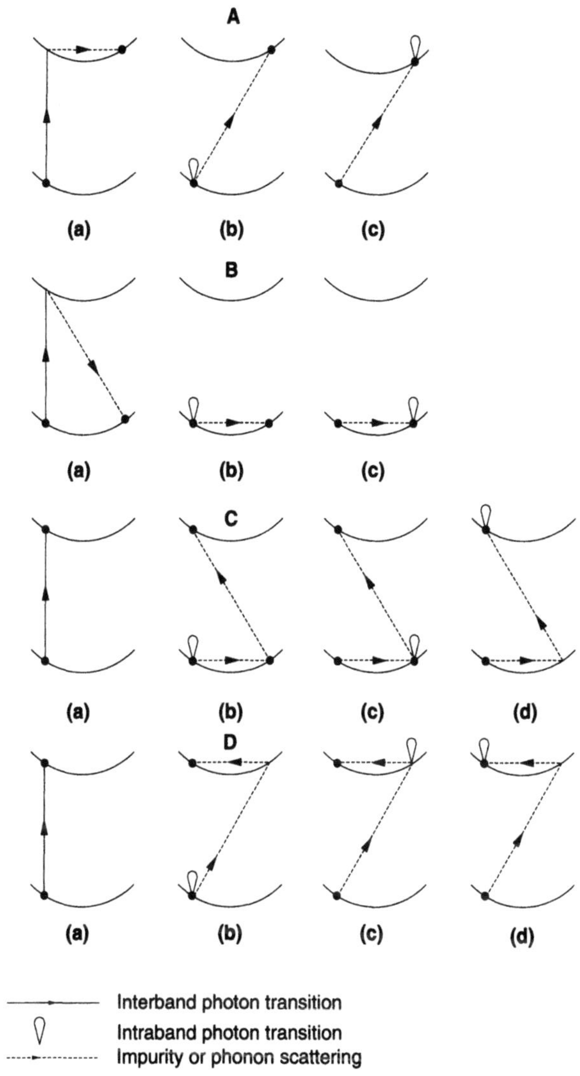

Fig. 10.5. Transitions contributing to linear PGE current under elastic electron scattering and Fermi degeneracy: **A,B** - interference of two second order matrix elements: **a** and **b** or **c**. **C, D** - interference of first order matrix element **a** and one of third order matrix elements **b**, **c** or **d**

where

$$Z_{ij} = \int_0^\infty f_i^*(z) z f_j(z) \mathrm{d}z.$$

The quantity L in (10.62) is the mean electron displacement in the x-direction (for $e_x = 1$):

$$L = -\frac{\pi}{4} N_s \frac{\hbar^2 m_{xz} m_{zz}^{1/2}}{m_\perp^2 m_\parallel^{3/2} \omega^2 Z_{21}} F(v), \tag{10.66}$$

where

$$F(v) = v_{21}\left[\frac{2v_{11}}{v_{11}^2+v_{21}^2}\frac{\hbar\omega}{\varepsilon_F}+\frac{2}{v_{11}}\right.$$
$$\left.+v_{11}\left(\frac{1}{v_{11}^2+v_{21}^2}+\frac{1}{v_{22}^2+v_{21}^2}\right)+\frac{2v_{22}}{v_{22}^2+v_{21}^2}\right]. \tag{10.67}$$

Here ε_F is the Fermi energy related to N_s through

$$\varepsilon_F = \frac{\pi\nu\hbar^2 N_s}{2\,m_\perp}\left(\frac{m_{zz}}{m_\parallel}\right)^{1/2}, \tag{10.68}$$

ν is the number of equivalent valleys (here $\nu = 2$). Equation (10.66) was derived taking into account that under the above assumptions and for $E \leqslant \varepsilon_F < E_{21}$ the relaxation time in the initial state

$$\tau_1^{-1} = v_{21}^2\frac{m_\perp}{\hbar^3}\left(\frac{m_\parallel}{m_{zz}}\right)^{1/2}. \tag{10.69a}$$

In the final state, for $E > E_{21}$, we have for branches 1 and 2

$$\frac{1}{\tau_2^{(1)}} = \left(v_{11}^2+v_{21}^2\right)\frac{m_\perp}{\hbar^2}\left(\frac{m_\parallel}{m_{zz}}\right)^{1/2},$$
$$\frac{1}{\tau_2^{(2)}} = \left(v_{22}^2+v_{21}^2\right)\frac{m_\perp}{\hbar^2}\left(\frac{m_\parallel}{m_{zz}}\right)^{1/2}. \tag{10.69b}$$

Equation (10.66) was derived assuming the scattering to occur primarily from surface roughness. In this case $v_{ij} \sim a_i a_j$, where

$$a_i = \left.\frac{\partial f_i}{\partial z}\right|_{z=0} \tag{10.70}$$

and, accordingly,

$$F(v) = \frac{2a_1a_2}{a_1^2+a_2^2}\left(\frac{\hbar\omega}{\varepsilon_F}+\frac{5}{2}+\frac{a_2^2}{a_1^2}+\frac{1}{2}\frac{a_1^2}{a_2^2}\right). \tag{10.71}$$

10.5.3 Displacement Contribution

To determine the contribution due to displacement, one has to calculate the nondiagonal component of Hamiltonian $\mathcal{H}_{21}$ in second order of perturbation theory

$$\mathcal{H}_{21}^{(2)} = -\frac{1}{2}\sum_{s\neq 1,2}\left[(E_s-E_1)^{-1}+(E_s-E_2)^{-1}\right]\mathcal{H}_{2s}^{(1)}\mathcal{H}_{s1}^{(1)} \tag{10.72}$$

where, by (10.53)

$$\mathcal{H}_{\alpha\beta}^{(1)} = \frac{\hbar^2 k_x}{m_{xz}}k_z^{(\alpha,\beta)}. \tag{10.73}$$

Using (10.65), one can transform (10.72) to

$$\mathcal{H}_{21}^{(2)} = -\frac{\mathrm{i}}{2}\frac{\hbar^2 k_x^2}{m_{xz}^2} m_{xx} \sum_{s\neq 1,2} \left(k_z^{(2,s)} Z_{s1} - Z_{2s} k_z^{(s,1)}\right). \tag{10.74}$$

From the commutation relation $\left[K_\alpha X_\beta\right] = -\mathrm{i}\delta_{\alpha\beta}$ it follows that

$$\sum_s \left(k_z^{(2,s)} Z_{s1} - Z_{2s} k_z^{(s,1)}\right) = 0. \tag{10.75}$$

In contrast to (10.74), in (10.75) the summation is done over all s including $s = 1, 2$. Using (10.75) and taking into account that $k_z^{(i,i)} = 0$, we finally come to

$$\mathcal{H}_{21}^{(2)} = \frac{\mathrm{i}}{2}\frac{\hbar^2 k_x^2}{m_{xz}^2} m_{zz} k_z^{(2,1)} \left(Z_{11} - Z_{22}\right). \tag{10.76}$$

Hence the transition matrix elements M_{21} including the first- and second-order terms will be

$$M_{21} = -\frac{e}{c}\frac{\hbar A_x}{m_{xz}} k_z^{(2,1)} \left[1 - \mathrm{i}k_z \frac{m_{zz}}{m_{xz}} \left(Z_{22} - Z_{11}\right)\right]. \tag{10.77}$$

According to (10.50), for the displacement current we obtain

$$j = -\frac{eKI}{\hbar\omega} R, \tag{10.78}$$

where, in accordance with (10.52) and (10.77),

$$R = \frac{m_{zz}}{m_{xz}} \left(Z_{22} - Z_{11}\right). \tag{10.79}$$

To evaluate the relative significance of the ballistic and displacement contributions, we use the variational functions

$$f_1(z) = C_1 z e^{-\alpha_1 z}, \tag{10.80}$$

$$f_2(z) = C_2 z \left(1 - \frac{\alpha_1 + \alpha_2}{3} z\right) e^{-\alpha_2 z}, \tag{10.81}$$

where the normalization factors were introduced in (7.78a). Assuming the field $\mathcal{I}$ in the inversion layer to be constant, minimization of the energy E_{11}, in accordance with (3.156), yields

$$\alpha_1 = 1.15 \left(e\mathcal{I} m_{zz}/\hbar^2\right)^{1/3}, \quad \alpha_2 = 0.935\alpha_1. \tag{10.82a}$$

Note that, according to Eq. (3.155)

$$E_1 = 1.97 \left[(e\mathcal{I}\hbar)^2 / m_{zz}\right]^{1/3}, \quad E_2 = 1.88 E_1, \quad \frac{a_2}{a_1} = 1.51, \tag{10.82b}$$

$$Z_{11} = 1.5\alpha_1^{-1}, \quad Z_{22} = 1.85 Z_{11}, \quad Z_{21} = -0.57 Z_{11}. \tag{10.82c}$$

To the experimentally obtained value [10.27] $\hbar\omega = E_2 - E_1 = 10.45$ meV corresponds the field $\mathcal{I} = 1.6 \times 10^4$ V/cm, $E_1 = 11.9$ meV, $Z_{11} = 0.51 \times 10^{-6}$ cm, $Z_{22} = 0.94 \times 10^{-6}$ cm, and $Z_{21} = -0.29 \times 10^{-6}$ cm.

The inclusion of partial field screening by free carriers in the inversion layer does not affect significantly these parameters and yields an estimate $N_s = 2 \times 10^{11}$ cm^{-2}. For the above values of the parameters, the displacement $R = 2.3 \times 10^{-7}$ cm, and the displacement L corresponding to the ballistic current is $L = 1.3 \times 10^{-6}$ cm, which shows that the ballistic contribution exceeds the displacement term by a factor five to six.[2]

10.5.4 Circular PGE

In contrast to the linear effect considered above, the circular PGE can be observed under excitation not only at the resonance frequency but, for instance, at frequencies $\hbar\omega < E_{21}$ as well. It is this case where light can be absorbed only in indirect transitions within one branch that we are going to consider now. According to (10.59, 60), the matrix element of an indirect transition from state k_1 to state k_2 within the same branch is

$$M_{21} = -\frac{M_{21}(\mathbf{k}_1)\, v_{21}}{E_2(\mathbf{k}_1) - E_1(\mathbf{k}_1)} + \frac{\mu(\mathbf{k}_1) - \mu(\mathbf{k}_2)}{\hbar\omega} v_{11} \tag{10.83}$$

with $\mu(\mathbf{k})$ defined by (10.60)

The first term in (10.83) describes the virtual transition via the second level, and the second, transitions within the first branch, virtual transitions involving the higher levels being neglected. The transition probability, i.e., the absorption coefficient, is determined by the squared moduli of both terms in (10.83), and the current, by their product,

Calculation by (10.47, 48) yields the following expression for the current

$$j_y = \frac{2\pi e^3 I Z_{21} N_s}{\hbar\omega^2 c n_\omega} \frac{E_{21}}{E_{21} - \hbar\omega} \frac{m_{zz}}{m_{xx}} \left(\frac{m_{zz}}{m_\parallel}\right)^{1/4} \left(1 + \frac{\varepsilon_F}{\hbar\omega}\right) P_{\text{circ}} \tag{10.84}$$

For the absorption coefficient of circularly polarized light for $\hbar\omega < E_{21}$, we have

$$K = \frac{2\pi e^2 N_s}{c\omega^2 n_\omega m_\perp \tau_1} \left[1 + \frac{m_{zz}}{m_\parallel} + \frac{a_2^2}{a_1^2} \frac{m_\perp \omega^2}{\varepsilon_F + \hbar\omega} \left(\frac{m_{zz}}{m_{xz}}\right)^2 \left(\frac{E_{21}}{E_{21} - \hbar\omega}\right)^2\right]. \tag{10.85}$$

For the above values of the parameters and $\hbar\omega = (2/3)E_{21}$, both terms in (10.85) are of the same order of magnitude, and the mean displacement $L = j\hbar\omega/eIK$ is 4×10^{-6} cm, which is larger than the corresponding displacement in the linear PGE, although the current in this case is smaller because of the smaller absorption coefficient. Close to the resonance, i.e., for $|\hbar\omega - E_{21}| \lesssim \Gamma$, where $2\Gamma = \hbar/\tau_1 + \hbar/\tau_2$ is the linewidth of resonant absorption caused by violation of the momentum or energy conservation law as a result of scattering,

[2] Assuming the field $\mathcal{I}$ to be screened only by free carriers, calculation with the variational function (10.81) shows that the corresponding level E_2 lies above the edge of the well whose height in this case is 1.94 E_1, where $E_1 = 5.21\,(e^2\hbar N_s/\kappa)^{2/3}\, m_{zz}^{-1/3}$.

the expression for the circular PGE current differs from (10.62) for j_y in the δ-function in (10.64) being replaced by $(1/\pi)(E_{21}-\hbar\omega)/[(E_{21}-\hbar\omega)^2+\Gamma^2]$, and $\mathrm{Re}\{e_x e_y^*\}$, by $1/2[\mathbf{e}\times\mathbf{e}^*]_z$ [10.44]. We see that the current in circular PGE reaches a maximum at $|E_{21}-\hbar\omega| \simeq \Gamma^{-1}$ and changes sign at $\hbar\omega = E_{21}$.

10.5.5 Linear PGE in Superlattices

In a noncentrosymmetric superlattice, linear PGE current appears also under illumination with unpolarized light. This effect was reported [10.28] to occur on a saw-tooth superlattice (Fig. 10.6a). Another example of a noncentrosymmetric SL is shown in Fig. 10.6b. A calculation was made [10.29] of the spectrum, wave functions and displacement contribution to the linear PGE current in such a superlattice produced by optical transitions from the first to second miniband. The displacement was shown to be the largest when it involved transitions from the bottom of the miniband, i.e., for small k_z. When expanding the transition matrix element M_{21} in this region, one can restrict oneself to the first two terms, $M_{21} \simeq M_0 + \mathrm{i}M_1 k_z$, which yields for the displacement, according to (10.52), $R = -M_1/M_0$. The magnitude of the displacement for a fixed k_z determined by the excitation frequency increases with increasing well asymmetry characterized by the quantity $\rho = (a-b)/(a+b)$ and reaches a maximum at $\rho \simeq 0.5$. The displacement increases also with increasing barrier asymmetry $\rho' = (c-d)/(c+d)$. Under weak localization, i.e., when the minibands formed by the levels of the two wells overlap, the displacement may exceed by far the superlattice period. In the case of strong localization, when the electrons of each miniband reside in their wells, the displacement is determined primarily by the terms Ω_1 and Ω_2 in (10.51) and is practically equal to the well separation. As already mentioned, the contribution associated with these terms under steady-state illumination is zero.

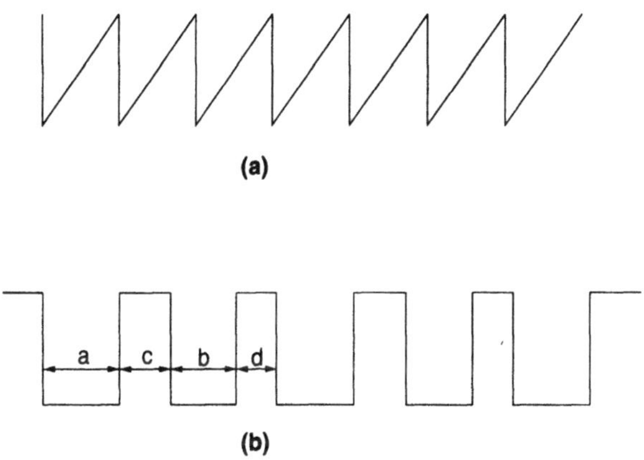

Fig. 10.6a,b. Noncentrosymmetric superlattices with **a** triangular and **b** rectangular wells

10.6 Current of Optically Oriented Electrons [10.30–32]

A peculiar photogalvanic effect is the current generation due to the spin relaxation or Larmour precession of optically oriented carriers. In quantum wells or superlattices, this effect can be revealed by producing carriers with circularly polarized light propagating along the normal to the layer surface, and by applying a transverse magnetic field **B** to turn the spin into the well plane.

As pointed out in Sect. 9.2, when carriers in A_3B_5 crystals are excited by polarized light with the photon momentum $m = +1$, the electrons transferring from the upper quantum level of the valence band turn out to be completely oriented, their spin S_0 at the moment of excitation being $-1/2$. If the relaxation time of the spin τ_s exceeds by far that of the momentum τ_p and of energy τ_e, then the thermalized electrons will preserve their orientation, and their steady state mean spin $\mathbf{S}_\perp$ in the presence of a transverse magnetic field will be:

$$S_\perp = \frac{T_\parallel T_\perp}{\tau_0} \frac{[\boldsymbol{\Omega}_B S_0]}{1 + \Omega_B^2 T_\parallel T_\perp}, \tag{10.86}$$

where $\hbar\boldsymbol{\Omega}_B = g\mu_B\mathbf{B}$, g is the g-factor, μ_B the Bohr magneton, $T_{\parallel,\perp}^{-1} = \tau_0^{-1} + \tau_{s\parallel,\perp}^{-1}$, τ_0 the lifetime and $\tau_{s\parallel,\perp}^{-1}$ are the components of the tensor τ_s^{-1} determining the spin relaxation rate. In (10.86) it is assumed that $\mathbf{S}_0 \parallel z$, the z axis being normal to the well plane. When the spin splitting of the conduction band is included, the equilibrium density matrix of the thermalized electrons will be defined by the expression

$$\rho_0 = nF_0, \tag{10.87}$$

where

$$F_0 = \left[f_0\left(E_\mathbf{k} + \mathcal{H}'\right)\left(\frac{1}{2} + \sigma \cdot \mathbf{S}\right)\right]_{\mathrm{sym}}. \tag{10.88}$$

Here n is the electron concentration;

$$[AB]_{\mathrm{sym}} = \frac{1}{2}(AB + BA),$$

$f_0(E)$ is the equilibrium distribution function assumed in what follows to be Maxwellian,

$$f_0\left(E_\mathbf{k} + \mathcal{H}'\right) \simeq f_0(E_\mathbf{k}) + \mathcal{H}' \frac{\partial f_0}{\partial E_\mathbf{k}}. \tag{10.89}$$

Here $E_\mathbf{k}$ is the kinetic energy and $\mathcal{H}'$ the term in the Hamiltonian $\mathcal{H}$ describing the spin splitting of the conduction band which can be written in the form

$$\mathcal{H}' = \frac{1}{2}\hbar\left(\sigma \cdot \boldsymbol{\Omega}_\mathbf{k}\right). \tag{10.90}$$

When only linear-in-k terms are included, we will have, according to (3.101, 102), for the (001) wells $\boldsymbol{\Omega}^{(1)} = (\beta/\hbar)(-k_x, k_y, 0)$, and for the (111) wells, $\boldsymbol{\Omega}^{(1)} = (\beta/\hbar)(k_y, -k_x, 0)$. It is assumed that the z-axis is directed along the normal to the well plane and that $\mathbf{k}_\perp$ lies in the latter. Two mechanisms of current generation were shown to exist [10.31]. The first of them, the relaxation current, is associated with a retardation in the distribution function relaxation when the spin is varying. At equilibrium, two contributions to the relaxation current, namely,

$$\mathbf{j}_1 = -en \sum_{\mathbf{k}} f_0(E_{\mathbf{k}})\, \Delta\mathbf{v}$$

where

$$\Delta\mathbf{v} = \nabla_{\mathbf{k}} (\mathbf{S}\boldsymbol{\Omega}^{(1)})$$

and

$$\mathbf{j}_2 = -en \sum_{k} \mathbf{v}_0 T_r(\Delta\rho),$$

where

$$\mathbf{v}_0 = \frac{\hbar\mathbf{k}}{m^*}$$

and $\Delta\rho$ is the component of ρ_0 determined by the second term in (10.89), cancel, the total current being $\mathbf{j} = \mathbf{j}_1 + \mathbf{j}_2 = 0$. When $\mathbf{S}$ is reversed, $\Delta\mathbf{v}$ changes instantly, while $\Delta\rho$ relaxes to the equilibrium value $\Delta\rho_0$ in a time τ_p. As a result, the sum of the two contributions no longer vanishes and a current will appear which, by (10.69), is described by the expression

$$\mathbf{j}(t) = eG(t)\tau_p \nabla_{\mathbf{k}}(\boldsymbol{\Omega}_{\mathbf{k}}^{(1)}, \mathbf{S}_0 - \mathbf{S}) - en(t)\tau_p \nabla_{\mathbf{k}} \left(\boldsymbol{\Omega}_{\mathbf{k}}^{(1)} \frac{\mathrm{d}\mathbf{S}}{\mathrm{d}t} \right), \tag{10.91}$$

where G is the generation rate.

In steady state, the second term vanishes, while the generation rate $G = n/\tau_0$, where τ_0 is the lifetime and the quantity $\mathbf{S}_0 - \mathbf{S}$ is defined by (10.86). After the switching off of illumination, i.e., at $G = 0$, the current relaxation is described by the second term in (10.91). Note that in a magnetic field such that $\Omega_B \tau_s \gg 1$ the current will oscillate at the Larmour frequency and decay in the time $T = (\tau_0^{-1} + \tau_s^{-1})^{-1}$.

The second contribution to current generation is of the kinetic nature and, in contrast to the relaxation one, depends on the actual mechanism of spin relaxation. When the Dyakonov-Perel mechanism involving random spin precession in an effective magnetic field $\mathbf{B}_{\text{eff}} = \hbar\Omega_k / g\mu_{\text{B}}$ is dominant, calculation of this contribution should include, besides the linear-in-$\mathbf{k}$, also the cubic-in-$\mathbf{k}$ contributions to $\boldsymbol{\Omega}_{\mathbf{k}}$:

$$\Omega_z^{(3)} = (2\gamma_c/\hbar)\, k_x k_y (k_x - k_y). \tag{10.92}$$

In the case where the scattering cross section does not depend on the scattering angle and electron energy, this contribution is

$$\mathbf{j} = \frac{4en\gamma_c}{\beta\hbar^3} m^* k_{\text{B}} T \frac{\tau_p}{\tau_{s\perp}} \nabla_{\mathbf{k}} \left(\Omega_{\mathbf{k}}^{(1)} \cdot \mathbf{S} \right). \tag{10.93}$$

For other scattering mechanisms, one should include in the collision integral describing the spin-flip process also the linear-in-**k** spin splitting. If the dominant mechanism of spin relaxation is the scattering from holes (the mechanism of *Bir, Aronov* and *Pikus*) or paramagnetic ions, the kinetic contribution to the current will be

$$\mathbf{j} = -\frac{2}{3}\frac{\tau_p}{\tau_s} en \nabla_{\mathbf{k}}(\boldsymbol{\Omega}_{\mathbf{k}}^{(1)}\mathbf{S}). \tag{10.94}$$

If the *Elliott-Yafet* mechanism involving conduction band spin state mixing in **k**-**p** interaction with the valence band predominates, then the ballistic and displacement contributions to the kinetic current will be equal, interference between transitions with and without spin flip playing the predominant role in the ballistic term. The total kinetic current is given by the expression

$$\mathbf{j} = -2en\delta\eta \frac{k_{\mathrm{B}}T}{E_{\mathrm{g}}} \operatorname{curl}_{\mathbf{k}} \left[\boldsymbol{\Omega}_{\mathbf{k}}^{(1)} \times \mathbf{S}\right], \tag{10.95}$$

where $\eta = \Delta/(E_g + \Delta)$, $\delta = (2 - \eta)/(3 - \eta)$, Δ is the spin-orbit splitting of the valence band.

Estimates show that the relaxation term is greater, as a rule, than the kinetic contribution [10.31]. Under optimum conditions, when $\Omega_B^{-2} = T_{\|} T_{\perp}$ and the photoconductivity exceeds the dark conductivity, the emf can reach as high as 10^{-2} V/cm. Note that besides the appearance of a photocurrent under optical orientation, one observes also the inverse effect, namely, carrier orientation under current [10.32].

10.7 Photon Drag Current

In the calculation of the linear and circular PGE currents, we always neglected the photon momentum $\hbar\mathbf{q}$. Taking into account that the photon momentum is finite, photocurrent can appear in crystals of any symmetry. Note that, in general, the direction of the current need not coincide with that of the momentum $\hbar\mathbf{q}$. This current was named the photon drag current. In bulk crystals the photon drag current was first observed experimentally in the microwave region [10.33] and in quantum transitions between the valence-band branches in Ge excited by a CO_2 laser [10.34, 37]. As long as classical theory holds, i.e., for $\omega \ll kT/\hbar$ (or $\omega \ll \varepsilon_F/\hbar$ for degenerate semiconductors) the photon drag current may be considered as a Hall current in crossed electric and magnetic fields of the light wave [10.33, 34]. Besides this contribution, there is another one, namely, the thermo-emf current caused by the microwave heating of the electrons [10.35]. This contribution may exceed the Hall current, however, as shown by *Normantas* and *Pikus* [10.36], it can be observed only if one uses a dc magnetic field and measures its Hall component. In the absence of magnetic field, the measuring electrodes should be in the illuminated region, in which case, however, the major contribution to the current will come from the contact emf.

For $\hbar\omega > k_B T$, the current has to be calculated by quantum theory. In the region $k_B T > \hbar\omega > \hbar/\tau_p$ the results obtained by quantum and classical theory coincide if in quantum theory one takes into account not only transitions involving absorption of photons but photoinduced emission as well.

In quantum transitions, the difference between the electron wave vectors in the final and initial states is $\mathbf{q}$, however, the magnitude of the momentum (quasimomentum) imparted to the electron and the remaining hole may exceed considerably $\hbar\mathbf{q}$. This effect manifests itself particularly strongly in two-dimensional systems in transitions between quantum levels in one band when the transverse effective masses for the two levels are close. As already mentioned, such direct transitions occur under excitation near the resonance frequency $\omega = E_{21}/\hbar$. When the finiteness of the photon momentum is included, the energy difference between the final and initial states will be

$$E_2(\mathbf{k}_2) - E_1(\mathbf{k}_1) = E_{21} + \frac{\hbar^2(\mathbf{k}_1+\mathbf{q})^2}{2m_\perp} - \frac{\hbar^2 k_1}{2m_\perp} \simeq E_{21} + \frac{\hbar^2 \mathbf{k}\cdot\mathbf{q}}{m_\perp}, \quad (10.96)$$

whence it is seen that for $\hbar\omega > E_{21}$ transitions will occur only from states with $(\mathbf{k}_1 \cdot \mathbf{q}) > 0$ and that, if the momentum relaxation time in the final state is less than that in the initial state, a current will appear in the direction of $\mathbf{q}$. On the contrary, for $\hbar\omega < E_{21}$, only transitions from states with $(\mathbf{k}_1 \cdot \mathbf{q}) < 0$ are possible, and the current will flow in the opposite direction. Therefore, in the absence of any other broadening mechanisms, the current will reach a maximum at the band edges, i.e., at $|\hbar\omega - E_{21}| = \Delta_\mathbf{q} = \hbar k_F q/m_\perp$. In this case, the magnitude of the current will not depend on q and can be written in the form

$$j = e\frac{KI}{\hbar\omega}\left[(\tau_1 - \tau_2)\frac{c\delta}{\hbar\omega} - \frac{\hbar q}{m_\perp}\tau_2\right], \quad (10.97)$$

where $\delta = \hbar\omega - E_{21}$. At the band edge $c|\delta|/\hbar\omega = v_F = \hbar k_F/\mathrm{m}_\perp$, where v_F and $\hbar k_F$ are the velocity and momentum at the Fermi surface. This peculiar feature of the photon drag phenomenon was first pointed out by *Dykhne* et al. [10.38].

In real structures the absorption line width exceeds by far 2 $\Delta_\mathbf{q}$ and is determined by scattering, i.e., by the magnitude of Γ, as well as by differences in the masses m_1 and m_2 at the first and second levels because of the band nonparabolicity. According to (3.103), in A_3B_5 crystals

$$\frac{m_2}{m_1} \simeq 1 + \frac{E_{21}}{E_g}, \quad (10.98)$$

where E_g is the band-gap width. For GaAs this evaluation gives for the coefficient of nonparabolicity $E_g^{-1} = 0.66\ \mathrm{eV}^{-1}$, whereas its experimental value is $0.69\ \mathrm{eV}^{-1}$. The band width due to nonparabolicity is $\Delta_F = \varepsilon_F m_1/\mu$, where

$$\frac{1}{\mu} = \frac{1}{m_1} - \frac{1}{m_2} \simeq \frac{1}{m_\perp}\frac{E_{21}}{E_g}. \quad (10.99)$$

To take into account the line broadening and the nonparabolicity, we replace the δ-function in (10.48) with the function

$$\Delta(E_2 - E_1) = \frac{1}{\pi} \frac{\Gamma}{(E_2 - E_1)^2 + \Gamma^2} \tag{10.100}$$

which becomes a δ-function as $\Gamma \to 0$. When calculating the photon drag current, the quantity $\Delta_{\mathbf{q}}$ may be considered small compared with Γ and Δ_{F}, and the Δ-function can be expanded and written in the form

$$\begin{aligned} &\Delta\left(\delta + \frac{\hbar^2 k^2}{2\mu} - \frac{\hbar \mathbf{k}\cdot\mathbf{q}}{m_\perp}\right) \\ &= \Delta\left(\delta + \frac{\hbar^2 k^2}{2\mu}\right) - \frac{\hbar \mathbf{k}\cdot\mathbf{q}}{m_\perp}\frac{\partial}{\partial\delta}\Delta\left(\delta + \frac{\hbar^2 k^2}{2\mu}\right). \end{aligned} \tag{10.101}$$

Then, according to (10.47, 48), for the number of absorbed photons we obtain

$$N = \frac{2m_\perp}{\hbar^3}|M_{21}|^2 \int_0^{\varepsilon_{\mathrm{F}}} \Delta\left(\delta + \frac{m}{\mu}E\right) \mathrm{d}E \tag{10.102}$$

and for the current

$$\begin{aligned} \mathbf{j} = -\frac{2e\mathbf{q}}{\hbar^2}|M_{21}|^2 &\left[(\tau_1 - \tau_2)\frac{\partial}{\partial\delta}\int_0^{\varepsilon_{\mathrm{F}}} \Delta\left(\delta + \frac{m}{\mu}E\right) E\,\mathrm{d}E\right. \\ &\left. + \tau_2 \int_0^{\varepsilon_{\mathrm{F}}} \Delta\left(\delta + \frac{m}{\mu}E\right) \mathrm{d}E\right]. \end{aligned} \tag{10.103}$$

It is assumed here that the relaxation times and the transition matrix element do not depend on energy. For an isotropic spectrum

$$M_{21} = \frac{\hbar e}{mc} A_z k_z^{(2,1)} = \mathrm{i}\frac{E_{21} Z_{21}}{\hbar}\frac{e}{c} A_z \tag{10.104}$$

and transitions can occur only for $e_z \neq 0$, i.e., when the light propagates in the quantum-well plane.

Integrating and taking into consideration (10.63), we obtain for the absorption coefficient

$$K(\omega) = \frac{4\pi m_\perp e^2 E_{21}^2 Z_{21}^2}{c n_\omega a \omega \hbar^4} F_1(\omega), \tag{10.105}$$

where a is the well width,

$$F_1(\omega) = \frac{\mu}{\pi m_\perp}\left(\arctan\frac{\delta + \frac{m}{\mu}\varepsilon_{\mathrm{F}}}{\Gamma} - \arctan\frac{\delta}{\Gamma}\right) \tag{10.106}$$

and for the current

$$\mathbf{j} = \frac{KI}{\hbar\omega}\frac{e\hbar\mathbf{q}}{m_\perp}\left[(\tau_1 - \tau_2)\frac{\mu}{m_\perp}\left(1 - \frac{F_2(\omega)}{F_1(\omega)}\right) - \tau_2\right], \tag{10.107}$$

where

$$F_2(\omega) = \varepsilon_F \Delta\left(\delta + \frac{m_\perp}{\mu}\varepsilon_F\right). \tag{10.108}$$

In the limit as $\Gamma \gg \Delta_F = \frac{m_\perp}{\mu}\varepsilon_F$

$$F_1(\omega) = \varepsilon_F \Delta(\delta) = \frac{1}{\pi}\varepsilon_F \frac{\Gamma}{\Gamma^2 + \delta^2}, \tag{10.109}$$

$$\frac{\mu}{m_\perp}\left(1 - \frac{F_2(\omega)}{F_1(\omega)}\right) = \varepsilon_F \frac{\delta}{\Gamma^2 + \delta^2}. \tag{10.110}$$

These expressions coincide with the relations given by *Lurye* [10.39], and *Grinberg* and *Lurye* [10.40]. In their consideration of the same limiting case, *Grinberg* and *Lurye* took into account the energy dependence of τ_1 and τ_2. However the dependence of Γ on energy was not included.

In the opposite limiting case, $\Delta_F \gg \Gamma$, we have over a broad frequency range, $E_{21} - (m_\perp/\mu)\varepsilon_F < \hbar\omega < E_{21}$ i.e., for $0 > \delta > -(m_\perp/\mu)\varepsilon_F$, $F_1(\omega) = \mu/m$, and $F_2(\omega) \ll F_1(\omega)$.

In this region, the absorption coefficient K and current do not depend on frequency. At its upper boundary, i.e., at $\hbar\omega = E_{21}$, the current and absorption coefficient decrease and fall off to zero for $\hbar\omega - E_{21} \gg \Gamma$. At the lower boundary, i.e., at $\hbar\omega \simeq E_{21} - (m_\perp/\mu)\varepsilon_F$, the absorption coefficient also decreases, whereas the current exhibits a spike of opposite polarity. The width of this spike is of the order of 2 Γ, the total areas of the positive and negative regions of the j (ω) curve being equal irrespective of the relative magnitude of Γ and Δ_F. As already pointed out, the cause of the sign reversal and of the abrupt increase of current near the long-wavelength edge of the absorption band may be found in the fact that for $\hbar\omega > E_{21} - (m_\perp/\mu)\varepsilon_F$, when light is capable of exciting electrons with an arbitrary direction of **k**, the main contribution to the current comes from electrons with $(\mathbf{k}\cdot\mathbf{q}) > 0$ which have a high velocity. For $\hbar\omega < E_{21} - (m_\perp/\mu)\varepsilon_F$ such states lie above the Fermi level, the contribution to the current comes from electrons with $(\mathbf{k}\cdot\mathbf{q}) < 0$ and with energies below ε_F, so that the current reverses the polarity.

The photon drag current in quantum wells was first observed by *Wieck* et al. [10.41] and *Stockman* et al. [10.42]. The study was carried out in GaAs/GaAlAs quantum wells at T = 10 K, under CO_2 laser excitation ($\hbar\omega = 124$ meV). In these experiments, the current was observed to grow near the resonance frequency and reverse the sign at $\hbar\omega = E_{21}$. The measured absorption line width 2 Γ was 5.2 meV. The carrier concentration was $N_s \simeq 10^{12} cm^{-2}$, which corresponds to $\varepsilon_F = 31$ meV and $\varepsilon_F m^*/\mu = \varepsilon_F E_{21}/E_g = 2.5$ meV, which is one half of 2 Γ. Therefore, in this case, (10.109, 110) provide a good approximation;

however, the asymmetry of the absorption line and of the j (ω) dependence suggest the significant role of nonparabolicity. Note that the actual experimental conditions [10.41] allowed multiple reflection of light from the sample boundaries. Under these conditions, the number of absorbed photons increases $(1-\mathrm{R}e^{-KL})^{-1}$ times, the drag current for the same incident flux decreasing by a factor $(1+\mathrm{R}e^{-KL})$. Here R is the reflection coefficient, L is the sample length. At the same time the linear or circular PGE current is always proportional to the number of absorbed photons.

References

Chapter 1

1.1 L. Chang, K. Ploog (eds.): *Molecular Beam Epitaxy and Heterostructures* (Nijhoff, Dorderecht 1985)
1.2 H. Jones, C. Zener: Proc. Roy. Soc. A **144**, 101 (1934)
1.3 C. Zener: Proc. Roy. Soc. A **145**, 521 (1934)
1.4 L.V. Keldysh: Fiz. Tverd. Tela **4**, 2265 (1962) [Sov. Phys. – Solid State **4**, 1658 (1963)]
1.5 L. Esaki, R. Tsu: IBM J. Res. Dev. **14**, 61 (1970)
1.6 R.F. Kazarinov, R.A. Suris: Fiz. Tekh. Poluprovodn. **5**, 797 (1971) [Sov. Phys. – Semicond. **5**, 707 (1971)]
1.7 R.H. Davis, H.H. Hosack: J. Appl. Phys. **34**, 864 (1963)
1.8 L.V. Iogansen: Zh. Exper. Teor. Fiz. **45**, 207 (1963) [Sov. Phys. – JETP **18**, 146 (1964)]
1.9 L.V. Iogansen: Zh. Exper. Teor. Fiz. **47**, 270 (1964) [Sov. Phys. – JETP **20**, 180 (1965)]
1.10 L. Esaki, L.L. Chang: Phys. Rev. Lett. **33**, 495 (1974)
1.11 L.L. Chang, L. Esaki, R. Tsu: Appl. Phys. Lett. **24**, 593 (1974)
1.12 V.N. Lutskii, D.N. Korneev, M.I. Elinson: Zh. Exper. Teor. Fiz. Pis'ma **4**, 267 (1966) [JETP Lett. **4**, 179 (1966)]
1.13 R. Dingle, W. Wiegmann, C.H. Henry: Phys. Rev. Lett. **33**, 827 (1974)

Additional Reading

Reviews

Ando T., A.B. Fowler, F. Stern: Electronic properties of two-dimensional systems. Rev. Mod. Phys. **54**, 437 (1982)
Esaki L.: The evolution of semiconductor superlattices and quantum wells. Int'l J. Mod. Phys. B **3**, 487 (1989)
Göbel E.O.: Fabrication and optical properties of semiconductor quantum wells and superlattices. Prog. Quant. Electron. **14**, 4 (1991)
Silin A.P.: Semiconductor superlattices. Usp. Fiz. Nauk, **147**, 485 (1985) [Sov. Phys. – Usp. **28**, 972 (1985)]

Chapter 2

2.1 I. Schur: J. Reine Angew. Mathem. **127**, 20 (1904)
2.2 I. Schur: J. Reine Angew. Mathem. **132**, 85 (1907)

2.3 I. Schur: J. Reine Angew. Mathem. **139**, 155 (1911)
2.4 H.A. Bethe: Ann. Phys. **3**, 133 (1929)
2.5 G.L. Bir, G.E. Pikus: *Symmetry and Strain-Induced Effects in Semiconductors* (Wiley, New York 1974)
2.6 O.V. Kovalev: *Irreducible and Induced Representations and Co-representations of the Fedorov Groups* (Nauka, Moscow 1986)
2.7 C. Herring: Phys. Rev. **52**, 361 (1937)

Additional Reading

Group Theory and its Physical Application

Bassani F., G. Pastori Parravicini: *Electronic States and Optical Transitions in Solids* (Pergamon, Oxford 1975)
Birman, J.L.: Theory of crystal space groups and infra-red and Raman lattice processes of insulating crystals, in *Encyclopedia of Physics* (Springer, Berlin, Heidelberg 1974) Vol.25/2b
Bradley C.J., A.P. Cracknell: *The Mathematical Theory of Symmetry in Solids: Representation Theory for Point Groups and Space Groups* (Clarendon, Oxford 1972)
Elliott, J.P., P.G. Dawber: *Symmetry in Physics* (MacMillan, London 1979) Vols.1,2
Evarestov R.A., V.P. Smirnov: *Site Symmetry in Crystals*, 2nd edn., Springer Ser. Solid-State Sci., Vol.108 (Springer, Berlin, Heidelberg 1996)
Inui T., Y. Tanabe, Y. Onodera: *Group Theory and Its Applications in Physics*, 2nd edn., Springer Ser. Solid-State Sci., Vol.78 (Springer, Berlin, Heidelberg 1996)
Knox R.S., A. Gold: *Symmetry in the Solid State* (Benjamin, New York 1964)
Ludwig W., C. Falter: *Symmetries in Physics*, 2nd edn., Springer Ser. Solid-State Sci., Vol.64 (Springer, Berlin, Heidelberg 1996)
Murnaghan F.D.: *The Theory of Group Representation* (Dover, New York 1963)
Nussbaum A.: Crystal symmetry, group theory and band structure. *Solid State Physics* **18**, 165 (Academic, New York 1966)
Tinkham M.: *Group Theory and Quantum Mechanics* (McGraw-Hill, New York 1964)
Tung W.K.: *Group Theory in Physics: Problems and Solutions* ed. by M. Aivazis (World Scientific, Singapore 1991)
Vainshtein B.K.: *Fundamentals of Crystals* (Springer, Berlin, Heidelberg 1994)
Wigner E.P.: *Group Theory and its Application to the Quantum Mechanics of Atomic Spectra* (Academic, New York 1959)

International Tables for Crystallography Vol.A. *Space Group Symmetry* (Reidel, Dordrecht 1983)

Time Reversal and Co-Representation Theory

Herring C.: Accidental degeneracy in the energy bands of crystals. Phys. Rev. **52**, 365 (1937)
Johnston D.F.: Space group operation and time-reversal for a Dirac electron in crystal field. Proc. Roy. Soc. A **243**, 546 (1958)
Kudryavtseva N.V.: Possible structure of the energy spectrum of electrons in crystals taking into account the operation of time reversal. Fiz.Tverd. Tela **7**, 998 (1965) [Sov. Phys. – Solid State **7**, 803 (1965)]

Selection Rules from Space Groups

Bradley C.J.: Space groups and selection rules. J. Math. Phys. **7**, 1145 (1966)

Elliott R.J., R. Loudon: Group theory of scattering processes in crystals. J. Phys. Chem. Sol. **15**, 146 (1960)

Karavaev G.F.: Selection rules for indirect transitions in crystals. Fiz. Tverd. Tela **6**, 3676 (1964) [Sov. Phys. – Solid State **6**, 2943 (1965)]

Lax M., Hopfield J.J.: Selection rules connecting different points in the Brillouin zone. Phys. Rev. **124**, 115 (1961)

Application of Group Theory to the Determination of Linearly Independent Components ot Tensors

Bhagavantan S., D. Suryanarayana: Crystal symmetry and physical properties – Application of the group theory. Acta Cryst. **2**, 21 (1949)

Fieschi R., F.G. Fumi: High-order matter tensors in symmetrical systems. Nuovo Cimento **10**, 865 (1953)

Jahn H.J.: Note on the Bhagavantan-Suryanarayana method of enumerating the physical constants of crystals. Acta Cryst. **2**, 30 (1949)

Nye J.F.: *Physical Properties of Crystals* (Clarendon, Oxford 1960)

Chapter 3

3.1 T. Ando, S. Wakahara, H. Akera: Phys. Rev. B **40**, 11609 (1989)

3.2 S.R. White, L.J. Sham: Phys. Rev. Lett. **47**, 879 (1981)

3.3 Q.-G. Zhu, H. Kroemer: Phys. Rev. B **27**, 3519 (1983)

3.4 K.B. Kahen, J.P. Leburton: Phys. Rev. B **33**, 5465 (1986)

3.5 T. Ando, H. Akera: Phys. Rev. B **40**, 11619 (1989)

3.6 R.A. Morrow, K.R. Brownstein: Phys. Rev. B **30**, 678 (1984)

3.7 G.L. Bir, G.E. Pikus: *Symmetry and Strain-Induced Effects in Semiconductors* (Wiley, New York 1974)

3.8 W. Shockley: Phys. Rev. **78**, 173 (1950)

3.9 R.J. Elliott: Phys. Rev. **96**, 266 (1954)

3.10 R.J. Elliott: Phys. Rev. **96**, 280 (1954)

3.11 G. Dresselhaus, A.F. Kip, C. Kittel: Phys. Rev. **98**, 368 (1955)

3.12 J.M. Luttinger: Phys. Rev. **102**, 1030 (1956)

3.13 E.O. Kane: J. Phys. Chem. Solids **1**, 82 (1957)

3.14 E.O. Kane: J. Phys. Chem. Solids **1**, 249 (1957)

3.15 R.A. Suris: Fiz. Tekh. Poluprovodn. **20**, 2008 (1986) [Sov. Phys. – Semicond. **20**, 1258 (1986)]
R.A. Suris, A.B. Sokolskii: Fiz. Tekh. Poluprovodn. **21**, 866 (1987) [Sov. Phys. – Semicond. **21**, 529 (1987)]

3.16 H.-R. Trebin, U. Rössler, R. Ranvaud: Phys. Rev. B **20**, 686 (1979)

3.17 D.L. Smith, C. Mailhiot: Rev. Mod. Phys. **69**, 173 (1990)

3.18 N.M. Ashkroft, N.D. Mermin: *Solid State Physics* (Holt, Rinehart and Winston, New York 1976)

3.19 I.A. Aleiner, E.L. Ivchenko: Fiz. Tekh. Poluprovodn. **27**, 594 (1993) [Semiconductors **27**, 330 (1993)]

Y. Fu, M. Willander, E.L. Ivchenko, A.A. Kiselev: Phys. Rev. B **47**, 13498 (1993)
3.20 G. Bastard: Phys. Rev. B **24**, 5693 (1981)
3.21 G. Bastard: Phys. Rev. B **25**, 7584 (1982)
3.22 D.L. Smith, C. Mailhiot: Phys. Rev. B **33**, 8345 (1986)
3.23 C. Mailhiot, D.L. Smith: Phys. Rev. B **33**, 8360 (1986)
3.24 R. Eppenga, M.F.H. Schuurmans, S. Colak: Phys. Rev. B **36**, 1554 (1987)
3.25 C.M. de Sterke, D.G. Hall: Phys. Rev. B **35**, 1380 (1987)
3.26 C. Mailhiot, D.L. Smith: Phys. Rev. B **35**, 1242 (1987)
3.27 Y.C. Chang, G.D. Sanders, D.Z.-Y. Ting: In *Excitons in Confined Systems*, ed. by R. Del Sole, A.D. Andrea, A. Lapiccirello. Springer Proc. Phys. **25**, 159 (Springer, Berlin, Heidelberg 1988)
3.28 S.S. Nedorezov: Fiz. Tverd. Tela **12**, 2269 (1970) [Sov. Phys. – Solid State **12**, 1814 (1971)]
3.29 D.A. Broido, L.J. Sham: Phys. Rev. B **31**, 888 (1985)
3.30 R. Wessel, M. Altarelli: Phys. Rev. B **40**, 12457 (1989)
3.31 A.M. Cohen, G.E. Marques: Phys. Rev. B **41**, 10608 (1990)
3.32 A. Matulis, K. Piragas: Fiz. Tekh. Poluprovodn. **9**, 220 (1975) [Sov. Phys. – Semicond. **9**, 1432 (1976)]
3.33 L.G. Gerchikov, A.V. Subashiev: Phys. Stat. Sol. (b) **160**, 443 (1990)
3.34 J.C. Maan: In *Two-Dimensional Systems. Heterostructures, and Superlattices*, ed. by G. Bauer, F. Kuchar, H. Heinrich, Springer Ser. Solid State Sci., Vol.53 (Springer, Berlin, Heidelberg 1984) p.183
3.35 G. Bastard, E.E. Mendez, L.L. Chang, L. Esaki: Phys. Rev. B **28**, 3241 (1983)
3.36 G.H. Wannier: Rev. Mod. Phys. **34**, 645 (1962)
3.37 W.V. Houston: Phys. Rev. **57**, 184 (1940)
3.38 L.V. Keldysh: Zh. Exper. Teor. Fiz. **33**, 994 (1957) [Sov. Phys. – JETP **6**, 763 (1958)]
3.39 A.G. Aronov, G.E. Pikus: Zh. Exper. Teor. Fiz. **51**, 281 (1966) [Sov. Phys. – JETP **24**, 188 (1967)]

Additional Reading

k·p Method

Bouckaert L.P., R. Smoluchowski, E. Wigner: Theory of Brillouin zones and symmetry properties of wave functions in crystals. Phys. Rev. **50**, 58 (1936)

Cardona M, F. Pollak: Energy-band structure of germanium and silicon: The **k·p** method. Phys. Rev. **142**, 530 (1966)

Gull F., M. Manas: The Dirac equation and integrable systems of **k·p** type. J. Phys. A **29**, 641 (1996)

Lage F.C. von der, H.A. Bethe: A method for obtaining electronic eigenfunctions and eigenvalues in solids with an application to sodium. Phys. Rev. **71**, 612 (1947)

Lew Yan Voon L.C., M. Willatzen, M. Cardona: Terms linear in **k** in the band structure of wurzite-type semiconductors. Phys. Rev. B **53**, 10703 (1996)

Loehr J.P.: Parameter consistency in multienergetic **k·p** models. Phys. Rev. B 52, 2374 (1995)

Krasovskii E.E., W. Schattke: The extended-LAPW-based **k·p** method for complex band structure calculations. Solid State Commun. **93**, 775 (1995)

Montasser S.S.: Analytical technique for extracting the eigenvalues of the **k**·**p** matrix that represent the band structure of semiconductors. Phys. Rev. B **42**, 7513 (1990)

Rashba E.I.: Symmetry of energy bands in wurtzite-type crystals. I. Symmetry of bands neglecting the spin-orbit interaction. Fiz. Tverd. Tela **1**, 407 (1959) [Sov. Phys. – Solid State **1**, 368 (1959)]

Rashba E.I., V.I. Sheka: Symmetry of energy bands in wurtzite-type crystals. II. Symmetry of bands taking into account the spin interactions. Fiz. Tverd. Tela, Sbornik II, p.162 (1959) (in Russian)

Wood D.M., A. Zunger: Successes and failures of the **k**·**p** method: A direct assessment for GaAs/AlAs quantum structures. Phys. Rev. B **53**, 7949 (1996)

Zak J.: The **k**·**p** representation in the dynamics of electron in solids. *Solid State Physics* **27**, 1 (Academic, New York 1972)

Effective Mass Theory

Brezini A., N. Zekri: Effective mass theory for abrupt heterojunctions. Solid State Commun. **86**, 613 (1993)

Foreman B.A.: Exact effective-mass theory for heterostructures. Phys. Rev. B **52**, 12241 (1995)

Geller M.R., W. Kohn: Quantum mechanics of electrons with graded composition. Phys. Rev. Lett. **70**, 3103 (1993)

Kittel C., A.H. Mitchell: The theory of donor and acceptor states in silicon and germanium. Phys. Rev. **96**, 1488 (1954)

Luttinger J.M., W. Kohn: Motion of electrons and holes in perturbed periodic fields. Phys. Rev. **97**, 869 (1955)

Pidgeon C.R., R.N. Brown: Interband magneto-absorption and Faraday rotation in InSb. Phys. Rev. **146**, 575 (1966)

Wannier G.H.: The structure of electronic excitation levels in insulating crystals. Phys. Rev. **52**, 191 (1937)

Deformation Potential Theory

Bir G.L., G.E. Pikus: Theory of deformation potential for semiconductors with the complicated band structure. Fiz. Tverd. Tela **2**, 2287 (1960) [Sov. Phys. – Solid State **2**, 2039 (1961)]

Blacha A., H. Presting, M. Cardona: Deformation potentials of k = 0 states in tetrahedral semiconductors. Phys. Status Solidi (b) **126**, 11 (1984)

Herring C.: Transport properties of a many-valley semiconductor. Bell. Syst. Techn. J. **34**, 237 (1955)

Herring C., E. Vogt: Transport and deformation-potential theory for many-valley semiconductors with anisotropic scattering. Phys. Rev. **101**, 944 (1956)

Shockley W., J. Bardeen: Energy bands and mobilities in monoatomic semiconductors. Phys. Rev. **77**, 407 (1950)

Method of Invariants

Koster G.F., H. Statz: Method of treating Zeeman splitting of paramagnetic ions in crystalline fields. Phys. Rev. **113**, 445 (1959)

Pikus G.E.: A new method of calculating of the energy spectrum of carriers in semiconductors. I. Neglecting spin-orbit interaction. Zh. Eksp. Teor. Fiz. **41**, 1258 (1961) [Sov. Phys. – JETP **14**, 898 (1962)]

Pikus G.E.: A New method for calculating the energy spectrum of carriers in semiconductors. II. Account of spin-orbit coupling. Zh. Eksp. Teor. Fiz. **41**, 1507 (1961) [Sov. Phys. – JETP **14**, 1075 (1962)]

Pikus G.E.: Polarization of exciton radiation of silicon in the presence of magnetic field or uniaxial strain. Fiz. Tverd. Tela **19**, 1653 (1977) [Sov. Phys. – Solid State **19**, 965 (1977)]

Rashba E.I., V.I. Sheka: Combined resonance on acceptor centers. Fiz. Tverd. Tela, **6**, 576 (1964) [Sov. Phys. – Solid State **6**, 451 (1964)]

Statz H., G.F. Koster: Zeeman splittings of paramagnetic atoms in crystalline fields. Phys. Rev. **115**, 1568 (1959)

Electron and Hole Spectrum in Cubic Crystals

Braunstein R., E.O. Kane: The valence band structure of the III-V compounds. J. Phys. Chem. Solids **23**, 1423 (1962)

Cardona M., N.E. Christensen, G. Fasol: Relativistic band structure and spin-orbit splitting of zinc-blende-type semiconductors. Phys. Rev. B **38**, 1806 (1988)

Dresselhaus G.: Spin-orbit coupling effects in zinc blende structures. Phys. Rev. **100**, 580 (1955)

Groves S.H., R.N. Brown, C.R. Pidgeon: Interband magnetoreflection and band structure of HgTe. Phys. Rev. **161**, 779 (1967)

Kalt H.: *Optical Properties of III-V Semiconductors. The Influence of Multi-Valley Band Structures*, Springer Ser. Solid-State Sci., Vol.120 (Springer, Berlin, Heidelberg 1996)

Montasser S.S.: Analytic approach to the inversion-asymmetry splitting of the valence band in zinc-blende-type semiconductor. Phys. Rev. B **48**, 12285 (1994)

Pikus G.E., V.A. Marushchak, A.N. Titkov: Spin splitting of energy bands and spin relaxation of carriers in cubic III-V crystals (Review). Fiz. Tekh. Poluprovodn. **22**, 185 (1988) [Sov. Phys. – Semicond. **22**, 115 (1988)]

Rashba E.I., V.I. Sheka: Combined resonance of band electrons in crystals with zinc-blende lattice. Fiz. Tverd. Tela **3**, 1735 (1961) [Sov. Phys. – Solid State **3**, 1257 (1961)]

Wetcel C., R. Winkler, M. Drechsler, B.K. Meyer, U. Rössler, J. Scriba, J.P. Kotthaus, V. Härle, F. Scholz: Electron effective mass and nonparabolisity in $Ga_{0.47}In_{0.53}As$/ InP quantum wells. Phys. Rev. B **53**, 1038 (1996)

Willatzen M., M. Cardona, N.E. Christensen: Linear muffin-tin-orbital and $\mathbf{k}\cdot\mathbf{p}$ calculations of effective masses and band structure of semiconducting diamond. Phys. Rev. B **50**, 18054 (1994)

Willatzen M., M. Cardona, N.E. Christensen: Spin-orbit coupling parameters and electron *g* factor of II-VI zinc-blende materials. Phys. Rev. B **51**, 17993 (1995)

Zawadski W., I.T. Yoon, C.L. Litter, X.N. Song, P. Pfeffer: Anisotropy of the conduction band of InSb: Orbital and spin properties. Phys. Rev. B **46**, 9469 (1992)

Electron Spectrum in Strained Cubic Crystals

Adams E.N.: Elastoresistance in *p*-type Ge and Si. Phys. Rev. **96**, 803 (1954)

Bahder T.B.: Eight-band $\mathbf{k}\cdot\mathbf{p}$ model of strained zinc-blende crystals. Phys. Rev. B **41**, 11992 (1990)

Bahder T.B.: Analytic dispersion relations near the Γ point in strained zinc-blende crystals. Phys. Rev. B **45**, 1629 (1992)

Bir G.L., G.E. Pikus: Effect of strain on energy spectrum and electrical properties of InSb-type semiconductors. Fiz. Tverd. Tela **3**, 3050 (1961) [Sov. Phys. – Solid State **3**, 2221 (1962)]

Blacha A., H. Presting, M. Cardona: Deformation potentials of k=0 states of tetrahedral semiconductors. Phys. Status Solidi (b) **126**, 11 (1984)

Enders P., A. Bärwolf,M. Woerner, D. Suisky: **k**·**p** theory of energy bands, wave functions, and optical selection rules in strained tetrahedral semiconductors. Phys. Rev. B **51**, 16695 (1995)

Montasser S.S.: Analytic approach to the inversion-asymmetry splitting of valence band in zinc-blende-type semiconductors. Phys. Rev. B **48**, 12285 (1994)

Pikus G.E., G.L. Bir: Effect of strain on energy spectrum and electrical properties of *p*-type germanium and silicon. Fiz. Tverd. Tela **1**, 1642 (1959) [Sov. Phys. – Solid State **1**, 1502 (1960)]

Pollak F.H., M. Cardona: Piezo-electroreflectance in Ge, GaAs and Si. Phys. Rev. **172**, 816 (1968)

Silver M., W. Batty, E.P.O. Reilly: Strain-induced valence subband splitting in III-V semiconductors. Phys. Rev. B **46**, 6781 (1992)

Spectrum of Electrons and Holes in Quantum Wells and Superlattices

a) *Nondegenerate Bands, Many-Band Models*

Ando T.: Valley mixing in short-period superlattices and the interface matrix. Phys. Rev. B **47**, 9621 (1993)

Balian R., D. Bessis, G.A. Mezincescu: Form of kinetic energy in effective-mass Hamiltonians for heterostructures. Phys. Rev. B **51**, 17624 (1995)

Burnett J.H., H.M. Cheong, W. Paul, E.S. Koteles, B. Elman: Γ-X mixing in GaAs/$Al_xGa_{1-x}As$ coupled double quantum wells under hydrostatic pressure. Phys. Rev. B **47**, 1991 (1993)

Burt M.G.: The justification for applying the effective-mass approximation to microstructures. J. Phys. Condens. Matter **4**, 6651 (1992)

Chu-liang Y., Y. Qing: Sublevels and excitons in GaAs-$Al_xGa_{1-x}As$ parabolic-quantum-well structures. Phys. Rev. B **37**, 1364 (1988)

Cnypers J.P., W. Van Haering: Coupling between Γ and X type envelope function at GaAs/Al(Ga)As interface. Phys. Rev. B **48**, 11469 (1993)

Dugaev V.K.: Electron energy spectrum and wave functions in quantum wells on the base of IV-VI narrow-gap semiconductors. Phys. Stat. Sol. (b) **184**, 347 (1994)

Einevoll G.T., L.T. Sham: Boundary conditions for envelope functions at interfaces between dissimilar materials. Phys. Rev. B **49**, 10533 (1994)

Elci A.: Effective band Hamiltonian in semiconductor quantum wells. Phys. Rev. B. **49**, 7432 (1994)

Ekenberg U.: Nonparabolicity effects in a quantum well: Sublevel shift, parallel mass, and Landau levels. Phys. Rev. B **40**, 7714 (1989)

Flatte M.E., P.M. Young, L.-H. Peng, H.E. Ehrenreich: Generalized superlattice theory and intersubband optical transitions. Phys. Rev. B **53**, 1963 (1996)

Forchhammer T., E. Veje, P. Tidemand-Petersson: Experimental determination of the conduction-band offset at GaAs/$Ga_{1-x}Al_xAs$ heterojunctions with the use of ballistic electrons. Phys. Rev. B **52**, 14693 (1995)

Foreman B.A.: Effective mass Hamiltonian and boundary conditions for the valence bands of semiconductor microstructures. Phys. Rev. B **48**, 4964 (1993)

Gershoni D., J. Oiknine-Schlesinger, E. Ehrenfreund, D. Ritter, R.A. Hamm, M.B. Panish: Minibands in the continuum of multi-quantum-well superlattices. Phys. Rev. Lett. **71**, 2975 (1993)

Hassenkam T., S. Pedersen, K. Balkanov, A. Kristensen, C.B. Sorensen, P.E. Lindenlof, F.G. Pikus, G.E. Pikus: Spin splitting and weak localization in (110) GaAs/AlGaAs quantum wells. Phys. Rev. B , to be published

Ivchenko E.L., A.A. Kiselev, Y. Fu, M. Willander: Fine structure of electron-transmission spectra across AlAs single barriers. Phys.Rev. B **50**, 7747 (1994)

Indjin D., V. Milanovic, Z. Ikonic: Bragg-confining structures with conventional and effective-mass superlattices. Phys. Rev. B **52**, 16762 (1995)

Knap W.,C. Skierbiszewski, A. Zduniak, E.Litvin-Staszevska, D. Berto, F. Kobbi, J.L. Robert, G.E. Pikus, F.G. Pikus, S.V. Iordanskii, V. Moser, K. Zekenes, Yu.B. Lyanda-Geller: Weak antilocalization and spin precession in quantum wells. Phys. Rev. B **53**, 3918 (1996)

Laikhtman B.: Boundary conditions for envelope functions in heterostructures. Phys. Rev. B **46**, 4769 (1992)

Lommer G., F. Malcher, U. Rössler: Spin splitting in semiconductor heterostructures for $B \rightarrow 0$. Phys. Rev. Lett. **60**, 728 (1988)

Luo J., H. Munekata, F.F. Fang, P.Y. Stilles: Effects of inversion asymmetry on electron energy band structures in GaSb/InAs/GaSb quantum wells. Phys. Rev. B **41**, 7685 (1990)

Pfeiffer P., W. Zawadzki: Spin-splitting of conduction subbands in GaAs/$Ga_{0.7}Al_{0.3}As$ heterostructures. Phys. Rev. B **41**, 1561 (1995)

Pikus F.G., G.E. Pikus: Conduction-band spin splitting and negative magnetoresistance in A_3B_5 heterostructures. Phys. Rev. B **51**, 16928 (1995)

Pötz W, D.K. Ferry: On the boundary conditions for envelope-function approaches for heterostructures. Superlatt. Microstruct. **3**, 57 (1987)

Rashba E.I., E.Ya. Sherman: Spin-orbital band splitting in symmetric quantum wells. Phys. Lett. A **129**, 175 (1988)

Santos P.V., M. Willatzen, M. Cardona, A. Cantarero: Tight-binding calculation of spin-splittings in semiconductor superlattices. Phys. Rev. B **51**, 5121 (1995)

Shi J., S. Pan: Envelope wave functions and subband energies in superlattices with complex bases: analytical solution and numerical examples. Phys. Rev. B **48**, 8136 (1993)

Ting D.Z.-Y., Y.-C. Chang: Γ-X mixing in GaAs/$Al_xGa_{1-x}As$ and $Al_xGa_{1-x}As$/AlAs superlattices. Phys. Rev. B **36**, 4359 (1987)

Vecris G., J.J. Quinn: Novel diagrammatic method for analysis of finite periodic and aperiodic multilayer structures. Solid State Commun. **76**, 1071 (1990)

Winkler R., U. Rössler: General approach to the envelope function approximation based on a quadrature method. Phys. Rev. B **48**, 8918 (1993)

b) *Degenerate Bands and Many-Band Models*

Altarelli M., U. Ekenberg, A. Fasolino: Calculations of hole subbands in semiconductor quantum wells and superlattices. Phys. Rev. B **32**, 5138 (1985)

Cuypers J.P., W. van Haeringen: Connection rules for envelope functions at semiconductor-heterostructure interfaces. Phys. Rev. B **47**, 10310 (1993)

Dyakonov M.I., A.V. Khaetskii: Size quantization of holes in a semiconductor with complicated valence band and of carriers in a gapless semiconductor. Zh. Eksp. Teor. Fiz. **82**, 1584 (1982) [Sov. Phys. – JETP **55**, 917 (1982)]

Edwards G., J.C. Inkson: Hole states in GaAs/AlAs heterostructures and the limitations of the Luttinger model. Solid State Commun. **89**, 595 (1994)

Edwards G., E.C. Valadares, F.W. Sheard: Hole-subband-states of GaAs/$Al_xGa_{1-x}As$ quantum wells within the 6×6 Luttinger model. Phys. Rev. B **50**, 8493 (1994)

Foreman B.A.: Effective-mass Hamiltonian and boundary conditions for the valence bands of semiconductor microstructures. Phys. Rev. B **48**, 4964 (1993)

Iconic Z., V. Milanovic, D. Tjapkin: Valence band structure of [100]-, [110]-, and [111]-grown GaAs/(Al, Ga)As quantum wells and the accuracy of the axial approximation. Phys. Rev. B **46**, 4285 (1992)

Ivchenko E.L., A.Yu. Kaminski, U. Rössler: Heavy-light hole mixing at zinc-blende (001) interfaces under normal incidence. Phys. Rev. B **54**, 5852 (1996)

Johnson N.F., H. Ehrenreich, P.M. Hui, P.M. Young: Electronic and optical properties of III-V and II-VI semiconductor superlattices. Phys. Rev. B **41**, 3655 (1990)

Kriechbaum M.: Envelope function calculation for superlattices. In *Two-Dimensional Systems and New Devices*, ed. by G. Bauer, F. Kuchar, H. Heinrich, Springer Ser. Solid-State Sci., Vol.67 (Springer, Berlin, Heidelberg 1986) p.120

Meney A.T., B. Gonul, E.P. O'Reily: Evaluation of various approximations used in the envelope-functions method. Phys.Rev. B **50**, 10893 (1994)

Nachev I.: Bound states in inversion layers on p-$Hg_{1-x}Cd_xTe$: Self-consistent results. Semicond. Sci. Technol. **3**, 29 (1988)

Nachev I.S.: Band-mixing and bound states in narrow-gap semiconductors. Physica Scripta **37**, 825 (1988)

Nojima S.: Anisotropy of optical transitions in [110] oriented quantum wells. Phys. Rev. B **47**, 13535 (1993)

Reboredo F.A., C.R. Proetto: Two-dimensional hole gas in acceptor δ-doped GaAs. Phys. Rev. B **47**, 4655 (1993)

Richards D., J. Wagner, H. Schneider, G. Hendorfer, M. Maier, A. Fischer, K. Ploog: Two-dimensional hole gas and Fermi-edge singularity in Be-δ-doped GaAs. Phys. Rev. B **47**, 9629 (1993)

Sobkowicz P.: Theory of n-inversion layers in narrow gap semiconductors: The role of the boundary conditions. Semicond. Sci. Technol. **5**, 183 (1990)

Szmulowicz F.: Derivation of a general expression for the momentum matrix element within the envelope-function approximation. Phys. Rev. B **51**, 1613 (1994)

c) *Strained Quantum Wells and Superlattices*

Anastassakis E.: Piezoelectric fields in strained heterostructures and superlattices. Phys. Rev. B **46**, 4744 (1992)

Baliga A., D. Tzivedi, N.G. Anderson: Tensile-strain effects in quantum well and superlattice band structure. Phys. Rev. B **49**, 10402 (1994)

Bertho D., J.-M. Jancu, C. Jouanin: [001] strain-induced band mixing in zinc-blende semiconductors: Intervalence versus upper-conduction – valence band effect. Phys. Rev. **50**, 16956 (1994)

Boring P., B. Gil, K.J. Moore: Optical properties and electronic structure of thin (Ga, In)As/AlAs multiple quantum wells and superlattices under internal and external strain fields. Phys. Rev. B **45**, 8413 (1992)

Caro De L., L. Tapfer: Strain and piesoelectric fields in orbitrary oriented semiconductor heterostructures: I. Multiple quantum wells. Phys. Rev. B **51**, 4374 (1995)

Chao C.Y.-P., S.L. Chuang: Spin-orbit-coupling effects on the valence-band structure of strained semiconductor quantum wells. Phys. Rev. B **46**, 4110 (1992)

Fantner E.J., G. Bauer: Strained layer IV-VI semiconductor superlattices. In *Two-Dimensional Systems, Heterostructures, and Superlattices*, ed. by G. Bauer, F. Kuchar, H. Heinrich, Springer Ser. Solid-State Sci., Vol.53 (Springer, Berlin, Heidelberg 1984) p.207

Fishman G.: Hole subbands in strained quantum-well semiconductors in [hhk] directions. Phys. Rev. B **52**, 11132 (1995)

Foreman B.A.: Analytic model for the valence band structure of a straining quantum well. Phys. Rev. B **49**, 1757 (1994)

Gil B., P. Lefebvre, P. Bonnel, H. Mathieu, C. Deparis, J. Massies, G. Neu, Y. Chen: Uniaxial-stress investigation of asymmetrical GaAs-(Ga, Al)As double quantum wells. Phys. Rev. B **47**, 1954 (1993)

Ilg M., K.H. Ploog, A. Trampert: Lateral piezoelectric field in strained semiconductor heterostructures. Phys. Rev. B **50**, 17111 (1994)

Jancu J.M., D. Bertho, C. Jouanin, B. Gil, N. Pelekanos, N. Magnea, H. Mariette: Upper-conduction-band effects in heavily strained low-dimensional zinc-blende semiconductor system. Phys. Rev. B **49**, 10802 (1994)

Kajikawa Y.: Anomaly in the in-plane polarization properties of (110)-oriented quantum wells under [110] uniaxial stress. Phys. Rev. B **47**, 3649 (1993)

Li T., H.J. Lozykowski, J.L. Reno: Optical properties of $CdTe/Cd_{1-x}Zn_xTe$ strained-layer single quantum wells. Phys. Rev. B **46**, 6961 (1992)

Los J., A. Fasalino, A. Catellanni: Generalization of the $\mathbf{k}\cdot\mathbf{p}$ approach for strained layered semiconductor structures grown of the high-index-planes. Phys. Rev. B **53**, 4960 (1996)

Moise T.S., L.J. Guido, R.C. Barker: Strain-induced heavy-hole-to-light-hole energy splitting in (111)B pseudomorphic $In_yGa_{1-y}As$ quantum wells. Phys. Rev. B **47**, 6758 (1993)

Shechter G., L.D. Shvartsman, J.E. Golub: Orientation as a key parameter in the valence-subband structure engineering of quantum wells. Phys. Rev. B **51**, 10857 (1995)

Sirenko Yu.M., J.-B. Jeon, K.W. Kim, M.A. Littlejohn: Envelope-function formalism for valence bands in wurzite quantum wells. Phys. Rev. B **53**, 1997 (1996)

Sugawaza M., N. Okazaki, T. Fujii, S. Yamazaki: Conduction-band and valence-band structures in strained $In_{1-x}Ga_xAs/InP$ quantum wells on (001) InP substrates. Phys. Rev. B **48**, 8161 (1993)

Talwar D.N., J.P. Loehr, B. Jogai: Comperative study of band-structure calculations for type-II $InAs/In_xGa_{1-x}Sb$ strained-layer superlattices. Phys. Rev. B **49**, 10345 (1994)

Valarades E.C.: Strong anisotropy of hole subbands in (311) GaAs/AlAs quantum wells. Phys. Rev. B **46**, 3935 (1992)

Voisin P.: Strained-layer superlattices. In *Two-Dimensional Systems, Heterostructures, and Superlattices*, ed. by G. Bauer, F. Kuchar, H. Heinrich, Springer Ser. Solid-State Sci., Vol.53 (Springer, Berlin, Heidelberg 1984) p.192

Weihofen R., G. Weiser, Ch. Starck, R.L. Simes: Energy gaps in strained $In_{1-x}Ga_xAs/In_{1-y}Ga_yAs_zP_{1-z}$ quantum wells, grown on (001)InP. Phys. Rev. B **51**, 4296 (1995)

Quantum Wells and Superlattices in Magnetic and Electric Fields

Agullo-Rueda F., E.E. Mendez, J.A. Brum, J.M. Hong: Coherence and localization in superlattices under electric fields. Surf. Sci. **228**, 80 (1990)

Austin E.J., M. Jaros: Electronic structure of an isolated GaAs-GaAlAs quantum well in a strong electric field. Phys. Rev. B **31**, 5569 (1985)

Bastard G., R. Ferreira, S. Chelles, P. Voisin: Interaction between Wannier-Stark states in semiconductor superlattices. Phys. Rev. B **50**, 4445 (1994)

Bereford R.: Envelope functions for a three-band semiconductor in a uniform electric field. Phys. Rev. B **49**, 13363 (1994)

Di Carlo A., P. Vogl, W. Pötz: Theory of Zener tunneling and Wannier-Stark states in semiconductors. Phys. Rev. B **50**, 8358 (1994)

Fasolino A., Altarelli, M.: Subband structure and Landau levels in heterostructures. In *Two-Dimensional Systems, Heterostructures, and Superlattices*, ed. by G. Bauer, F. Kuchar, H. Heinrich, Springer Ser. Solid-State Sci., Vol.53 (Springer, Berlin, Heidelberg 1984) p.176

Ferreira R., G. Bastard: Wannier-Stark levels in the valence band of semiconductor multiple quantum wells. Phys. Rev. B **38**, 8406 (1988)

Fukuyama H., R.A. Bari, H.C. Fogedby: Tightly bound electrons in a uniform electric field. Phys. Rev. B **8**, 5579 (1973)

Goldoni G., A. Fasalino: Hole states in quantum wells in high in-plane magnetic fields: Implications for resonant magnetotunneling spectroscopy. Phys. Rev. B **48**, 4948 (1993)

Greene S.K., J. Singelton, P. Sobkowicz, T.D. Golding, M. Pepper, J.A.A.J. Perenboom, J. Dinan: Subband occupancies and zero-field spin splitting in InSb-CdTe heterojunctions: Magnetotransport and self-consistent calculations. Semicond. Sci. Technol. **7**, 1377 (1992)

Hagon J.P., M. Jaros: Stark shifts in GaAs/$Ga_{1-x}Al_xAs$ finite-length superlattices. Phys. Rev. B **41**, 2900 (1990)

Hembree C.E., B.A. Mason, A. Zhang, J.A. Slinkman: Subband spectrum of a parabolic quantum well in a perpendicular magnetic field. Phys. Rev. B **46**, 7588 (1992)

Ivchenko E.L., A.A. Kiselev: Electron g-factor of quantum wells and superlattices. Fiz. Techn. Poluprovodn. **26**, 1471 (1992) [Sov. Phys. – Semicond. **26**, 827 (1992)]

Kajikawa Y.: Level anticrossing and related giant optical anisotropy caused by the Stark effect in a strained (110) quantum well. Phys. Rev. B **49**, 8136 (1994)

Krieger J.B., G.J. Iafrate: Time evolution of Bloch electrons in a homogeneous electric field. Phys. Rev. B **33**, 5494 (1986)

Matsuura M., T. Kamizato: Subbands and excitons in a quantum well in an electric field. Phys. Rev. B **33**, 8385 (1986)

Merlin R.: Subband-Landau-level coupling in tilted magnetic fields: Exact results for parabolic wells. Solid State Commun. **64**, 99 (1987)

Morifuji M., M. Yamaguchi, K. Taniguchi, C. Hamaguchi: Electric-field-induced Γ-X mixing between Stark ladders in short-period GaAs/AlAs supperlattices. Phys. Rev. B **50**, 8722 (1994)

Müller W., H.T. Grahn, K. von Klitzing, K. Ploog: Resonant tunneling in crossed electric and magnetic fields in GaAs-AlAs superlattices. Phys. Rev. B **48**, 11176 (1993)

Nakayama M., M. Ando, I. Tanaka, H. Nishimura, H. Schneider, K. Fujiwara: Electric-field effect on above-barrier states in a GaAs/$Al_xGa_{1-x}As$ superlattice. Phys. Rev. B **51**, 4236 (1995)

Pacheco M., Z. Barticevic, F. Claro: Optical response of a superlattice in parallel magnetic and electric fields. Phys. Rev. B **46**, 15200 (1993)
Platero G., M. Altarelli: Valence-band levels and optical transitions in quantum wells in a parallel magnetic field. Phys. Rev. B **39**, 3758 (1989)
Ritze M., N.J.M. Horing, R. Enderlein: Density of states and Wannier-Stark levels of superlattices in an electric field. Phys. Rev. B **47**, 10437 (1993)
Schmidt K.H., N. Linder, G.H. Döhler, H.T. Grahn, K. Ploog, H. Schneider: Coexistence of Wannier-Stark transitions and miniband Franz-Keldysh oscillations in strongly coupled GaAs/AlAs superlattices. Phys. Rev. Lett. **72**, 2769 (1994)
Schwedler R., F. Brüggemann, A. Ziebell, B. Opitz, B., A. Kohl, H. Kurz: Valence band structure of fractional Wannier-Stark-effect shallow InGaAs/InGaAs superlattices. Superlatt. Microstruct. **15**, 145 (1994)
Smith T.P., F.F. Fang: *g*-Factor of electrons in an InAs quantum well. Phys. Rev. B **35**, 7729 (1977)
Wollrab R., R. Sizmann, F. Koch, J. Ziegler, H. Maier: Spin splitting of electron subbands in the electrostatic interface potential on HgCdTe. Semicond. Sci. Technol. **4**, 491 (1989)
Wong S.L., R.W. Martin, M. Lakrimi, R.J. Nicholas, T.-Y. Seong, N.J. Mason, P.J. Walker: Optical and transport properties of piezoelectric [111]-oriented strained $Ga_{1-x}In_xSb/GaSb$ quantum wells. Phys. Rev. B **48**, 17885 (1993)
Xu W.: Electronic structure of $Al_xGa_{1-x}As$-GaAs-$Al_yGa_{1-y}As$ single quantum wells subjected to in-plane magnetic field. Phys. Rev. B **51**, 9770 (1995)
Zawadski W., S. Klahn, H. Merkt: Inversion electrons on narrow-band-gap semiconductors in crossed electric and magnetic fields. Phys. Rev. B **33**, 6916 (1986)

Chapter 4

4.1 Yu.E. Kitaev, R.A. Evarestov: Fiz. Tverd. Tela **30**, 2970 (1988) [Sov. Phys. – Solid State **30**, 1712 (1988)]
4.2 M.I. Alonso, M. Cardona, G. Kanellis: Solid State Commun. **69**, 479 (1989)
4.3 P. Molinas i Mata, M.I. Alonso, M. Cardona: Solid State Commun. **74**, 374 (1990)
4.4 B.H. Bairamov, R.A. Evarestov, I.P. Ipatova, Yu.E. Kitaev, A.Yu. Maslov, M. Delaney, T.A. Gant, M.V. Klein, D. Levi, J. Klem, H. Morkoc: Superlatt. Microstruct. **6**, 227 (1989)
4.5 B. Jusserand, M. Cardona: In *Light Scattering in Solids V*, ed. by M. Cardona, G. Güntherodt, Topics Appl. Phys., Vol.66 (Springer, Berlin, Heidelberg 1989) p.49
4.6 J. Menendez: J. Lumin. **44**, 285 (1989)
4.7 M. Cardona: Superlatt. Microstruct. **4**, 27 (1989)
4.8 E. Molinari, S. Baroni, P. Giannozzi, D. Gironcoli: *Proc. 20th Int'l Conf. Phys. Semicond.* (Thesaloniki, Greece 1990), ed. by E.M. Anastassakis, J.D. Joannopoulos (World Scientific, Singapore 1990) p.1429
4.9 M. Rytov: Acust. Zh. **2**, 71 (1956) [Sov. Phys. – Acoust. **2**, 63 (1956)]
4.10 C. Colvard, T.A. Gant, M.V. Klein, R. Merlin, R. Fischer, H. Morkoc, A.C. Gossard: Phys. Rev. B **31**, 2080 (1985)
4.11 E.P. Pokatilov, S.I. Beril: Phys. Stat. Sol. (b) **118**, 567 (1983)
4.12 R.E. Gamley, D.L. Mills: Phys. Rev. B **29**, 1695 (1984)

4.13 K. Huang, B. Zhu: Phys. Rev. B **38**, 13377 (1988)
4.14 H. Chu, S.F. Ren, Y.-C. Chang: Phys. Rev. B **37**, 10476 (1988)
4.15 H. Sato, Y. Cory: Phys. Rev. B **39**, 10192 (1989)
4.16 V.L. Gurevich, K.E. Shtengel: J. Phys. Cond. Matter. **2**, 6323 (1990)

Additional Reading

Vibrational Spectra of Superalattices, Electron-Phonon Interaction

Ahn D.: Calculations of hole-phonon scattering in strained-layer quantum wells. J. Appl. Phys. **78**, 4505 (1995)

Bannov N., V. Aristov, V. Mitin, M.A. Stroscio: Electron relaxation times due to the deformation-potential interaction of electrons with confined acoustic phonons in a free-standing quantum well. Phys. Rev. B **51**, 9931 (1995)

Chamberlain M.P., M. Cardona, B.K. Ridley: Optical modes in GaAs/AlAs superlattices. Phys. Rev. B **48**, 14356 (1993)

Chamberlain M.P., C. Trallerro-Giner, M. Cardona: Anisotropy effects on optical phonon modes in GaAs/AlAs quantum wells. Phys. Rev. B **50**, 1611 (1994)

Cruz de la R.M., S.W. Teitsworth, M.A. Stroscio: Interface phonons in spherical GaAs/$Al_xGa_{1-x}As$ quantum dots. Phys. Rev. B **52**, 1489 (1995)

El Boudouti E.H., B. Djafari-Rouhani: Acoustic waves in finite superlattices. Phys. Rev. B **49**, 4586 (1994)

Enderlein R.: Macroscopic dynamic theory of polar interface modes of superlattices. Phys. Stat. Sol. (b) **150**, 85 (1988)

Foreman B.A.: Foundations of the envelope-function theory for phonons in heterostructures. Phys. Rev. B **52**, 12260 (1995)

Hai G.Q., F.M. Peeters, J.T. Devreese: Electron optical-phonon coupling in GaAs/$Al_xGa_{1-x}As$ quantum wells due to interface, slab and half-space modes. Phys. Rev. B **48**, 4666 (1993)

Jusserand B., D. Paquet, A. Regreny: "Folded" optical phonons in GaAs/$Ga_{1-x}Al_xAs$ superlattices. Phys. Rev. B **30**, 6245 (1984)

Jusserand B., M. Cardona: In *Light Scattering in Solids V*, ed. by M. Cardona, G. Güntherodt, Topics Appl. Phys., Vol.66 (Springer, Berlin, Heidelberg 1989) Chap.3

Klein M.V.: Phonons in semiconductor superlattices: IEEE J. QE-**22**, 1760 (1986)

Knipp P.A., T.L. Reinecke: Coupling of electrons to interface phonons in semiconductor quantum wells. Phys. Rev. B **48**, 12338 (1993)

Knipp P.A., T.L. Reinecke: Effect of boundary conditions on confined optical phonons in semiconductor nanostructures. Phys. Rev. B **48**, 18037 (1993)

Knipp P.A., T.L. Reinecke: Coupling between electrons and acoustic phonons in semiconductor nanostructures. Phys. Rev. B **52**, 5923 (1995)

Michl F.C., R. Winkler, U. Rössler: Exciton transfer in asymmetrical double quantum wells: The one-LO phonon process. Solid State Commun. **99**, 13 (1996)

Popovic Z.V., M.B. Vukmirovic, Y.S. Raptis, E. Anastassakis, R. Nötzel, K. Ploog: Folded phonons from lateral periodicity in (311) GaAs/AlAs corrugated supperlattices. Phys. Rev. B **52**, 5789 (1995)

Rey-Gonzalez R., L. Quiroga, A. Camacho: Phonon spectra of (001), (011) and (111) surfaces and interfaces of the CdTe/HgTe system. J. Phys., Condensed Matter **7**, 1979 (1995)

Ridley B.K.: Electron-hybridon interaction in a quantum well. Phys. Rev. B **47**, 4592 (1993)

Sela I., V.V. Gridin, R. Beserman, H. Morkoç: Resonant Raman study of the LO-phonon energy fluctuations in III-V alloy semiconductors. Phys. Rev. B **37**, 6393 (1988)

Shanabrook B.V., B.R. Bennett, R.J. Wagner: Planar vibrational modes in superlattices. Phys. Rev. B **48**, 17172 (1993)

Shi J., S. Pan: Surface and interface optical-phonon modes in a finite double heterostructure of polar crystals. Phys. Rev. B **46**, 4265 (1992)

Shi J., S. Pan: Electron-interface-phonon interaction and scattering in asymmetric semiconductor quantum-well structures. Phys. Rev. B **51**, 17681 (1995)

Sun G., L. Friedmann: Heavy-hole scattering by confined nonpolar optical phonons in a single $Si_{1-x}Ge_x/Si$ quantum well. Phys. Rev. B **53**, 3966 (1996)

Tsuchiya T., T. Ando: Electron-phonon interaction in GaAs/AlAs superlattices. Phys. Rev. B **47**, 7240 (1993)

Zhu B.: Optical phonon modes in superlattices. Phys. Rev. B **38**, 7694 (1988)

Zi J., K. Zhang, X. Xie: Vibrational properties of Si/Ge and α-Sn/Ge superlattices with intermixed interfaces. Phys. Rev. **47**, 9937 (1993)

Zucker J.E., A. Pinczuk, D.S. Chemla, A.C. Gossard, W. Wiegmann: Optical vibrational modes and electron-phonon interaction in GaAs quantum wells. Phys. Rev. Lett. **53**, 1280 (1984)

Chapter 5

5.1 G. Bastard: Phys. Rev. B **24**, 4714 (1981)
5.2 C. Mailhiot, Y.-C. Chang, T.C. McGill: Phys. Rev. B **26**, 4449 (1982)
5.3 W.T. Masselink, Y.-C. Chang, H. Morkoc: Phys. Rev. B **32**, 5190 (1985)
5.4 R.L. Greene, K.K. Bajaj: Phys. Rev. B **31**, 913 (1985)
5.5 S. Chaudhuri: Phys. Rev. B **28**, 4480 (1983)
5.6 H. Chen, S. Zhou: Phys. Rev. B **36**, 9581 (1987)
5.7 E.L. Ivchenko, A.V. Kavokin: Fiz. Tekh. Poluprovodn. **25**, 1780 (1991) [Sov. Phys. – Semicond. **25**, 1070 (1991)]
5.8 N.F. Gashimzade, E.L. Ivchenko, V.A. Kosobukin: Fiz. Tekh. Poluprovodn. **23**, 839 (1989) [Sov. Phys. – Semicond. **23**, 529 (1989)]
5.9 R.L. Greene, K.K. Bajaj: Solid State Commun. **45**, 831 (1983)
5.10 U. Ekenberg, M. Altarelli: Phys. Rev. B **35**, 7585 (1987)
5.11 A. Chomette, B. Lambert, D. Deveaud, F. Clerot, A. Regreny, G. Bastard: Europhys. Lett. **4**, 461 (1987)
5.12 D.A. Kleinman: Phys. Rev. B **28**, 871 (1983)
5.13 H.W. van Kesteren, E.C. Cosman, W.A.J.A. van der Poel, C.T. Foxon: Phys. Rev. B **41**, 5283 (1990)

Additional Reading

Shallow Impurities

Greene R.L., K.K. Bajaj: Energy levels of hydrogenic impurity states in GaAs-$Ga_{1-x}Al_xAs$ quantum well structures. Solid State Commun. **45**, 825 (1983)

Holtz P.O., Q.X. Zhao, A.C. Ferreira, B. Monemar, M. Sundaram, J.L. Merz, A.C. Gossard: Excited states of shallow acceptors confined in $GaAs/Al_xGa_{1-x}As$ quantum wells. Phys.Rev. B **48**, 8872 (1993)

Lane P., R.L. Greene: Shallow donors in multiple-well GaAs-$Ga_{1-x}Al_xAs$ heterostructures. Phys. Rev. B **33**, 5871 (1985)
Latgé A., N. Porras-Montenegro, L.E. Oliveira: 1s-$2p_{\pm}$ infrared-absorption spectra of donor-doped quantum wells under electric and magnetic fields, Phys. Rev. B **51**, 13344 (1995)
Liu Z., D. Ma: Energy spectra of donors in GaAs-$Ga_{1-x}Al_xAs$ superlattices. J. Phys. C **19**, 2757 (1986)
MacDonald A.H., D.S. Ritchie: Hydrogenic energy levels in two dimensions at arbitrary magnetic fields. Phys. Rev. B **33**, 8336 (1986)
Pasquarello A., L.C. Andreani, R. Buczko: Binding energies of excited shallow acceptor states in GaAs/$Ga_{1-x}Al_xAs$ quantum wells. Phys. Rev. B **40**, 5602 (1989)
Tanaka K., M. Nagaoka, T. Yamabe: Binding energy of the impurity level in the $Ga_{1-x}Al_xAs$-GaAs-$Ga_{1-y}Al_yAs$ supperlattice. Phys. Rev. B **28**, 7068 (1983)
Zhu J.-L.: Coupling between a donor potential and quantum wells: effect on binding energies. Phys. Rev. B **40**, 10529 (1989)

Excitons

Andreani L.C., A. Pasquarello: Accurate theory of excitons in GaAs-$Ga_{1-x}Al_xAs$ quantum wells. Phys. Rev. B **42**, 8928 (1990)
Bastard G., E.E. Mendez, L.L. Chang, L. Esaki: Exciton binding energy in quantum wells. Phys. Rev. B **26**, 1974 (1982)
Bauer G.E.W., T. Ando: Exciton mixing in quantum wells. Phys. Rev. B **38**, 6015 (1988)
Blackwood E., M.J. Snelling, R.T. Harley, S.R. Andrews, C.T.B. Foxon: Exchange interaction of excitons in GaAs heterostructures. Phys. Rev. B **50**, 14246 (1994)
Chao C.Y.-P., S.L. Chuang: Momentum-space solution of exciton excited states and heavy-hole-light-hole mixing in quantum wells. Phys.Rev. B **48**, 8210 (1993)
Dawson P., K.J. Moore, G. Duggan, H.I. Ralph, C.T.B. Foxon: Unambiguous observation of the 2s state of the light- and heavy-hole excitons in GaAs-(AlGa)As multiple-quantum-well structures. Phys. Rev. B **34**, 6007 (1986)
Del Sole A., A. D'Andrea, A. Lapiccirella (eds.): *Excitons in Confined Systems*, Springer Proc. Phys., Vol.25 (Springer, Berlin, Heidelberg 1988)
Dingle R., W. Wiegmann, C.H. Henry: Quantum states of confined carriers in very thin $Al_xGa_{1-x}As$-GaAs-$Al_xGa_{1-x}As$ heterostructures. Phys. Rev. Lett. **33**, 827 (1974)
Greene R.L., K.K. Bajaj, D.E. Phelps: Energy levels of Wannier excitons in GaAs-$Ga_{1-x}Al_xAs$ quantum-well structures. Phys. Rev. B **29**, 1807 (1984)
Greene R.L., K.K. Bajaj: Binding energies of Wannier excitons in GaAs-$Ga_{1-x}Al_xAs$ quantum-well structures in a magnetic field. Phys. Rev. B **31**, 6498 (1985)
Jiang T.-F.: An alternative approach to exciton binding energy in a GaAs-$Al_xGa_{1-x}As$ quantum well. Solid State Commun. **50**, 589 (1984)
Kavokin A.V., A.I. Nesvizhskii, R.P. Seisyan: Exciton in a semiconductor quantum well subjected to a strong magnetic field. Fiz. Tekh. Poluprovodn. **27**, 977 (1993) [Sov. Phys. – Semiconductors **27**, 530 (1993)]
Koteles E.S., J.Y. Chi: Experimental exciton binding energies in GaAs/$Al_xGa_{1-x}As$ quantum wells as a function of well width. Phys. Rev. B **37**, 6332 (1988)
Matsuura M., Y. Shinozuka: Excitons in type-II quantum-well systems: binding of the spatially separated electron and hole. Phys. Rev. B **38**, 9830 (1988)

Miller R.C., D.A. Kleinman, W.T. Tsang, A.C. Gossard: Observation of the excited level of excitons in GaAs quantum wells. Phys. Rev. B **24**, 1134 (1981)

Pikus F.G.: Exciton in quantum wells with a two dimensional electron gas. Fiz. Tekh. Poluprovodn. **26**, 43 (1992) [Sov. Phys. – Semicond. **26**, 26 (1992)]

Priester C., G. Allan, M. Lannoo: Wannier excitons in GaAs-$Ga_{1-x}Al_xAs$ quantum-well structures. Influence of the effective-mass mismatch. Phys. Rev. B **30**, 7302 (1984)

Rashba E.I., M.D. Sturge (eds.): *Excitons* (North-Holland, Amsterdam 1982)

Weisbusch C., R.C. Miller, R. Dingle, A.C. Gossard, W. Wiegmann: Intrinsic radiative recombination from quantum states in GaAs-$Al_xGa_{1-x}As$ multi-quantum well structures. Solid State Commun. **37**, 219 (1981)

Song K.S., R.T. Williams: *Self-Trapped Excitons*, 2nd edn., Springer Ser. Solid-State Sci., Vol.105 (Springer, Berlin, Heidelberg 1996)

Zhu B., K. Huang: Effect of valence-band hybridization on the exciton spectra in GaAs-$Ga_{1-x}Al_xAs$ quantum wells. Phys. Rev. B **36**, 8102 (1987)

Biexcitons

Charbonneau S., T. Steiner, M.L.W. Thewalt, E.S. Koteles, I.Y. Chi, B. Elman: Optical investigation of biexcitons and bound excitons in GaAs quantum wells. Phys. Rev. B **38**, 3583 (1988)

Miller R.C., D.A. Kleinman, A.C. Gossard, O. Munteanu: Biexcitons in GaAs quantum wells. Phys. Rev. B **25**, 6545 (1982)

Reynolds D.C., K.K. Bajaj, C.E. Stutz, R.L. Jones, W.M. Theis, P.W. Yu, K.R. Evans: Binding energies of biexcitons in $Al_xGa_{1-x}As$/GaAs multiple quantum wells. Phys. Rev. B **40**, 3340 (1989)

Chapter 6

6.1 R.C. Miller, A.C. Gossard, G.D. Sanders, Y.-C. Chang, J.N. Schulman: Phys. Rev. B **32**, 8452 (1985)

6.2 H. Chu, Y.-C. Chang: Phys. Rev. B **39**, 10 861 (1989)

6.3 S. Schmitt-Rink, D.S. Chemla, D.A.B. Miller: Adv. Phys. **38**, 89 (1989)

6.4 H. Iwamura, H. Kobayashi, H. Okamoto: Jpn. J. Appl. Phys. **23**, L795 (1984)

6.5 E.L. Ivchenko, V.A. Kosobukin: Fiz. Tekh. Poluprovodn. **22**, 24 (1987) [Sov. Phys. – Semicond. **22**, 15 (1988)]

6.6 E.L. Ivchenko, V.P. Kochereshko, I.N. Uraltsev: In *Semiconductors and Insulators: Optical and Spectroscopic Research*, ed. by Yu.I. Koptev, Ioffe Physicotechnical Institute Research Series, Vol.12 (Nova Science, New York 1992) p.21

6.7 E.L. Ivchenko, A.V. Kavokin: Fiz. Tekh. Poluprovodn. **25**, 1780 (1991) [Sov. Phys. – Semicond. **25**, 1070 (1991)]

6.8 I.N. Uraltsev, E.L. Ivchenko, P.S. Kop'ev, V.P. Kochereshko, D.R. Yakovlev: Phys. Stat. Solidi (b) **150**, 673 (1988)

6.9 E.E. Mendez, F. Agullo-Rueda: J. Lumin. **44**, 223 (1989)

6.10 A.P. Thorn, A.J. Shields, P.C. Klipstein, N. Apsley, T.M. Kerr: J. Phys. C **20**, 4229 (1987)

6.11 S. Tarucha, H. Okamoto, Y. Ywasa, N. Miura: Solid State Commun. **52**, 815 (1984)
6.12 H. Chu, Y.-C. Chang: Phys. Rev. B **40**, 5497 (1989)

Additional Reading

Optical Superlattices

Agranovich V.M., V.E. Kravtsov: Notes on crystal optics of superlattices. Solid State Commun. **55**, 85 (1985)
Gashimzade N.F.: Guided polariton waves in semiconductor superlattices. Phys. Status Solidi (b) **160**, K113 (1990)
Shi H., C. Tsai: Polariton modes in superlattice media. Solid State Commun. **52**, 953 (1984)

Interband Transitions in Heterostructures

Atanasov R., F. Bassani, A. D'Andrea, N. Tomassini: Exciton properties and optical responce in $In_xGa_{1-x}As/GaAs$ strained quantum wells. Phys. Rev. B **50**, 14381 (1994)
Andreani L.C., F. Tassone, F. Bassani: Radiative lifetime of free excitons in quantum wells. Solid State Commun. **77**, 641 (1991)
Belousov M.V., V.L. Berkovets, A.O. Gusev, E.L. Ivchenko, P.S. Kop'ev, N.N. Ledentsov, A.I. Nesvizhskii: Optical anisotropy of GaAs/AlAs superlattices grown in the [113] direction, Fiz. Tverd. Tela **36**, 1098 (1994) [Phys. Solid State **36**, 596 (1994)]
Belousov M.V., E.L. Ivchenko, A.I. Nesvizhskii: Biaxial optical anisotropy in GaAs/AlAs(hhl) superlattices. Fiz. Tverd. Tela **37**, 1408 (1995) [Phys. Solid State **37**, 763 (1995)]
Belousov M.V., N.N. Ledentsov, M.V. Maximov, P.D. Wang, I.N. Yasievich, N.N. Faleev, I.A. Kozin, V.M. Ustinov, P.S. Kop'ev, C.M. Sotomayor Torres: Energy levels and exciton oscillator strength in submonolayer InAs-GaAs heterostructures, Phys. Rev. B **51**, 14346 (1995)
Cingolani R., K. Ploog, L. Baldassarre, M. Ferrara, M. Lugara, C. Moro: Spectroscopic studies of real space indirect symmetric GaAs/AlAs short period superlattices. Appl.Phys. A **50**, 189 (1990)
Cingolani R., P. Prete, D. Greco, P.V. Giugno, M. Lomascolo, R. Rinaldi, L. Calcagnile, L. Vanzetti, L. Sorba, A. Franciosi: Exciton spectroscopy in $Zn_{1-x}Cd_xSe/ZnSe$ quantum wells, Phys. Rev. B **51**, 5176 (1995)
D'Andrea A., N. Tomassini, L. Ferrari, M. Righini, S. Selci, M.R. Bruni, M.G. Simeone, N. Gambacorti: Normalized reflection spectra of $GaAs/In_xGa_{1-x}As$ single quantum wells: Structure characterizations and excitonic properties. Phys. Rev. B **52**, 10713 (1995)
Dignam M.M., J.E. Sipe: Exciton states in type-I and type-II $GaAs/Ga_{1-x}Al_xAs$ superlattices. Phys. Rev. B **41**, 2865 (1990)
Efros Al.L.: Excitons in the structures with quantum wells. Fiz. Tekh. Poluprovodn. **20**, 1281 (1986) [Sov. Phys. – Semicond. **20**, 808 (1986)]
Fainstein A., P. Etchegoin, P.V. Santos, M. Cardona, K. Tötemeyer, K. Eberl: In-plane birefringence of GaAs/AlAs multiple quantum wells. Phys. Rev. B **50**, 11850 (1994)

Fu Y., K.A. Chao: Subband structures of GaAs/$Al_xGa_{1-x}As$ multiple quantum wells. Phys. Rev. B **40**, 8349 (1989)

Fujiwara K., K. Kawashima, T. Yamamoto, K. Ploog: Optical transitions in GaAs/AlAs superlattices with different miniband widths. Solid State Electron. **37**, 889 (1994)

Glembovskii, O.J., B.V. Shanabrook, W.T. Beard: Temperature dependence of photoreflectance in GaAs-AlGaAs multiple quantum wells. Surf. Sci. **174**, 206 (1986)

Houdre R., R.P. Stanley, U. Oesterle, M. Ilegems, C. Weisbuch: Room-temperature cavity polaritons in a semiconductor microcavity. Phys. Rev. B **49**, 16761 (1994)

Humlicek J., F. Lukes, K. Navratil, M. Carriga, K. Ploog: Ellipsometric and reflectance studies of GaAs/AlAs superlattices. Appl.Phys. A **49**, 407 (1989)

Ivchenko E.L.: Exciton polaritons in periodic quantum-well structures. Fiz. Tverd. Tela, **33**, 2388 (1991) [Sov. Phys. – Solid State **33**, 1344 (1991)]

Ivchenko E.L., A.V. Kavokin: Light reflection from structures with quantum wells, quantum wires and quantum dots. Fiz. Tverd. Tela **34**, 1815 (1992) [Sov. Phys. – Solid State **34**, 968 (1992)]

Ivchenko E.L., P.S. Kop'ev, V.P. Kochereshko, I.N. Uraltsev, D.R. Yakovlev, S.V. Ivanov, B.Ya. Meltser, M.A. Kalitievskii: Reflection in the excitonic spectral region from a single quantum well structure. Oblique and normal incidence of light. Fiz. Tekh. Poluprovodn. **22**, 784 (1988) [Sov. Phys. – Semicond. **22**, 495 (1988)]

Ivchenko E.L., V.P. Kochereshko, G.R. Pozina, D.R. Yakovlev, A. Waag, W. Ossau, G. Landwehr, R. Hellmann, E.O. Göbel: Giant exciton resonance reflectance in Bragg MQW structures. Superlatt. Microstruct. **15**, 471 (1994)

Ivchenko E.L., V.P. Kochereshko, P.S. Kop'ev, V.A. Kosobukin, I.N. Uraltsev, D.R. Yakovlev: Exciton longitudinal-transverse splitting in GaAs/AlGaAs superlattices and quantum wells. Solid State Commun. **70**, 529 (1989)

Ivchenko E.L., A.N. Nesvizhskii, S. Jorda: Bragg reflection of light from quantum-well structures. Fiz. Tverd. Tela **36**, 2118 (1994) [Phys. Solid State **36**, 1156 (1994)]

Ivchenko E.L., A.N. Nesvizhskii, S. Jorda: Resonant Bragg reflection from quantum-well structures. Superlatt. Microstruct. **16**, 17 (1994)

Ivchenko E.L., M.A. Kaliteevski, A.V. Kavokin, A.I. Nesvizhskii: Reflection and absorption spectra from microcavities with resonant Bragg quantum wells. J. Opt. Soc. Am. B **13**, 1061 (1996)

Jorda S.: Exciton polaritons in quantum wells embedded in waveguides and microcavities. J. Opt. Soc. Am. B **13**, 1054 (1996)

Kahen K.B., J.P. Leburton: Index of refraction of GaAs-$Al_xGa_{1-x}As$ superlattices and multiple quantum wells. Superlatt. Microstruct. **3**, 251 (1987)

Kajikawa Y.: Well-width dependence of the optical anisotropies in (001) and (110) semiconductor quantum wells: The effect of spin-orbit split-off bands. Phys. Rev. B **51**, 16790 (1995)

Livescu G., D.A.B. Miller, D.S. Chemla, M. Ramaswamy: Free carrier and many-body effects in absorption spectra of modulation-doped quantum wells. IEEE J. QE-**24**, 1677 (1988)

Lorusso G.F., V. Capazzi, J.L. Staehli, C. Flesia, D. Martin, P. Favia: Absorption spectra of GaAs/$Al_xGa_{1-x}As$ random superlattices at 2 K. Phys. Rev. B **53**, 1018 (1996)

Masselink W.T., P.J. Pearah, J. Klem, C.K. Peng, H. Morkoc, G.D. Sanders, Y.-C. Chang: Absorption coefficients and exciton oscillator strengths in AlGaAs-GaAs superlattices. Phys. Rev. B **32**, 8027 (1985)

Masumoto Y., M. Matsuura, S. Tarucha, H. Okamoto: Two-dimensional shrinkage of the exciton wavefunction in quantum wells probed by optical absorption. Surf. Sci. **170**, 635 (1986)

Norris T.B., J.-K. Rhee, C.-Y. Sung, Y. Arakawa, M. Nishioka, C. Weisbuch: Time-resolved vacuum Rabi oscillations in a semiconductor quantum microcavity. Phys. Rev. B **50**, 14 663 (1994)

Pau S., G. Björk, J. Jacobson, H. Cao, Y. Yamamoto: Stimulated emission of a microcavity dressed exciton and suppression of phonon scattering. Phys. Rev. B **51**, 7090 (1995)

Pau S., G. Björk, J. Jakobson, H. Cao, Y. Yamamoto: Microcavity exciton-polariton splitting in the linear regime. Phys. Rev. B **51**, 14437 (1995)

Pereira M.F.: Analytical solutions for the optical absorption of semiconductor superlattices. Phys. Rev. B **52**, 1978 (1995)

Santos P.V., A. Cantarero, M. Cardona, R. Nötzel, K. Ploog: Optical properties of (311)-oriented GaAs/AlAs superlattices. Phys. Rev. B **52**, 1970 (1995)

Savona V., L.C. Andreani, P. Schwendimann, A. Quattropani: Quantum well excitons in semiconductor microcavities: Unified treatment of weak and strong coupling regimes. Solid State Commun. **93**, 733 (1995)

Schubert M.: Polarization-dependent optical parameters of arbitrarily anisotropic homogeneous layered systems. Phys. Rev. B **53**, 4265 (1996)

Schultheis L., K. Ploog: Reflectance in two-dimensional excitons in GaAs-AlGaAs quantum wells. Phys. Rev. B **30**, 1090 (1984)

Stroucken T., A. Knorr, P. Thomas, S.W. Koch: Coherent dynamics of radiatively coupled quantum-well excitons. Phys. Rev. B **53**, 2026 (1996)

Tang Z.K., A. Yanase, T. Yasui, Y. Segawa, K. Cho: Optical selection rule and oscillator strength of confined exciton system in CuCl thin films. Phys. Rev. Lett. **71**, 1431 (1993)

Tsitsishvili E.G.: Intrinsic optical anisotropy of quantum wells in cubic crystals. Phys. Rev. B **52**, 11172 (1995)

Voliotis V., R. Grousson, P. Lavallard, E.L. Ivchenko, A.A. Kiselev, R. Planel: Absorption coefficient in type-II GaAs/AlAs short-period superlattices. Phys. Rev. B **49**, 2576 (1994)

Weisbuch C., M. Nishioka, A. Ishikawa, Y. Arakawa: Observation of the coupled exciton-photon mode splitting in a semiconductor quantum microcavity. Phys. Rev. Lett. **69**, 3314 (1992)

Wen G., Y.-C. Chang: Theory of absorption spectra for above-barrier excitons in semiconductor superlattices, Phys. Rev. B **49**, 16585 (1994)

Yamanaka K., T. Fukunaga, T. Tsukada, K.L.I. Kobayashi, M. Ishii: Photocurrent spectroscopy in GaAs/AlGaAs multiple quantum wells under a high electric field perpendicular to the heterointerface. Appl. Phys. Lett. **48**, 840 (1986)

Yuan S., N. Frank, G. Bauer, M. Kriegbaum: Optical properties of $PbSe/Pb_{1-x}Mn_xSe$ multiple-quantum-well structures: Phys. Rev. B **50**, 5286 (1994)

Zhu B.: Oscillator strength and optical selection rule of excitons in quantum wells. Phys. Rev. B **37**, 4689 (1988)

Electro-Optics

Barticevic Z., M. Pachecho, F. Claro: Exciton Stark and Landau ladders in a $GaAs/Al_xGa_{1-x}As$ superlattice. Phys. Rev. B **51**, 14414 (1995)

Bastard G., E.E.Mendez, L.L. Chang, L. Esaki: Variational calculations on a quantum well in an electric field. Phys. Rev. B **28**, 3241 (1983)

Bleuse J., G. Bastard, P. Voisin: Electric-field-induced localization and oscillatory electro-optical properties of semiconductor superlattices. Phys. Rev. Lett. **60**, 220 (1988)

Cao S.M., M. Willander, A.I. Nesvizhskii, E.L. Ivchenko, A.A. Toropov: Electrons and excitons in an imperfect superlattice in electric fields. Superlatt. Microstruct. **17**, 97 (1995)

Cao S.M., M. Willander, A.A. Toropov, T.V. Shubina, B.Ya. Mel'tser, S.V. Shaposhnikov, P.S. Kop'ev, P.O. Holtz, J.P. Bergman, B. Monemar: Resonant coupling of electrons and excitons in an aperiodic superlattice under electric fields studied by photoluminescence spectroscopy. Phys. Rev. B **51**, 17267 (1995)

Chang C.P., Y.-T. Lu: Electroabsorption of thin AlAs/GaAs quantum well: Effect of Γ-X valley mixing. Solid State Commun. **89**, 949 (1994)

Chang Y.-C., J.N. Schulman, U. Efron: Electro-optic effect in semiconductor superlattices. J. Appl. Phys. **62**, 4533 (1987)

Chtanov A., T. Baars, M. Gal: Excitation-intensity-dependent photoluminescence in semiconductor quantum wells due to internal electric fields. Phys. Rev. B **53**, 4704 (1996)

Collins R.T., L. Vina, W.I. Wang, L.L. Chang, L. Esaki, K. von Klitzing, K. Ploog: Mixing between heavy-hole and light-hole excitons in GaAs/$Al_xGa_{1-x}As$ quantum wells in an electric field. Phys. Rev. B **36**, 1531 (1987)

Dignam M.M., J.E. Sipe: Exciton Stark ladder in GaAs/$Ga_{1-x}Al_xAs$ superlattices. Phys. Rev. Lett. **64**, 1797 (1990)

Dignam M.M., J.E. Sipe, J. Shah: Coherent excitations in the Stark ladder: Excitonic Bloch oscillations. Phys. Rev. B **49**, 10502 (1994)

Hogg R.A., T.A. Fisher, A.R.K. Willcox, D.M. Whittaker, M.S. Skolnick, D.J. Mowbray, J.P.R. David, A.S. Pabla, G.J. Rees, R. Grey, J. Woodhead, J.L. Sanchez-Rojas, G. Hill, M.A. Pate, P.N. Robson: Piezoelectric-field effects on transition energies, oscillator strengths, and level widths in (111)B-grown (In, Ga)As/GaAs multiple quantum wells. Phys. Rev. B **48**, 8491 (1993)

Hunsperger R.G.: *Integrated Optics*, 4th edn. (Springer, Berlin, Heidelberg 1995)

Kavokin A.V., A.I. Nesvizhskii: Stark effect near the type-I--type-II transition point in semiconductor quantum wells. Phys. Rev. B **49**, 17055 (1994)

Mendez E.E., G. Bastard, L.L. Chang, L. Esaki, H. Morkoc, R. Fisher: Effect of an electric field on the luminescence of GaAs quantum wells. Phys. Rev. B **26**, 7101 (1982)

Miller D.A.B., D.S. Chemla, T.C. Damen, A.C. Gossard, W. Wiegmann, T.H. Wood, C.A. Burrus: Band-edge electroabsorption in quantum well structures: the quantum-confined Stark effect. Phys. Rev. Lett. **53**, 2173 (1984)

Schmeller A., W.Hansen, J.P. Kotthaus, G. Tränke, G. Weimann: Franz-Keldysh effect in a two-dimensional system. Appl. Phys. Lett. **64**, 330 (1994)

Schmidt K.H., N. Linder, G.H. Döhler, H.T. Grahn, K. Ploog, H. Schneider: Coexistence of Wannier-Stark transitions and miniband Franz-Keldysh oscillations in strongly coupled GaAs-AlAs superlattices. Phys. Rev. Lett. **72**, 2769 (1994)

Tanaka I., M. Nakayama, H. Nishimura, K. Kawashima, K. Fujiwara: Electric-field dependence of oscillator strengths of Stark-ladder transitions in a GaAs/AlAs superlattice. Solid State Commun. **92**, 385 (1994)

Zhu B.: Exciton spectra in GaAs/$Ga_{1-x}Al_xAs$ quantum wells in an externally applied electric field. Phys. Rev. B **38**, 13316 (1988)

Magneto-Optics

Ancilotto F., A. Fasolino, J.C. Maan: Effect of band mixing of the hole subbands in quantum wells on the optical transition intensities in a magnetic field. Superlatt. Microstruct. **3**, 187 (1987)

Ancilotto F., A. Fasolino, J.C. Maan: Hole-subband mixing in quantum wells: a magneto-optical study. Phys. Rev. B **38**, 1788 (1988)

Bychkov Yu.A., S.V. Iordanski, G.M. Eliashberg: Two dimensional electrons in a strong magnetic field. Zh. Exper. Teoz. Fiz. Pisma **33**, 152 (1981) [JETP Lett. **33**, 143 (1981)]

Kallin C., B.I. Halperin: Excitation from a filled Landau level in the two dimensional electron gas. Phys. Rev. B **30**, 5655 (1984)

Maan J.C., G. Belle, A. Fasolino, M. Altarelli, K. Ploog: Magneto-optical determination of exciton binding energy in GaAs/$Ga_{1-x}Al_xAs$ quantum wells. Phys. Rev. B **30**, 2253 (1984)

Oettinger K., Al.L. Efros, B.K. Meyer, C. Woelk, H. Brugger: Diamagnetic shift as a measure of the penetration of a quasi-two-dimensional exciton into quantum-well barriers. Phys. Rev. B **52**, R5531 (1995)

Truchseß M. von, V. Latussek, F. Goschenhofer, C.R. Becker, G. Landwehr, E. Batke: Magneto-optics and valence-band discontinuity in a HgTe-$Hg_{1-x}Cd_xTe$ superlattice. Phys. Rev. B **51**, 17618 (1995)

Yang S.-R.E., L.J. Sham: Theory of magnetoexcitons in quantum wells. Phys. Rev. Lett. **58**, 2598 (1987)

Yasui T., Y. Segawa, Y. Aoyagi, Y. Iimura, G.E.W. Bauer, I. Mogi, G. Kido: Exciton states in two-dimensional systems of GaAs/AlAs multi-quantum-well structures under high magnetic fields. Phys. Rev. B **51**, 9813 (1995)

Chapter 7

7.1 N.F. Gashimzade, E.L. Ivchenko: Fiz. Tekh. Poluprovodn. **25**, 323 (1991) [Sov. Phys. – Semicond. **25**, 195 (1991)]

7.2 T. Duffield, B. Bhat, M. Koza, F. de Rosa, D.M. Hwang, P. Grable, S.J. Allen Jr.: Phys. Rev. Lett. **56**, 2724 (1986)

7.3 B.F. Levine, R.J. Malik, J. Walker, K.K. Choi, C.G. Bethea, D.A. Kleinman, J.M. Vandenberg: Appl. Phys. Lett. **50**, 273 (1987)

7.4 C. Hermann, C. Weisbuch: In *Optical Orientation* ed. by F. Meier, B.P. Zakharchenya (North-Holand, Amsterdam 1984) p.463

7.5 M. Dobers, K. von Klitzing, G. Weimann: Solid State Commun. **70**, 41 (1989)

7.6 E.L. Ivchenko, A.A. Kiselev: Fiz. Tekh. Poluprovodn. **26**, 1471 (1992) [Sov. Phys. – Semicond. **26**, 827 (1992)]

7.7 B. Lou, S. Sudharsanan, S. Perkowitz: Phys. Rev. B **38**, 2212 (1988)

7.8 O.K. Kim, W.G. Spitzer: J. Appl. Phys. **50**, 4362 (1979)

Additional Reading

Cyclotron Resonance

Bass F., V.A. Lykakh, A.P. Tetervov: Cyclotron resonance in a semiconductor with superlattice. Fiz. Tekh. Poluprovodn. **14**, 2314 (1980) [Sov. Phys. – Semicond. **14**, 1372 (1980)]

Bluyssen H., J.C. Maan, P. Wyder, L.L. Chang, L. Esaki: Cyclotron resonance in an InAs-GaSb superlattice. Solid State Commun. **31**, 35 (1979)

Schlesinger Z., S.J. Allen, J.C.M. Hwang, P.M. Platzman, N. Tzoar: Cyclotron resonance in two dimensions. Phys. Rev. B **30**, 435 (1984)

Song S.-H., D.C. Tsui, F.F. Fang: Cyclotron mass of two-dimensional holes in strained $Si/Si_{0.88}Ge_{0.12}/Si$ heterostructures. Solid State Commun. **96**, 61 (1995)

Winkler R., M. Merkler, T. Darnhofer, U. Rössler: Theory for the cyclotron resonance of holes in strained asymmetric Ge-SiGe quantum wells. Phys. Rev. B **53**, 10858 (1996)

Wixforth A., M. Kaloudis, C. Rocke, K. Enslin, M. Sundaram, J.H. English, A.C. Gossard: Dynamic response of parabolically confined electron systems. Semicond. Sci. Technol. **9**, 215 (1994)

Wong S.L., D. Kinder, R.J. Nicholas, T.E. Whall, R. Kubiak: Cyclotron-resonance measurements on *p*-type strained-layer $Si_{1-x}Ge_x/Si$ heterostructures. Phys. Rev. B **51**, 13499 (1995)

Yuh P., K.L. Wang: Intersubband optical absorption in coupled quantum wells under an applied electric field. Phys. Rev. B **38**, 8377 (1988)

Electron Spin Resonance, g-Factor

Chen Y.-F., M. Dobrowolska, J.K. Furdyna: *g*-factor anisotropy of conduction electrons in InSb. Phys. Rev. B **31**, 7989 (1985)

Dobers M., K. von Klitzing, G. Weimann: Electron-spin resonance in the two-dimensional electron gas of $GaAs-Al_xGa_{1-x}As$ heterostructures. Phys. Rev. B **38**, 5453 (1988)

Gudmundsson V., J.J. Palacios: Spin effects in a confined two-dimensional electron gas: Enhancement of the *g*-factor, spin-inversion states, and their far-infrared absorption. Phys. Rev. B **52**, 11266 (1995)

Hannak R.M., M. Oestreich, A.P. Heberle, W.W. Rühle, K. Köhler: Electron *g* factor in quantum wells determined by spin quantum beats. Solid State Commun. **93**, 313 (1995)

Heberle A.P., W.W. Rühle, K. Ploog: Quantum beats of electron Larmor precession in GaAs wells. Phys. Rev. Lett. **72**, 3887 (1994)

Hermann C., C. Weisbuch: Optical detection of conduction electron spin resonance in semiconductors and its application to **kp** perturbation theory. In *Optical Orientation*, ed. by F. Meier, B.P. Zakharchenya (North-Holland, Amsterdam 1984) p.463

Lommer F., F. Malcher, U. Rössler: Reduced **g**-factor of subband Landau levels in AlGaAs/GaAs heterostructures. Phys. Rev. B **32**, 6965 (1985)

Ogg N.R.: Conduction-band *g* factor anisotropy in indium antimonide. Proc. Phys. Soc. **89**, 431 (1966)

Smith III T.P., F.F. Fang: *g*-Factor of electrons in an InAs quantum well. Phys. Rev. B **35**, 7729 (1987)

Snelling M.J., E. Blackwood, C.J. McDonagh, R.T. Harley: Exciton, heavy-hole and electron *g* factors in type-I GaAs/$Al_xGa_{1-x}As$ quantum wells. Phys. Rev. B **45**, 3922 (1992)

Snelling M.J., G.P. Flinn, A.S. Plaut, R.T. Harley, A.C. Tropper, R. Eccleston, C.C. Phillips: Magnetic *g* factor of electrons in GaAs/$Al_xGa_{1-x}As$ quantum wells. Phys. Rev. B **44**, 11345 (1991)

Intersubband Transitions and IR Spectroscopy

Ahn D., S.L. Chuang: Intersubband optical absorption in a quantum well with an applied electric field. Phys. Rev. B **38**, 4149 (1987)

Berezhkovskii A.M., R.A. Suris: Absorption of electromagnetic radiation by carriers in semiconductors with a superlattice in the transverse magnetic field. Fiz. Tekh. Poluprovodn. **18**, 1224 (1984) [Sov. Phys. – Semicond. **18**, 764 (1984)]

Brown L.D.L., M. Jaroc, D.C. Herbert: Large intersubband infrared transitions in GaAs-$Ga_{1-x}Al_xAs$ superlattices. Phys. Rev. B **40**, 1616 (1989)

Golub L.E., E.L. Ivchenko, R.Ya. Rasulov: Intersubband absorption of light in quantum wells for semiconductors with the complicated band structure. Fiz. Tekh. Poluprovodn. **29**, 1093 (1995) [Semiconductors **29**, 566 (1995)]

Ikonic Z., V. Milanovic, D. Tjapkin: On the linewidths of intersubband transitions in GaAs-$Al_xGa_{1-x}As$ quantum wells in electric field. Solid State Commun. **72**, 835 (1989)

Kono J., S.T. Lee, M.S. Salib, G.S. Herold, A. Petrou, B.D. McCombe: Optically detected far-infrared resonances in doped GaAs quatum wells. Phys. Rev. B **52**, 8654 (1995)

Levine B.F., K.K. Choi, C.G. Bethea, J. Walker, R.J. Malik: New 10 μm infrared detector using intersubband absorption in resonant tunneling GaAlAs superlattices. Appl. Phys. Lett. **50**, 1092 (1987)

Miyatake T., S. Horihata, T. Ezaki, H. Kubo, N. Mori, K. Taniguichi, C. Hamaguchi: GaAs/AlGaAs quantum well infrared photodetectors. Solid State Electron. **37**, 1187 (1994)

Piro O.E.: Anisotropy and infrared response of the GaAs-AlAs superlattice. Phys. Rev. B **36**, 3427 (1987)

Sengers A.J., L. Tsang, K.J. Kuhn: Optical properties due to intersubband transitions in n-type quantum wells including the effects of the exchange interaction. Phys. Rev. B **48**, 15116 (1993)

Warburton R.J., C. Gauer, A. Wixforth, J.P. Kotthaus, B. Brar, H. Kroemer: Intersubband resonances in InAs/AlSb quantum wells: Selection rules, matrix elements, and the depolarization field. Phys. Rev. B **53**, 7903 (1996)

West L.E., S.J. Eglash: First observation of an extremely large-dipole infrared transition within the conduction band of a GaAs quantum well. Appl. Phys. Lett. **46**, 1156 (1985)

Xu W., Y. Fu, M. Willander, S.C. Shen: Theory of normal-incidence absorption for the intersubband transition in n-type indirect-gap semiconductor quantum wells. Phys. Rev. B **49**, 13760 (1994)

Yuh P., K.L. Wang: Intersubband optical absorption in coupled quantum wells under an applied electric field. Phys. Rev. B **38**, 8377 (1988)

Zanier S., J.M. Berroir, Y. Guldner, J.P. Vieren, I. Sagnes, F. Glowacki, Y. Campidelli, P.A. Badoz: Free-carrier and intersubband infrared absorption in *p*-type $Si_{1-x}Ge_x/Si$ multiple quantum wells. Phys. Rev. B **51**, 14311 (1995)

Chapter 8

8.1 A. Mooradian: In *Light Scattering Spectra of Solids*, ed. by G.B. Wright (Springer, Berlin, Heidelberg 1969) p.285

8.2 B.H. Bairamov, V.A. Voitenko, I.P. Ipatova, A.V. Subashiev, V.V. Toropov, E. Yane: Fiz. Tverd. Tela **28**, 754 (1986) [Sov. Phys. – Solid State **28**, 420 (1986)]

8.3 G. Abstreiter, M. Cardona, A. Pinczuk: In *Light Scattering in Solids IV*, ed. by M. Cardona, G. Güntherodt, Topics Appl. Phys., Vol.54 (Springer, Berlin, Heidelberg 1984) p.5

8.4 A. Pinczuk, S. Schmitt-Rink, G. Danan, J.P. Valladares, L.N. Pfeiffer, K.W. West: Phys. Rev. Lett. **63**, 1633 (1989)

8.5 C. Colvard, T.A. Gant, M.V. Klein, R. Merlin, R. Fischer, H. Morkoc, A.C. Gossard: Phys. Rev. B **31**, 2080 (1985)

8.6 A.K. Sood, J. Menendez, M. Cardona, K. Ploog: Phys. Rev. Lett. **54**, 211 (1985); ibid. **54**, 2115 (1985)

8.7 V.F. Sapega, M. Cardona, K. Ploog, E.L. Ivchenko, D.N. Mirlin: Phys. Rev. B **45**, 4320 (1992)

8.8 D.N. Mirlin, A.A. Sirenko: Fiz. Tverd. Tela **34**, 205 (1992) [Sov. Phys. – Solid State **34**, 108 (1992)]

8.9 E.L. Ivchenko: Fiz. Tverd. Tela **34**, 476 (1992)[Sov. Phys. – Solid State **34**, 254 (1992)]

8.10 V.F. Sapega, T. Ruf, M. Cardona, K. Ploog, E.L. Ivchenko, D.N. Mirlin: *Proc. 20th Int'l Conf. on Raman Spectroscopy*, ed. by W. Kiefer, M. Cardona, G. Schaack, F.W. Schneider, M.W. Schroetter (Wiley, New York 1992) p.856

Additional Reading

Scattering in Bulk Semiconductors

Cantarero A,, C. Trallero-Giner, M. Cardona: Excitons in one-phonon resonant Raman scattering: Deformation-potential interaction. Phys. Rev. B **39**, 8388 (1989); Excitons in one-phonon resonant Raman scattering: Fröhlich and interface effects. Phys. Rev. B **40**, 12290 (1989)

Garcia-Cristobal A., A. Cantarero, G. Trallero-Giner, M. Cardona: Excitonic model for second-order resonant Raman scattering. Phys. Rev. B **49**, 13430 (1994)

Hamilton D.C., A.L. McWhorter: Raman scattering from spin-density fluctuations in *n*-GaAs. In *Light Scattering Spectra of Solids*, ed. by G.B. Wright (Springer, Berlin, Heidelberg 1969) p.309

Klein M.V.: Electronic Raman Scattering. In *Light Scattering in Solids I*, 2nd edn., ed. by M. Cardona, Topics Appl. Phys., Vol.8 (Springer, Berlin, Heidelberg 1983) p.147

Mooradian A., A.L. McWhorter: Light scattering from plasmons and phonons in GaAs. In *Light Scattering Spectra of Solids*, ed. by G.B. Wright (Springer, Berlin, Heidelberg 1969) p.297

Pinczuk A., E. Burstein: Fundamentals of inelastic light scattering in semiconductors and insulators. In *Light Scattering in Solids I*, 2nd edn., ed. by M. Cardona, Topics Appl. Phys., Vol.8 (Springer, Berlin, Heidelberg 1975) p.23

Zemski V.I., E.L. Ivchenko, D.N. Mirlin, I.I. Reshina: Spatial dispersion and damping of plasmon-phonon modes in semiconductors. *Proc. 1st Joint USA-USSR Symp. Theory of Light Scattering in Condensed Matter* (Plenum, Moscow 1975) p.341; Dispersion of plasmon-phonon modes in semiconductors: Raman scattering and infrared spectra. Solid State Commun. **16**, 221 (1975)

Scattering by Free Carriers in Heterostructures

Abstreiter G., R. Merlin, A. Pinczuk: Inelastic light scattering by electronic excitations in semiconductor heterostructures. IEEE J. QE-**22**, 1771 (1986)

Bajema K., R. Merlin, F.-Y. Juang, S.-C. Hong, J. Singh, P.K. Bhattacharya: Stark effect in GaAs-$Al_xGa_{1-x}As$ quantum wells: light scattering by intersubband transitions. Phys. Rev. B **36**, 1300 (1987)

Burstein E., A. Pinczuk, D.L. Mills: Inelastic light scattering by charge carrier excitations in two-dimensional plasmas: Theoretical considerations. Surf. Sci. **98**, 451 (1980)

Katayama S., T. Ando: Light scattering by electronic excitations in *n*-type GaAs-$Al_xGa_{1-x}As$ superlattices. J. Phys. Soc. Jpn. **54**, 1615 (1985)

Menéndez J., A. Pinczuk, A.C. Gossard, M.G. Lamont, F. Cerdeira: Light scattering in GaAs parabolic quantum wells. Solid State Commun. **61**, 601 (1987)

Pinczuk A., G. Abstreiter: Spectroscopy of free carrier excitations in semiconductor quantum wells. In *Light Scattering of Solids V*, ed. by M. Cardona, G. Güntherodt, Topics Appl. Phys., Vol.66 (Springer, Berlin, Heidelberg 1989) p.153

Pinczuk A., S. Schmitt-Rink, G. Danan, J.P. Valladares, L.N. Pfeiffer, K.W. West: Large exchange interactions in the electron gas of GaAs quantum wells. Phys. Rev. Lett. **63**, 1633 (1989)

Pinczuk A., H.L. Störmer, R. Dingle, J.M. Worlock, W. Wiegmann, A.C. Gossard: Observation of intersubband excitations in a multilayer two dimensional electron gas. Solid State Commun. **32**, 1001 (1979)

Shanabrook B.V., T. Comas, T.A. Perry, R. Merlin: Raman scattering from electrons bound to shallow donors in GaAs-$Al_xGa_{1-x}As$ quantum-well structures. Phys. Rev. B **29**, 7096 (1984)

Scattering by Lattice Vibrations in Heterostructures

Chamberlain M.P: Theory of Raman scattering from interface phonons in GaAs/AlAs superlattices. Phys. Rev. B **48**, 14356 (1993)

Colvard C., R. Merlin, M.V. Klein, A.C. Gossard: Observation of folded acoustic phonons in a semiconductor superlattice. Phys. Rev. Lett. **45**, 298 (1980)

Fainstein A., P. Etchegoin, M.P. Chamberlain, M. Cardona, K. Tötemeyer, K. Eberl: Selection rules and dispersion of GaAs/AlAs multiple-quantum-well optical phonons studied by Raman scattering in right-angle, forward, and backscattering in-plane geometries. Phys. Rev. B **51**, 14448 (1995)

Gammon D., R. Merlin, H. Morkoc: Magnetic-field-enhanced Raman scattering by confined and interface phonons in semiconductor superlattices. Phys. Rev. B **35**, 2552 (1987)

Huang K., B. Zhu, H. Tang: Microscopic theory of optic-phonon Raman scattering in quantum-well systems. Phys. Rev. B **41**, 5825 (1990)

Jusserand B., M. Cardona: Raman spectroscopy of vibrations in superlattices. In *Light Scattering of Solids V*, ed. by M. Cardona, G. Güntherodt, Topics Appl. Phys., Vol.66 (Springer, Berlin, Heidelberg 1989) p.49

Jusserand B., D. Paquet, A. Regreny: "Folded" optical phonons in GaAs/$Ga_{1-x}Al_xAs$ superlattices. Phys. Rev. B **30**, 6245 (1984)

Kauschke W., A.K. Sood, M. Cardona, K. Ploog: Resonance Raman scattering in GaAs-$Al_xGa_{1-x}As$ superlattices: Impurity-induced Frölich-interaction scattering. Phys. Rev. B **36**, 1612 (1987)

Meynadier M.H., E. Finkman, M.D. Sturge, J.M. Worlock, M.C. Tamargo: High-order resonant Raman scattering by combinations and overtones of interface phonons in GaAs-AlAs short-period superlattices. Phys. Rev. B **35**, 2517 (1987)

Mirlin D.N., A.A. Sirenko, R. Planel: Resonance Raman scattering by interface phonons in the magnetic field in GaAs/AlAs superlattices. Solid State Commun. **91**, 545 (1994)

Mirlin D.N., I.A. Merkulov, V.I. Perel, I.I. Reshina, A.A. Sirenko, R. Planel: Magnetic field enhancement of Raman scattering on folded acoustic phonons in GaAs-AlGaAs superlattice, Solid State Commun. **82**, 305 (1992); Temperature dependence of Raman scattering by acoustic phonons in a superlattice GaAs/AlGaAs in strong magnetic fields. Solid State Commun. **84**, 1093 (1992)

Popovic Z.V., M. Cardona, E. Richter, D. Strauch, L. Tapfer, K. Ploog: Raman scattering of $(GaAs)_{n_1}/(AlAs)_{n_2}$ superlattices grown along the [110] direction. Phys. Rev. B **40**, 3040 (1989); Phonons in GaAs/AlAs superlattices grown along the [111] direction. Phys. Rev. B **41**, 5904 (1990)

Popovic Z.V., M.B. Vukmirovic, Y.S. Raptis, E. Anastassakis, R. Nötzel, K. Ploog: Folded phonons from lateral periodicity in (311) GaAs/AlAs corrugated superlattices. Phys. Rev. B **52**, 5789 (1995)

Ruf T., V.I. Belitsky, J. Spitzer, V.F. Sapega, M. Cardona, K. Ploog: Raman scattering from folded phonon dispersion gaps. Phys. Rev. Lett. **71**, 3035 (1993)

Ruf T., J. Spitzer, V.F. Sapega, V.I. Belitsky, M. Cardona, K. Ploog: Interface roughness and homogeneous linewidths in quantum wells and superlattices studied by resonant acoustic-phonon Raman scattering. Phys. Rev. B **50**, 1792 (1994)

Sapega V.F., V.I. Belitsky, T. Ruf, H.D. Fuchs, M. Cardona, K. Ploog: Secondary emission and acoustic-phonon scattering induced by strong magnetic fields in multiple quantum wells. Phys. Rev. B **46**, 16005 (1992)

Shields A.J., M. Cardona, K. Eberl: Resonant Raman line shape of optic phonons in GaAs/AlAs multiple quantum wells. Phys. Rev. Lett. **72**, 412 (1994)

Shields A.J., M.P. Chamberlain, M. Cardona, K. Eberl: Raman scattering due to interface optical phonons in GaAs/AlAs multiple quantum wells. Phys. Rev. B **51**, 17728 (1995)

Silva da S.W., Yu.A. Pusep, J.C. Galzerani, M.A. Pimenta, D.I. Lubyshev, P.P. Gonzalez Borrero, P. Basmaji: Raman study of interface roughness in $(GaAs)_n/(AlAs)_n$ superlattices grown on tilted surfaces: Evidence of corrugation of the (113) interface. Phys. Rev. B **53**, 1927 (1996)

Viña L., J.M. Calleja, A. Cros, A. Cantarero, T. Berendschot, J.A.A.J. Perenboom, K. Ploog: Role of excitons in double Raman resonances in GaAs quantum wells. Phys. Rev. B **53**, 3975 (1996)

Yakovlev D.R., W. Ossau, A. Waag, G. Landwehr, E.L. Ivchenko: Double 2s-1s resonance in LO-assisted secondary emission of quantum-well structures. Phys. Rev. B **52**, 5773 (1995)

Zucker J.E., A. Pinczuk, D.S. Chemla, A.C. Gossard, W. Wiegmann: Raman scattering resonant with quasi-two-dimensional excitons in semiconductor quantum wells. Phys. Rev. Lett. **51**, 1293 (1983); Optical vibrational modes and electron-phonon interaction in GaAs quantum wells. Phys. Rev. Lett. **53**, 1280 (1984)

Zucker J.E., A. Pinczuk, D.S. Chemla, A.C. Gossard: Resonant Raman study of low-temperature exciton localization in GaAs quantum wells. Phys. Rev. B **35**, 2892 (1987); also Surf. Sci. **196**, 563 (1988)

Spin-Flip Scattering

Geschwind S., R. Romestain: High resolution spin-flip Raman scattering in CdS. In *Light Scattering in Solids IV*, ed. by M. Cardona, G. Güntherodt, Topics Appl. Phys., Vol.54 (Springer, Berlin, Heidelberg 1984) p.151

Gubarev S.I., T. Ruf, M. Cardona: Resonant spin-flip Raman scattering on photoexcited carriers in *p*-type $Cd_{0.95}Mn_{0.05}Te$ crystals. Phys. Rev. B **43**, 14564 (1991)

Meyer R., N. Hirsch, G. Schaack, A. Waag, R.-N. Bicknell-Tassius: Resonant spin-flip Raman scattering of excitons and magnetic polarons in $CdTe/Cd_{1-x}Mn_xTe$ quantum-well-structures. Superlatt. Microstruct. **9**, 165 (1991)

Sapega V.F., T. Ruf, M. Cardona, K. Ploog, E.L. Ivchenko, D.N. Mirlin: Resonant Raman scattering due to bound-carrier spin flip in $GaAs/Al_xGa_{1-x}As$ quantum wells. Phys. Rev. B **50**, 2510 (1994)

Scott J.F.: Spin-flip Raman scattering in *p*-type semiconductors. Rep. Prog. Phys. **43**, 951 (1980)

Sirenko A.A., T. Ruf, N.N. Ledentsov, A.Yu. Egorov, P.S. Kop'ev, V.M. Ustinov, A.E. Zhukov: Resonant spin-flip Raman scattering and localized exciton luminescence in submonolayer InAs-GaAs structures. Solid State Commun. **97**, 169 (1996)

Thomas D.G., J.J. Hopfield: Spin-flip Raman scattering in cadmium sulfide. Phys. Rev. **175**, 1021 (1968)

Chapter 9

9.1 P.D. Altukhov, A.V. Ivanov, Yu.N. Lomasov, A.A. Rogachev: Zh. Exper. Teor. Fiz. Pisma: **38**, 5 (1983) [JETP Lett. **38**, 4 (1983)]

9.2 I.V. Kukushkin, V.B. Timofeev: Zh. Exper. Teor. Fiz. Pis'ma **40**, 413 (1984) [JETP Lett. **40**, 1231 (1984)]

9.3 I.V. Kukushkin, S.V. Meshkov, V.B. Timofeev: Ups. Fiz. Nauk **155**, 219 (1988) [Sov. Phys. – Usp. **31**, 511 (1988)]

9.4 I.V. Kukushkin, V.B. Timofeev, K. von Klitzing, K. Ploog: *Festkörperprobleme / Adv. Solid State Phys.* **28**, 21 (Viewg, Braunschweig 1988)

9.5 G.L. Bir, G.E. Pikus: *Symmetry and Strain-Induced Effects in Semiconductors* (Wiley, New York 1974)

9.6 I.V. Kukushkin, V.B. Timofeev: Zh. Exper. Teor. Fiz. Pis'ma **44**, 179 (1986) [JETP Lett. **44**, 228 (1986)]

9.7 D. Heiman, B.B. Goldberg, A. Pinczuk, C.W. Tu, A.C. Gossard, J.H. English: Phys. Rev. Lett. **61**, 605, (1988)

9.8 A.J. Turberfield, S.R. Haynes, P.A. Wright, R.A. Ford, R.G. Clark, J.F. Ryan, J.J. Harris, C.T. Foxon: Phys. Rev. Lett. **65**, 637 (1990)
9.9 B.B. Goldberg, D. Heiman, A. Pinczuk, L. Pfeiffer, K. West: Phys. Rev. Lett. **65**, 641 (1990)
9.10 V.M. Apal'kov, E.I. Rashba: Zh. Exper. Teor. Fiz. Pis'ma **53**, 420 (1991) [JETP Lett. **53**, 442 (1991)]
9.11 H. Buhmann, W. Joss, K. von Klitzing, I.V. Kukushkin, A.S. Plaut, G. Martinez, K. Ploog, V.B. Timofeev: Phys. Rev. Lett. **66**, 926 (1991)
9.12 N.S. Averkiev, G.E. Pikus, M.L. Shmatov: Fiz. Tverd. Tela **30**, 3276 (1988) [Sov. Phys. – Solid State **30**, 1844 (1988)]
9.13 P.D. Altukov, A.A. Bakun, A.V. Krutitskii, G.P. Rubtsov: Zh. Tekh. Fiz. Pis'ma **15**, 17 (1989) [Sov. Tekh. Phys. Lett. **15**, 127 (1989)]
9.14 B.M. Apal'kov, E.I. Rashba: Zh. Exper. Theor. Fiz. Pis'ma **54**, 160 (1991) [JETP Lett. **54**, 155 (1991)]
9.15 H.W. Jiang, R.L. Willett, H.L. Stormer, D.C. Tsui, L.N. Pfeiffer, K.W. West: Phys. Rev. Lett. **65**, 633 (1990)
9.16 M.G.W. Alexander, M. Nido, W.W. Rühle, K. Köhler: Phys. Rev. B **41**, 12295 (1990)
9.17 M.G.W. Alexander, W.W. Rühle, R. Sauer, W.T. Tsang: Appl. Phys. Lett. **55**, 885 (1989)
9.18 K. Leo, E.O. Göbel, T.C. Damen, J. Shah, S. Schmitt-Rink, W. Schäfer, J.F. Müller, K. Köhler, P. Ganser: *Proc. 20th Int'l Conf. Phys. Semicond.* (Thessaloniki, Greece 1990), ed. by E.M. Anastassakis, J.D. Joannopoulos (World Scientific, Singapore 1990) p.1481
9.19 B.F. Feuervbacher, J. Kuhl, R. Eccleston, K. Ploog: *Proc. 20th Int'l Conf. Phys. Semicond.* (Thessaloniki, Greece 1990), ed. by E.M. Anastassakis, J.D. Joannopoulos (World Scientific, Singapore 1990) p.1485
9.20 K. Leo, T.G. Damen, J. Shah, E.O. Göbel, K. Köhler: Appl. Phys. Lett. **57**, 19 (1990)
9.21 W.A.J.A. van der Poel, A.L.G.J. Severens, C.T. Foxon: Optics Commun. **76**, 116 (1990)
9.22 M.I. Dyakonov, V.I. Perel': Zh. Exper. Theor. Fiz. **60**, 1954 (1971) [Sov. Phys. – JETP **33**, 1053 (1971)]
9.23 V.D. Dymnikov, M.I. Dyakonov, V.I. Perel': Zh. Exper. Theor. Fiz. **71**, 2373 (1976) [Sov. Phys. – JETP **44**, 1252 (1976)]
9.24 G. Lampel: Phys. Rev. Lett. **20**, 491 (1968)
9.25 R.R. Parsons: Phys. Rev. Lett. **23**, 1152 (1969)
9.26 I.Ya. Karlik, D.N. Mirlin, L.P. Nikitin, D.G. Polyakov, V.F. Sapega: Zh. Exper. Theor. Fiz. Pis'ma **36**, 155 (1982) [JETP Lett. **36**, 192 (1982)]
9.27 M.I. Dyakonov, V.I. Perel: *Optical Orientation*, ed. by F. Meier, B.P. Zakharchenya (North-Holland, Amsterdam 1984) p.11
9.28 D.N. Mirlin: *Optical Orientation*, ed. by F. Meier, B.P. Zakharchenya (North-Holland, Amsterdam 1984) p.133
9.29 V.G. Fleisher, I.A. Merkulov: *Optical Orientation*, ed. by F. Meier, B.P. Zakharchenya (North-Holand, Amsterdam 1984) p.173
9.30 G.E. Pikus, A.N. Titkov: Applications of the Hanle effects in solid state physics. In *The Hanle Effect and Level-Crossing Spectroscopy*, ed. by G. Morussi, F. Strumia (Plenum, New York 1991) p.283
9.31 A. Twardowski, G. Herrmann: Phys. Rev. B **35**, 8144 (1987)

9.32 I.A. Merkulov, V.I. Perel', M.E. Portnoi: Zh. Exper. Theor. Fiz. **99**, 1202 (1991) [Sov. Phys. – JETP **72**, 669 (1991)]
9.33 C. Weisbuch, R.C. Miller, R. Dingle, A.C. Gossard, W. Wiegmann: Solid State Commun. **37**, 219 (1981)
9.34 E.L. Ivchenko, P.S. Kop'ev, V.P. Kochereshko, I.N. Uraltsev, D.R. Yakovlev: Zh. Exper. Theor. Fiz. Pis'ma **47**, 407 (1988) [JETP Lett. **4**, 486 (1988)]
9.35 G.L. Bir, G.E. Pikus: In *Proc. 7 Int'l Conf. Phys. Semicond.* (Dunod, Paris 1964) p.789
9.36 M.I. D'yakonov, V.I. Perel: Zh. Exp. Teor. Fiz. **60**, 1954 (1971) [Sov. Phys. – JETP **33**, 1053 (1971)]
9.37 M.I. D'yakonov, V.I. Perel: Fiz. Tverd. Tela **13**, 3581 (1971) [Sov. Phys. – Solid State **13**, 3023 (1972)]
9.38 G.L. Bir, A.G. Aronov, G.E. Pikus: Zh. Ekspr. Teor. Fiz. **69**, 1382 (1975) [Sov. Phys. – JETP **42**, 705 (1975)]
9.39 G.E. Pikus, A.N. Titkov: *Optical Orientation*, ed. by F. Meier, B.P. Zakharchenya (North Holland, Amsterdam 1984) p.73
9.40 R. Ferreira, G. Bastard: Phys. Rev. B **49**, 9687 (1991)
9.41 T. Uenoyama, L.J. Sham: Phys. Rev. Lett. **64**, 3070 (1990)
9.42 T. Uenoyama, L.J. Sham: Phys. Rev. B **42**, 7114 (1990)
9.43 E.L. Ivchenko: Fiz. Tverd. Tela **15**, 1566 (1973) [Sov. Phys. – Solid State **15**, 1048 (1973)]
9.44 J. Wagner, H. Schneider, D. Richards, A. Fischer, K. Ploog: Phys. Rev. B **47**, 4786 (1993)
9.45 V.K. Kalevich, V.L. Korenev, O.M. Fedorova: Zh. Exper. Theor. Fiz. Pis'ma **52**, 964 (1990) [JETP Lett. **52**, 349 (1991)]
9.46 M.E. Portnoi: Fiz. Tekh. Poluprovodn. **25**, 2150 (1991) [Sov. Phys. – Semicond. **25**, 1294 (1990)]
9.47 B.P. Zakharchenya, P.S. Kop'ev, D.N. Mirlin, D.G. Polyakov, I.I. Reshina, V.F. Sapega, A.A. Sirenko: Solid State Commun. **66**, 203 (1989)
9.48 P.S. Kop'ev, D.N. Mirlin, D.G. Polyakov, I.I. Reshina, V.F. Sapega, A.A. Sirenko: Fiz. Tekh. Semicond. **24**, 1200 (1990) [Sov. Phys. – Semicond. **24**, 757 (1990)]
9.49 E.F. Gross, A.I. Ekimov, B.S. Razbirin, V.I. Safarov: Zh. Exper. Theor. Fiz. Pis'ma **14**, 108 (1971) [JETP Lett. **14**, 70 (1971)]
9.50 G.L. Bir, G.E. Pikus: Zh. Exper. Theor. Fiz Pis'ma **15**, 730 (1972) [JETP Lett. **15**, 516 (1972)]
9.51 G.L. Bir, G.E. Pikus: *Proc. 11 Int'l Conf. Phys. Semicond.* (Warsaw, Poland 1972) (Elsevier, Amsterdam 1972) p.1341
9.52 A. Bonnot, R. Planel, C. Benoit a la Guillaume: Phys. Rev. B **9**, 690 (1974)
9.53 G.E. Pikus, E.L. Ivchenko: *Excitons*, ed. by E.I. Rashba, M.D. Sturge (North-Holland, Amsterdam 1972) p.205
9.54 S. Permogorov: *Excitons*, ed. by E.I. Rashba, M.D. Sturge (North-Holland, Amsterdam 1972) p.177
9.55 R. Planel, C. Benoit a la Guillaume: *Optical Orientation*, ed. by F. Meier, B.P. Zakharchenya (North-Holland, Amsterdam 1984) p.353
9.56 W.T. Masselink, Y.L. Sun, R. Fischer, T.J. Drummond, Y.C. Chang, M.V. Klein, H. Morkoc: J. Vac. Sci. Technol. B **2**, 117 (1984)
9.57 I.N. Uraltsev, E.L. Ivchenko, P.S. Kop'ev, D.R. Yakovlev: Phys. Status Solidi (b) **150**, 673 (1988)

9.58 H. Stolz, D. Schwarre, W. von der Osten, G. Weimann: Superlatt. Microstruct. **6**, 271 (1989)

9.59 E.L. Ivchenko, V.P. Kochereshko, I.N. Uraltsev, D.R. Yakovlev: In *High Magnetic Fields in Semiconductor Physics III*, ed. by G. Landwehr, Springer Ser. Solid-State Sci., Vol.101 (Springer, Berlin, Heidelberg 1992) p.533

9.60 E.L. Ivchenko, V.P. Kochereshko, A.Yu. Naumov, I.N. Uraltsev, P. Lavallard: Supperlatt. Microstruct. **10**, 497 (1991)

9.61 T. Koda, D.W. Langer: Phys. Rev. Lett. **20**, 50 (1968)

9.62 O. Akimoto, H. Hasegawa: Phys. Rev. Lett. **20**, 916 (1968)

9.63 G.L. Bir, G.E. Pikus, L.G. Suslina, D.L. Fedorov: Fiz. Tverd. Tela **12**, 1187 (1970) [Sov. Phys. – Solid State **12**, 926 (1970)]; Fiz. Tverd. Tela **12**, 3218 (1970) [Sov. Phys. – Solid State **12**, 2602 (1971)]

9.64 C. Gourdon, P. Lavallard, R. Planel: Proc. Int'l Meet. Optics of Excitons in Conf. System (IoP, London 1992) Preprint

9.65 C. Gourdon, P. Lavallard: Phys. Rev. B **46**, 4644 (1992)

9.66 I.L. Aleiner, E.L. Ivchenko: Pis'ma Zh. Exper. Teor. Fiz. **55**, 662 (1992) [JETP Lett. **55**, 692 (1992)]

9.67 E.I. Ivchenko, A.Yu. Kaminskii, I.L. Aleiner: Zh. Exper. Teor. Fiz. **104**, 3401 (1993) [Sov. Phys. – JETP **77**, 609 (1993)]

9.68 M.Z. Maialle, E.A. de Andrada e Silva, L.J. Cham: Phys. Rev. B **47**, 15776 (1993)

9.69 G.E. Pikus, G.L. Bir: Zh. Exp. Teor. Fiz. **60**, 195 (1971) [Sov. Phys. – JETP **33**, 108 (1971)]

9.70 S. Permogorov, A. Naumov, C. Gourdon, P. Lavallard: *Proc. 20th Int'l Conf. Phys. Semicond.* (Thesaloniki, Greece 1990), ed. by E.M. Anastassakis, J.D. Joannopoulos (World Scientific, Singapore 1990) p.1517

9.71 G.E. Pikus, F.G. Pikus: Solid State Commun. **89**, 319 (1994)

9.72 G.E. Pikus, F.G. Pikus: J. Lumin. **54**, 279 (1993)

9.73 P.S. Kop'ev, D.N. Mirlin, V.F. Sapega, A.A. Sirenko: Zh. Exper. Theor. Fiz. Pis'ma **51**, 624 (1990) [JETP Lett. **51**, 708 (1990)]

9.74 I.A. Merkulov, V.I. Perel: J. Lum. **60/61**, 293 (1994)

9.75 H.W. van Kesteren, E.C. Cosman, F.J.A.M. Greindanus, P. Dawson, K.J. Moore, C.T. Foxon: Phys. Rev. Lett. **61**, 129 (1988)

9.76 H.W. van Kesteren, E.C. Cosman, P. Dawson, K.J. Moore, C.T. Foxon: Phys. Rev. B **39**, 13426 (1989)

9.77 H.W. van Kesteren, E.C. Cosman, W.A.J.A. van der Poel, C.T. Foxon: Phys. Rev. B **41**, 5283 (1990)

9.78 J.M. Trombetta, T.A. Kennedy, D. Gammon, V.B. Shanabrook, S.M. Prokes: *Proc. 20th Int'l Conf. Phys. Semicond.* (Thesaloniki, Greece 1990), ed. by E.M. Anastassakis, J.D. Joannopoulos (World Scientific, Singapore 1990) p.1361

9.79 P.S. Kop'ev, V.P. Kochereshko, I.N. Uraltsev, D.R. Yakovlev: Fiz. Tekh. Poluprovodn. **22**, 597 (1988) [Sov. Phys. – Semicond. **22**, 373 (1988)]

9.80 P.S. Kop'ev, V.P. Kochereshko, I.N. Uraltsev, D.R. Yakovlev: *Proc. Int'l Conf. Shallow Impur. in Semicond.* (Linköpling, Sweden 1988). Inst. Phys. Conf. Ser. **95**,39 (IoP, Bristol)

9.81 I.N. Uraltsev, V.P. Kochereshko, V.S. Vikhnin, D.R. Yakovlev: Materials Sci. Forum **65 & 66**, 111 (1990) Proc. Int'l. Conf. Shallow Impur. in Semicond. (London 1990)

Additional Reading

Luminescence Under the Condition of Ordinary and Fractional Quantum Hall Effect and Wigner Crystallization

Apalkov V.M., E.I. Rashba: Interaction of excitons with an incompressible quantum liquid. Phys. Rev. B **46**, 1628 (1992)

Apalkov V.M., E.I. Rashba: Doublet structure of emission spectra from fractional quantum Hall states. Phys. Rev. B **48**, 18312 (1993)

Apalkov V.M., F.G. Pikus, E.I. Rashba: Theory of exciton spectra of incompressible quantum liquids. Phys. Rev. B **52**, 6111 (1995)

Butov L.V., A. Zrenner, M. Shayegan, G. Abstreiter, H.C. Manoharan: Magneto-optics of two-dimensional hole systems in the extreme quantum limit. Phys. Rev. B **49**, 14054 (1994)

Chakraborty T., P. Pietiläinen: *The Quantum Hall Effects*, 2nd edn., Springer Ser. Solid-State Sci., Vol.85 (Springer, Berlin, Heidelberg 1995)

Chen X.M., J.J. Quinn: Anyonic ions, energy bands, and photoluminescence of fractional quantum Hall systems. Phys. Rev. B **50**, 2354 (1994)

Chen X.M., J.J. Quinn: On the interpretation of the photoluminescence spectrum of fractional quantum Hall systems. Solid State Commun. **90**, 303 (1994)

Goldus E.M., S.A. Brown, R.B. Dunford, A.G. Davies, R. Newbury, R.G. Clark, P.E. Simmonds, J.J. Harris, C.T. Foxon: Magneto-optical probe of two-dimensional electron liquid and solid phase. Phys. Rev. B **46**, 7957 (1992)

Hawrylak P., N. Pulsford, K. Ploog: Magneto-optics of acceptor-doped GaAs/$Ga_{1-x}Al_xAs$ heterostructures in the quantum Hall regime: Resonant magnetoexcitons and many-electron effects. Phys. Rev. B **46**, 15193 (1992)

Heiman A., A. Pinczuk, M. Dahl, B.S. Dennis, L.N. Pfeiffer, K.W. West: Time-resolved photoluminescence in the fractional quantum Hall regime. Surf. Sci. **305**, 50 (1994)

Kukushkin I.V.: Magneto-optics in one- and two-dimensions. *Proc. 20th Int'l Conf. Phys. Semicond.* (Thessaloniki, Greece 1990), ed. by E.M. Anastassakis, J.D. Joannopoulos (World Scientific, Singapore 1990) p.1266

Kukushkin I.V., V.I. Fal'ko, R.J. Haug, K. von Klitzing, K. Eberl, K. Tötemayer: Evidence of the triangular lattice of crystallized electrons from time resolved luminescence. Phys. Rev. Lett. **72**, 3594 (1994)

Kukushkin I.V., R.J. Haug, K. von Klitzing, K. Ploog: Hierarchy of the fractional quantum Hall effect states studied by time-resolved magnetoluminescence. Phys. Rev. Lett. **72**, 736 (1994)

Kukushkin I.V., R.J. Haug, K. von Klitzing, K. Eberl, K. Tötemeyer: Influence of the electron-hole spatial separation on the magnetoluminescence measurements of the fractional-quantum-Hall-effect energy gaps. Phys. Rev. B **50**, 11259 (1994)

MacDonald A.H., E.H. Rezayi, D. Keller: Photoluminescence in the fractional quantum Hall regime. Phys. Rev. Lett. **68**, 1939 (1992)

Price R., X. Zhu, P.M. Platzman, S.G. Louie: Freezing of the quantum Hall liquid at $\nu = 1/7$ and 1/9. Phys. Rev. B **48**, 11473 (1993)

Rashba E.I., M.E. Portnoi: Anyon excitons. Phys. Rev. Lett. **70**, 3315 (1994)

Tatarinova T.V., E.I. Rashba, A.L. Efros: Continuum excitons in the fractional-quantum-Hall-effect regime: Steps in the filling-factor dependence of energy. Phys. Rev. B **50**, 17349 (1994)

Turberfield A.J., R.A. Ford, I.N. Harris, J.F. Ryan, C.T. Foxon, J.J. Harris: Incompressible electron liquid states studied by optical spectroscopy. Phys. Rev. B **47**, 4794 (1993)

Zang J., J.L. Birman: Structure of the magnetoexciton and optical properties in fractional quantum Hall systems. Phys. Rev. B **51**, 5574 (1995)

Transient Luminescence, Quantum Beats

Baumberg J.J., D.D. Awschalom, N. Samarth, H. Luo, J.K. Furdyna: Spin beats and dynamic magnetization in quantum structures. Phys. Rev. Lett. **72**, 717 (1994)

Carmel O., H. Shtrikman, T. Bar-Joseph: Quantum-beat spectroscopy of the Zeeman splitting of heavy- and light-hole excitons in GaAs/As_xGa_{1-x}As quantum wells. Phys. Rev. B **48**, 1995 (1993)

Gourdon C., D.Yu. Rodichev, P. Lavallard, G. Bacquet, R. Planel: Anisotropic exciton states in GaAs/AlAs superlattices in zero and non-zero magnetic field. J. de Physique IV-**4**, Supple. JPII, C5-183 (1993)

Heberle A.P., W.W. Rühle, K. Ploog: Quantum beats of electron Larmor precession in GaAs wells. Phys. Rev. Lett. **72**, 3887 (1994)

Leisching P., P.H. Bolivar, W. Beck, Y. Dhaibi, F. Brüggemann, R. Schwedler, H. Kurz, K. Leo, K. Köhler: Bloch oscillations of excitonic wave packets in semiconductor superlattices. Phys. Rev. B **50**, 14389 (1994)

Leo K., J. Shah, J.P. Gordon, T.C. Damen, D.A.B. Miller, C.W. Tu, J.E. Cunningham: Effect of collisions and relaxation on coherent resonant tunneling: hole tunneling in GaAs/Al_xGa_{1-x}As double-quantum-well structures. Phys. Rev. B **42**, 7065 (1990)

Luo S.M.C., S.L. Chuang, P.C.M. Planken, I. Brener, M.C. Nuss: Coherent double-pulse control of quantum beats in a coupled quantum well. Phys. Rev. B **48**, 11043 (1993)

Mayer E.J., G.O. Smith, V. Heuckeroth, J. Kuhl, K. Bott, A. Schulze, T. Meier, S.W. Koch, P. Thomas, R. Hey, K. Ploog: Polarization dependence of beating phenomena at the energetically lowest exciton transition in GaAs quantum wells. Phys. Rev. B **51**, 10909 (1995)

Pantke K.-H., D. Oberhauser, V.G. Lyssenko, J.M. Hwam, G. Weimann: Coherent generation and interference of excitons and biexcitons in GaAs/Al_xGa_{1-x}As quantum wells. Phys. Rev. B **47**, 2413 (1993)

Pantke K.-H., J.M. Hvam: Nonlinear quantum beat spectroscopy in semiconductors. Int' J. Modern Physics B **8**, 73 (1994)

Paul A.E., W. Sha, S. Patkar, A.L. Smirl: Excitation-induced phase shifts of heavy- and light-hole quantum beats in GaAs/Al_xGa_{1-x}As multiple quantum wells. Phys. Rev. B **51**, 4242 (1995)

Rühle W.W., M.G.W. Alexander, M. Nido: Picosecond spectroscopy on tunneling between quantum wells. *Proc. 20th Int'l Conf. Phys. Semicond.* (Thessaloniki, Greece 1990), ed. by E.M. Anastassakis, J.D. Joannopoulos (World Scientific, Singapore 1990) p.1226

Schmitt-Rink S., D. Bennhardt, V. Heuckeroth, P. Thomas, P. Haring, G. Maidorn, H. Bakker, K. Leo, D.-S. Kim, J. Shah, K. Köhler: Polarization dependence of heavy- and light-hole quantum beats. Phys. Rev. B **46**, 10460 (1992)

Shah J.: *Ultrafast Spectroscopy of Semiconductors and Semiconductor Nanostructures*, Springer Ser. Solid-State Sci., Vol.115 (Springer, Berlin, Heidelberg 1996)

Srinivas V., Y.J. Chen, G.E.C. Wood: Spin relaxation of two-dimensional electrons in GaAs quantum wells. Phys. Rev. B **47**, 1090 (1993)

Vinattieri A., J. Shah, T.C. Damen, D.S. Kim, L.N. Pfeiffer, M.Z. Maialle, L.J. Sham: Exciton dynamics in GaAs quantum wells under resonant excitation. Phys. Rev. B **50**, 10868 (1994)

Weckendrup D., S. Saschek, U. Neukirch, J. Gutowski, S.O. Ferreira, H. Sitter: Quantum beats of excitons in $Zn_{1-x}Cd_xSe/ZnSe$ multiple quantum well structures. J. Appl. Phys. **77**, 4145 (1995)

Optical Spin Orientation and Momentum Alignment of Electrons and Holes

Barrett S.E., R. Tycko, L.N. Pfeiffer, K.W. West: Directly detected nuclear magnetic resonance of optically pumped GaAs quantum wells. Phys. Rev. Lett. **72**, 1368 (1994)

Bastard G.: Spin flip relaxation time of quantum-well electrons in a strong magnetic field. Phys. Rev. B **46**, 4253 (1992)

Baylac B., T. Amand, X. Marie, B. Dareys, M. Brousseau, G. Bacquet, V. Thierry-Mieg: Hole spin relaxation in *n*-modulation doped quantum wells. Solid State Commun. **93**, 57 (1995)

Crookman H.C., E.R. Glaser, R.L. Henry, T.A. Kennedy: Optically detected magnetic resonance in zinc-doped indium phosphide under uniaxial stress. Phys. Rev. B **48**, 14157 (1993)

Hackenberg W., H.P. Hughes: Directional resolution of the GaAs heavy-hole band dispersion and photoelectron-momentum orientation from hot-electron luminescence. Phys. Rev. B **49**, 7990 (1994)

Kawazoe T., Y. Masumoto, T. Mishina: Spin-relaxation process of holes in type-II $Al_{0.34}Ga_{0.66}As/AlAs$ multiple quantum wells. Phys. Rev. B **47**, 10452 (1993)

Maruyama T., E.L. Garwin, R. Prepost, G.H. Zapalac: Electron-spin polarization in photoemission from strained GaAs grown on $GaAs_{1-x}P_x$. Phys. Rev. B **46**, 4261 (1992)

Mirlin D.N., V.I. Perel: Hot photoluminescence in quantum-well structures under continuous wave pumping. Semicond. Sci. Technol. **7**, 1221 (1992)

Munoz L., E. Pérez, L. Viña, K. Ploog: Spin relaxation in intrinsic GaAs quantum wells: Influence of excitonic localization. Phys. Rev. B **51**, 4247 (1995)

Roussignol Ph., P. Rolland, R. Ferreira, D. Delalande, G. Bastard, A. Vinattieri, J. Martinez-Pastor, L. Garraresi, M. Colocci, J.F. Palmer, B. Ettiene: Hole polarization and slow hole-spin relaxation in a *n*-doped quantum well structure. Phys. Rev. B **46**, 7292 (1992)

Sapega V.F., M.P. Chamberlain, T. Ruf, M. Cardona, D.N. Mirlin, K. Tötemeyer, A. Fisher, K. Eberl: Optical-phonon emission in GaAs/AlAs multiple-quantum-well structures determined by hot-electron luminescence. Phys. Rev. B **52**, 14144 (1995)

Srivinas V., Y.J. Chen, C.C. Wood: Spin relaxation of two-dimensional electrons in GaAs quantum wells. Phys. Rev. B **47**, 10907 (1993)

Wagner J., H. Schneider, D. Richards, A. Fischer, K. Ploog: Observation of extremely long electron-spin-relaxation times in *p*-type δ-doped $GaAs/Al_xGa_{1-x}As$ double heterostructures. Phys. Rev. B **47**, 4786 (1993)

Warburton R.J., J.G. Michels, R.J. Nicholas, J.J. Harris, C.T. Foxon: Optically detected cyclotron resonance of GaAs quantum wells: effective-mass measurements and offset effects. Phys. Rev. B **46**, 13394 (1992)

Zhitomirskii V.E., V.E. Kirpichev, A.I. Filin, V.B. Timofeev, B.N. Shepel, K. von Klitzing: Optical detection of spin relaxation of 2D electrons during photoexcitation. Pis'ma Zh. Eksp. Teor. Fiz. **58**, 429 (1993) [JETP Lett. **58**, 439 (1993)]

Optical Orientation and Alignment of Excitons, Exchange Splitting of Exciton Levels

Amand T., B. Dareys, B. Baylac, X. Marie, J. Barrau, M. Brousseau, D.J. Dunstan, R. Planel: Exciton formation and hole-spin relaxation in intrinsic quantum wells. Phys. Rev. B **50**, 11624 (1994)

Amand T., X. Marie, D. Robart, P. Le Jeune, M. Brousseau, J. Barrau: Spin repolarization in bidimensional gas of interacting excitons. Solid State Commun. **100**, 445 (1996)

Bacquet G., F. Hassen, M.J. Alonso, M. Garriga: Spin orientation by optical pumping in $In_xGa_{1-x}As/AlAs$ quantum wells. Solid State Commun. **91**, 703 (1994)

Baranov P.G., I.V. Mashkov, N.G. Romanov, P. Lavallard: Optically detected magnetic resonance of excitons and carriers in pseudodirect GaAs/AlAs superlattices. Solid State Commun. **87**, 649 (1993)

Blackwood E., M.J. Snelling, R.T. Harley, S.R. Andrews, C.T.B. Foxon: Exchange interaction of excitons in GaAs heterostructures. Phys. Rev. B **50**, 14246 (1994)

Frommer A., Arza Ron, E. Cohen, J.A. Kash, L.N. Pfeiffer: Dynamics and spin relaxation of excitons in $GaAs/Al_xGa_{1-x}As$ quantum wells. Phys. Rev. B **50**, 11833 (1994)

Frommer A., E. Cohen, Arza Ron, L.N. Pfeiffer: Linear and circular polarizations of exciton luminescence in $GaAs/Al_xGa_{1-x}As$ quantum wells. Phys. Rev. B **48**, 2803 (1993)

Harley R.T., M.J. Snelling: Magnetic-field dependence of exciton spin relaxation in $GaAs/Al_xGa_{1-x}As$ quantum wells. Phys. Rev. B **53**, 9561 (1996)

Hassen F., G. Bacquet, N. Lauret, J. Barrau, L.K. Howard, D.J. Dunstan: Spin orientation by optical pumping in strained $In_xGa_{1-x}As/GaAs$ quantum wells. Solid State Commun. **87**, 889 (1993)

Ivchenko E.L., G.E. Pikus, L.V. Takunov: Alignment and orientation of hot excitons in semiconductors. Fiz. Tverd. Tela **20**, 2598 (1978) [Sov. Phys. – Solid State **20**, 1502 (1978)]

Ivchenko E.L.: Spectroscopy of spin-polarized excitons in semiconductors, Pure & Appl. Chem. **67**, 463 (1995)

Ivchenko E.L., A.Yu. Kaminskii: Optically-detected localized-exciton level anticrossing in semiconductors. Fiz. Tverd. Tela **37**, 1418 (1995) [Phys. Solid State **37**, 768 (1995)]

Jorda S., U. Rössler, D. Broido: Fine structure of excitons and polariton dispersion in quantum wells. Phys. Rev. B **48**, 1669 (1993)

Kochereshko V.P., G.R. Pozina, G.L. Sandler, I.N. Uraltsev: Effect of a magnetic field on polarized luminescence of superlattices. Semicond. Sci. Technol. **10**, 1269 (1995)

Ota K., H. Yaguchi, K. Onabe, R. Ito, Y.Takahashi, K. Muraki, Y. Shiraki: Observation of the valence-subband level crossing in GaAs/GaAsP strained-barrier quantum well structures using circularly polarized photoluminescence excitation spectroscopy. Appl. Phys. Lett. **63**, 946 (1993)

Robart D., T. Amand, X. Marie, M. Brousseau, J. Barrau, G. Bacquet: Exciton-exciton interaction under elliptically polarized light excitation. J. Opt. Soc. Am. B **13**, 1000 (1996)

Wimbauer Th., K. Oettinger, Al.L. Efros, B.K. Meyer, H.Brugger: Zeeman splitting of the excitonic recombination in $In_xGa_{1-x}As/GaAs$ single wells. Phys. Rev. B **50**, 8889 (1994)

Chapter 10

10.1 S.B. Arifzhanov, E.L. Ivchenko: Fiz. Tverd. Tela **17**, 81 (1975) [Sov. Phys. – Solid State **17**, 46 (1975)]

10.2 K. Tai, A. Myzyrovicz, R.J. Fischer, R.E. Sluscher, A.Y. Cho: Phys. Rev. Lett. **62**, 1784 (1989)

10.3 D. Frölich, R. Wille, W. Schlapp, G. Weiman: Phys. Rev. Lett. **61**, 1878 (1988)

10.4 E.L. Ivchenko, V.P. Kochereshko, I.N. Uraltsev, D.R. Yakovlev: Phys. Status Solidi (b) **161**, 217 (1990)

10.5 E.L. Ivchenko, V.P. Kochereshko, I.N. Uraltsev: In *Semiconductors and Insulators: Optical and Spectroscopic Research*, ed. by Yu.I. Koptev, Ioffe Physico-technical Institute Research Series, Vol.12 (Nova Science, New York 1992)

10.6 G. Coopermann, L. Friedman, W.L. Bloss: Appl. Phys. Lett. **44**, 977 (1984)

10.7 A. Shimizu: Phys. Rev. Lett. **61**, 613 (1988)

10.8 M. Bass, P.A. Franken, J.F. Ward: Phys. Rev. A **138**, 534 (1965)

10.9 P. Bois, E. Rosencher, J. Nagle, B. Vinter, E. Martinet, S. Delaitre, E. Costard, P. Boucaud, J.-M. Lourtioz, F.H. Julien, D.D. Yang: *Proc. 20th Int'l Conf. Phys. Semicond.* (Thesaloniki, Greece 1990), ed. by E.M. Anastassakis, J.D. Joannopoulos (World Scientific, Singapore 1990) p.1681

10.10 A.M. Glass, D. von der Linde, T.J. Negran: Appl. Phys. Lett. **25**, 233 (1974)

10.11 A.M. Glass, D. von der Linde, D.H. Auston, T.J. Negran: J. Electron. Mater. **4**, 915 (1975)

10.12 V.I. Belinicher, B.I. Sturman: Usp. Fiz. Nauk **130**, 415 (1980) [Sov. Phys. – Usp. **23**, 199 (1980)]

10.13 E.L. Ivchenko, G.E. Pikus: *Semiconductor Physics*, ed. by V.M. Tuchkevich, V.Ya. Frenkel (Cons. Bureau, New York 1986) p.427

10.14 K.H. Herrmann, R. Vogel: *Proc. 11 Int'l Conf. Phys. Semicond.* (Warsaw, Poland 1972) (Elsevier, Amsterdam 1972) p.870

10.15 C.R. Hammond, J.R. Jenkins, C.R. Stanley: Optoelectronics **4**, 189 (1972)

10.16 A.V. Andrianov, E.L. Ivchenko, G.E. Pikus, R.Ya. Rasulov, I.D. Yarochetskii: Zh. Exper. Teor. Fiz. **81**, 2080 (1981) [Sov. Phys. – JETP **54**, 1105 (1981)]

10.17 E.L. Ivchenko, G.E. Pikus: Zh. Exper. Teor. Fiz. Pis'ma **27**, 640 (1978) [JETP Lett. **27**, 604 (1978)]

10.18 V.I. Belinicher: Phys. Lett. A **66**, 213 (1978)

10.19 V.M. Asnin, A.A. Bakun, A.M. Danishevskii, E.L. Ivchenko, G.E. Pikus, A.A. Rogachev: Zh. Exper. Teor. Fiz. Pis'ma **28**, 80 (1978) [JETP Lett. **28**, 74 (1978)]

10.20 V.M. Asnin, A.A. Bakun, A.M. Danishevskii, E.L. Ivchenko, G.E. Pikus, A.A. Rogachev: Solid State Commun. **30**, 565 (1979)
N.S. Averkiev, V.M. Asnin, A.A. Bakun, A.M. Danishevskii, E.L. Ivchenko, G.E. Pikus, A.A. Rogachev: Fiz. Tekh. Poluprovodn. **18**, 639, 648 (1984) [Sov. Phys. – Semicond. **18**, 397, 402 (1984)]
10.21 V.I. Belinicher, B.I. Sturman: Fiz. Tverd. Tela **20**, 821 (1978) [Sov. Phys. – Solid State **20**, 476 (1978)]
10.22 N. Kristoffel, A. Gulbis: Izv. A.N. Est. SSR **28**, 268 (1979)
10.23 R. von Baltz, W. Kraut: Phys. Lett. A **79**, 364 (1980)
10.24 V.I. Belinicher, E.L. Ivchenko, B.I. Sturman: Zh. Exper. Teor. Fiz. **83**, 649 (1982) [Sov. Phys. – JETP **56**, 359 (1982)]
10.25 E.L. Ivchenko, Yu.B. Lyanda-Geller, G.E. Pikus, R.Ya. Rasulov: Fiz. Tekh. Poluprovodn. **18**, 93 (1984) [Sov. Phys. – Semicond. **18**, 55 (1984)]
10.26 L.I. Magarill, M.V. Entin: Fiz. Tverd. Tela **31** (8), 37 (1989) [Sov. Phys. – Solid. State **31**, 1299 (1989)]
10.27 G.M. Gusev, Z.D. Kvon, L.I. Magarill, A.M. Palkin, V.I. Sozinov, O.A. Shegay, M.V. Entin: Zh. Exper. Teor. Fiz. Pis'ma **46**, 28 (1987) [JETP Lett. **46**, 33 (1987)]
10.28 F. Capasso, S. Luryi, W.T. Tzang, C.G. Bethea, B.F. Levine: Phys. Rev. Lett. **51**, 2318 (1983)
10.29 F.G. Pikus: Fiz. Tekh. Poluprovodn. **22**, 940 (1988) [Sov. Phys. – Semicond. **22**, 594 (1988)]
10.30 Yu.B. Lyanda-Geller, G.E. Pikus: Fiz. Tverd. Tela **31** (12), 77 (1989) [Sov. Phys. – Solid State **31**, 2068 (1989)]
10.31 E.L. Ivchenko, Yu.B. Lyanda-Geller, G.E. Pikus: Zh. Exper. Teor. Fis **98**, 989 (1990) [Sov. Phys. – JETP **71**, 550 (1990)]
10.32 A.G. Aronov, Yu.B. Lyanda-Geller, G.E. Pikus: Zh. Exper. Teor. Fiz. **100**, 973 (1991) [Sov. Phys. – JETP **73**, 537 (1991)]
10.33 H.E.M. Barlow: Proc. IRE **46**, 1411 (1958)
10.34 A.M. Danishevskii, A.A. Kastalskii, S.M. Ryvkin, I.D. Yaroshetskii: Zh. Exper. Teor. Fiz. **58**, 544 (1970) [Sov. Phys. – JETP **31**, 292 (1970)]
10.35 V.I. Perel', Ya.M. Pinskii: Zh. Exper. Teor. Fiz. **54**, 1889 (1968) [Sov. Phys. – JETP **27**, 1014 (1968)]
10.36 E. Normantas, G.E. Pikus: Zh. Exper. Teor. Fiz. **94**, 150 (1988) [Sov. Phys. – JETP **67**, 1169 (1988)]
10.37 A.F. Gibson, M.F. Kimmitt, A.C. Walker: Appl. Phys. Lett. **17**, 75 (1970)
10.38 A.M. Dykhne, V.A. Roslyakov, A.N. Starostin: Dokl. A.N. USSR **254**, 559 (1980) [Sov. Phys. – Dokl. **25**, 741 (1980)]
10.39 S. Luryi: Phys. Rev. Lett. **58**, 2263 (1987)
10.40 A.A. Grinberg, S. Luryi: Phys. Rev. B **38**, 87 (1988)
10.41 A.D. Wieck, H. Sigg, K. Ploog: Phys. Rev. Lett. **64**, 463 (1990)
10.42 M.I. Stockman, L.N. Pandey, T.E. George: Phys. Rev. Lett. **65**, 3433 (1990)

Additional Reading

Schmitt-Rink S., D.S. Chemla, D.A.B. Miller: Linear and nonlinear optical properties of semiconductor quantum wells. Adv. Phys. **38**, 89 (1989)

Two-Photon Absorption, Biexitons, Trions

Bar-Ad S., I. Bar-Joseph, G. Finkelstein, Y. Levinson: Biexcitons in short-pulse optical expertiments in strong magnetic fields in GaAs quantum wells. Phys. Rev. B **50**, 18375 (1994)

Buhmann H., L. Mansouri, J. Wang, P.H. Beton, N. Mori, L. Eaves, M. Henini, M. Potemski: Electron-concentration-dependent quantum-well ce: Evidence for a negatively charged exciton. Phys. Rev. B **51**, 7969 (1995)

Catalano I.M., A. Cingolani, M. Lepore, R. Cingolani, K. Ploog: Polarization dependence of the excitonic two-photon absorption spectra of GaAs/AlGaAs quantum wells. Solid State Commun. **71**, 217 (1989)

Cingolani R., Y. Chen, K. Ploog: Biexiton formation in GaAs/$Al_xGa_{1-x}As$ multiple quantum well: An optical investigation. Phys. Rev. B **38**, 13478 (1988)

Finkelstein G., S. Bar-Ad, O. Carmel, I. Bar-Joseph: Biexcitonic effects in transient nonlinear optical experiments in quantum wells. Phys. Rev. B **47**, 12964 (1993)

Finkelstein G., H. Shtrikman, I. Bar-Joseph: Optical spectroscopy of a two-dimensional gas near the metal-insulator transition. Phys. Rev. Lett. **74**, 976 (1995)

Kheng K., R.T. Cox, Y. Merle d'Aubigné, F. Bassani, K. Saminadayar, S. Tatarenko: Observation of negatively charged excitons X^{-} in semiconductor quantum wells. Phys. Rev. Lett. **71**, 1752 (1993)

Kuroda T., K. Inoue, R. Kuribayashi, F. Minami, I. Suemune, A. Myzyrowich: Observation of biexciton in ZnSe/ZnSSe quantum wells through transient four-wave mixing. Solid State Commun. **98**, 951 (1996)

Mayer E.J., G.O. Smith, V. Heuckeroth, J. Kuhl, K. Bott, A. Schulze, T. Meier, D. Bennhardt, S.W. Koch, P. Thomas, R. Hey, K. Ploog: Evidence of biexcitonic contribution to four-wave mixing in GaAs quantum wells. Phys. Rev. B **50**, 14731 (1994)

Pasquarello A., A. Quattropani: Gauge-invariant two-photon transitions in quantum wells. Phys. Rev. B **38**, 6206 (1988)

Reynolds D.C, K.K. Bajaj, C.E. Stutz, R.L. Jones, W.M. Theis, P.W. Yu, K.R. Evans: Binding energies of biexcitons in $Al_xGa_{1-x}As$/GaAs multiple quantum wells. Phys. Rev. B **40**, 3340 (1989)

Shields A.J., M. Pepper, D.A. Ritchie, M.Y. Simmons, G.A.C. Jones: Quenching of excitonic optical transitions by excess electrons in GaAs quantum wells. Phys. Rev. B **51**, 18049 (1995)

Shields A.J., J.L. Osborne, M.Y. Simmons, M. Pepper, D.A. Ritchie: Magneto-optical spectroscopy of positively charged excitons in GaAs quantum wells. Phys. Rev. B **52**, R5523 (1995)

Spector H.N.: Two-photon absorption in semiconducting quantum-well structures. Phys. Rev. B **35**, 5876 (1987)

Yamada Y., T. Mishina, Y. Masumoto, Y. Kawakami, J. Suda, Shizuo Fujita, Shigeo Fujita: Time-resolved nonlinear luminescence of biexcitons in ZnSe-$Zn_xMg_{1-x}S_ySe_{1-y}$ single quantum wells. Phys. Rev. B **52**, 2289 (1995)

Diffraction from a Light-Induced Spatial Grating

Bakker H.J., K. Leo, J. Shah, K. Köhler: Time-resolved study of dephasing mechanisms of excitons in GaAs/$Al_xGa_{1-x}As$ quantum-well structures. Phys. Rev. B **49**, 8249 (1994)

Carmel O., I. Bar-Joseph: Four-wave-mixing studies of quantum-well excitons in a magnetic field. Phys. Rev. B **47**, 7606 (1993)

Carmel O., H. Shtrikman, I. Bar-Joseph: Quantum-beat spectroscopy of the Zeeman splitting of heavy- and light-hole excitons in GaAs/$Al_xGa_{1-x}As$ quantum wells. Phys. Rev. B **48**, 1955 (1993)

Erland J., I. Balslev: Theory of quantum beat and polarization interference in four-wave mixing. Phys. Rev. A **48**, R1765 (1993)

Erland J., K.-H. Pantke, V. Mizeikis, V.G. Lyssenko, J.M. Hvam: Spectrally resolved four-wave mixing in semiconductors: Influence of inhomogeneous broadening. Phys. Rev. B **50**, 15047 (1994)

Hübner M., J. Kuhl, T. Stroucken, A. Knorr, S.W. Koch, R. Hey, K. Ploog: Collective effects of excitons in multiple-quantum-well Bragg and anti-Bragg structures. Phys. Rev. Lett. **76**, 4199 (1996)

Leo K., M. Wegener, J. Shah, D.S. Chemla, E.O. Göbel,T.C. Damen, S. Schmitt-Rink, W. Schäfer: Effects of coherent polarization interactions on time-resolved degenerate four-wave mixing. Phys. Rev. Lett. **65**, 1340 (1990)

Paul A.E., J.A. Bolger, A.L. Smirl, J.G. Pellegrino: Time-resolved measurements of the polarization state of four-wave mixing signals from GaAs multiple quantum wells. J. Opt. Soc. Am. B **13**, 1016 (1996)

Schmitt-Rink S., S. Mukamel, K. Leo, J. Shah, D.S. Chemla: Stochastic theory of time-resolved four-wave mixing in interacting media. Phys. Rev. A **44**, 2124 (1991)

Schultheis L., A. Honold, J. Kuhl, K. Köhler, C.T. Tu: Optical dephasing of homogeneously broadened two-dimensional exciton transitions in GaAs quantum wells. Phys. Rev. B **34**, 9027 (1986)

Wegener M., D.S. Chemla, S. Schmitt-Rink, W. Schäfer: Line shape of time-resolved four-wave mixing. Phys. Rev. A **42**, 5675 (1990)

Higher-Harmonics Generation

Atanasov R., F. Bassani, V.M. Agranovich: Second-order nonlinear optical susceptibility of asymmetric quantum wells, Phys. Rev. B **50**, 7809 (1994)

Bloss W.L., L. Friedman: Theory of optical mixing by mobile carriers in superlattices. Appl. Phys. Lett. **41**, 1023 (1982)

Khurgin J.: Second-order nonlinear effects in asymmetric quantum-well structures. Phys. Rev. B **38**, 4056 (1988)

Qu X.H., D.J. Bottomley, H. Ruda, A.J. Springthorpe: Second-harmonic generation from GaAs/$Al_{1-x}Ga_xAs$ asymmetric quantum-well structure. Phys. Rev. B **50**, 5703 (1994)

Shaw M.J., K.B. Wong, E. Corbin, M. Jaros: GaAs-AlAs and Si-SiGe quantum well structures for applications in nonlinear optics. Solid-State Electron. **37**, 1303 (1994)

Xu Z., P.M. Fauchet, C.W. Rella, B.A. Richman, H.A. Schwettman, G.W. Wicks: Second-harmonic generation near 4 μm in *p*-type asymmetric GaAs/AlGaAs/AlAs quantum wells. Solid State Commun. **93**, 903 (1995)

Photoreflection and Photoabsorption

Glembocki O.J., B.N. Shanabrook, N. Bottka, W.T. Beard, J. Comes: Photoreflectance characterization of interband transmissions in GaAs/AlGaAs multiple quantum wells and modulation-doped heterojunctions. Appl. Phys. Lett. **46**, 970 (1985)

Hoof C.Van, D.J. Arent, K. Deneffe, J. De Boeck, G. Borghs: Photomodulated absorption spectroscopy on AlGaAs-GaAs heterostructures. J. Appl. Phys. **64**, 4233 (1988)

Liu C.P., J. Lee, E.S. Koteles, B. Elman, K.T. Hsu, G.J. Jan, P.K. Tseng, I.F. Chang: Photoreflectance studies of quantum wells at oblique incidence. Solid State Commun. **76**, 1229 (1990)

Sanders G.D., Y,-C. Chang: Theory of photoabsorption in modulation-doped semiconductor quantum wells. Phys. Rev. B **35**, 1300 (1987)

Shanabrook B.N., O.J. Glembocki, W.T. Beard: Photoreflectance modulation mechanisms in GaAs-$Al_xGa_{1-x}As$ multiple quantum wells. Phys. Rev. B **35**, 2540 (1987)

Shen H., S.H. Pan, Z. Hang, J. Leng, F.H. Pollak, J.M. Woodall, R.N. Sacks: Photorefelctance of GaAs and $Ga_{0.82}Al_{0.18}As$ at elevated temperatures up to 60°C. Appl. Phys. Lett. **53**, 1080 (1988)

Tang Y., B. Wang, D. Jiang, W. Zhuang, J. Liang: Photoreflectance spectroscopy of GaAs doping superlattices. Solid State Commun. **63**, 793 (1987)

Worsley R.S., N.J. Traynor, T. Grevatt, R.T. Harley: Transient linear birefringence in GaAs quantum wells: Magnetic field dependence of coherent exciton spin dynamics. Phys. Rev. Lett. **76**, 3224 (1996)

Photogalvanic Effects

Agranovich V.M., G.C. La Rocca, F. Bassani: Photovoltaic effect in asymmetrical donor-acceptor organic superlattices. Zh. Eksper. Teor. Fiz., Pis'ma **62**, 407 (1995) [JETP Lett. **62**, 418 (1995)]

Liu J.M., L.P. Chen, C.T. Liu, C.A. Burrus, A.Y. Cho: Picosecond response of steady-state photovoltaic effect in an asymmetrical graded semiconductor superlattice structure. Appl. Phys. Lett. **62**, 1179 (1993)

Luo C.P, D.S. Jiang, W.H. Zhuang, F. Li, Y.Z. Li: Photovoltaic effect and its polarity in Si doping superlattices. Appl. Phys. Lett. **63**, 1777 (1993)

Zhu W., Q. Shen, S. Liu: Photovoltage spectroscopy study of Ge_xSi_{1-x}/Si superlattices. Superlatt. Microstruct. **13**, 275 (1993)

Photon Drag Current

Sigg H., S. Graf, M.H. Kwakernaak, B. Margotte, D. Erni, P. Van Son, K.Köhler: Ultrafast photon drag detector for intersubband spectroscopy. Superlatt. Microstruct. **19**, 105 (1996)

Vasko F.T.: Photon drag effect in tunnel-coupled quantum wells. Phys. Rev. B **53**, 9576 (1996)

Subject Index

Springer Series in Solid-State Sciences

Editors: M. Cardona P. Fulde K. von Klitzing H.-J. Queisser

1 **Principles of Magnetic Resonance**
3rd Edition By C. P. Slichter

2 **Introduction to Solid-State Theory**
By O. Madelung

3 **Dynamical Scattering of X-Rays in Crystals** By Z. G. Pinsker

4 **Inelastic Electron Tunneling Spectroscopy**
Editor: T. Wolfram

5 **Fundamentals of Crystal Growth I**
Macroscopic Equilibrium and Transport Concepts
By F. E. Rosenberger

6 **Magnetic Flux Structures in Superconductors** By R. P. Huebener

7 **Green's Functions in Quantum Physics**
2nd Edition
By E. N. Economou

8 **Solitons and Condensed Matter Physics**
Editors: A. R. Bishop and T. Schneider

9 **Photoferroelectrics** By V. M. Fridkin

10 **Phonon Dispersion Relations in Insulators** By H. Bilz and W. Kress

11 **Electron Transport in Compound Semiconductors** By B. R. Nag

12 **The Physics of Elementary Excitations**
By S. Nakajima, Y. Toyozawa, and R. Abe

13 **The Physics of Selenium and Tellurium**
Editors: E. Gerlach and P. Grosse

14 **Magnetic Bubble Technology** 2nd Edition
By A. H. Eschenfelder

15 **Modern Crystallography I**
Fundamentals of Crystals
Symmetry, and Methods of Structural Crystallography
2nd Edition
By B. K. Vainshtein

16 **Organic Molecular Crystals**
Their Electronic States By E. A. Silinsh

17 **The Theory of Magnetism I**
Statics and Dynamics
By D. C. Mattis

18 **Relaxation of Elementary Excitations**
Editors: R. Kubo and E. Hanamura

19 **Solitons** Mathematical Methods for Physicists
By. G. Eilenberger

20 **Theory of Nonlinear Lattices**
2nd Edition By M. Toda

21 **Modern Crystallography II**
Structure of Crystals 2nd Edition
By B. K. Vainshtein, V. L. Indenbom, and V. M. Fridkin

22 **Point Defects in Semiconductors I**
Theoretical Aspects
By M. Lannoo and J. Bourgoin

23 **Physics in One Dimension**
Editors: J. Bernasconi and T. Schneider

24 **Physics in High Magnetics Fields**
Editors: S. Chikazumi and N. Miura

25 **Fundamental Physics of Amorphous Semiconductors** Editor: F. Yonezawa

26 **Elastic Media with Microstructure I**
One-Dimensional Models By I. A. Kunin

27 **Superconductivity of Transition Metals**
Their Alloys and Compounds
By S. V. Vonsovsky, Yu. A. Izyumov, and E. Z. Kurmaev

28 **The Structure and Properties of Matter**
Editor: T. Matsubara

29 **Electron Correlation and Magnetism in Narrow-Band Systems** Editor: T. Moriya

30 **Statistical Physics I** Equilibrium Statistical Mechanics 2nd Edition
By M. Toda, R. Kubo, N. Saito

31 **Statistical Physics II** Nonequilibrium Statistical Mechanics 2nd Edition
By R. Kubo, M. Toda, N. Hashitsume

32 **Quantum Theory of Magnetism**
2nd Edition By R. M. White

33 **Mixed Crystals** By A. I. Kitaigorodsky

34 **Phonons: Theory and Experiments I**
Lattice Dynamics and Models of Interatomic Forces By P. Brüesch

35 **Point Defects in Semiconductors II**
Experimental Aspects
By J. Bourgoin and M. Lannoo

36 **Modern Crystallography III**
Crystal Growth
By A. A. Chernov

37 **Modern Chrystallography IV**
Physical Properties of Crystals
Editor: L. A. Shuvalov

38 **Physics of Intercalation Compounds**
Editors: L. Pietronero and E. Tosatti

39 **Anderson Localization**
Editors: Y. Nagaoka and H. Fukuyama

40 **Semiconductor Physics** An Introduction
6th Edition By K. Seeger

41 **The LMTO Method**
Muffin-Tin Orbitals and Electronic Structure
By H. L. Skriver

42 **Crystal Optics with Spatial Dispersion, and Excitons** 2nd Edition
By V. M. Agranovich and V. L. Ginzburg

43 **Structure Analysis of Point Defects in Solids**
An Introduction to Multiple Magnetic Resonance Spectroscopy
By J.-M. Spaeth, J. R. Niklas, and R. H. Bartram

44 **Elastic Media with Microstructure II**
Three-Dimensional Models By I. A. Kunin

45 **Electronic Properties of Doped Semiconductors**
By B. I. Shklovskii and A. L. Efros

46 **Topological Disorder in Condensed Matter**
Editors: F. Yonezawa and T. Ninomiya

Springer Series in Solid-State Sciences

Editors: M. Cardona P. Fulde K. von Klitzing H.-J. Queisser

47 **Statics and Dynamics of Nonlinear Systems**
Editors: G. Benedek, H. Bilz, and R. Zeyher

48 **Magnetic Phase Transitions**
Editors: M. Ausloos and R. J. Elliott

49 **Organic Molecular Aggregates**
Electronic Excitation and Interaction Processes
Editors: P. Reineker, H. Haken, and H. C. Wolf

50 **Multiple Diffraction of X-Rays in Crystals**
By Shih-Lin Chang

51 **Phonon Scattering in Condensed Matter**
Editors: W. Eisenmenger, K. Laßmann, and S. Döttinger

52 **Superconductivity in Magnetic and Exotic Materials** Editors: T. Matsubara and A. Kotani

53 **Two-Dimensional Systems, Heterostructures, and Superlattices**
Editors: G. Bauer, F. Kuchar, and H. Heinrich

54 **Magnetic Excitations and Fluctuations**
Editors: S. W. Lovesey, U. Balucani, F. Borsa, and V. Tognetti

55 **The Theory of Magnetism II** Thermodynamics and Statistical Mechanics By D. C. Mattis

56 **Spin Fluctuations in Itinerant Electron Magnetism** By T. Moriya

57 **Polycrystalline Semiconductors**
Physical Properties and Applications
Editor: G. Harbeke

58 **The Recursion Method and Its Applications**
Editors: D. G. Pettifor and D. L. Weaire

59 **Dynamical Processes and Ordering on Solid Surfaces** Editors: A. Yoshimori and M. Tsukada

60 **Excitonic Processes in Solids**
By M. Ueta, H. Kanzaki, K. Kobayashi, Y. Toyozawa, and E. Hanamura

61 **Localization, Interaction, and Transport Phenomena** Editors: B. Kramer, G. Bergmann, and Y. Bruynseraede

62 **Theory of Heavy Fermions and Valence Fluctuations** Editors: T. Kasuya and T. Saso

63 **Electronic Properties of Polymers and Related Compounds**
Editors: H. Kuzmany, M. Mehring, and S. Roth

64 **Symmetries in Physics** Group Theory Applied to Physical Problems 2nd Edition
By W. Ludwig and C. Falter

65 **Phonons: Theory and Experiments II**
Experiments and Interpretation of Experimental Results By P. Brüesch

66 **Phonons: Theory and Experiments III**
Phenomena Related to Phonons
By P. Brüesch

67 **Two-Dimensional Systems: Physics and New Devices**
Editors: G. Bauer, F. Kuchar, and H. Heinrich

68 **Phonon Scattering in Condensed Matter V**
Editors: A. C. Anderson and J. P. Wolfe

69 **Nonlinearity in Condensed Matter**
Editors: A. R. Bishop, D. K. Campbell, P. Kumar, and S. E. Trullinger

70 **From Hamiltonians to Phase Diagrams**
The Electronic and Statistical-Mechanical Theory of sp-Bonded Metals and Alloys By J. Hafner

71 **High Magnetic Fields in Semiconductor Physics**
Editor: G. Landwehr

72 **One-Dimensional Conductors**
By S. Kagoshima, H. Nagasawa, and T. Sambongi

73 **Quantum Solid-State Physics**
Editors: S. V. Vonsovsky and M. I. Katsnelson

74 **Quantum Monte Carlo Methods in Equilibrium and Nonequilibrium Systems** Editor: M. Suzuki

75 **Electronic Structure and Optical Properties of Semiconductors** 2nd Edition
By M. L. Cohen and J. R. Chelikowsky

76 **Electronic Properties of Conjugated Polymers**
Editors: H. Kuzmany, M. Mehring, and S. Roth

77 **Fermi Surface Effects**
Editors: J. Kondo and A. Yoshimori

78 **Group Theory and Its Applications in Physics**
2nd Edition
By T. Inui, Y. Tanabe, and Y. Onodera

79 **Elementary Excitations in Quantum Fluids**
Editors: K. Ohbayashi and M. Watabe

80 **Monte Carlo Simulation in Statistical Physics**
An Introduction 2nd Edition
By K. Binder and D. W. Heermann

81 **Core-Level Spectroscopy in Condensed Systems**
Editors: J. Kanamori and A. Kotani

82 **Photoelectron Spectroscopy**
Principle and Applications 2nd Edition
By S. Hüfner

83 **Physics and Technology of Submicron Structures**
Editors: H. Heinrich, G. Bauer, and F. Kuchar

84 **Beyond the Crystalline State** An Emerging Perspective By G. Venkataraman, D. Sahoo, and V. Balakrishnan

85 **The Quantum Hall Effects**
Fractional and Integral 2nd Edition
By T. Chakraborty and P. Pietiläinen

86 **The Quantum Statistics of Dynamic Processes**
By E. Fick and G. Sauermann

87 **High Magnetic Fields in Semiconductor Physics II**
Transport and Optics Editor: G. Landwehr

88 **Organic Superconductors**
By T. Ishiguro and K. Yamaji

89 **Strong Correlation and Superconductivity**
Editors: H. Fukuyama, S. Maekawa, and A. P. Malozemoff